Evolutionary Biology

VOLUME 24

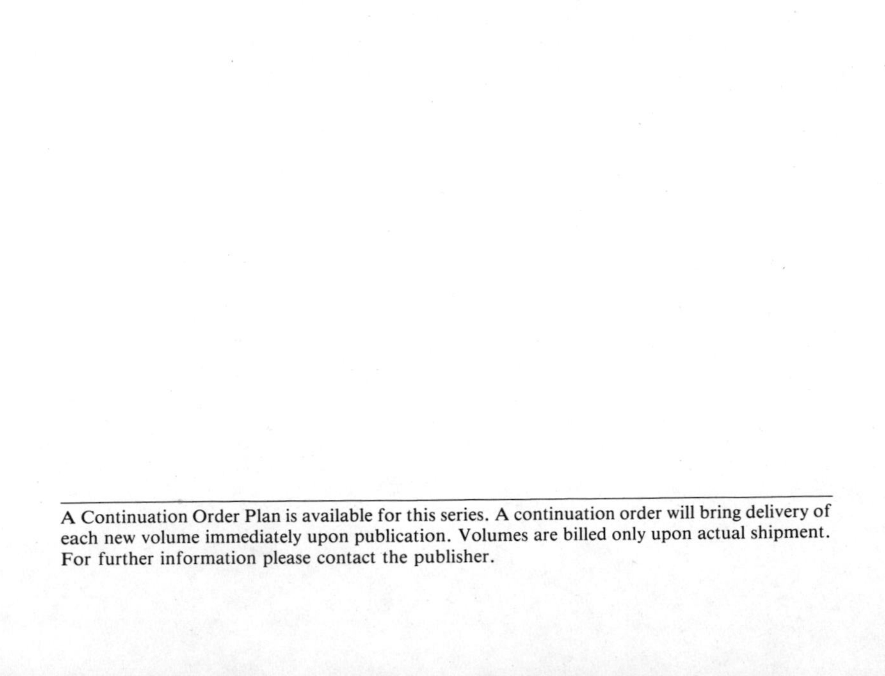

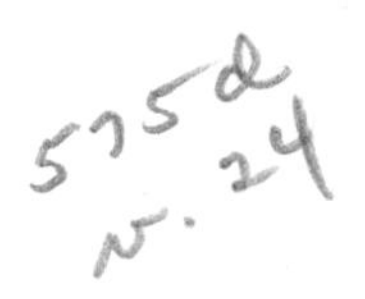

Evolutionary Biology

VOLUME 24

Edited by

MAX K. HECHT
Queens College of the City University of New York
Flushing, New York

BRUCE WALLACE
Virginia Polytechnic Institute and State University
Blacksburg, Virginia

and

ROSS J. MACINTYRE
Cornell University
Ithaca, New York

PLENUM PRESS • NEW YORK AND LONDON

The Library of Congress cataloged the first volume of this title as follows:

Evolutionary biology. v. 1- 1967-
New York, Appleton-Century-Crofts.
v. illus. 24 cm annual.
Editors: 1967- T. Dobzhansky and others.
1. Evolution—Period. 2. Biology—Period. I. Dobzhansky, Theodosius Grigorievich, 1900-
QH366.A1E9 575'.005 67-11961

ISBN 0-306-43299-4

A Division of Plenum Publishing Corporation
233 Spring Street, New York, N.Y. 10013

Printed in the United States of America

Contributors

Peter W. Atkinson • *Department of Biology, Syracuse University, Syracuse, New York 13244*

Michael J. Benton • *Department of Geology, The Queen's University of Belfast, Belfast BT7 1NN, Northern Ireland, United Kingdom; present address: Department of Geology, University of Bristol, Bristol BS8 1RJ, England, United Kingdom*

P. Bernhardt • *Department of Biology, St. Louis University, St. Louis, Missouri 63103*

A. Dafni • *Institute of Evolution, Haifa University, Haifa 31999, Israel*

B. David • *CNRS Associate Unit 157, Earth Sciences Center, F-21100 Dijon, France*

Jean Gayon • *Department of Philosophy, University of Bourgogne, F-21000 Dijon, France*

Marjorie Grene • *Department of Philosophy, Virginia Polytechnic Institute and State University, Blacksburg, Virginia 24061*

Liisa Halkka • *Department of Genetics and Tvärminne Zoological Station, University of Helsinki, SF-00100 Helsinki, Finland*

Olli Halkka • *Department of Genetics and Tvärminne Zoological Station, University of Helsinki, SF-00100 Helsinki, Finland*

William B. Heed • *Department of Ecology and Evolutionary Biology, University of Arizona, Tucson, Arizona 85721*

Marc-Andre Lachance • *Department of Plant Sciences, University of Western Ontario, London, Ontario N6A 5B7, Canada*

Herman J. Phaff • *Department of Food Science and Technology, University of California at Davis, Davis, California 95616*

Siegfried Scherer • *Faculty for Biology, University of Konstanz, D-7750 Konstanz, Federal Republic of Germany*

Ulf Sorhannus • *Department of Biology, Queens College of the City University of New York, Flushing, New York 11367*

William T. Starmer • *Department of Biology, Syracuse University, Syracuse, New York 13244*

David T. Sullivan • *Department of Biology, Syracuse University, Syracuse, New York 13244*

Preface

Evolutionary Biology, of which this is the twenty-fourth volume, continues to offer its readers a wide range of original articles, reviews, and commentaries on evolution, in the broadest sense of that term. The topics of the reviews range from anthropology to molecular evolution and from paleobiology to principles of systematics.

In recent volumes, a broad spectrum of chapters has appeared on such subjects as biochemical systematics, comparative morphology and embryology, evolutionary ecology, biogeography, paleobiology, and the history and philosophy of science. We have also attempted to provide a forum for conflicting ideas. Articles such as these, often too long for standard journals, are the material for *Evolutionary Biology.*

The editors continue to solicit manuscripts on an international scale in an effort to see that every one of the many facets of biological evolution is covered. Manuscripts should be sent to any one of the following: Max K. Hecht, Department of Biology, Queens College of the City University of New York, Flushing, New York 11367; Bruce Wallace, Department of Biology, Virginia Polytechnic Institute and State University, Blacksburg, Virginia 24061; or Ross J. MacIntyre, Department of Genetics and Development, Cornell University, Ithaca, New York 14853.

The editors of *Evolutionary Biology* wish to express their regret at having lost a good friend and valued colleague in the passing of William Campbell Steere (1907–1989), a founder of this series and co-editor of the first thirteen volumes.

The Editors

Contents

1

Critics and Criticisms of the Modern Synthesis

The Viewpoint of a Philosopher

JEAN GAYON

INTRODUCTION

I must be clear at the outset about the intentions of the present essay: I have been asked to comment on a significant sample of books from the recent and abundant critical literature on the synthetic theory of evolution. I have focused my attention on six books, namely: *Evolution at a Crossroads: The New Biology and The New Philosophy of Science* (Depew and Weber, 1985); *Unfinished Synthesis: Biological Hierarchies and Modern Evolutionary Thought* (Eldredge, 1985); *Dimensions of Darwinism: Themes and Counterthemes in Twentieth-Century Evolutionary Theory* (Grene, 1983); *Evolutionary Theory: Paths into the Future* (Pollard, 1984); *The Structure of Biological Science* (Rosenberg, 1985); *Evolving Hierarchical Systems: Their Structure and Representation* (Salthe, 1985).

On the occasion of such a review, the editors of *Evolutionary Biology* have given a formidable challenge to philosophers. In a correspondence inviting this review, the editors asked the philosopher to provide "his evaluation of the status of modern evolutionary biology." More specifically, it was suggested (1) to examine "the underlying assumptions implicit in classical and current evolutionary theory," (2) to clarify "the major limitations and contradictions posed by the modern synthesis," and (3) to suggest "alternative assumptions [which] will best guide future research on evolution into new and fruitful directions." I find

JEAN GAYON • Department of Philosophy, University of Bourgogne, F-21000 Dijon, France.

these questions somewhat terrifying.* They assign to the philosopher a role that makes one think of Auguste Comte's characterization of the philosopher's position within the framework of "positivist" science and "positivist" society: He will be "the specialist of scientific generalities" (Comte, 1975, p. 32; originally published 1830) and will assume "the presidency of knowledge." I submit that it is not clear at all that the three questions mentioned above might receive any specifically "philosophical" answer.

However, these questions appear within a particular context. One of the most conspicuous features of the current controversies on the synthetic theory of evolution is the general suspicion, among the syntheticists as well as their critics, that each of the basic rival concepts involves its own philosophical prejudice. All the texts I have examined confirm this suspicion. A closely related observation is the following: Evolutionary literature of the 1980s has a stylistic look that A. Comte, the founding father of all "positivisms," might have characterized by "the historical mode of presentation" as opposed to the "dogmatic mode of presentation" (Comte, 1975, p. 50; originally published 1830). When one compares the evolutionary literature of the past decade with that of 1940–1970, one cannot but be struck by the increasing importance of the recital of the discipline's past history and by the detailed narration of the earlier disputes. The questions raised by the editors of *Evolutionary Biology* do occur within a theoretical context where the borderline between philosophy and science is unclear. Let me explain in a few simple words how I view this situation, which, while it may appear perplexing, is nevertheless quite common in the actual history of science.

There is never any single criterion which allows an *a priori* decision as to whether a problem is of a scientific or a philosophical nature. Nonetheless, it is reasonable to say that the spirit in which the "scientist" and the "philosopher" treat a given problem is not the same. Piaget has effectively described the difference: Problems of a scientific nature are those which "have been successfully isolated in such a way that their solution does not question everything," while problems of a philosophical nature "stand as bound up with an indefinite sequence of preliminary questions which require the adoption of a position on the totality of reality" (Piaget, 1973, p. 14). It goes without saying that such an opposition is never possible but *a posteriori*. The distinction has no strict operative rigor, although it provides a guide for the process of locating a number of fundamental problems debated in current evolutionary biology. For this reason, in the themes I will be discussing, I will try to determine what is methodologically delimited (or delimitable) and what belongs to a totalizing ambition typical of the philosophical attitude.

*These quotations are taken from a series of letters which Max Hecht sent to several philosophers. It must be clear that the sentences I quote were not intended to be direct instructions to the authors. In fact, I have picked them out of a list of open suggestions. However, reacting as a philosopher, I could not be insensitive to a scientist's evocation of the foundations, contradictions, and possible alternatives of a theory.

Before indicating the themes I shall deal with in this chapter, I must give a minimal characterization of what is meant by the term "modern synthesis" in connection with the critical literature I will analyze. It is very difficult to define the modern synthesis as a "theory" with a clear-cut thesis; it is indeed questionable whether the modern synthesis should be considered as one single theory (see, however, Ruse, 1973, and Caplan, 1980). However, opponents and advocates of the "synthetic theory of evolution" seem to understand what they speak of when they dispute over the validity and the possible decline of the aforesaid "theory."

I see three strata of signification in the expression "modern synthesis" as it is commonly used among evolutionary biologists. The first meaning of the "modern synthesis" consists in a more or less extensive list of authors and books which function as a reference list of problems, beliefs, and theoretical possibilities for all those evolutionists who recognize themselves as "syntheticists." This kind of definition has two interesting implications. One is that the synthesis is more or less clearly defined as a historical fact: most evolutionary biologists would no doubt admit that the modern synthesis developed between 1930 and 1950, that its name originated in Julian Huxley's (1942) *Evolution: The Modern Synthesis,* and that its crucial doctrines and methodologies were elaborated by Fisher (1930, 1958), Wright (1931), Ford (1931), Haldane (1932), Dobzhansky (1937), Huxley (1940, 1942), Mayr (1942), Simpson (1944), Schmalhausen (1949), Stebbins (1950), and Waddington (1957) [on this presentation of the synthesis see Provine (1980)]. The other implication is that, if this list of shared references (or something like that) is the only reasonable characterization of the synthetic theory, then the synthesis is not so much a structured theory as a "paradigm" in the original sense given by Thomas Kuhn to the word, that is, a tradition of research sufficiently defined by a tacit agreement on some "exemplars" as legitimate ways of doing science.*

Beyond this rather circular way of viewing the synthesis, there is a second layer of signification: the word connotes a "synthesis" of several (virtually all) biological sciences involved in the study of organic evolution. Originally, this idea was formulated in a very eclectic manner, as it appears in Huxley's preface of the first edition of *Evolution: The Modern Synthesis:* "The time is ripe for a rapid advance in our understanding of evolution. Genetics, developmental physiology, ecology, systematics, paleontology, mathematical analysis, have all provided new facts or new tools of research: the need to-day is for concerted attack and synthesis" (Huxley, 1942, cited in Huxley, 1973, p. xii).† In spite of this eclectic project, it became clear at the end of the 1940s that the new theory of

*A very interesting application of the notion of "paradigm" to Darwinism and to the modern synthesis can be found in Burian (1988). The notion of "paradigm" as an epistemological category was originally proposed by Kuhn (1962).

†Note that the list given by Huxley would by no means seem evident today; it has often been said that ecology and developmental physiology have not been really integrated into the synthesis.

evolution was an attempt to articulate three fields of inquiry: genetics (particularly population genetics), speciation, and macroevolutionary phenomena. This is particularly clear in the organization of the Princeton Conference (Jepsen *et al.*, 1949), especially in Jepsen's introduction.* The important point here is that in such a representation of the synthesis, the obvious question is: Is one of the three fields of research more "fundamental" than the others, and possibly able to absorb the others? If one takes seriously Dobzhansky's assertion that "Evolution is a change in the genetic composition of populations" (Dobzhansky, 1937, p. 11) and Mayr's idea that "all the processes and phenomena of macroevolution . . . can be traced back to intraspecific variation" (Mayr, 1942, p. 298), one understands why a "speciation theory of evolution" (as advocated by Gould and Eldredge) may be said to be antisyntheticist. This does not mean that speciation was not important for the modern synthesis (obviously not: perhaps it was the main *problem*); it means that speciation perhaps was not the hard core of a causal theory of evolution.

There is finally a third common meaning of the "synthetic theory of evolution." It is interesting to observe that in the 1950s, when the term "synthetic theory" was not as conventional as it became in the 1960s, people used to mention as obvious synonyms the "selection theory of evolution," "neo-Darwinism," or sometimes "neo-Mendelism" (Olson, 1960, p. 524). Such a terminological hesitation clearly indicates that there is indeed a major theoretical commitment in the synthetic theory: It is fundamentally a general consensus on a *genetic theory of natural selection*. Mayr clearly stated this consensus in his opening text of the historical volume on *The Evolutionary Synthesis*. To recall Mayr's formula:

> The term evolutionary synthesis was introduced by Julian Huxley . . . to designate the general acceptance of two conclusions: gradual evolution can be explained in terms of small genetic changes ('mutations') and recombination, and the ordering of this genetic variation by natural selection; and the observed evolutionary phenomena, particularly macroevolutionary processes and speciation, can be explained in a manner that is consistent with the known genetic mechanisms. (Mayr, 1980, p. 1.)

From the three strata of signification I have distinguished in the expression "modern synthesis," it is easy to understand what a "criticism of the modern synthesis" can mean (if such a criticism is not simply a reflection on such or such particular contribution among the many which compose the whole synthetic industry). Obviously, the first meaning (the synthesis as a set of "exemplars") cannot be a serious target for a criticism, just because it would signify criticizing the whole work of Fisher + Wright + Ford + Haldane + Dobzhansky + ... One may imagine choosing other models of reference, but that is all. Now, the two other meanings are possible targets for a "criticism." There are two ways of

*Most of Jepsen's Introduction is a justification of the title of the Conference (Genetics, Systematics, and Paleontology), as expressing the structure of modern evolutionary theory, as fundamentally concerned with the genetic state of an evolving population, speciation, and a morphological approach to phylogenies (Jepsen *et al.*, 1949, pp. ix–x).

criticizing the synthetic theory as a whole. (1) To attack its most obvious central claim, that is, natural selection as the ultimate factor controlling all evolutionary processes; (2) to contest the tacit subordination between the diverse fields of research.

The four themes I treat fall into these two categories. The section following the next one is devoted to critical reflections on the concept of natural selection, as elaborated since the 1930s. The section after that is devoted to possible alternative paradigms. The section following this introduction, as well as the final section, are concerned with criticisms bearing on the general structure of the synthetic theory as a research program. I begin in the section following this introduction with a characterization of the rhetorical structure of punctuationism because, as an explicit challenge to the synthesis, it forces evolutionary biologists to reflect on the fundamental assumptions of the synthesis. The final section is devoted to various recent epistemological discussions on the structure of evolutionary theory, with strong emphasis on the "ontological" commitment of some of them.

It must be made clear that, while this chapter is intended to be a critical review, its scope is actually broader, in that the books are taken as indications of present dominant strategies for questioning the validity of neo-Darwinism. This point being clarified, I now give a brief description of the six books considered in this review.

Unfinished Synthesis (Eldredge, 1985) and *Evolving Hierarchical Systems* (Salthe, 1985) are both arguments in favor of a hierarchical structure of the theory of evolution. In spite of a common central doctrine (the distinction of two hierarchies of evolutionary entities, the "genealogical" and the "ecological," each provides a different reflection. Eldredge devotes a good half of his book to a reconstitution of the theoretical framework of the synthesis. This is viewed as a rather confused anticipation of a hierarchical view of evolution. Salthe's book is largely devoted to a general characterization of hierarchical theory, and provides valuable information and proposals on the subject. The three collective books (Depew and Weber, 1985; Grene, 1983; Pollard, 1984) are a mixture of classical evolutionary discussions, proposals of alternative paradigms to the synthesis, historical retrospectives, and philosophy of biology. Grene (1983) provides a remarkable array of historical and philosophical analyses on Darwinism. Depew and Weber (1985) and above all Pollard (1984) are more interested in understanding what a theoretical alternative to neo-Darwinism might be. Rosenberg's (1985) book on *The Structure of Biological Science* is the only book whose objective is not specifically to enter into the controversies over the synthetic theory. The essay is a philosopher's brilliant reflection on the old question of the meaning of "autonomy" for biological sciences. Rosenberg provides an exemplary clear discussion of two obsessions of modern evolutionary theory, the concept of fitness and that of species.

PUNCTUATIONISM: AN ATTRACTOR FOR CRITICAL DISCOURSE ON THE MODERN SYNTHESIS

Since its first formulation in 1972 (Eldredge and Gould, 1972), the theory of punctuated equilibria has been one of the major axes of criticism of the modern synthesis. After 17 years of active debate, the status of punctuationism in relation to synthetic orthodoxy remains extremely equivocal. Sometimes it appears as a fundamental theoretical alternative to the synthesis. For example, Turner (1983) or Maynard Smith (1983) take seriously Gould's proclamation (Gould, 1980, p. 120) that the synthetic theory, "as a general proposition, is effectively dead." On the other hand, sometimes the distance between punctuationism and the synthesis is minimized. There is an abundance of classical literature that tries to demonstrate the compatibility of the two theories [for a complete recapitulation, see Hecht and Hoffman (1986)]. Ayala (1985) provides a typical example of such a position. However, even Eldredge and Gould (1977, p. 117) occasionally say that punctuated equilibria are a development of an aspect of the synthesis.* In the same vein, Salthe, whose sympathy for the punctuationists is beyond doubt, writes: "It seems clear enough that the problems raised by the challenge of the punctuated equilibria idea can be assimilated by the Synthetic Theory of Evolution without much difficulty, if that is what one wishes to do" (Salthe, 1985, p. 187). Therefore, the status of punctuationism relative to the synthesis is ambiguous, both for the proponents of the theory as well as for its critics. In this section, I will show that this ambiguity depends on the very structure of the punctuationist argument. Furthermore, punctuationism will appear in its effects as a rhetorical configuration which functions as a recapitulation of criticisms and alternatives to the synthesis. I repeat that I am not describing the theory of punctuated equilibria and objections to this theory. I am discussing a hypothesis about the rhetorical structure, including its reasons, motives, and perceptions.

Punctuationism as an Argument

An examination of the punctuationist literature has convinced me that its mode of argumentation may be characterized by using Lakotos' distinction between the "hard core" and the "heuristic periphery" of a theory (Lakatos, 1976). In the present case, the opposition between an inalterable central claim and its plastic environment is exceptionally clear.

*"For all the hubbub it engendered, the model of punctuated equilibria is scarcely a revolutionary proposal. As Simpson, with his unfailing insight, recognized in three lines . . . , our model tries to clarify and emphasize ideas nascent studies of the Synthetic Theory" (Gould and Eldredge, 1977, p. 117).

Central Claim

Basically, the theory of punctuated equilibria consists of two inseparable theses which are the *de facto* invariant in all its presentations. The two theses concern the "tempo" and the "mode" of evolution. The question of tempo is typically subordinated to that of mode. The most suggestive formulation I have found appears in Gould:

> [The theory] holds, speaking of mode, that significant evolutionary change arises in coincidence with events of branching speciation. . . . It maintains, speaking of tempo, that the proper geological scaling of speciation renders branching events geologically instantaneous and that, following this rapid origin, most species fluctuate mildly in morphology during a period of stasis. (Gould, 1982, p. 88.)

The interdependence of the two theses is essential. In the original paper (Eldredge and Gould, 1972), the Simpsonian vocabulary of "tempo" and "mode" does not appear, but the famous diagram of rectangular evolution fully expresses the important decision to concentrate morphological change on the events of speciation. This diagram shows that there is in fact only *one* essential thesis: The theory of punctuated equilibria is a theory that makes speciation "the fundamental event of evolution" (Gould, 1983*a*).* It is only from the point of view of empirical tests that two aspects (interdependence of morphological change and speciation, alternation of periods of stasis and periods of crisis) may be distinguished. It is worth noting that the central thesis is presented as an empirical fact, a feature which makes the theory of punctuated equilibria epistemologically asymmetrical with the synthetic theory, sometimes known in the past as the "selection theory of evolution" (Olson, 1960, pp. 523–524). Natural selection is primarily a causal principle rather than a "fact." A rigorous symmetry between the synthetic theory and the theory of punctuated equilibria would require that the latter be named a "speciation theory of evolution" (that is, a theory of evolution by speciation). The proponents of the theory are not very far from saying this. In fact, they only speak of a "speciation theory of MACROevolution" (Gould and Eldredge, 1977, p. 140), and they present this causal theory with great care as a possible extension of the initial thesis.

This characterization of the hard core of the theory leads to some observations on current descriptions found in the literature. Eldredge and Gould's model

*In a certain sense, this thesis has nothing original: speciation has always been the fundamental *event* of evolution, because it is the source of all diversity. However, when punctuationists use this formula, they seem to have at least two ideas in mind. First, they mean that morphological change is mainly concentrated in events of speciation. They claim it is a *fact* (being patently clear that this is not clear for all evolutionists). Second, there is a strong tendency in punctuationist literature to expand this empirical claim into a causal one. In this perspective, speciation is the major cause of evolutionary change, and not only of diversity. This second claim involves: deemphasizing adaptation, refusing the "extrapolationist strategy of explanation," and stressing macroevolution for itself.

must be distinguished from various saltationist conceptions of the past. In its central claim the theory of punctuated equilibria is not a theory of macromutation, as Turner (1983) or Maynard Smith (1983) would seem to suggest. Similarly, the analogy with Schindewolf's "typostrophism" is more nominal than conceptual: Schindewolf was speaking of saltations from type to type, at all levels of the taxonomic hierarchy, so that the notions of "typogenesis," "typostasis," and "typolysis" apply to all ranks of the hierarchy, not only to species (Reif, 1983).

Furthermore, many authors distinguish a "weak version" and a "strong version" of punctuationism. This distinction sometimes coincides with the two theses on "tempo" and "mode" of evolution (e.g., Hecht and Hoffman, 1986), but this is not always the case. Thus, Maynard Smith (1983) identifies a "minor claim" (the alternation of stasis and crisis) and a "major claim" (the decoupling of macroevolution from microevolution). This interpretation is confusing. "Decoupling" does not belong directly to the hard core of punctuationism. It is an auxiliary hypothesis which makes possible the transformation of a factual assertion (speciation is the major mode of evolution) into a causal one (speciation is the main cause of evolutionary change). There is in fact only one "version" of the central theoretical commitment of punctuated equilibria. This is the assertion that evolutionary change is given in what will be called, according to the context, "speciation," "branching evolution," "splitting of lineage," "ramification," or "cladogenesis."

Periphery

The second aspect of the punctuationist argument consists of a network of explanatory suggestions and heuristic extensions woven around the central nucleus. This network spreads out in an indefinite number of directions. The remarkable feature of this sum of auxiliary hypotheses and programs is a plasticity nearly as great as the central claim is inalterable. Indeed, everything happens as if the intrinsic validity of each causal suggestion were more or less indifferent, the important point being the network of conceptual solidarities which emerge from the whole construction. Let me illustrate this assertion. Punctuationists have proposed numerous hypotheses in order to explain the alternation of speciation events and stasis. The initial explanations used typically syntheticist concepts such as allopatric speciation (Mayr), canalization of development (Waddington), or genetic homeostasis (Lerner). Alternative models of speciation were later invoked (Gould, 1982, p. 87), and also arguments taken from ecology (Gould and Eldredge, 1977), embryology, and recent developments of molecular biology (Gould, 1980).

Meanwhile, the theory of punctuated equilibria expands into a number of heuristic proposals, such as discussions of species selection or species drift, and various paleobiological programs. Other extensions were the debates on developmental constraints (Gould, 1980, p. 128), or heterochrony, two aspects which contribute to locate punctuationism as a way of rehabilitating a morphological and embryological approach to evolution. At this point, punctuationism becomes a well-known criticism of panadaptationist explanations to which it opposes the two notions of constraint of history and developmental architecture (Gould, 1980, p. 129; Gould and Lewontin, 1979, pp. 265–269).

All these heuristic propositions are open to debate. Therefore, one cannot honestly discredit the central thesis of punctuated equilibria by attacking such or such an aspect of the heuristic periphery. On the other hand, from its very beginnings, the punctuationist argumentary strategy has always been based on an indefinite series of preliminary questions. This gives it the style of a biological philosophy. Moreover, philosophical expansion is fully assumed, as may be seen in the two canonical texts of 1972 and 1977. In this sense, it may be useful to distinguish "punctuationism" from the "theory of punctuated equilibria." The latter is a precise claim about the tempo and mode of evolution, which is open to standard empirical and methodological discussion. Punctuationism is a philosophical hyperbole of that theory. The rhetorical transition from one aspect to the other is explicit in Gould's reflections on the emergence of a "new and general theory of evolution." In Gould (1980, p. 119), the author pleads the cause of a hierarchical theory of evolution based on the acceptance of a "punctuational change at all levels." At this point, punctuationism becomes a deliberate reconstruction of the epistemology (Eldredge, 1985) and the metaphysics (Salthe, 1985) of evolution.

Punctuationism as a Complex Fact of the History of Science

Now that I have clarified the structure of the punctuationists' argument, I can discuss the way the argument itself has been perceived by the larger community of evolutionary biologists. Within this perspective, it is clear that punctuationism has filled two roles or functions in evolutionary discourse.

The first function of punctuationism is a consequence of its central paleobiological thesis on the tempo and mode of evolution. N. Eldredge and S. J. Gould continually repeat that they are addressing a question of fact: Stasis and concentration of morphological change in speciation events are factual assertions, they are not, strictly speaking, principles or causal laws. In other words, the central thesis of the "theory of punctuated equilibria" has the epistemological status of an empirical generalization. This inductive aspect of the "theory" stimulated an extensive and useful debate on the interpretation of the fossil record. Although it would be beyond my competence and irrelevant to this chapter to enter into this discussion, I nevertheless see the necessity for two epistemological observations.

One observation is related to the "testability" of punctuated equilibria. The proponents of the theory claim that it is testable (Gould and Eldredge, 1977). A majority of evolutionary biologists conclude that it is not. Their major objection is a methodological one, which comes from the neontologists. It consists in saying that the paleontologist has access only to the morphological aspect of evolution, and not to the evolution of species proper. As a matter of fact, (1) in most cases the paleontologist is not able to distinguish a cladogenetic event from a migration event; (2) the fossil record is such that paleontologists are compelled to use a morphological notion of species (morphospecies), which does not take the complexes of sibling species, which are very frequent in nature, into account; (3) neontology demonstrates that there is a considerable discontinuous and stable polymorphism in biological species, another aspect of "species" which is not easily accessible to paleontologists (Hoffman, 1983; Hecht and Hoffman, 1986). When taken to its logical conclusion, this kind of argument tends to deny the paleontologist any competence for making inferences on the units that evolve. It also leads to the recognition of a definitional circularity in the very concept of speciation used by punctuationists (speciation defined by morphological change, and vice versa) (Gould and Eldredge, 1977). This typically syntheticist objection is indeed a very serious one for a theory which presents itself as a speciation theory of evolution. However, this is not a conceptual objection, but one that focuses more on the technical limitations of paleontology as such.

Another epistemological remark is that the theory of punctuated equilibria organizes itself around basic dogmas which are given as statistical approximations. This is indeed a very ordinary situation, particularly in the biological and human sciences. In paleobiology, this feature is expressed by the overuse of the word "pattern." Eldredge (1985, p. 85) defines "patterns" as "classes of historical events" (e.g., allopatric speciation), as opposed to "causative processes" (e.g., natural selection) and to "entities" (e.g., genes or species). These examples are explicitly given by Eldredge (see also Eldredge and Cracraft, 1980). It might be of some interest to the English-speaking reader to learn that this sense of "pattern" has no satisfactory translation in Latin languages in its scientific use. This is a puzzling fact, if one considers that "pattern" is another form of "patron," a very common word in French, which means "something to be copied" (by a craftsman). In fact, this meaning is also the primary one for the English "pattern." However, a translation of the English "pattern" into the French "patron" inevitably suggests "model," deliberate artifact, or *a priori* construction, in other words, the exact epistemological opposite of what English-speaking scientists mean when they define "pattern" as "a class of events." If one remembers the insistence with which paleobiologists, especially punctuationists, use the term "pattern," and moreover give the impression of having access only to "patterns," one notices a symptom of theories which do not have a well-defined causal status (Hoffman, 1983). This observation is confirmed by the repeated claims of punctuationists that their central theses on the tempo and mode of evolution are assertions of fact.

The second function of punctuationism derives from its heuristic periphery. Because of its extreme plasticity, and also because of the undeniable visionary know-how of its advocates, punctuationism came to function in practice as an attractor for a large array of critical discourse on the synthesis. There have been many kinds of conceptual convergences between those critical discourses and the theory of punctuated equilibria. I will now describe them as "attraction effects." By this physical metaphor, I mean that punctuationist argumentation has been a very efficient theoretical framework for gathering many different kinds of criticisms of the modern synthesis.

The most conspicuous attraction effects are on paleobiology, some classical topics of which agree well with a concept of evolution as punctuated by speciation events. This is obviously the case for the notion of macroevolution as "decoupled" from microevolution. As popularized by Stanley, "decoupling" means that *there is* a transpecific evolution, founded on a *process* of species selection, distinct from natural selection: "In this higher-level process, species become analogous to individuals, and speciation replaces reproduction" (Stanley, 1975, p. 648). Clearly, this claim is a causal one. It is, moreover, formulated as a direct refutation of typical neo-Darwinian themes (the all-sufficiency of natural selection, and "extrapolationism"). The attraction between this causal theory and punctuationism has been so strong that many people now identify them (e.g., Ayala, 1985; Maynard Smith, 1983). Furthermore, the theory of punctuated equilibria is in resonance with research programs on the diversity of the biosphere and on paleoecology. Studying diversity for its own sake more or less presupposes that change occurs through speciation, while paleoecology may provide an interpretation of stasis. This first category of attraction effects may be interpreted as the basis for claims by paleontologists for disciplinary autonomy, as seen in a number of authors (Eldredge, 1985; Salthe, 1985; Hoffman, 1983; Ayala, 1985).

A more conceptual attraction effect brings into play what Goodwin (1984) calls the "generative paradigm," suggesting it as an alternative neo-Darwinism. Here (see below, p. 24), Goodwin advocates a rehabilitation of an old and venerable tradition, which has two main features: (1) the conviction that "logical principles of biological order" exist; and (2) a basic interest in morphogenesis. One must recognize here a transcendental conception of evolution, which refuses to consider the "form" of living beings as a mere product of a cumulative and contingent history. Goodwin says that such a conceptual framework has some relation to Gould and Eldredge's idea of evolution, which "leads to the proposition that species are real entities existing as natural kinds'' (Goodwin, 1984, p. 100), from which it follows that the speciation process must be interpreted as the ''generation of separate and distinct forms *ab initio,*'' and not ''as a result of natural selection operating on a continuum of forms'' (Goodwin, 1984, p. 117). This kind of statement is highly significant of what I call an attraction effect. Everybody knows that punctuationists do not see species as ''kinds,'' but as ''individuals''; furthermore, they are most often satisfied with the classical synthetic models of speciation. Therefore, one cannot infer that they consider species to be ''natural kinds.'' Yet it is also true that punctuationists emphasize morphological and embryological constraints. But these themes appear once more to be auxiliary and not decisive arguments, as seen in the texts where Gould suggests that the famous internal constraints are in fact historical constraints (e.g., Gould, 1980, p. 129). In any case, such ambiguities have favored an association between punctuated equilibria and an imprecise constellation of radical alternatives to the synthesis: ''epigenetic'' theories of evolution,* resurgences of transcendental or typological interpretations (e.g., Riedl, 1983), a systemic interpretation (Salthe, 1985), and neocatastropic theories (Saunders and Ho, 1984; Ho and Saunders, 1984; Thom, 1985).† None of these concepts is punctuationist *per se*. But all can present themselves as possible explanations of the ''fact'' of punctuated equilibria. The point they have in common is that all are variants of formalist or ''*Gestaltist*'' conceptual schemes, neatly playing on all the connotations of the word ''form'' (such as figure, structure, organization, type, and species, the last term being the Latin translation of the Greek *eidos*). Such conceptual schemes claim that they explain the patterns of change advocated by punctuationism, in the sense that these patterns are explicitly ''morphological.''

A third, more subtle attraction effect deserves to be analyzed. Janvier (1984, pp. 63–64) sees a close connection between cladistic analysis of phylogenies and models of punctuated equilibria. The latter would be ''a result of cladistic analyses of various fossil invertebrate groups previously interpreted in the light of phyletic speciation.'' In the same vein, Maynard Smith (1983, p. 275) writes: ''in the debate about macroevolution, an alliance has been found between Punctuationalists and Cladists.'' At first sight, such statements are surprising. There is no reference to Hennig's methodology in the canonical presentation of punctuated equilibria. Moreover, while Eldredge seems to have been interested rather early in the method (Eldredge, 1979*b*), this is not so for Gould. More fundamentally, cladistic analysis is not a theory of processes, but a method for testing phylogenetic inferences. Therefore, there seems to be no clear

*Epigenetic theories of evolution are well characterized in their relation to punctuationism in Bowler (1983, pp. 222–223).

†Hoffman (1983, p. 259) notes that, curiously, catastrophe has not been much considered by proponents of punctuated equilibria.

historical relationship, nor is there any rigorous conceptual or methodological connection. However, what we have here is a fascinating case from the viewpoint of the history of science, a case which illustrates the difference between a scientific fact and a fact in the history of science.

The relationship between punctuationism and cladistics might be reconstructed historically as follows: The history begins before the emergence of punctuationist concepts. The ideas of phylogenetic systematics [baptized "cladistics" by Mayr (1969, p. 211)] were severely criticized by several founding fathers of the synthetic theory (Mayr, 1969, 1974; Simpson, 1975). Their major argument was that cladistic analysis "confuses genealogical with genetic relationship" (Mayr, 1969, p. 230), the consequence being that cladistics did not take into account the biological concept of species. Therefore, cladistics was suspected of restoring a typological kind of thinking, incompatible with the synthesis. In reply to these criticisms, numerous doctrines of the synthetic theory were attacked by the cladists (Janvier, 1986). Rosen (1984) is an example of the ferocity of these attacks. Among the criticisms of the cladists, there is one which clarifies the "alliance" with punctuated equilibria. As noted by Janvier, cladistic analysis, when applied to stratigraphic series of fossils, may result in contesting the commonest anagenetic interpretation. As a matter of fact, cladistic analysis invites the consideration of several traits, a procedure which often confronts the paleontologist with several incompatible phylogenies. The cladistic methodology arrives at this point, and suggests the most "parsimonious" genealogy, which happens to be a tree more often than a series. The neat gradual anagenetic series are then replaced by cascades of ramifications. The convergence with the punctuationists' vocabulary and images is striking.

This third attraction effect of punctuationism does not have the same epistemological meaning as the previous two. In the two former cases, convergences occur in the heuristic periphery, and the possibility that they might not be relevant does not necessarily jeopardize the punctuationists' central thesis, which is given as an assertion of fact. This is not the case with cladistic methodology, for two reasons: (1) The model of punctuated equilibria states that "the fundamental event of evolution is speciation (ramification of lineages)" (Gould, 1983*a*, p. 39). Logically, this statement implies that the model of punctuated equilibria can apply only if the phylogenetic structure is elucidated (Goujet *et al.*, 1983, p. 142). A theory that makes ramification the major mode of evolution depends on a methodology capable of establishing the effective structure of ramifications. (2) Correlatively, one may have some doubts about what punctuationists mean by "speciation" and "species." As noted by Rieppel (1984), punctuationists recognize species by *differences* (cf. Eldredge, 1979*a*, p. 16). This confers an essentially formalistic, not biological, significance to their "rectangular" ramification diagrams. In such diagrams, the horizontal axis simultaneously connotes phenotypic change and geographic separation, so that one is driven back to a time when evolutionary biologists did not know how to distinguish species from varieties (Turner, 1983). It would surely be unfair to reproach paleontology for not having access, or at best having minimal access, to the biological reality of species. However, it is possible that punctuationists might have abused the terms "species" and "speciation." In spite of many references to the synthetic conception of species, and despite "ontological" claims (species as individuals), methodologically speaking, punctuationalists' species are forms. Would the impact of the theory have been the same if they had spoken of "morphostasis" and "morphocrisis" in relation to phylogenies? A theory of "rectangular" cladomorphic *patterns* is not necessarily synonymous with a theory of evolution by speciation.

Let us conclude on punctuationism. The main feature of its argumentary structure is that the central theses are factual claims, not explanatory principles. These claims (on tempo and mode) are not hopelessly incompatible with the explanatory principles of the synthesis. The general scenario, where there is only one mode, is simply less eclectic than Simpson's. On the other hand, their central empirical thesis generates a systematic exploration of many kinds of antisynthetist argument. In this respect, eclecticism and plasticity are the rule. This suggests that the punctuationists' discourse is not so much an alternative paradigm as a "philosophical" mediation in a time of theoretical crisis. The theory of punctuated equilibria has always claimed this style of reflection, which is in accordance with an epistemological position that might be formulated as follows: Scientific theories are not relays between facts, facts are relays between theories. To be fair, punctuationists have made the synthesis less naive.

NATURAL SELECTION, FITNESS, AND ADAPTATION: A BLIND SPOT IN THE MODERN SYNTHESIS

It is impossible not to be struck by the amount of discussion on the very meaning of the concept of natural selection in the evolutionary literature of the past two decades. An important crisis persists, which is abundantly commented on in the six volumes under examination. Although I cannot say that I have noticed radically new elements on the question, I have found a large sample of classical epistemological questions on the relationships among the concepts of natural selection, fitness, and adaptation. I will first give a schematic description of these questions, then follow with a few historical remarks, which will prove useful in a debate which I believe is too often considered exclusively in epistemological terms.

The Spectrum of Classical Questions

Let us begin with the most troublesome question, that of the "tautological" character of natural selection. The presumed tautology appears when the Darwinian principle is formulated in Spencer's terms as "survival of the fittest." It seems impossible to avoid the circular conclusion that the "fittest" are those who survive, while "those who survive" are "the fittest." Yet this kind of remark is often attributed to dishonest anti-Darwinians, or to parasitic epistemologists, entangled in a sterile neopositivist reflection on the structures of scientific theories. The historical reality is quite different. Riddiford Penny (1984) gives an impressive list of neo-Darwinian biologists who, ever since Haldane (1935), have labeled the principle of natural selection as a "tautology" (e.g., Waddington, 1960; Dobzhansky, 1968; Stebbins, 1974). I believe another reference must be added, although it is always omitted: Spencer himself, in the very text where he proposed a reformulation of Darwin's principle of natural selection as "survival of the fittest," presented it as a certain *a priori* truth, which does not need to be proven by facts (Spencer, 1864, §165).* One can see, therefore, that the tautological characteristic of natural selection is not a novelty; it is a deliberate attribute of its formulation by Spencer. Under these conditions, it comes as no surprise that so many circular definitions of natural selection, fitness, and adaptation are to be found in current Darwinian literature.

Now there is a difficulty, which must be examined. A. Rosenberg, developing Hull's (1974) analysis, strikes the balance with exemplary clarity: "The charge [of tautology] of course rests on the assumption held by nearly everyone that a real theory must be testable and cannot be merely a set of definitions, disguised or undisguised, without bearing on the way things actually are" (Rosenberg, 1985, p. 127). Yet the significance of this requirement should not be misunderstood. The issue is *not* that a theory might be viewed as a set of definitions; the problem is to determine whether the system of definitions is applicable to the real world or not. Thus, Newtonian mechanics may be considered as a system of definitions of terms such as "force," "gravity," and "mass," with obvious circularities. But this same theory also treats a series of empirical assertions about "force," "gravity," and "mass" in such a way that the theory as a whole enables predictions to be made, and thus it has an explanatory power.

In the case of the theory of natural selection, the difficulties are concentrated on the term "fitness." In its commonest use within evolutionary theory this term connotes both differential reproduction and a vague notion of the cause

*Darwin incorporated Spencer's formula into the fifth (1859) and sixth (1862) editions of *The Origin of Species* (Darwin, 1959).

responsible for the reproductive performance (the "general aptitude"). Therefore, the same term applies both to a phenomenon and to its causes, to the *explanandum* (differential reproduction) and to the *explanans* (fitness *senso stricto*). Rosenberg observes that there is a reason for this typically operationalist confusion that can easily be understood. In reality, fitness is measurable only by its effects, and the only means (or "currency" as Medawar said) by which these effects are expressed is differential reproduction. As opposed to what happens in physics, this measurement cannot be corrected in biology. For instance, when one measures a temperature with an alcohol thermometer, corrections are possible by two means: either by using other kinds of measuring instruments, or by using a physiochemical theory which enables the behavior of the alcohol thermometer to be understood. There is no similar general theory for fitness. There obviously are causal factors affecting the level of fitness, but these factors are so numerous and so variable that the only thing one can measure is their overall effect on reproduction. For this reason Rosenberg suggests a new way of considering this problem of tautology: First, one must stop defining fitness by the effects which measure it. Second, one must accept the fact that no general interpretation of fitness is provided by evolutionary theory. One solution might be to give up the term. Another would be to abandon any definition of fitness, and to consider it as a primitive term in the theory of natural selection. "Evolutionary biologists who feel constrained to define their central theoretical terms sometimes incautiously state that fitness is differential reproduction. This turns into a tautology. Their error is to suppose that every term in a theory is defined within this theory" (Rosenberg, 1985, p. 159).

This conclusion will perhaps seem excessive, or appear to belong to an attitude of terminological hygiene, which does not have much to do with evolutionists' practice. Nevertheless, it has the advantage of calling attention to the causal status of the natural selection principle. There are of course other ways of questioning the status of the natural selection concept, and they provide other possible solutions to its paradoxes. Most of these solutions belong to what may be called semantic interrogations about the term and its related concepts, in other words, classical analysis of content. These analyses developed as a result of Williams' (1966) *Adaptation and Natural Selection*. They led to a proliferation of distinctions that lend the current discussions on selection the style of a scholastic specialty. As they are very close to current evolutionary practice, these reflections concern the evolutionists themselves as well as the philosophers or historians. I will content myself here with enumerating the major axes of semantic discussions, making clear that all of these analyses are ultimately concerned with the causal status of the selectionist explanation. Two general categories may be distinguished.

Clarification of Adaptive Language

Williams (1966, p. 158) apparently convinced the evolutionist community that it was reasonable to dissociate the statistical demographic aspect of fitness (reproduction performance of a given entity) from its properly adaptive connotation (effective design for reproductive survival). The first concept is a bookkeeping one, in the sense that all evolutionary effects cannot avoid being expressed in its language as recalled by Dyke (1985). However, this is a very old story (Medawar, 1960; Wimsatt, 1980; Kingsland, 1985). The second concept (fitness as adaptation *sensu stricto*) is a causal concept. As has often been remarked by ecological geneticists, this concept is methodologically onerous because it necessarily requires a detailed demonstration of the way a specific characteristic affects survival in a given context.

Furthermore, it has become common to distinguish the adapted state (''adaptedness'') from the causal historical process that constructs the adapted state (''adaptation'' in a strong sense). Burian (1983) provides a very useful and refined restatement of these two distinctions and their many difficulties.

A third class of distinctions must be mentioned. A number of authors try to rehabilitate the old debate on adaptation version adaptability as a fundamental aporia in evolutionary theory. For example, see Eldredge's interesting discussion of this problem as it appears in Dobzhansky (1968), or the more direct approach in Salthe (1985). Finally, there is Gould and Vrba's (1982) distinction between ''aptation'' and ''exaptation.'' Frankly, I do not see the point of restoring such an overly-debated and confusing problem. Recall that: (1) There is at least one reputable and specific answer to the question in Darwin himself. Darwin used to distinguish ''infinitesimal advantage'' and ''adaptation'' (always a result). Relating the two notions leads to the concept of natural selection as an opportunistic process. (2) If one does not accept this theme, the alternative is that of a concept of predaptation (Cuénot, 1914). This concept inevitably leads to the assumption that the adapted *state* preexists the adaptive *process*.

Clarification of Selective Language

Williams' (1966) essay also generated a huge controversy over the kind of *objects* affected by the selective process. Let me simply outline the relationship among the various topics which are connected with this dispute.

Although it was not considered to be one of its most important aspects, the controversy over the units and levels of selection appears to have existed even in the early days of the history of the theory of natural selection. Wallace allowed group selection as well as individual selection, while Darwin accepted only the latter (Ruse, 1980). The question, however, turned into a fundamental one with the appearance of a genetic theory of natural selection, and the legendary quarrel which opposed Fisher to Haldane and to Wright. While Fisher only recognized a genic selection, Haldane insisted on the necessity of restricting the attribution of selection coefficients to genotypes. Wright advocated an ''organismic'' selection (1980), that is, an interdemic selection acting on approximately homozygous genomes. The debate was further enlarged and generalized with the concept of group selection (Wynne-Edwards, 1962), followed by the paleobiological proposal of a ''species selection'' (Stanley, 1975). The multiplication of ''units'' that might possibly be involved in a selective process finally produced a subtle change in the original question: Disputes over the relevant units of selection led to an examination of the logical structure of the concept, and thence to the idea of its ''generalization'' (Lewontin, 1970). As a result of the parallel analyses of Dawkins (1978) and Hull (1980, 1981), a standard terminology now seems to be well established. Two kinds of entities are seen to operate in any selective process: ''replicators'' (a replicator is ''an entity that passes on its structures directly in replication'') and ''interactors'' (an interactor is ''an entity that directly interacts as a cohesive whole with its environment in such a way that replication is differential'' (Hull, 1980, p. 318).*

As I see it, the main interest of this distinction is that it clarifies the causal aspect of the principle of natural selection. The dissociation of replication and interaction may be usefully understood with Sober's illuminating distinction between ''selection OF'' and ''selection FOR'' (Sober, 1984, pp. 97–102), in other words, selection of *things* (selection as sorting, a sieve effect) and selection for *properties* (selection as a causal process, the effect of which is a sorting.† The second aspect is too often neglected, as if the sorting effect were the full reality of natural selection. However, Sober's distinction does not necessarily coincide with that of replicators and interactors. Let me clarify this point with an example. Genes are the clearest case of replicators. But in Sober's terminology they may be seen in two ways: (1) as the *things* (among possible others) that are sorted; or (2) as causal determinants, or as physiological units responsible for some property affecting the viability or fertility of the organisms to which they belong. On the other hand, an organism (if sexual) is not the kind of thing which is sorted; nevertheless, it can be the subject of these properties which make it interact with its environment in such a way that some replicating entities are sorted. The problem here is that we see the replicating entity (the gene) as both

*The term ''replication,'' in the particular context of natural selection, was first proposed by Dawkins (1978). ''Interaction'' was proposed by Hull (1980, 1981).

†Rigorously speaking, the expressions ''selection OF'' and ''selection FOR'' (as well as ''selection BY'') were first used by Waddington (1957, Chapter 3, Selection of, for and by). However, Sober's epistemological analysis is based on arguments which wholly belong to him.

cause and effect, while the interacting entity (the organism) is also both effect and cause. Thus, the distinction between replicators and interactors is a subtle way of getting around the causal problem. In this context, Brandon's (1985) deserves particular attention. Brandon fully assumes a causal presentation of replicators and interactors: "Interactors, by definition, are proximate causes, while *active* replicators are remote causes of differential reproduction" (Brandon, 1985, p. 90; emphasis added.) Therefore, it is equally irrelevant to consider replicators (e.g., genes) as no more than "passengers" (a current holistic view) and interactors (e.g., organisms) as "vehicles" (Dawkins' vocabulary). Both replicators and interactors are causes. However, a marked asymmetry appears as soon as adaptive considerations are taken into account. Provided that it is understood as design for survival, adaptation is intrinsically associated with the notion of interactor. This is why Brandon concludes that "any natural selectionist *explanation* will have to be given in terms of properties of interactors" (Brandon, 1985).

I will not go further into this kind of discussion. The main conclusion to be drawn from all the epistemological discussions on natural selection that I have examined is that they clearly reflect a rehabilitation of the concept of adaptation in the original sense of engineering solutions to environmental problems. Moreover, this rehabilitation appears to be closely connected with the epistemological interrogation on natural selection as a causal principle.

The Same Questions within a Historical Perspective

I think that epistemological questions on natural selection deserve to be placed within a historical perspective. I will give two examples showing how the central concept of modern evolutionism came to be an epistemological blind spot, first in Darwinian thinking, then in the synthesis.

"Natural Selection" versus "Survival of the Fittest"

Historians have been prolix on the conditions under which Darwin adopted Spencer's expression of "survival of the fittest" and finally integrated it into the last two editions of *The Origin of Species*. As already indicated, Spencer proposed the expression "survival of the fittest" in his *Principles of Biology* (Spencer, 1864, §165). In his mind, this expression had the advantage of manifesting the character of *a priori* truth he saw in Darwin's principle. Moreover, Spencer thought Darwin's "natural selection" was dangerously exposed to the objection of anthropomorphism. Wallace, who shared the same conviction, tried to convince Darwin to abandon the formula "natural selection." According to Darwin, it seems clear enough that for him that was simply a matter of vocabulary [see the Wallace–Darwin correspondence of 1866, in Wallace (1916)]. This is why Darwin finally introduced Spencer's expression next to his own "natural selection."

Nevertheless, I believe that a more interesting question would be to understand why Spencer proposed an alternative formulation. There is a very eloquent text which clarifies the opposition between Darwin and Spencer on natural selection. Although this text was written in 1893, or 10 years after Darwin's death, it makes clear something which was already present in Darwin's and Spencer's respective thinking in the 1860s, although not clearly formulated as an opposition:

> Artificial selection can pick out a particular *trait* and, regardless of other traits of the individuals displaying it, can increase it by selective breeding in successive generations. . . . On the other hand, if we regard Nature . . . we see that there is no selection of this or that *trait;* but there is a selection only of *individuals,* which are, by the aggregates of their traits, best fitted for living. And here, I may note an advantage of the expression of "survival of the fittest," since this does not tend to raise the thought of any one *character* which, more than others, is to be maintained or increased, but tends to raise the thought of *a general adaptation* for all purposes. (Spencer, 1893, p. 160; emphasis added.)

It is sufficiently instructive to compare this text with the definition Darwin gave of natural selection in Chapter 4 of *The Origin of Species:*

> Can we doubt . . . that *individuals* having *any advantage,* however slight, over others, would have the best chance of surviving and creating their kind? On the other hand, we may feel sure that *any variation* in the last degree injurious would be rigidly destroyed. This preservation *of favoured variations,* and the rejection of *injurious variations* I call Natural Selection or [6th ed.] the Survival of the fittest. (Darwin, 1859, pp. 80–81; emphasis added.)

The comparison shows that, in spite of Darwin's adoption of Spencer's formula, Spencer and Darwin had two heterogeneous concepts of "natural selection." Spencer sees selection as bearing upon individuals, Darwin upon differential properties. Spencer's conception is an ontological one, inseparable from the sieve metaphor. On the contrary, Darwin reasons on properties that vary in "advantage": Each individual is viewed as n-dimensional space where a set of forces simultaneously acts on n characters presenting infinitesimal variations in advantage. Hence the subsequent interpretations in the language of differential calculus, and the search for a maximizing function: Fisher's "fundamental theorem of natural selection" is a differential equation. The Darwinian schema cannot be represented in ontological terms. It does not consist of speculating on preadapted *entities,* but on the dynamic idea of an opportunistic construction of adaptations. In such a schema, adaptation is never a starting point, but is the result of a process. This schema is a causal one. Unlike that of Spencer, it is not circular.*

It is important to emphasize that the emergence of a particular theory of inheritance (Mendelism) did not negate the epistemological opposition I have just described. On the contrary, Mendelism brought out its basic importance. Although Mendelism favored a Spencerian way of representing natural selection as a selection of things (genes), this attitude—with its characteristic tendency to put the causal dimension of natural selection into parentheses—has led to the operationalist dead ends described above (pp. 15–16).

The Choice of the Modern Synthesis: Genetical Theory versus Ecological Theory of Natural Selection

It has often been said that ecology was not really integrated into the synthetic theory of evolution. Nevertheless, one must understand what this means from a historical point of view, so that one can evaluate whether or not this omission affects the synthesis in its fundamental thesis.

Kimler (1983) clearly states the historical issue. Darwin's theory of natural selection confers a definite importance on arguments of an ecological nature:

> The general sensitivity of the ecological balance that is required for the control of the evolutionary dynamic by natural selection is found in high levels of competition and predation, which act on the fruits of Malthusian fecundity. It is the constraint of these selective pressures that gives the utilitarian argument its meaning and its fecundity. (Grene, 1983, p. 112.)

The synthesis presents a noticeably different perspective. In its Darwinian aspect, the synthesis did not directly rely upon the ecologists' work on populations. The synthesis was constructed on the periphery of a *genetical* theory of natural selection, whose main objective was to demonstrate the plausibility of Darwinism within the context of the Mendelian theory of inheritance. Hence a strategy of research concentrating on the existence of minor variations and the *a priori* demonstration of the efficiency of weak selective pressures. As Kimler says, "Ecology contributed in the sense of being there with ready support for selectionist theories" (Kimler, 1983, p. 115), not in the sense of a theoretically integrated element.

*In the same perspective, there is another interesting question: Why did Darwin maintain the expression "natural selection," although he sometimes acknowledged that it was, in the literal sense, "a false term" (*Origin,* 6th ed., introductory paragraph; Darwin, 1959)? As I see it, the reason is that the parallel between artificial selection and natural selection is absolutely crucial in Darwin's argumentation. While Wallace (1859) or Spencer (1893) think they are very different processes, Darwin insists on the similarities. What makes the parallel crucial is that Darwin has no direct evidence in favor of natural selection. But he knows there is much evidence in favor of the efficiency of artificial selection. This is particularly important if natural selection is not an *a priori* truth (as in Spencer), but a hypothesis with *some* kind of empirical support.

The consequence of this (relative) absence of ecologists may be described in the following way. As a *theory* of natural selection, the synthesis mainly appears as a theory of the Mendelian *effects* of natural selection within populations. This makes it the kinetics rather than the dynamics of natural selection, in the sense that it does not incorporate considerations on the nature of selective forces within its theoretical framework. Everything happens indeed as if selective forces were always contingent, so that they can appear only in case studies, never in the theory.

One might object to this presentation that the existence of ''ecological genetics'' was a major component of the synthesis. There is no doubt that without Dobzhansky, Ford, Dowdeswell, Cain, Sheppard, Teissier, and a few others (among them Fisher, Haldane, and Wright in their empirical work), the synthesis would hardly be recognizable. Yet most of the field work of the various schools of ''ecological genetics'' confirms my analysis. As noted by Lamotte (1988), most of the investigations remain within the limits of the monospecific framework. In this context, ecology is nearly always an autoecology, which leaves the biotic environment of populations outside its theoretical field. When the biotic environment intervenes in the explanation, it does so as a contingent and empirical periphery, not as an element of the theory of natural selection. Therefore, the modern theory of natural natural selection, in the common form it takes in ''ecological genetics,: is open to the objection that it is mainly a theory of effects (not a true causal theory of natural selection).

This characteristic tendency finds its paradigmatic expression in one of the most obscure, although fundamental and permanent, discussions in contemporary neo-Darwinism. I am referring to the discussion of Fisher's ''fundamental theorem of natural selection'' (Fisher, 1930, 1958).

It would be beyond the scope of this chapter to examine either the history of this theorem or that of the concept of ''fitness'' (Fox Keller, 1987; Gayon, 1988, 1989*a*). A single observation will suffice. I believe this fundamental theorem sums up the difficulties of the synthetic theory as a theory of natural selection. Anyone who takes the trouble to read the puzzling Chapter 2 of *The Genetical Theory of Natural Selection* (Fisher, 1930, 1958) from beginning to end (and not only the middle) must conclude that there are two very different aspects to the theorem.

The most obvious aspect is that it is a ''manifesto'' in favor of a purely genic concept of selection, as illustrated by Fisher's commentary: ''The only evolutionary effect, in increased fitness, or in anything else, that I recognize as such, is constituted by the changes in gene ratio'' [Fisher to Kempthorne, 18 February 1955, in Bennett (1983, p. 229)]. Within this perspective, Fisher's theorem presents evolution as a statistically-described cloud of genes, with the genes being exposed to a field of forces specified only in their effects on the particles of the system.

The second aspect of this theorem is that Fisher associated it with considerations on the demographic dynamics of populations. These considerations are of a typically ecological nature: instantaneous increase of population size (in absolute, not relative value), the interaction between fitness and population density, and the "deterioration of the environment." Although they were generally ignored by the many authors who "refuted" the theorem, these considerations are vital to its algorithmic meaning.* At any rate, the point I wish to stress here is that Fisher's attitude was probably unique in the history of the synthesis; it was a timid (and obscure) attempt to build a bridge between theoretical population genetics and theoretical ecology. For two decades a number of people (e.g., Roughgarden) have been trying to explore this theoretical pathway again. It is possible that this kind of investigation might lead to a real dynamical theory of natural selection. It is also possible that such an attempt might reveal itself to be chimerical. Were this to be the case, one would have to admit, to borrow Lotka's expression, that the theory of natural selection can be no more than a "stoichiometry of evolution" (Lotka, 1956, p. 50),† the selective *forces* being accessible only to case studies. This is precisely the situation which obviously characterized the synthetic approach to the theory of natural selection.

IS THERE SOME ALTERNATIVE PARADIGM TO THE SYNTHESIS? THE RETURN OF ORGANIZATION

A precise description of the antisyntheticist nebula would be a difficult task. As seen above, punctuationism provides a useful landmark. It functions as an attractor for a variety of critical propositions. Yet punctuationism does not stand as an alternative paradigm, because it does not address itself to problems that are different from those considered by the synthesis. Punctuationism is part of the Darwinian industry in its largest sense. Much more confusing is the constellation of openly anti-Darwinian attitudes. There are, for instance, clear signs of a

*Refutations of the theorem always rely upon Wright's interpretation of it. This interpretation assumes separate generations, and presumes that the theorem speaks of the mean relative fitness of the population (Wright's "peak"). There is a misunderstanding on those two points [see Fisher's letter to Kimura, 3 May 1956, in Bennett (1983, p. 229); see also Edwards (1971, 1977)].

†Lotka distinguished two branches in the "mechanics of evolution." One was "the stoichiometry of the systems in evolutionary transformation," that is, "the science which concerns itself with . . . the *masses* of the components." The other was the energetics or dynamics of evolution (Lotka, 1956, p. 50). The stoichiometry of evolution, Lotka said, was mainly a "kinetics." As an example of the "kinetics of intra-group evolution," Lotka gave Haldane's first paper on the mathematical theory of evolution, written in 1924 (Lotka, 1956, pp. 122–127). It is worth recalling that Lotka's *Elements of Mathematical Biology* was first published in 1926.

revival of transcendental morphology, or of neo-Lamarckianism, whether mechanicist (e.g., Steele *et al.*, 1984) or psychologically oriented (e.g., Campbell 1985). One recognizes the old theoretical oppositions to Darwinism.

A recurring conceptual configuration may be observed among these and other revivals, that of "organization." The same theme appears in a variety of disciplinary fields: experimental embryology, molecular biology (genome structure), thermodynamics of the evolutionary process, or population biology modeling. I have been impressed by the number of texts which emphasize terms such as "self-organization," "complexity," "form," "Gestalt," "structure," "morphology," "type," and "emergence." The question is whether this organicist epidemic reflects more than a metaphysical sensibility, and in what sense "an organizational interpretation of evolution" [to borrow Campbell's (1985) title] might be an alternative to the neo-Darwinian synthesis. I will briefly evoke two texts which provide suggestions within this perspective.

Goodwin (in Pollard, 1984) accuses the synthesis of having neglected the important lessons of developmental mechanics for evolution. This neglect, Goodwin argues, dissimulates an original sophism in neo-Darwinism. Neo-Darwinism treated embryogenesis as an aspect of inheritance, considering that genes ultimately determine form. Coupled with the principle of natural selection, this assumption leads to a type of explanation which systematically emphasizes contingencies, genealogies, and history. The construction of forms is reduced to an accumulation of survival contingencies, and there is no room left for the general principle of biological organization. To this paradigm Goodwin opposes the notion of morphogenetic potential: A given morphological arrangement is only one solution among a spectrum of other possibilities. The effective orientation toward such or such a solution is a function of the initial and limiting condition, which includes both genetic and external parameters. This is why Goodwin denounces a fundamental sophism in neo-Darwinism. The sophism consists in giving a certain category of occasional causes (genes) the status of primary causes. This criticism suggests an alternative paradigm: Basically, the process of evolution consists in "the exploration of the potential set of forms defined by [the] laws of organization, by changes in genes and in the environment." In other words, the process of natural selection acting on genes is subordinated to an "intrinsic dynamics of the living state."

Let us now compare Goodwin's speculation with Kauffman's (1985) in his remarkable paper, "Self-organization, selective adaptation and its limits." Kauffman proposes a modeling of the possible action of natural selection on an autoregulating genomic system. In its formal and visionary rigor this text makes one think of the seminal papers of population genetics in the 1920s and 1930s. The question raised by Kauffman is whether or not selection is able to produce and maintain any kind of arbitrary and unique regulatory diagram in a given population. The model makes a certain number of initial assumptions on the number of structural and regulatory genes, and on the rates of chromosomal mutations (in other words, mutations capable of modifying the topology of the system: transposition and duplication). There is no doubt about the highly idealized nature of the model. Nevertheless, Kauffman provides a strategy for theorizing a problem that seemed well beyond the possibilities of theoretical population genetics: How can one evaluate *a priori* the effects of selection on a system of more than two loci? As I understand it, Kauffman's proposal would abandon an individual characterization of genotypic fitness, and reason in terms of the topology of the system. At this point the question turns into an evaluation of the degree of precision attainable by selection. Kauffman's conclusion is that selection cannot act with precision on a complex diagram of regulation; the system indeed manifests "generic properties" which impose serious limitations on the possible action of selection.

Goodwin's and Kauffman's speculations are typical of a major shift in today's evolutionary thinking, which is abundantly illustrated in a number of articles and books. It is important to qualify this shift, avoiding the stereotyped traps of the organizational discourse.

One may characterize this major conceptual breakthrough as the theoretical decision to think of selection in terms of its *limits*. This claim should not be simply discredited by arguing that Darwinian biologists knew for a long time that selection cannot produce just anything. Evolutionary biologists such as Waddington, Schmalhausen, Dobzhansky, Mayr, and Wright obviously knew that living beings were organized things, and often advocated a holistic presentation of their object. But we need to distinguish the fact of "being conscious of" the organizational dimension of living beings from the decision to begin a strategy of research commanded by this idea. As a strategy of research, the neo-Darwinian synthesis may be characterized as a systematic exploration of the theory of natural selection as an "extremal theory." An extremal theory is one that "treats the objects in its domain as behaving in such a way as to maximize and/or minimize the values of certain variables" (Rosenberg, 1985, p. 238). This feature of the modern synthesis may be clearly seen in the theoretical constructions of Fisher and Wright: Both the "fundamental theorem" and "adaptive topographies" are schemata of maximization of fitness (Turner, 1969). Furthermore, other followers of the synthetic theory confirm this tendency. Ecological genetics, the new systematics, Simpson's paleontology, all carefully explored the possibility of natural selection as an unavoidable aspect of any evolutionary phenomenon whatsoever. This did not mean that natural selection explained everything, but rather that nothing was explainable without taking natural selection into account. A comparison may be made with Newtonian physics. After Newton, any mass had to be affected at any moment and in any circumstance by the principle of gravity. Similarly, in evolutionary biology, any living thing that reproduces had to be affected by the principle of natural selection. We may carry this comparison a little further. Eighteenth century physicists believed that ultimately all phenomena in the universe would be understandable in terms of the gravitation principle. Their hope was largely realized for macroscopic objects (tides, structure and origin of the solar system, form of the galaxies). However, the principle could not explain why the sun is hot and revealed itself to be almost totally sterile for explaining chemical properties or, more recently, atomic structures. Physicists had to discover other forces than that of gravity.

The neo-Darwinian synthesis faced a similar problem. It clearly recognized other forces than natural selection, such as mutation, population size, etc. The syntheticists measured them, and constructed a delicate and rigorous theory describing their interaction. The well-known conclusion, in *a priori* deductions as well as in field work, was that natural selection is strong enough to overcome the other forces. However, these evaluations are restricted to a particular conceptual framework: They operate within the parametric space of gene frequencies. The validity of synthetic claims is based on this restriction.

If we now return to the "organizational" attacks against the synthetic theory, we see that they break with this strategy of research. Instead of exploring the phenomena that illustrate the theory of natural selection as an extremal theory, they systematically look for phenomena that would indicate the limits within which the theory operates. Goodwin or Kauffman do not say that natural selection does not act on genomic or morphogenetic systems, they say those systems have a certain logic (generic properties) which is not sufficiently explained by the historical and cumulative action of natural selection. I see no reason to disqualify this attitude *a priori* as a *research program.*

One must, however, be conscious of the implications of such a program. If evolutionary theory turns into an investigation of properties of living systems that are not reducible to an opportunistic tinkering, it will increasingly be a theory of what *does not* change in evolution (at least at a given scale of observation). In other words, it will increasingly be a structuralist science, and decreasingly a historical science. It will indeed be a transcendental science, that is, a science of the most general conditions of possibility for any evolutionary process. Once again, there is no *a priori* reason for such a program to fail. It is a very old cliche that the more you move in the direction of formal constraints, the more you lose in information about the history of the system you examine. In that sense, one may have some doubts that a genuine transcendental science of evolution (one that would discover "biological universals" or "laws of biological order") could provide an alternative solution to the kind of problems the synthesis has tried to disentangle, such as intraspecific variation, speciation, and paleobiological patterns. If this general science can exist, the problem will more likely be that of conciliating it with the vast empirical construction of the synthesis, rather than that of replacing the latter. However that may be, there are obvious signs of an alternative organizational paradigm, not in the sense of an alternative truth, but in the sense of an alternative research program.

ONTOLOGY

Modern Evolutionary Theory and Ontology: A Brief Historical Survey

We now come to a puzzling debate to which the philosopher cannot be indifferent. For several years the epistemology of evolutionary theory has largely turned into an ''ontological'' reflection. Two books among those examined here are explicitly devoted to an examination of the ontological structure of evolutionary theory (Eldredge, 1985; Salthe, 1985). In general it appears that a good deal of current critical literature on the synthesis is similarly concerned with ontology through two themes: species as individuals, and hierarchical structure of evolutionary theory. Before examining these two themes, I want to clarify the relationship between ontological infatuation and critical attitudes toward the synthesis.

It has often been argued that the synthetic theory ''suffers from a lack of articulation between its parts (Salthe, 1985, p. 187). The three major parts of the synthetic theory are: (1) a theory of gene frequency dynamics in populations; (2) speciation theory; and (3) a conglomeration of patterns and interpretations relating to macroevolution. Now let us ask whether the whole construction may claim to constitute one single unified theory of evolution. From a purely epistemological point of view, this looks like a rather daunting task (see, however, Caplan, 1980). Things are perhaps clearer from a historical point of view. In fact, 50 years of internal controversies within the synthesis show that the unity of the ''theory'' has been seriously jeopardized on two critical fronts. The first difficulty arises from the theory of speciation. It is well known that the founding fathers of the synthesis, in spite of many conciliatory arguments, seriously disagreed on the meanings of ''species'' and ''speciation.'' One of the conflicts was that between Dobzhansky and Mayr concerning the adaptive (or nonadaptive) significance of speciation. This conflict involved the ''genetic side'' of the theory. The other conflict was between Mayr and Simpson about the very concept of species—biological versus evolutionary. This controversy concerned the ''paleontological'' side of the speciation theory. These two conflicts are very clearly stated in Eldredge's (1985) book and show that speciation theory is anything but a unifying doctrine in the synthesis. It is a central doctrine, but in the sense that it has created a permanent controversy, a controversy which I see as an essential feature of the synthesis.* The second difficulty lies in the ''ex-

*It would be beyond the scope of this chapter to enter into the detailed history of controversies over species and speciation within the synthesis. It is fairly well-established today that the three major actors of the synthesis (Dobzhansky, Mayr, and Simpson) *did not* agree on the species concept, in spite of many efforts for presenting it as a central piece of the synthetic consensus. Dobzhansky and Mayr disagreed on the role of natural selection in speciation. For Dobzhansky, isolating mecha-

trapolationist'' argument. Mayr's argument stated that all the processes and phenomena of macroevolution were nothing but an ''extrapolation'' of those of microevolution, that is, intraspecific variations and speciation (Mayr, 1942, p. 298). Nearly half a century after its formulation, this argument looks like an idealized expression of the unity of the synthetic theory. As a matter of fact, this assumption was sufficiently ambiguous to evoke the suspicion that the theoretical unity of the synthesis was based on purely rhetorical grounds.

At this point it is easy to understand how a critical reflection on the theoretical structure of the synthesis turned into an ''ontological'' inquiry. First, the ambiguity of the species concept gave rise to a direct examination of the kind of entities species were. Thus appeared, in a purely orthodox Darwinian context, the idea of species as individuals (Ghiselin, 1969, 1974, 1987*a,b;* Hull, 1976, 1978). This idea quickly became enmeshed with controversies over the units of selection, as well as with the punctuationists' attacks against the synthesis, and finally resulted in a general investigation into the nature of the entities involved in the evolutionary process. Although at first confined to a debate on selection, this ontological investigation recently surfaced as a conceptual tool for casting discredit on the synthetic ''extrapolationist'' argument as reductionist as well as for hierarchically restructuring evolutionary theory. The books by Eldredge (1985) and Salthe (1985) are examples of this strategy, which they had already described in a joint article (Eldredge and Salthe, 1984).

In the following pages, I will develop some critical observations about the idea of the individuality of species. I will also discuss the hierarchical structure of evolutionary theory. A great deal of intellectual confusion has accumulated in both cases.

What Does It Mean That Species Are "Individuals"?

Rosenberg (1985), in his chapter devoted to ''species,'' recalls the origin and context of the notion of species as individuals. The synthetic theory did not succeed in developing a species concept which would incorporate the whole array of phenomena commonly covered by the word ''species.'' All of the

---

nisms are the result of a special selective process; for Mayr, they are incidental effects of adaptation in different environments. Another fundamental disagreement bears on the dimensionality of the species concept. Dobzhansky's and Mayr's species concept is a nondimensional one, because it is based on *potential* interfecundity; Simpson's concept is based on the notion of the continuity of lineage (Simpson, 1961) and is thus dimensional. More recently, Van Valen's (1976) ecological concept or Paterson's (1978, 1979) recognition concept reinforce an approach to the species problem in positive terms, species being thought of as cohesive wholes. Thus, it is not obvious that any novelty in the species debate might be a serious challenge to the synthesis. To a historian of science, it seems that there has never been one single clear-cut ''synthetic'' concept of species.

concepts of species that have been proposed as alternatives to the original essentialist notion (Mayr's biological concept, Simpson's evolutionary concept, and more recently Van Valen's ecological concept; to this list I would like to add Paterson's "recognition concept"), are both methodologically ambiguous and nonexhaustive. For instance, the biological concept of species is not operational in its use of the notion of potential interfecundity, it does not apply to asexual organisms, nor does it have any clear meaning for a paleontologist because it is nondimensional. The other modern concepts face similar difficulties, which are familiar to readers of *Evolutionary Biology*. Therefore, the modern synthesis may be said to have failed in its attempt to construct a species concept as a natural kind: "There is no single type of event properly called speciation, nor is there a single type of collection of organisms properly called species" (Rosenberg, 1985, p. 202).

To put it briefly, modern evolutionary theory is not far from reaching the conclusion that the very nature of species is a verbal fiction. Naturally things are rarely if ever stated this baldly, for the simple reason that speciation theory is an essential part of the modern synthesis (see footnote to pp. 27–28).

Within this context Ghiselin (1969, 1974, 1987*a,b*) proposed his famous "radical solution." The solution is based on arguments of a logical nature. Species are not classes, but individuals, in the meaning of "singular things." They are not classes because there are no instances of a given species, but only parts or constituents thereof. They are "individuals" because they have a spatially and temporally determined existence; for that reason they receive a proper name. These logical arguments are inspired in some degree by a well-known metaphysical essay of P. F. Strawson on "individuals" (Strawson, 1959). Ghiselin also observed that treating species as individuals conforms to three fundamental evolutionary notions: first, the notion that species evolve (classes do not change, only individuals transform themselves); second, the importance of intraspecific variation (a fact which makes many species indefinable); finally, the fact that species interact (e.g., competition, predation).

Ghiselin's proposal must be appreciated within its own specific context. It is an attempt to solve a problem of definition. Conscious that the very word "species" was first a technical term in philosophy and that it carries within itself some traces of the old problem at universals, Ghiselin proposes a clarification of the biologists' language by confronting it with some basic notions of logic. The biologists' "species" does not have the status of a class, that is to say: a collection of entities which have some property (predicate) in common. In other words, biologists are concerned with "species" which are not "species" in the philosophical sense, but the exact opposite: They are "individuals," that is, logical subjects which admit predicates (properties), while they cannot themselves be predicates for any other subjects (Leibniz, 1686). Finally, the thesis that "species are individuals" means neither more nor less than that species are real, an assertion which expresses the conviction of all syntheticists over and above their disagreements.

One more comment on Ghiselin's analysis. I have characterized this analysis as based on arguments of a ''logical'' nature. This is only an approximation. In fact, Ghiselin uses some current notions in logic. For instance, the opposition between terms of ''properties'' (predicates) and terms designating ''individuals'' is a traditional one, in ancient logic, as well as in modern predicate logic. Similarly, ''class,'' ''instance,'' and ''part'' are obvious borrowings from formal logic. However, it would be more rigorous to say that Ghiselin uses ''logical'' notions within an argumentation of a philosophical nature. This is particularly important in the notion of ''individual.'' For a modern logician, there is no necessity whatsoever that an ''individual'' (as opposed to a ''property'') should be a singular being, with spatiotemporal limits. In most standard presentations of predicate logic, integers (1, 2, 3, . . .) are classical examples of ''individuals,'' in reference to a certain domain of discourse. Within this context, ''individual'' is a formal notion, independent of any metaphysical thesis on ''reality.'' Clearly enough, Ghiselin's use of the concept of ''individual'' is not purely formal: it must be placed in the context of the old philosophical problem of the ''universals.'' This is, moreover, perfectly explicit (Ghiselin, 1987*b,* p. 207). Consequently, my presentation of Ghiselin's thesis, with reference to Strawson's essay (an essay of ''descriptive metaphysics'') and to Leibniz, should make clear that the thesis of ''species as individuals'' can only be a metaphysical thesis, although based on elegant logical arguments. This consideration largely accounts for the confusing debate that resulted from Ghiselin's proposal. In the following pages, I propose three remarks on this debate: first, I examine the very different meanings the ''species qua individuals'' may have for evolutionary biologists, according to various classical metaphysical connotations of the term ''individual''; second, I submit that the formula ''species as individuals'' is able to express two radically opposed philosophies of evolutionary biology; third, I contest the formula on specifically logical grounds.

Remark 1. There are many possible meanings of ''individual,'' and correlatively many meanings of the individuality-of-species thesis.

The word ''individual'' now has many very strong meanings, which rapidly swamped the rigorous logical-metaphysical one that Ghiselin gave it. ''Individual'' (*individuum*) is the exact Latin translation of *atomos* (atom); in other words, that which cannot be divided. This negative notion was always interpreted in two ways: The individual may be thought of either as an elementary particle (the Greek materialist's ''atom''), or an irreducible whole, that is, something which cannot be divided without losing its own being (from the second perspective, the paradigmatic individual is an organism). In other words, the same word is an expression of both atomistic and holistic thinking. To these philosophical connotations must be added a popular one which is of a bio-anthropomorphic nature: An individual is a being who was born, who lives, and who will die. It does not take much imagination to guess the possible effects of such connotations on modern evolutionary theory. If species are individuals in the sense of irreducible wholes, they are superorganisms; this is an idea which is involved in some aspects of the synthesis (e.g., the notion of integrated gene pool or genetic homeostasis). If species are "individuals" in the sense of particles, one may transpose to their level such notions as mutation (speciation), natural selection (species selection), and genetic drift (species drift). This is what happens in the theory of punctuated equilibria. In this context, if species are considered as individuals, a clear hierarchical argument against the extrapolationist strategy of the modern synthesis appears. We are confronted with the paradox of an argument (that of Ghiselin) which was primarily conceived in a typical Darwinian and syntheticist atmosphere, but which finally turns into an argument against the neo-Darwinian synthesis.

Remark 2. The thesis that species are individuals is capable of expressing two radically opposed philosophies of evolutionary biology.

In the precise sense advocated by Ghiselin, the thesis meant that species are historical singularities. Therefore, one must not expect to discover any general laws about them (Hull, 1976).* As Rosenberg (1985, Chapter 7, §9) says, evolutionary biology—and for that matter biology as a whole—is a natural science that cannot go beyond case studies. In opposition to physico-chemical sciences, which can claim to be about natural classes of objects, biology never succeeds in formulating laws without exceptions. In this way, treating species as individuals makes one think of Jacob's conclusion that life, even at the molecular level, is always "tinkering" (Jacob, 1981). The consequence for evolutionists is that they must be conscious that they are studying a historical science within which all events affecting species are unique, so that the notion of a transpecific evolution is not a rigorous one. This is what Hoffman (1983, p. 251) calls "the individualistic view of life."

On the other hand, if the notion of species as individuals means that at a certain level of observation species can be seen to behave as elementary particles, a more uniformitarian and less historical perspective inevitably follows. Equilibrium laws will be conceivable at this level, and a less contingent vision of the history of life will seem reasonable. Species' adaptive opportunism will

*It is somewhat rhetorical to say simply that, species being historical singularities (individuals), there are no laws for them, nor for taxa. In fact, Hull and Ghiselin seem to have evolved toward a more subtle formulation of this thesis. They now admit that, even though one cannot establish laws for an individual species or taxon, one can conceive of laws for classes of species, as long as those classes are not confounded with the taxa. For instance, asexual species perhaps constitute a natural class, while mammals, or primates, or *Homo sapiens* are taxa, that is, historical singularities. I am indebted to both authors for this observation.

appear as canalized by a dynamics operating on a higher level. Macroevolution will therefore be understood as a genuine transpecific evolution, with its own regularities. The ''species-as-individuals'' dictum is therefore capable of attracting what Schopf (1982) calls two fundamentally different approaches to evolutionary events: the equilibrium approach, which focuses on micro- and macroevolutionary laws underlying the diversity of life; and the historical approach, which considers evolutionary events as unique results of microevolutionary laws operating under particular conditions, both physical and biological (see also Hecht and Hoffman, 1986).

Remark 3. An alternative view of the logical structure of the Darwinian concept of species is possible.

Finally, I want to suggest that, even on purely logical grounds, the notion of species as individuals may be open to some criticism. It is not at all clear that the logical structure of the Darwinian concept of species is correctly expressed by the notion ''individual.'' Ghiselin's analysis may be compared to a proposal by Nowinski (1967). Nowinski argues that the Darwinian notion of species must be analyzed logically in terms of an original structure of ''class relation.'' The Darwinian notion of species implies a fusion of two fundamentally distinct logical categories, classification and ordering. In a certain sense species are classes because the objects of which they are composed manifest a certain number of similarities. However, these similarities are never more than indications of a certain position in a genealogical system, which is a system of order relations. A fusion between classes and order relations is nothing extraordinary in itself. The idea of ''number'' is another classical example of such a fusion. Yet the class-relation structure underlying the Darwinian species concept has something very original: While the idea of number requires that the qualities of objects be left out of account, the notion of species requires that the qualities of the object be considered, both in their similarities *and* differences. In other words, differentiation is the key word in a logical description of Darwinian species.

The implication for our subject is straightforward. If evolutionary theory is a theory of descent with differentiation, it forbids a consideration of the continuity of transformations separate from the discontinuity of forms (the ''relation-of-order'' aspect from the ''class'' aspect of the groups named species). There are two ways of neglecting the Darwinian methodological principle. One is to ignore the genealogical aspect of species: Species then appear as forms or types, making it possible to view them as classes of objects which share the same set of definite attributes. A more subtle temptation is to neglect the class aspect, and to focus exclusively on the genealogical one. This is precisely what happens when one tries to represent species in terms of individuals. Putting differentiation aside, species appear as individuated segments in the course of the evolutionary process, ''spatiotemporally *bounded* entities'' as many people call them. However, this vocabulary stressing limits is questionable. Species certainly can be described in a spatiotemporal manner; they have a spatiotemporal *determined* existence. But this does not mean that they necessarily have limits, particularly temporal limits. Species are processes, not entities. One could make this comment in more general terms. Recently there has been a tendency to structure the whole of evolutionary theory in ontological terms. This attitude leads to identifying ''individuals'' of many sorts and many levels. I will return later to one of these theoretical constructions (see p. 39–62). Suffice it to say that it is rather puzzling to see modern evolutionists trying to organize their theory as a speculation on *things* rather than on *phenomena.**

Synthetic Theory and Hierarchical Thinking

It would be difficult to imagine a more confusing debate than this one. ''Hierarchy'' has become a very fashionable word in current evolutionary theory. Among the six books examined here, three—Eldredge (1985), Pollard (1984), and Salthe (1985)—are concerned primarily with a hierarchical approach to evolution. In the following pages, I will analyze three aspects of the current controversies on hierarchies and evolution. A preliminary analysis of the notion of hierarchy itself will also be given.

What Does ''Hierarchy'' Mean?

From a philosopher's vantage point, I see three different ways of approaching the question. I do not aim here at anything other than outlining a general conceptual framework. For more details, see the classic literature cited in the footnotes.

Hierarchical Thinking as an Epistemological Preference. Hierarchical thinking is a conceptual tool which has long been used in virtually all scientific disciplines. There has been a tremendous amount of literature on the subject since the beginning of the century.† The idea common to all supporters of hierarchical thinking is that in any scientific field there are certain organizational properties that can come into operation only in connection with specific stages of

*For a more detailed elaboration of the present argumentation, see Gayon (1988*b*).

†For a panoramic view of hierarchical thinking, see Koestler and Smythies (1969). Whatever one may think of Koestler, this conference provides a fairly good perception of the conceptual and disciplinary constellation underlying hierarchical thinking. For recent developments, Salthe (1985) provides much useful information.

complexity (Needham, 1936, p. 165). This idea implies that theories are structured in terms of "levels," each level being both a part and a whole. For this reason, Koestler (Koestler and Smythies, 1969; Koestler, 1978) proposed replacing the old word "hierarchy," full of religious and political connotations, by the term "holarchy," which is supposed to express the idea of "holons," or intermediate structures in an increasing order of complexity. In its most refined aspects, the theory of hierarchy is closely related to systems theory. More commonly, it is a strategy of research which finds its typical expression in a dendrogrammatic representation of phenomena.

This epistemological aspect of hierarchy has already been popularized; it is not necessary to emphasize it. I simply add that although often used in many fields of research (documentation sciences, computer technology, psycholinguistics, social sciences), hierarchical thinking has a particularly strong affinity with biology. It is reasonable to say that since the beginning of the century most of the major debates in biology have been the occasion for repeated questions about the possibility and the meaning of a hierarchical structuring of theories. Experimental embryology, ecology, molecular biology, and neurophysiology would best illustrate this assertion. After much disagreement over the reducibility of higher level theories to lower level theories, the most common outcome of this kind of debate is that investigating the higher level in and of itself is at any event a wise heuristic strategy (Nagel, 1961). As will be shown, modern evolutionary controversies are no exception to this rule.

Hierarchy and Metaphysics. If one speaks in terms of metaphysics, there are two topics that are clearly related to the hierarchical theme. It will be enough simply to name them.

Among other things (soul, God), metaphysics is concerned with the ultimate meaning of the word "world" (or "nature" if we consider it in a dynamic way). The concept of "world" is that of the totality of all possible phenomena accessible to human experience. Given that concept, one might ask whether or not anything simple can exist in the world. If such absolutely simple things exist, any composite thing whatsoever would be a mere aggregate of simples; the world itself would be such an aggregate. In other words, the only thing that really exists is the simple. The alternative view consists in saying that nothing simple exists in the world. Therefore, any composed thing is made up of parts, which themselves are made of parts, *ad infinitum*. In other words, organization, that is, the whole–part relationship, is the ultimate structure of the world. Hierarchical schema obviously belong to the second alternative. The opposition between these two concepts of world has been debated over and over for centuries. Immanuel Kant (1781) devoted a good deal of his *Critique of Pure Reason* to showing that questions of that sort were beyond any experience, and therefore are constitutive "antinomies" of reason. Consequently, Kant said, they cannot pretend to structure any scientific inquiry, if science is concerned with what is accessible to us in experience. I strongly recommend that any naturalist who would believe that being "for" or "against" hierarchy is of primordial importance read Kant's "Second Antinomy of Pure Reason" and its "solution" as given in *The Critique of Pure Reason* (Kant, 1781).

The second metaphysical issue is transcendence, as clearly appears in the etymology of the world (*hieron* means "sacred"). In this sense, hierarchical is defined as any reality the principle of which is sacred. Here we find the original meaning of "levels": a "higher" level is of another order of reality, in such a way that the "lower" level proceeds from the higher, while being incommensurable with it.

To sum up, from a metaphysical point of view, hierarchical schema belong to the related realms of cosmological and religious questions.

Hierarchy and Anthropology. We now come to the anthropological roots of "hierarchy." In Western culture, hierarchy most often appears as a political category, the clearest expression of which may be found in the sociological notion of "social stratification," with its connotation of levels of complexity, each defined by its own political-economic structures and rules. However, as can be seen in anthropological studies of the ancient tradition of castes in India, this is an oversimplified perspective: As a matter of fact, a nonhierarchical culture sees hierarchy in terms of political and economic power. In other words, our own cultural context limits us to seeing only some products of hierarchical schema (ranking, levels, primacy of whole or parts); we are blind to the underlying mental structure. Hence the interest of examining social structures which in their very principle are inspired by a hierarchical conception of man, society, and the universe. I will content myself here with providing an example of the kind of general conclusions to which such an approach to hierarchy might lead.

One of the most fascinating and controversial anthropological explanations of the hierarchical mentality is probably Dumont's (1966, 1979) *Homo Hierarchicus*. Dumont's main conclusions are stated as follows: (1) The main point in hierarchy is the omnipresence of a relationship that can be described as "the embodiment of the opposite" (*englobement du contraire*). This relationship is the ultimate logical root of the notion of levels. In the Bible, the creation of Eve from Adam provides a good example: At a first level, man and woman are identical (Eve is an element of the human genus, the prototype of which is Adam; this is rendered symbolically by saying that Eve is materially embodied in Adam as one of his ribs); at a second level, woman is the opposite or the contrary of man. (2) In a hierarchical scheme, levels can be multiplied without the generative law being altered (e.g., in India, the opposition purity/impurity generates an indefinite series of caste distinctions). (3) Hierarchy is basically a cosmological and religious schema rather than a political one.

Is the Synthetic Theory of Evolution a Hierarchical Theory?

The preceding analysis of hierarchy enables us to provide a concise answer to this question.

If we look at the pioneers of the synthesis, we see that some of them explicitly assumed a "hierarchical" representation of evolution. This was associated with a hierarchical epistemology, and ultimately with a hierarchical representation of reality. Wright is the best example, as clearly appears in his philosophical writings (Wright, 1964), as well as in his evolutionary practice. Wright's evolutionary thinking is paradigmatically "hierarchical" in all the meanings distinguished above, particularly the third, or anthropological, meaning. If we transpose Dumont's anthropological description, we see as a pure matter of fact that: (1) As a statistician, Wright formalizes the notion of genic selection in such a way that it embodies its own opposite (genetic drift). This strategy results in the production of two levels of selection (intra- and interdemic selection). (2) There is a strong tendency in Wright's thinking to multiply the "levels," according to a common scheme of opposition between deterministic and stochastic processes: What is deterministic at one level is stochastic at another (e.g., speciation is equated with mutation). This suggestion opened the road to the species selection concept (Moorehead and Kaplan, 1967; Gould and Eldredge, 1977). (3) Wright does not find philosophical justification in political metaphors, but rather in a panpsychical cosmological mediation (Wright, 1964).

Of course there have been other pioneers of the synthesis who assumed the exact opposite of a hierarchical conception of evolution in any sense of the term. Fisher is a typical case: atomistic schemes of explanation, refusal of "levels," metaphors and motivations mainly drawn from political interests (eugenics, etc.). Between these two extremes, all the imaginable intermediates are easily recognizable. The conclusion is that the modern synthesis was neither a consensus in favor of, nor a consensus against, a hierarchical concept of evolution. In other words, hierarchy is not crucial for the synthetic doctrines, it is rather a matter of personal epistemological and metaphysical preference.

The "Extrapolationist" Argument: nth Manifestation of the Problem of Reductionism

The preceding conclusion must be tempered by taking into account the objection often addressed to the synthesis by those who advocate a restructuring of evolutionary theory. Their objection focuses on the "extrapolationist" conception of macroevolution shared by most—if not all—syntheticists. The extrapolationist strategy of explanation was first denounced by Gould (1980); it is widely discussed in several of the books examined here (Gould, 1983*b;* Ayala, 1985; Eldredge, 1985; Salthe, 1985). Although familiar to contemporary evolutionists, the notion of extrapolation deserves a rigorous epistemological discussion, in the sense that it is indeed one of the clearly identifiable fundamental assumptions of the synthesis.

As mentioned by Eldredge (1985), Mayr used the term "extrapolation" as early as 1942 in the concluding sentences of *Systematics and the Origin of Species:*

> All the available evidence indicates that the origin of the higher categories is a process which is nothing but an extrapolation of speciation. All the processes and phenomena of macroevolution and the origin of the higher categories can be traced back to intraspecific variations, even though the first steps of such processes are usually very minute. (Mayr, 1942, p. 298.)

This assertion was later refined to express the greatest possible consensus within the synthesis: Macroevolution is nothing but an extrapolation of microevolution, provided that in "microevolution" are included: (1) all known processes and patterns of anagenetic phenotypic change; (2) speciation, in its diverse possible modes; (3) underlying genomic change (Bock, 1979; Hecht and Hoffman, 1986). Actually, the problem is understanding what "extrapolation" means. Gould (1980) interprets it as "a strong faith in reductionism," an attitude sometimes fully assumed by syntheticists themselves. (e.g., Bock, 1979) However, this is not the usual case, and the term "extrapolation" or "amplification" appears to be a mysterious compromise. The oddness of the word is readily apparent if one replaces it by a Greek equivalent. Such an equivalent might be *"hyperbola,"* in both its rhetorical (hyperbole: exaggeration) and geometrical (hyperbola: plane curve) meaning. Does the microevolutionist know the equation of the curve well enough to be able to predict *a priori* what happens at its limit? This metaphor indicates the kind of ambiguity affecting the notion of "extrapolation," that macroevolution, although not deducible from microevolution, is not strictly speaking irreducible to it.

It seems reasonable to treat the problem by using the classical epistemological notion of reduction. Ayala (1985) provides a good example of this approach. He distinguishes three issues: (1) Do microevolutionary processes underlie macroevolutionary phenomena? The answer is very likely yes. It is reasonable to believe that presently observed microevolutionary processes did apply to the organisms and populations which made up the taxa or communities studied by macroevolutionists. (2) Can microevolutionary processes account for macroevolution, or is it necessary to postulate different processes, such as Goldschmidt's system mutations? Ayala answers that microevolutionary processes *can* account for macroevolutionary change, in its observed tempo. I think that both this question and its answer are rather ambiguous. They strongly resemble Mayr's "extrapolation." (3) Can macroevolutionary theory be derived from microevolutionary knowledge? This question is that of reducibility in the precise meaning that word has in epistemology (Nagel, 1961).

An important remark must be added at this point. As noted by Nagel (1961), there is never an *a priori* solution to reducibility questions, and the answer is largely independent of metaphysical representations of reality. This means that there is a difference if one speaks of a hierarchical organization of theories (which means that they are not "reducible" or of a hierarchical organization of things and processes in themselves. A practical consequence of this is that hierarchical proclamations can most often be interpreted as a claim for autonomy in research:

> A wise strategy of research may indeed require that a given discipline be cultivated as a relatively independent branch of science, at least during a certain period of its development, rather than as an appendage to some other discipline, even if the theories of the latter are more inclusive and better established than the explanatory principles of the former. (Nagel, 1961, p. 445; see also Hempel, 1966.)

The reducibility question bears on scientific statements, not on things. There are two conditions for one theory to be reducible to another: First, the concepts of the reduced theory must be defined in terms of the reducing theory (condition of connectability); second, one must be able to deduce the laws of the reduced theory from the laws of the reducing theory (condition of derivability). In the case we are considering it is clear that the condition of derivability is not satisfied: Macroevolutionary theories cannot seriously be presented as logical implications of microevolutionary theory. In that sense, Ayala justly concludes that there is an obvious autonomy of paleontological theories.

These epistemological remarks clarify the troublesome question of "extrapolationism." Mayr's dictum carries the conviction that "there is no transpecific evolution" (Bock, 1979, p. 27). Such an assertion means that there is no need for a special theory of evolution above the species level, because there are no macroevolutionary *processes*. Heuristically speaking, at least, in retrospect this position appears to be exaggerated. Whatever may be the outcome of the debate on macroevolutionary processes, a paleobiologist is heuristically justified in claiming methodological autonomy. If indeed there are macroevolutionary *processes,* the justification of the paleobiologist will be to find them. If there are none, the paleobiologist's work is that of a historian, comparable in its relation with microevolutionary theory to the historical work of the astronomer in relation to nomothetic research of physicochemical sciences.

I now combine my previous assertions (p. 36) that hierarchy is not crucial for the synthetic doctrines with the present discussion on the "extrapolationist" representation of macroevolution. As a theory of microevolution, the synthesis recognized a certain number of levels of *description* that nobody ever contested. These levels of description stand between gene and species; they vary in number and nature according to the context, but the important point is that they are operationally interconnected. This is why it is a matter of philosophical preference to consider the whole array of *micro*evolutionary phenomena as being either of a "hierarchical" or "atomist" nature. In the case of macroevolution, we only have a loose conglomeration of "patterns," which according to the syntheticists are not operationally connected with the microevolutionary corpus in a clear way (they are only compatible with it). In this context, the extrapolationist argument is a rhetorical one in favor of a unified theory of evolution. It is as clear—neither more nor less—as was the claim in the 1700s that Newton's principle of gravity explained not only the present behavior of the solar system, but also its origin, a claim that was finally revealed to be partly truc (cf. Laplace) and partly false (cf. nuclear physics). Correlatively, the "hierarchical" pleading of many paleobiologists resembles an efficient argument in favor of heuristic autonomy. Therefore, one may think that it does not concern so much the structure of evolutionary *theory* as the control of research programs.

"Ontological" Evolutionary Hierarchies

I now turn to the arguments of those who advocate a fully hierarchical structure of evolutionary theory. This recently took the form of an enquiry into the kind of entities involved in the evolutionary process. The principal objective of the books of both Eldredge (1985) and Salthe (1985) is to propose a general ontological framework for evolutionary theory. Both authors undertake to structure evolutionary thinking along two axes representing two hierarchies of evolutionary entities, the axis of "genealogical" entities and the axis of "ecological" entities. Genealogical entities are those which transmit and transform information (in Eldredge's schema: gene, organism, deme, species, monophyletic taxa). Ecological entities are defined in terms of matter and energy transfer (Eldredge: proteins, organisms, populations, communities, and higher-level ecological entities). The relationship between the two hierarchies is that "the genealogical hierarchy simply supplies the players in the ecological arena" (Eldredge, 1985, p. 180). All entities are thought of as "Individuals." Only one of them (organism) belongs to the two hierarchies, and stands as the paradigmatic individual. Dissociation of the genealogical and the ecological aspects in the other entities requires the acceptance of a counterintuitive ontology, according to Hull's (1980) expression (although with a very different perspective on the synthesis). This is the price paid for surpassing the extrapolationist simplifications of the modern synthesis.

I would like to express some fundamental epistemological and philosophical reservations on this fashionable schema. I begin with the epistemological objections.

The distinction between "genealogical" and "ecological" entities obviously derives from Hull's distinction between "replicators" and "interactors" (Hull, 1980, 1981). In other words, as noted by Eldredge himself, it takes root in a typically Darwinian discussion of natural selection. Hull's distinction may be viewed as a generalization of the traditional genotype–phenotype distinction (Brandon, 1985, p. 82) or even more broadly as a generalization of Weismann's distinction between germen and soma (Eldredge, 1985, p. 8). We are therefore back at the starting point in a neo-Darwinian atmosphere. The "genealogical–ecological" duality itself results from a generalization of the replicators–interactors distinction. A genealogical entity (Eldredge, 1985, p. 186) is an entity that "makes additional entities of like kind." This definition makes it possible to consider not only genes, but also organisms, demes, and species as "genealogical entities." But we must observe that in so doing, Eldredge rids himself of the most characteristic constraint of the replication idea, or the persistence of a *structure*. In Eldredge's schema, "structure" is replaced by "information"; genealogical entities are "packages of information" (Eldredge, 1985, pp. 107, 137). Such a concept is obviously much less restrictive than that of a replication of structure.

This conceptual shift is essential and also questionable. From this, one might infer that the revolution brought about by Mendel in his theory of inheritance is of minor importance. Eldredge's genealogical entities imply a concept of inheritance in which Mendelian determinants are but one possible case among others. This is simply amazing: An entire century of experimental investigation on heredity is presented as merely a detail. All of the methodological constraints that little by little formed the notion of heredity as distinct from reproduction suddenly disappear in the "genealogical entities." Genealogical entities are not defined as units of transmission (Mendel); they are not causal determinants (early genetics); they are not physical entities located somewhere and capable of replicating (chromosomal theory of inheritance); finally, genealogical entities are not required to have a structure (molecular biology). What does remain then? Only filiation and resemblance, the two notions that defined "information" within pre-Mendelian concepts of inheritance ever since Aristotle. This is why Eldredge uses "genealogical" and "reproduction," carefully avoiding the term "heredity." My impression, however, is that the idea of "genealogical entities" appears to be a subtle rehabilitation of the old notion of a multistoried heredity. At the beginning of the century it was common to oppose two modes of inheritance, "special inheritance" and "general inheritance" or sometimes even "special inheritance" (that is, individual plus social inheritance), "specific inheritance," and "supraspecific inheritance." These notions were commonly used against Mendelism (cf. Guyénot, 1924, pp. 13–15, 287). I have the impression that "genealogical entities" (genes, organisms, demes, species, monophyletic taxa) look a little like this pre-Mendelian representation of heredity in terms of various ancestral "influences" (individual, racial, specific, supraspecific legacy).

Moreover, it must be observed that the generalization of "replicators" in "genealogical entities" produces a similar ambiguity with reference to "interactors.'' According to Hull (1980, p. 318), the interactor concept is integrally associated with that of replicator. An interactor is ''an entity that directly acts as a cohesive whole with its environment in such a way that replication is differential.'' Eldredge (1985, p. 183) and Salthe (1985, p. 288) think this concept is too restrictive and that it must be freed of its historical dimension. Hence the definition of ecological entities as ''primarily spatially construed.'' This definition is tantamount to emancipating the interactor concept from natural selection. Very logically indeed, natural selection becomes a particular use of a more general process called ''sorting'' (Vrba and Eldredge, 1984; Eldredge, 1985, pp. 202–205). Therefore, the whole theoretical construction of genealogical and ecological entities appears to be a meticulous dismantling of the neo-Darwinian fortress by pulling out the operational content of its two components, genetics and selection theory.

My second reservation with respect to the genealogical–ecological hierarchies is more philosophical. A philosophical observer cannot help being impressed by the naturalist's determination to reconstruct the epistemological foundations of the discipline in a resolute language of individuality. Salthe's and Eldredge's hierarchical ontology of evolution resembles a general metaphysics of life. Viewed as a metaphor, the genealogical–ecological epic may be presented in the following way: It is a reconstruction of nature based on the utilization of two remarkable and venerable dualistic schema. The first duality is that of time and space, succession order and coexistence order. Genealogical entities are characterized by their temporal continuity, ecological entities by spatial structure. A second duality is superimposed on the first: form and matter. Genealogical entities transmit and transform ''information.'' They inform in the Aristotelian sense. Ecological entities are nothing but places traversed by matter and energy flows. It is true in a certain sense that an ontology which reconstructs the notion of biological individuality by dissociating time and space, form and matter, challenges common sense (Eldredge, 1985, p. 124). In fact, such an ontology challenges the most powerful connotations of individuality: (1) the notion that an individual is a spatiotemporal singularity, and (2) the representation of the individual as a compromise between form (which does not admit division) and matter (which accepts division) (Canguilhem, 1975, p. 62).*

However, Eldredge's and Salthe's axes are not completely dissociated. They meet in one—and only one—point: the organism, which therefore is the paradigmatic individual. Moreover, as it is the intersection of two axes, the individual organism generates a representation of the entire universe. From this central meeting point, it is indeed possible to contemplate both the ecological (or material) hierarchy, which expands indefinitely in the cosmos, and the genealog-

*''L'individu, c'est ce qui ne peut être divisé quant à la forme, alors même qu'on sent la possibilité de division quant à la matière'' (Canguilhem, 1975, p. 62).

ical (or formal) hierarchy, which is "truncated" (Salthe) and does not extend beyond the biosphere.

Such underlying metaphysical and symbolic schemata are powerful enough to account for the seductiveness of the genealogical–ecological hierarchy. There is no reason why it would not stimulate the scientific imagination. If it does so, however, it will be with the creative power of myth.

However, this evaluation should not be taken too negatively. Eldredge's and Salthe's books are fascinating, in that they show the effort of scientists to face the philosophical implications of their own field. If I take off the polemical aspects of my argument, the main thing is: I am skeptical about the idea of construing the epistemology of a whole field of research on the basis of the ontological notion of individual.

CONCLUSION

In the introduction of this essay, I mentioned my skepticism concerning the philosopher's ability to behave as an "evaluator of the status of modern evolutionary biology" (or of any other scientific theory). However, it is always in periods of crisis that philosophers are invited to participate in their debates: in these periods, the boundary between science and other kinds of speculation is less clear than ever. The considerable number of philosopher's contributions in books or journals devoted to evolution seems to indicate such a state of crisis. The role of the philosopher in a situation of this sort is less to pronounce on the validity of the theories than to settle as clearly as possible the issues. This was the strategy of Bishop Berkeley, himself following the advice of Father Malebranche: When there is some obscurity in any problem, they say, you must try to formulate an alternative in the crudest possible way. This is precisely what I will do in this conclusion, my purpose being to provide landmarks rather than solutions.

Let us suppose, not without some sort of naivety, that the question is: Is the synthetic theory of evolution seriously challenged by the many criticisms presently developed against it? I see two categories of answers.

1. The synthesis is not a well-defined "scientific theory." The major argument in favor of this assertion would be that the so-called theory is a loose conglomeration of subtheories, with no central doctrine. In this perspective, the identity of the synthesis is mainly sociological: It is a tacit agreement on a number of problems which are worth being explored in so many academic specialties. Moreover, the cohesiveness of the synthesis relies mainly on a common reference to a series of exemplary figures and texts (Dobzhansky, Mayr, etc.), not on clear-cut doctrines.

Anyone who accepts such a representation of the "modern synthesis" must accept the following consequence: There do not exist anything like *criticisms* of the synthetic *theory. Simply, there are many contestations of the authority of a certain lobby.*

2. The synthesis is a well-defined theory. The main tenet of this theory is that it is plainly Darwinian, in the sense that it assumes that natural selection is the dominant factor in organic evolution. This characterization may seem naive, in that it does not take into account the numerous differences between Darwin's and the syntheticists' representation of "natural selection." It is interesting, however, to recall what the thesis of the predominance of natural selection meant in Darwin's mind. Darwin conceived natural selection as explaining both the adaptation of species and the diversity of life [for this second aspect, see Darwin (1859, Chapter 4, section Divergence of character, pp. 111–126)]. If we consider the modern synthesis and its critics through this grid, a series of fundamental issues can be raised.

a. It is rather obvious today that the synthetic theory of evolution was autoecological and not synecological. In other words, the synthesis was mainly a "genetical theory of natural selection," not a true "ecological theory of natural selection." As I tried to show in the present review, this has been the source of many paradoxes in the theory of natural selection, all more or less summarized in Lewontin's comment on population genetics as an "auto-mechanics of evolution."

b. There is a crucial cleavage within the modern synthesis itself between those who think that speciation is an effect of some adaptive process (e.g., Dobzhansky), and those who disconnect adaptation and speciation (e.g., Mayr). In a sense, this cleavage (or ambiguity) may be seen as a fundamental "contradiction" of the synthesis. This, among other things, makes understandable why so many theoreticians are presently involved in disputes over the ultimate signification of the term "species." One may think here of Darwin saying that a major interest of his conceptions was that "we shall at least be freed from the vain search for the undiscovered and undiscoverable essence of the term species" (Darwin, 1859, p. 485).

c. Behind the many controversies on the concept of species, one may discern a more and more urgent questioning on the causes of the diversity of life. The current literature shows that four solutions are commonly considered. The first solution, certainly in best agreement with the beliefs of many syntheticists, is to consider that diversity formation is under the control of biotic mechanisms (such as competition, coevolution, etc.). Such a vision of the history of life fits nicely with the "extrapolationist" atmosphere, this history being "nothing more than" the result of a complex interaction between many evolving units (the populations) (e.g., Bock, 1979; Mayr, 1988). A second solution consists in a rehabilitation of the old schemata of "transcendental morphology," with the idea that "not everything is possible," and that the diversification of life is largely channeled by a limited number of developmental possibilities. In this perspective, embryology is the key to the understanding of evolution. A third solution is to postulate some macroevolutionary mechanism, which "supervenes" on the microadaptive processes (for instance, "species selection" or "species drift"). The three former solutions amount to the philosophical postulate that the diversity formation is under biological control. A fourth solution, which is not met in the six books discussed [except for some allusions of Hoffman (1983)], is that "much of the formation of new biota is due to major environmental changes caused by a dynamic earth resulting in major environmental change beyond biological control or limitation" (M. Hecht, personal communication). This last solution might well be the most "Darwinian" of all: It comes down to saying that microevolution is a utilitarian tinkering in a changing and hazardous physical world.

It is probable that Darwin would have been very much interested in problems (a) and (c), and skeptical concerning problem (b). Problem (b) is one chapter among many others in one of the oldest debates in the philosophy of nature, the debate on the kinds of objects that are serious candidates for the status of "individuals." It may finally be suggested that an assertion such as "species are individuals" has in addition some anthropological signification: it is not equivalent to saying that I am a "part" of the human species, and to saying that I am a "representative" of the human genus, as well as any other human being. This is particularly clear in the context of ethics.

Acknowledgments

This chapter is very much indebted to R. M. Burian, M. Delsol, M. Ghiselin, M. Grene, D. Hull, B. Hecht, M. Hecht, and E. Mayr, for their encouragements and fruitful criticisms. I also thank V. Lyons, R. Merrill, and D. Beaune for their patient linguistic help.

REFERENCES

Ayala, F. J., 1985, Reduction in biology: A recent challenge, in: *Evolution at a Crossroads: The New Biology and the New Philosophy of Science* (D. J. Depew and B. H. Weber, eds.), pp. 65–79, MIT Press, Cambridge, Massachusetts.

Bennet, J. H., 1983, *Natural Selection, Heredity and Eugenics, Including Correspondance of R. A. Fisher with Leonard Darwin and others,* Clarendon Press, Oxford.

Bock, W. J., 1979, The synthetic explanation of macroevolutionary change: A reductionist approach, *Bull. Carnegie Mus. Nat. Hist.* **13:**20–69.

Bowler, P. J., 1983, *The Eclipse of Darwinism,* Johns Hopkins University Press, Baltimore, Maryland.

Brandon, R. N., 1985, Adaptive explanations: Are adaptations for the good of replicators or interactors? (D. J. Depew and B. J. Weber, eds.), pp. 81–96, MIT Press, Cambridge, Massachusetts.

Burian, R. M., 1983, "Adaptation", in: *Dimensions of Darwinism: Themes and Counterthemes in Twentieth-Century Evolutionary Theory* (M. Grene, ed.), pp. 287–314, Cambridge University Press, Cambridge.

Burian, R. M., 1988, The influence of the evolutionary paradigm (in preparation).

Campbell, J. H., 1985, An organizational interpretation of evolution, in: *Evolution at a Crossroads: The New Biology and the New Philosophy of Science* (D. J. Depew and B. H. Weber, eds.), pp. 133–167, MIT Press, Cambridge, Massachusetts.

Canguilhem, G., 1975, *La Connaissance de la Vie,* Vrin, Paris.

Caplan, A. L., 1980, Philosophical issues concerning the modern synthetic theory of evolution, Ph.D. dissertation, Columbia University, New York (unpublished).

Comte, A., 1975, *Cours de Philosophie Positive* (M. Serres, F. Dagognet, and A. Sinaceur, eds.), Hermann, Paris.

Cuénot, L., 1914, Théorie de la préadaptation, *Scientia* **8:**60–73.

Darwin, C., 1859, *On the Origin of Species,* John Murray, London.

Darwin, C., 1868, *The Variation of Animals and Plants under Domestication,* John Murray, London.

Darwin, C., 1959, *On the Origin of Species. A Variorum Text* (M. Peckham, ed.), University of Pennsylvania Press, Philadelphia, Pennsylvania.

Dawkins, R., 1978, Replication selection and the extended phenotype, *Z. Tierpsychol.* **47:**61–76.

Depew, J., and Weber, B. H., eds., 1985, *Evolution at a Crossroads: The New Biology and the New Philosophy of Science,* MIT Press, Cambridge, Massachusetts.

Dobzhansky, T., 1937, *Genetics and the Origin of Species,* Columbia University Press, New York.

Dobzhansky, T., 1968, Adaptedness and fitness in: *Population Biology and Evolution* (R. C. Lewinton, ed.), pp. 109–121, Syracuse University Press, Syracuse, New York.

Dumont, L., 1966, *Homo Hierarchicus, Le système des castes et ses implications,* Gallimard, Paris.

Dumont, L., 1972, *Homo Hierarchicus, Le système des castes et ses implications,* 2nd ed., Gallimard, Paris.

Dumont, L., 1979, *Homo Hierarchicus, Le système des castes et ses implications,* 2nd ed. (revised), Gallimard, Paris.

Dyke, C., 1985, Complexity and closure, in: *Evolution at a Crossroads: The New Biology and the New Philosophy of Science* (D. J. Depew and B. H. Weber, eds.), pp. 97–131, MIT Press, Cambridge, Massachusetts.

Edwards, A W. F., 1971, Review of S. Wright: Evolution and the Genetics of Populations, vol. 2., The Theory of Gene Frequencies, *Heredity* **26:**332–337.

Edwards, A. W. F., 1977, *Foundations of Mathematical Genetics,* Cambridge University Press, Cambridge.

Eldredge, N., 1979*a,* Alternative approaches to evolutionary theory, *Bull. Carnegie Mus. Nat. Hist.* **13:**7–19.

Eldredge, N., 1979*b,* Cladism and common sense, in: *Phylogenetic Analysis and Paleontology* (J. Cracraft and N. Eldredge, eds.), pp. 165–198, Columbia University Press, New York.

Eldredge, N., 1985, *Unfinished Synthesis: Biological Hierarchies and Modern Evolutionary Thought,* Oxford University Press, New York.

Eldredge, N., and Cracraft, J., eds., 1980, *Phylogenetic Pattern and the Evolutionary Process,* Columbia University Press, New York.

Eldredge, N., and Gould, S. J., 1972, Punctuated equilibria: An alternative to phyletic gradualism, in: *Models in Paleobiology* (T. J. M. Schopf, ed.), pp. 82–115, Freeman, San Francisco.

Eldredge, N., and Salthe, S. N., 1984, Hierarchy and evolution, *Oxf. Surv. Evol. Biol.* **1:**82–206.

Fisher, R. A., 1930, *The Genetical Theory of Natural Selection,* Clarendon Press, Oxford.

Fisher, R. A., 1958, *The Genetical Theory of Natural Selection,* 2nd ed., Dover, New York.
Ford, E. B., 1931, *Mendelism and Evolution,* Methuen, London.
Fox Keller, E., 1987, Reproduction and the central project of evolutionary theory, *Biol. Philos.* **2:** 383–396.
Gayon, J., 1988, Le "Théorème fondamental" de la sélection naturelle de R. A. Fisher: Une approche historique, in: *Biologie théorique,* Solignac, pp. 45–72. CNRS, Paris.
Gayon, J., 1989*a, Elementos para una historia del concepto de fitness,* Universidad autonoma de Mexico, Mexico.
Gayon, J., 1989*b, L'individualité de l'espèce: Une thèse transformiste?,* Colloque Buffon, Dijon.
Ghiselin, M., 1969, *The Triumph of the Darwinian Method,* Chicago University Press, Chicago.
Ghiselin, M., 1974, A radical solution to the species problem, *Syst. Zool.* **23:**536–544.
Ghiselin, M., 1987*a,* Species concepts, individuality, and objectivity, *Biol. Philos.* **2:**127–143.
Ghiselin, M., 1987*b,* Response to commentary on the individuality of species, *Biol. Philos.* **2:**207–212.
Goodwin, B. C., 1984, Changing from an evolutionary to a generative paradigm in biology, in: *Evolutionary Theory: Paths into the Future* (J. W. Pollard, ed.), pp. 99–120, Wiley, New York.
Goujet, D., Janvier, P., Rage, J. C., and Tassy, P., 1983, Structures ou modalités de l'évolution: Point de vue sur l'apport de la paléontologie, in: *Modalités, Rythmes et Mécanismes de l'Evolution Biologique* (J. Chaline, ed.), pp. 137–143, CNRS, Paris.
Gould, S. J., 1980, Is a new and general theory of evolution emerging?, *Paleobiology* **6:**119–130.
Gould, S. J., 1982, The meaning of punctuated equilibria and its role in validating a hierarchical approach to macroevolution, in: *Perspectives in Evolution* (R. Milkman, ed.), pp. 83–104, Sinauer, Sunderland, Massachusetts.
Gould, S. J., 1983*a,* Dix-huit points au sujet des équilibres ponctués, in: *Modalités, Rythmes et Mécanismes de l'évolution Biologique* (J. Chaline, ed.), pp. 39–41, CNRS, Paris.
Gould, S. J., 1983*b,* The hardening of the modern synthesis, in: *Dimensions of Darwinism: Themes and Counterthemes in Twentieth-Century Evolutionary Theory* (M. Grene, ed.), pp. 71–93, Cambridge University Press, Cambridge.
Gould, S. J., and Eldredge, N., 1977, Punctuated equilibria: The tempo and mode of evolution reconsidered, *Paleobiology* **3:**115–151.
Gould, S. J., and Lewontin, T. C., 1979, The spandrels of San Marco and the Panglossian paradigm: A critique of the adaptationist programme, *Proc. R. Soc. Lond., B* **205:**581–598.
Gould, S. J., and Vrba, E. S., 1982, Exaptation—a missing term in the science of form, *Paleobiology* **8:**115–151.
Grene, M., ed., 1983, *Dimensions of Darwinism: Themes and Counterthemes in Twentieth-Century Evolutionary Theory,* Cambridge University Press, Cambridge, Massachusetts.
Guyénot, E., 1924, *L'hérédité,* Doin, Paris.
Haldane, J. B. S., 1932, *The Causes of Evolution,* Longmans, London.
Haldane, J. B. S., 1935, Darwinism under revision, *Rationalist Annu.,* **1965:**19–29.
Hecht, M. K., and Hoffman, A., 1986, Why not neo-Darwinism? A Critique of paleobiological challenge, *Oxf. Rev. Evol. Biol.* **3:**1–47.
Hempel, C. G., and Oppenheim, F., 1936, *Der Typus begriff im Lichte der neuen Logik,* A. W. Sijthoff, Leiden.
Hempel, K., 1966, *Philosophy of Natural Science,* Prentice-Hall, Englewood Cliffs, New Jersey.
Ho, M. W., and Saunders, P. T., eds., 1984, *Beyond Neo-Darwinism,* Academic Press, London.
Hoffman, A., 1983, Paleobiology at the Crossroads: A critique of some modern paleobiological research programs, in: *Dimensions of Darwinism: Themes and Counterthemes in Twentieth-Century Evolutionary Theory* (M. Grene, ed.), pp. 241–271, Cambridge University Press, Cambridge.

Hull, D. L., 1974, *The Philosophy of Biological Science,* Prentice-Hall, Englewood Cliffs, New Jersey.

Hull, D. L., 1976, Are species really individuals?, *Syst. Zool.* **25:**174–191.

Hull, D. L., 1978, A matter of individuality, *Philos. Sci.* **45:**335–360.

Hull, D. L., 1980, Individuality and selection, *Annu. Rev. Ecol. Syst.* **11:**311–332.

Hull, D. L., 1981, The units of evolution, in: *Studies in the Concept of Evolution* (U. J. Jensen and R. Harré, eds.), pp. 23–44, Harvester Press, London.

Huxley, J., ed., 1940, *The New Systematics,* Oxford University Press, Oxford.

Huxley, J., 1942, *Evolution: The Modern Synthesis,* Allen and Unwin, London.

Huxley, J., 1973, *Evolution: The Modern Synthesis,* 3rd ed., Allen and Unwin, London.

Jacob, F., 1981, *Le jeu des possibles,* Fayard, Paris.

Janvier, P., 1984, Cladistics: Theory, purpose and evolutionary implications, in: *Evolutionary Theory: Paths into the Future* (J. W. Pollard, ed.), pp. 39–75, Wiley, New York.

Janvier, P., 1986, L'impact du cladisme sur la recherche dans les sciences de la vie et de la terre, in: *L'ordre et la diversité du vivant* (P. Tassy, ed.), pp. 99–120, Fayard, Paris.

Jepsen, G. L., Mayr, E., and Simpson, G. G., eds., 1949, *Genetics, Paleontology and Evolution,* Columbia University Press, New York.

Kant, I., 1781, *Kritik der Reinen Vernunft.*

Kauffman, S. A., 1985, Self-organization, selective adaptation, and its limits: A new pattern of inference in evolution and development, in: *Evolution at a Crossroads: The New Biology and the New Philosophy of Science* (D. J. Depew and B. H. Weber, eds.), pp. 169–207, MIT Press, Cambridge, Massachusetts.

Kimler, W. C., 1983, Mimicry: Views of naturalists and ecologists before the modern synthesis, in: *Dimensions of Darwinism: Themes and Counterthemes in Twentieth-Century Evolutionary Theory* (M. Grene, ed.), pp. 97–127, Cambridge University Press, Cambridge.

Kingsland, S., 1985, *Modeling Nature: Episodes in the History of Population Ecology,* University of Chicago Press, Chicago.

Koestler, A., 1978, *Janus: A Summing Up,* Vintage Books, New York.

Koestler, A., and Smythies, J. R., eds., 1969, *Beyond Reductionism: New Perspectives in the Life Sciences,* London.

Kuhn, T., 1962, *The Structure of Scientific Revolutions,* Chicago University Press, Chicago.

Lakatos, J., 1976, History of science and its national reconstruction, in: *Method and Appraisal in the Physical Sciences* (C. Howson, ed.), Cambridge University Press, Cambridge.

Lamotte, M., 1988, Phénomènes fortuits et évolution, in: *L'évolution dans sa réalité et ses diverses modalités* (M. Marois, ed.), pp. 241–268, Masson, Paris.

Leibniz, G. W. F., 1686, *Discours de Métaphysique.*

Lewontin, R. C., 1970, The units of selection, *Annu. Rev. Ecol. Syst.* **1:**1–18.

Lotka, A. J., 1956, *Elements of Mathematical Biology,* Dover, New York.

Maynard Smith, J., 1983, Current controversies in evolutionary biology, in: *Dimensions of Darwinism: Themes and Counterthemes in Twentieth-Century Evolutionary Theory* (M. Grene, ed.), pp. 273–286, Cambridge University Press, Cambridge.

Mayr, E., 1942, *Systematics and the Origin of Species,* Columbia University Press, New York.

Mayr, E., 1969, *Principles of Systematic Zoology,* McGraw-Hill, New York.

Mayr, E., 1974, Cladistic analysis or cladistic classification?, *Z. Zool. Syst. Evolutionsforsch.* **12:** 94–128.

Mayr, E., 1980, Prologue: Some thoughts on the history of the evolutionary synthesis, in: *The Evolutionary Synthesis: Perspectives on the Unification of Biology* (E. Mayr and W. B. Provine, eds.), Harvard University Press, Cambridge, Massachusetts.

Mayr, E., 1988, *Toward a New Philosophy of Biology,* Harvard University Press, Cambridge, Massachusetts.

Medawar, P., 1960, *The Future of Man,* Basic Books, New York.
Moorehead, P., and Kaplan, M., 1967, *Mathematical Challenge to the Neo-Darwinian Interpretation of Evolution,* Wistar Institute Press, Philadelphia, Pennsylvania.
Nagel, E., 1961, *The Structure of Science,* Routledge and Keagan Paul, London.
Needham, J., 1936, *Order and Life,* Yale University Press, New Haven, Connecticut.
Nowinski, Cz., 1967, Biologie, théories du développement et dialectique, in: *Logique et connaissance scientifique* (J. Piaget, ed.), pp. 862–892, Gallimard, Paris.
Olson, E. C., 1960, Morphology, paleontology and evolution, in: *The Evolution of Life* (S. Tax, ed.), pp. 523–545, University of Chicago Press, Chicago.
Paterson, H. E. H., 1978, More evidence against speciation by reinforcement, *S. Afr. J. Sci.* **74:** 369–371.
Paterson, H. E. H., 1979, A comment on mate recognition systems, *Evolution* **34:**330–331.
Piaget, J., 1973, *Introduction à l'épistémologie génétique,* 2nd ed., Vol. 1, Presses Universitaires de France, Paris.
Pollard, J. W., ed., 1984, *Evolutionary Theory: Paths into the Future,* Wiley, Chichester.
Provine, W. B., 1980, Epilogue, in: *The Evolutionary Synthesis: Perspectives on the Unification of Biology* (E. Mayr and W. B. Provine, eds.), Harvard University Press, Cambridge, Massachusetts.
Reif, W. E., 1983, Evolutionary theory in German paleontology, in: *Dimensions of Darwinism: Themes and Counterthemes in Twentieth-Century Evolutionary Theory* (M. Grene, ed.), pp. 173–203, Cambridge University Press, Cambridge.
Riddiford, A., and Penny, D., 1984, The scientific status of modern evolutionary theories, in: *Evolutionary Theory: Paths into the Future* (J. W. Pollard, ed.), pp. 1–38, Wiley, New York.
Riedl, R., 1983, The role of morphology in the theory of evolution, in: *Dimensions of Darwinism: Themes and Counterthemes in Twentieth-Century Evolutionary Theory* (M. Grene, ed.), pp. 205–238, Cambridge University Press, Cambridge.
Rieppel, O., 1984, Schindewolf's typostrophism in light of the punctuated equilibria, *Neues Jahrb. Geol. Evol. Paleontol.* **8:**491–496.
Rosen, D. E., 1984, Hierarchies and history, in: *Evolutionary Theory: Paths into the Future* (J. W. Pollard, ed.), pp. 77–97, Wiley, New York.
Rosenberg, A., 1985, *The Structure of Biological Science,* Cambridge University Press, Cambridge.
Roughgarden, J., 1979, *Theory of Population Genetics and Evolutionary Ecology: An Introduction,* Macmillan, New York.
Ruse, M., 1973, *Philosophy of Biology,* Hutchinson, London.
Ruse, M., 1980, Charles Darwin and group selection, *Ann. Sci.* **37:**615–630.
Salthe, S. N., 1985, *Evolving Hierarchical Systems: Their Structure and Representation,* Columbia University Press, New York.
Saunders, P. T., and Ho, M.-W., 1984, The complexity of organisms, in: *Evolutionary Theory: Paths into the Future* (J. W. Pollard, ed.), pp. 121–139, Wiley, New York.
Schmalhausen, I. I., 1949, *Factors of Evolution,* Blakiston, Philadelphia.
Schopf, T. J. M., 1982, Historical approaches versus equilibrium approaches to evolutionary data in: *Biochemical Aspects of Evolutionary Biology* (M. H. Nitecki, ed.), pp. 1–8, Chicago University Press, Chicago.
Simpson, G. G., 1944, *Tempo and Mode in Evolution,* Columbia University Press, New York.
Simpson, G. G., 1961, *The Principles of Animal Taxonomy,* Columbia University Press, New York.
Simpson, G. G., 1975, Recent advances in methods of phylogenetic inference, in: *Phylogeny of the Primates* (W. P. Lucket and F. S. Szaley, eds.), pp. 3–19, Plenum Press, New York.
Sober, E., 1984, *The Nature of Selection: Evolutionary Theory in Philosophical Focus,* MIT Press, Cambridge, Massachusetts.
Spencer, H., 1864, *Principles of Biology,* Vol. 1, Williams and Norgate, London.

Spencer, H., 1893, The inadequacy of natural selection, *Contemp. Rev.* **43:**153–166, 439–456.

Stanley, S. M., 1975, A theory of evolution above the species level, *Proc. Natl. Acad. Sci. USA* **72:** 646–650.

Stebbins, G. L., 1950, *Variation and Evolution in Plants,* Columbia University Press, New York.

Stebbins, G. L., 1974, Adaptive shifts and evolutionary novelty: A compositionalist approach, in: *Studies in the Philosophy of Biology* (F. J. Ayala and T. Dobzhansky, eds.), pp. 285–306, Macmillan, London.

Steele, E. J., Gorczynski, R. M., and Pollard, J. W., 1984, The somatic selection of acquired characters, in: *Evolutionary Theory: Paths into the Future* (J. W. Pollard, ed.), pp. 217–237, Wiley, New York.

Strawson, P. F., 1959, *Individuals,* Methuen, London.

Thom, R., 1985, Pour ou contre Darwin, *Science* (Paris) **1**(4)**:**51–57.

Turner, J. R. G., 1969, The basic theorems of natural selection: A naive approach, *Heredity* **24:**76–84.

Turner, J. R. G., 1983, "The hypothesis that explains mimetic resemblance explains evolution": The gradualist–saltationist schism, in: *Dimensions of Darwinism: Themes and Counterthemes in Twentieth-Century Evolutionary Theory* (M. Grene, ed.), pp. 129–169, Cambridge University Press, Cambridge.

Van Valen, L., 1976, Ecological species, multispecies, and oaks, *Taxon* **25:**233–239.

Vrba, E. S., and Eldredge, N., 1984, Individuals, hierarchies and processes: Towards a more complete evolutionary theory, *Paleobiology* **10:**146–171.

Waddington, C. H., 1957, *The Strategy of the Genes,* Allen and Unwin, London.

Waddington, C. H., 1960, Evolutionary adaptation, in: *The Evolution of Life* (S. Tax, ed.), pp. 381–402, University of Chicago Press, Chicago.

Wallace, A. R., 1859, On the tendency of varieties to depart infinitely from the original type, *J. Proc. Linn. Soc. (Zool.)* **3.** [Reprinted in *The Collected Papers of Charles Darwin,* Vol. 2, (P. H. Barrett, ed.), pp. 10–19, University of Chicago Press, Chicago (1977)].

Wallace, A. R., 1916, *Alfred Russel Wallace Letters and Reminiscences* (J. Marchant, ed.), Vol. 1, Cassel, London.

White, M. J. D., 1945, *Animal Cytology and Evolution,* Cambridge University Press, Cambridge.

Williams, G. C., 1966, *Adaptation and Natural Selection: A Critique of Some Current Evolutionary Thought,* Princeton University Press, Princeton, New Jersey.

Wimsatt, W. C., 1980, Reductionist research strategies and their biases in the units of selection controversy, in: *Scientific Discoveries: Case Studies* (T. Nickles, ed.), pp. 213–259, Reidel, Dordrecht.

Wright, S., 1931, Evolution in Mendelian populations, *Genetics* **16:**97–159.

Wright, S., 1964, Biology and philosophy of science, in: *The Hartshorme Festschrift: Process and Divinity* (W. R. Freese and E. Freeman, eds.), pp. 101–125, Open Court Publishing, La Salle, Illinois.

Wright, S., 1980, Genic and organismic selection, *Evolution* **34:**825–843.

Wynne-Edwards, V. C., 1962, *Animal Dispersion in Relation to Social Behavior,* Oliver and Boyd, Edinburgh.

2

Is Evolution at a Crossroads?

MARJORIE GRENE

INTRODUCTION

Is evolution at a crossroads, as Depew and Weber's title suggests (Depew and Weber, 1985)? It would lie far beyond the knowledge or competence of this writer to review the literature on this much debated—and allegedly often definitively answered—question. What follows is rather a handful of reflections occasioned (at the suggestion of the editors of the present volume) by a series of relatively recent volumes that appear, at first sight, to suggest, at least in part, a positive reply. My answer will be more ambiguous: "Yes" and "No," but on the whole more "No" than "Yes."*

To ask whether "evolution" (sc. "evolutionary biology") is at a crossroads is, obviously, to ask whether the evolutionary synthesis of the middle decades of this century calls for radical revision or even, as some authors suggest, replacement. Exactly what is the synthesis? On most accounts, it has been understood to entail primarily the reconciliation of Darwinian natural history with Mendelian genetics effected through the influence of the great mathematical geneticist-evolutionists Fisher, Haldane, and Wright (Provine, 1971). According to Mayr, however, it was the synthesis of evolutionary theory with systematics that was definitive (Mayr, 1942, 1982; Mayr and Provine, 1980). But again, on

*As proposed by the editors of *Evolutionary Biology*, the present chapter considers some of the papers in recent collections, as well as the argument of a recent book. The titles under consideration are Buss (1987), Depew and Weber (1985), Eldredge (1985), Ho and Fox (1988*a*), Ho and Saunders (1984), Pollard (1984*a*), and Weber *et al.* (1988). I have of course ignored my own purely philosophical contribution to Depew and Weber, although I shall want to say something, if briefly, about the impact of some of the other arguments in these volumes on developments in the philosophy of science. I should also add that since I was able to consider the Ho and Fox volume only after I had written this essay, I could work in some of its contents only rather sketchily.

MARJORIE GRENE • Department of Philosophy, Virginia Polytechnic Institute and State University, Blacksburg, Virginia 24061.

many occasions, the synthesis has been celebrated as bringing under the all-enfolding wing of Darwinism not only genetics and systematics, but paleontology, embryology (via Schmalhausen), and possibly even comparative anatomy [through the reinterpretation of the concept of homology?—see, e.g., Gould (1986)]. Given Mayr's emphasis on the role of geographical isolation in the origin of species, biogeography would also seem to be included; and although the discipline of ecology developed, at least in part, in relative independence of evolutionary theory, any Darwinian theory is founded on the conception of organisms adapting, or adapted, to their environments, and so, at bottom "ecological" in its orientation [for the relations between ecology and evolution, see, e.g., McIntosh (1985, pp. 256–266), Kingsland (1985), Collins (1986), and Kimler (1986)].

In view of this complex and many-sided situation, therefore, it seems reasonable to ask our question about the synthesis in the light of the proposals made for its reform or radical revision in, or involving, these several areas. Thus I shall look, if sketchily, at the arguments raised in the literature before us with respect to: (1) genetics, (2) systematics, (3) paleontology, (4) the problem of form (problems raised in connection with morphological considerations, including morphogenesis), (5) embryology, (6) biogeography, and (7) ecology. In addition, (8) there have been proposals for new programs that appear to alter the classic synthesists' view of the relation of biology to the exact sciences; and finally (9) some of the contributions to the volumes in question carry implications for the philosophy of science that are worth noting.*

GENETICS

Whatever the importance of other fields, it can scarcely be denied that genetics, and in particular population genetics, played a major, if not the major, role in bringing the synthetic theory to its authoritative position in evolutionary biology. The signal difficulties of relating theory to practice even here (Lewontin, 1974) were far from deterring the advance both of experimental research and

*In view of the inclusion of two papers on animal behavior in Ho and Fox (Bateson, 1988; Gordon, 1988), one would like to ask also why the field of ethology has not been considered as such in connection with the synthetic theory. "Sociobiology," after all, is an extrapolation from some genetic, and evolutionary, explanations to a genetic (and evolutionary) explanation of all animal behavior. Indeed, from the start there have been explanations of animal behavior in the Darwinian tradition, and such explanations have in turn supported Darwinian approaches to evolution, if, at different times and places, in different styles. However, rather than adding yet another field of consideration here, I shall refer briefly to the contributions of Bateson (1988) and Gordon (1988) in discussing development.

of theorizing in the bonding of population genetics and evolutionary theory. What, if anything (on the very partial but probably representative evidence of the books here under review), has happened to alter this aspect of the synthesis? Obviously, genetics has developed explosively since 1930, if we want to date the origin of the synthesis with Fisher's *Genetical Theory,* or since 1937 and the first edition of Dobzhansky's classic text (Fisher, 1930; Dobzhansky, 1937). Has the explosion burst the bonds of Darwinism? Several of the authors in the Pollard, Ho and Saunders, and Ho and Fox collections argue that it has (Pollard, 1984*a*; Ho and Saunders, 1984; Ho and Fox, 1988*a*). Two neo-neo-Lamarckian papers, if cogent, would of course radically undercut a Darwinian approach; they are based, however, on experimental work that has not, so far I have been able to ascertain, proved reproducible (Steele *et al.*, 1984; Pollard, 1984*c*; cf. Pollard, 1988). Genetic assimilation, in its day, was taken by some to support Lamarckism, but was easily enough incorporated in a more orthodox theory. Steele's work, the chief experimental basis for this recent revival, seems, so far, less credible in itself and hence even less likely to overturn the Darwinian craft (Brent *et al.*, 1981, 1982; Howard, 1981; Maynard Smith, 1982).

The papers of Fox (1984) and of Matsuno (1984) appear, on different grounds, to insist on internal sources of variation that contradict the "randomness" of variations demanded by the synthetic theory (Fox, 1984; Matsuno, 1984; cf. Ho and Fox, 1988*b*). What Fox's "holistic determinism" amounts to is not wholly clear to this reader; one suspects a certain self-important insistence on the uniqueness of the writer's own work as distinct from other suggestions for the origin of life. Moreover, his interpretation of "randomness" in the context of evolutionary theory seems to suffer from the same double misunderstanding as that of Eden and others in the well-known Wistar Institute debate [both Fox and Matsuno cite Eden with approval; see Moorhead and Kaplan (1967)]. The "mathematical challenge" to the Darwinian tradition posed by these writers in the 1960s depended on the argument that, given the conditions on this planet before life's origin, the existence of life as we now know it was too wildly improbable to be explained in Darwinian, microevolutionary terms. It was clear on the occasion of the Wistar Institute confrontation in 1966, however, and it is entirely plain from the literature, that natural selection works from generation to generation, not all of a sudden from then to now. Thus, that form of "improbability" challenge was easily defeated. Moreover, the Darwinian use of "random" refers unambiguously to the fact that mutation occurs without relation to the needs of the organism, and that is all it means. Mathematical problems about "pure randomness" have nothing to do with the case. To raise these old issues once again seems futile. However, what both Fox and Matsuno may have to contribute is perhaps best understood in the context of the problem of form and its origin rather than of genetics.

Campbell (1985) treats, as many have done, of the complexity, harmony, and hierarchical organization of the genome. Temin and Engels (1984) discuss the role of movable elements in the genome, and Vrba (1984) refers in passing to Dover's concept of molecular drive. Similar "counterexamples" are included in some of the contributions to Ho and Fox (1988*a*), especially in Pollard's (1988) review of recent developments in genetics. When one considers, however, how easily orthodox evolutionists have assimilated the complexity of the genome to Darwinian thinking, none of these lines of research appears to undermine on principle the basic reasoning of the synthetic theory. Indeed, Kauffman's work on genetic regulatory networks [discussed in Kauffman (1985)], while contributing to the complexification of theory in genetics, is explicitly intended to complement rather than to replace Darwinian theory. "Self-organization" (Fox, 1988) here supplements or supports selection rather than replacing it. Similarly, Brandon (1985), while opposing genetic reductionism, argues in a thoroughly Darwinian spirit, insisting, as does Buss (1987), on the significance of multiple levels of selection; neither of these writers, however, is attempting to escape from the basic adaptationist framework of Darwinian theory. Further, Gordon is plainly proposing methodological changes in the study of natural selection, not its replacement (Gordon, 1988, 1989). Nor does the work of Fitch and Upper (1988) on the origin of the genetic code appear to furnish a fundamental challenge to orthodoxy.

One more case to complete our review; consider Cullis's (1988) presentation of the effect of repeated "shocks" on variation in higher plants (in particular in flax). It is worth looking at his conclusion:

> In summary . . . , it appears that higher plants have a genetically controlled variation generating system. This can be activated by a number of "shocks" to the organism and it generates variation in a specific subset of the genome. This genomic variation can be manifest as phenotypic variation from which better adapted lines can be selected. In the absence of any "shocks" the variation system is not active, or active at a very low level, and it is the exposure of the organism to the "shock" which causes the variation. The limitation of the variation to a subset of the genome, which is controlled by the physiological state of the cell, gives the variation a Lamarckian dimension in that repeated exposures to the same shocks generate the same range of variants. However, this is not caused by a preknowledge of what would be advantageous but simply by the generation of a subset of variants which are subject to selection by the same environment each time the experiment is repeated. (Cullis, 1988, p. 60.)

Here is another instance that may look Lamarckian. Yet surely this is rather another case of complexification within a recognizably Darwinian framework, not a revolutionary shift to a thoroughly anti-Darwinian point of view. Variation whose genesis is irrelevant, in any direct way, to the needs of the organism; differential adaptedness of such variation; heritability, in particular, of such adaptive differences; and consequently the occurrence of natural selection: these elements of the Darwinian evolutionary process readily accommodate both the complexification of genomic organization and its molecularization. There has always been, and still is in some areas, a tendency to atomization in the Darwinian tradition; but the synthesis never rested on "beanbag genetics," and if it ever had, its most credible spokesmen long ago recovered from such reductionistic tendencies. What matters is not isolated items, but programs, and these can be instantiated at a number of levels. Of course, as Wimsatt (1980) has argued, the evolutionary books can be kept in terms of single genes, but nobody—perhaps not even Richard Dawkins—believes that evolutionary causality is wholly reducible to such simplistic terms.

What then do the advocates of "a new paradigm" (whatever that much debated word really means!) in fact propose? Temin and Engels (1984), for example, conclude that work on movable elements in the genome will produce "a change in emphasis," thus contradicting the notion that a radically new theory is called for, or that the synthesis, as Pollard (1984*b*) quotes Gould as saying, is "effectively dead." "Changes in emphasis" may well be signs of healthy growth in theories; they are not symptomatic of death or replacement. Perhaps the synthesis in its "hardened" phase (Gould, 1983) should be rechristened, as G. L. Stebbins has suggested (in a lecture at Cornell University in 1987), "the synthesis of the mid-twentieth century," while the Darwinian tradition in its now developing form, with its new, but readily assimilable, molecular foundation, would need a modified title. In short, when it comes to the relation of evolutionary biology to genetics, the answer to our leading question is a resounding "No!" There are marvelous advances along the broad highway, but no bewildering crossroads is in sight.

SYSTEMATICS

If genetics still fits smoothly into its Darwinian niche, systematics is another, and more complicated, story. Of the two schools that have resisted evolutionary taxonomy, phenetics, indeed, appears to hold no further threat, but cladistics (erstwhile phylogenetic systematics) is not so easy to assimilate or to ignore. [For a history of these two movements, especially the social or personal interactions involved, see Hull (1988).] Paradoxically, "cladistics" began with Hennig's *magnum opus,* which certainly made no effort to contradict current orthodoxy. Hennig accepted the biological species concept, and in the second edition of his book (since he had not had access to the English language literature when writing it just after the war), he made an effort to take account of the classics of the synthesis that had not previously penetrated to Nazi or post-Nazi Germany (Hennig, 1950, 1966). His disciples, however, have rebelled constantly and vociferously against what they hold to be the arbitrariness or subjectivity of evolutionary taxonomists' weighting of characters. Thus, Janvier's essay (1984), which provides a clear, if slanted, introduction to cladistic taxonomy, concludes: ''In comparative biology and paleontology, the future will belong to those who understand the structure of nature, who concentrate their efforts on what is knowable instead of creating new mysteries in the form of untestable hypotheses'' (Janvier, 1984, p. 70). And ''testability'' in this context means Popperian falsifiability—presumably because the test of congruence can be failed, whereas conformity to selection theory can always be saved by the invention of yet another just-so story. Sometimes it seems that cladists, rebelling against the authority of population genetics in recent evolutionary theory, take refuge in some other discipline: Rosen (1984) apparently wanted to substitute embryology for genetics as an auxiliary field: after all, whatever has been the fate of ''recapitulation,'' it *is* the case that phylogeny is a sequence of ontogenies and nothing more. If we knew all the ontogenies of all the organisms that ever inhabited this planet, one after another and one alongside the other, we would know it all—more than if, as orthodox evolutionists have to dream of doing, we knew all the details of every genome. [Attempts by Goodwin (1984 *a, b,* 1988) or Oyama (1988) to assimilate development to phylogeny appear rather subtler, and in any case they are not, I believe, so plainly motivated by the need to find a foundation for a particular position in systematics.] Nelson and Platnick (1984), on the other hand, want Croizatian panbiogeography (Croizat, 1964) to do their work for them; here [as also to some extent in Janvier's (1984) argument] it seems to be dispersal that is the villain—everywhere and always, vicariance must take its place. [For Croizat's objection to Nelson and Platnick's use of his ideas, see Hull (1988). For a more reasoned discussion of vicariance biogeography and its role in evolutionary theory, see e.g., Cracraft (1989*a,b*), Kluge (1989), Sober (1989).] Craw and Page (1988) continue the reliance on Croizat's panbiogeography, going so far as to contrast his ''relative space-time'' with an alleged use by Darwin of Newtonian absolute space. Now admittedly, as far as methodology goes, there is a clear ''Newtonian'' theme in Darwin, if mediated by Herschel (Depew and Weber, 1988; Hodge, 1977). But to elicit from the *Origin,* in all its biological complexity and, indeed, ambiguity, a doctrine of absolute space and time is an outright absurdity. Thus, these debates appear to the present writer, unfortunately, rather faddish. It seems, I am afraid, somewhat outlandish to take a writer like Croizat so seriously.

Yet, alas, that is by no means all there is to it. In fact, there is no possible doubt—no possible doubt whatever—that cladism, whether of the original phylogenetic genre or ''transformed,'' has exerted a very powerful and widespread influence in systematics (see, e.g., Eldredge and Cracraft, 1980). And in the hands of a thinker as incisive as, say, Colin Patterson, it affords a very sharp and powerful taxonomic instrument indeed (see, e.g., Patterson, 1981, 1982). It does make one think (and that is usually salutary) to be told that many standard groups of standard evolutionary "history" are paraphyletic, that ancestors are plesiomorphic and so not much use, if any—and so on. It is also wholesome to become somewhat critical of the biological species concept, and to recognize, as indeed some of its own supporters have explicitly done—and done with pride—that this "evolutionists'" species concept holds only synchronically and hence indubitably NOT for evolution (e.g., Bock, 1979). Nor are all practitioners of cladism as extreme in their antisynthesis views as those just mentioned. Vrba's (1984) essay takes off from cladistic taxonomy, but her "effect hypothesis" has, so far as I can tell, nothing radically "anti-Darwinian" about it. Eldredge's (1985) book, moreover, which is also sympathetic to cladism, purports to "complete" or supplement the synthesis, certainly not to replace it. I will have a little more to say about these workers in the next section: on paleontology. The point here is that the cladistic movement has developed an impetus of its own, which sometimes seems to supplement, but sometimes to contradict, the basic structure of Darwinian thinking. Ernst Mayr is as ingenious as any one well could be at absorbing all possible rebellions into the Darwinian camp; yet in this case his distinction between "cladistic analysis" and "cladistic classification" does not seem to put the issue quite to rest—as, for example, his "How to carry out the adaptationist program" seemed to silence the Panglossian protest against extreme forms of selectionism (Mayer, 1974, 1983). Nevertheless, if there are bets to be taken, the present writer will put her money on something still akin to Darwinism to win out over its more radical cladistic opponents. For one thing, the opposition is still too strikingly a house divided against itself to defeat a position, itself admittedly less than monolithic, but sufficiently flexible and adaptable to withstand even a more single-minded attack than cladists of so many different stripes can mount. With respect to systematics, therefore, we may conclude, tentatively and superficially perhaps, but still with a certain confidence, that for evolutionary biology as a whole there is no significant crossroad here either of which we can say: here the synthesis stopped and evolution went another way. Thanks to computerization and some ferocious personalities, cladistics is a flourishing business, but it is not about to take over, let alone to abolish, the study of evolution.

PALEONTOLOGY

Although descent with modification was admitted by nearly all, and by all reasonable, biologists not long after 1859, the fossil record—to a layman the most obvious evidence that evolution happened—resisted assimilation to the Darwinian mechanism with its attendant gradualism. Single-handedly, Simpson brought his discipline into the synthetic fold, where it stood rather like a reformed drunkard at a temperance meeting. Even in Tax's celebratory centennial volumes, however, a paleontologist like Olson could be found giving less than whole-hearted assent (Olson, 1960). The trouble is that, like many of Darwin's own arguments (or so his critics complained), this reconciliation rests on a "could-possibly" foundation. As that arch-conservative synthesist Ayala (1985) demonstrates once more, the claim of contemporary Darwinism is that macroevolution—the broad sweep of events through geological time—*could* be explained in terms of the within-species, short-order mechanisms that have been explored on a microevolutionary time scale. If, as Hodge (1977) has convincingly argued, there are three parts in an argument based on the *Vera Causa* principle—existence, competence, and responsibility—the micro–macro claim has never gone further than the competence stage.*

Here, then, one could expect, if not "a new paradigm," or a return to the "-ogenesis" fashions of an earlier period, at least some radical objections to the established program. In the works here considered, however, no such dramatic claim occurs—unless one counts the attempt by Fox (1984) to revive orthogenesis or Pollard's (1988) separation of micro- from macroevolution; neither of these writers, however, is a paleontologist. Vrba works on a foundation which claims the occurrence of punctuated equilibrium; and of course Eldredge does, too: he invented (or discovered) it, a year before the famous joint paper usually cited (Eldredge, 1971; Eldredge and Gould, 1972). A textbook like that of Dobzhansky *et al.* (1977), however, easily admits the existence of such phenomena—as it easily admits the phenomena celebrated in the "neutral theory": the synthesis has room for many bedfellows.

And what did punctuated equilibrium amount to? It was, it appears, and both the 1972 authors have agreed to this in conversation, an attempt by two young invertebrate paleontologists to face the phenomena presented by their discipline and eliminate, or at least alleviate, the apologies for "gaps" that have haunted paleontology ever since Darwin's chapter on "difficulties" in the record. As Gould, like T. H. Huxley long before him, has recurrently argued, selection does not entail constant and gradual change; why, even in Darwinian terms, should not the long-time stabilities often seen in the fossil evidence be data rather than embarrassments? It is not so much the seemingly shocking notion that nature "jumps" that matters here: punctuated equilibrium's "abrupt" transitions are no saltations, but *relatively* rapid changes—at the kind of rate, one presumes, needed for such wholly kosher phenomena as adaptive radiation. What matters most, however, is that gradual change in a given direction does not have to be going on all the time—and that is consistent with Darwinism, too. The point here, in other words, is that sometimes, perhaps even often, a species or even a genus or family

*The claim to decouple micro- from macroevolution is at any event an ambiguous one; it is not clear from the literature whether macroevolution begins with or beyond speciation. Cf., e.g., Stanley (1975, 1979).

hangs around for a very long time without much major change: and one can *look* at these persistent species as *entities* existent through time. The species-are-individuals idea, first proposed (amid immense logical and philosophical confusion) as a defense of pure Darwinism and the BSC (Ghiselin, 1974; Hull, 1976) is taken over by both Vrba and Eldredge as an ally of cladism in systematics and a prop for the recognition of the independent existence of macroevolutionary patterns (sometimes of punctuated equilibrium character) in paleontology.

Further, the acceptance of species as ontological individuals—as real historical entities—supports a hierarchical conception of evolutionary theory not in itself contradictory to the synthesis, but, one hopes, enriching it. There is nothing radically ''anti-Darwinian'' either about Vrba's (1984) ''effect hypothesis'' [in fact, she cites Williams (1966) in its support] or about Eldredge's (1985) ''finishing'' of the synthesis by injecting a hierarchical dimension into its sometimes less than hierarchical construction (cf. Lewontin, 1970; Brandon and Burian, 1984). If Simpson by 1953 narrowed and exaggerated his previously rather tentative harmonization of paleontology and neo-Darwinism (Simpson, 1944, 1953; Gould, 1983), a rather subtler and more open treaty between paleontology and the reigning theory, or metatheory, demands no sharp turn aside to a new stance—certainly not a reintroduction of Simpson's *bête noire,* orthogenesis. What it does demand—and this comes from cladistics as much as from paleontology—is a rejection of the omnicompetence of phyletic gradualism to sort out the patterns of evolutionary change through geological time. This means a declaration of independence of ''macroevolution'' at the level of pattern; it does not *entail* any new ''process,'' that is, causal mechanism, that contradicts those acceptable to more traditional evolutionary biologists.*

THE PROBLEM OF FORM

The field next on my list was embryology. The resistance of many embryologists to the synthesis rests, however, in my view, on a deeper or more general ground: that is, the incompatibility between any Darwinian vision on the one hand and on the other any view, whether of individual development or of evolution, that rests on the acceptance of the significance of form. It seems appropriate, therefore, to postpone an explicit (if brief) treatment of embryology to the next section and consider first, if not morphology or comparative anatomy as ''fields,'' a more general methodological and even metaphysical question that is fundamental to some of the issues some of our authors deal with.

*Neither punctuated equilibrium nor a hierarchical view of evolution necessarily entails ''species selection,'' which appears to have been a coinage by Stanley (1975, 1979). See Gilinsky (1986).

''Natural selection,'' Darwin wrote, ''can act only by the preservation and accumulation of infinitesimally small inherited modifications, each profitable to the preserved being'' (Darwin, 1859, p. 95). This is of course the turn to population thinking so loudly vaunted by evolutionists and recently also by philosophers, and it entails, equally of course, the defeat of that evil demon typology. Darwin does not deny the existence of what he calls Unity of Type, but declares that of the two laws on which ''all organic beings have been formed,'' Unity of Type and Conditions of Existence, it is the latter that is ''the higher law,'' . . . ''as it includes, through the inheritance of former adaptations, that of unity of type.'' Or, as he had put it earlier in the same paragraph, ''on my theory unity of type is explained by unity of descent'' and since natural selection, the shaper of adaptations, is the chief motor of descent (with modification), it follows that unity of type is subordinated to ''conditions of life'' (Darwin, 1859, p. 206). Yet in Chapter Thirteen, in the section on Morphology, Darwin writes:

> We have seen that the members of the same class, independently of their habits of life, resemble each other in the general plans of their organization. This resemblance is often expressed by the term ''unity of type;'' or by saying that the several parts and organs in the different species of the class are homologous. The whole subject is included under the general name of Morphology. This is the most interesting department of natural history, and may said to be its very soul. What can be more curious than that the hand of a man, formed for grasping, that of a mole for digging, the leg of the horse, the paddle of the porpoise, and the wing of the bat, should all be constructed on the same pattern, and should include the same bones, in the same relative position? (Darwin, 1859, p. 434.)

Homology, then, the identity of pattern across dissimilar instances, the ''soul of natural history,'' is explained through unity of descent, with the modifications entailed by natural selection's working on infinitesimally small inherited differences. But to say that *a* is explained by *b* is not to deny the existence of *a;* indeed, effect and cause must be distinct, or the explanation is no explanation. On the other hand, to declare that unity of type is *included* in ''the law of conditions of existence'' does seem to suggest, at least, that *a* is a part of *b*, assimilated to it. Indeed, the explanation of homology through evolution has by now become so ingrained in our thinking that earlier nonevolutionary accounts seem no accounts at all: homology seems to *mean* unity of descent, so that the sameness of pattern evaporates and only the continuous process of slight and gradual selective change remains. Up with population thinking, down with typology! The synthesis, especially in its more rhetorical expressions, reinforced this contrast. At the 1947 conference (see Mayr and Provine, 1980, p. 42) there was one comparative anatomist, D. Dwight Davis, who does not seem to fit wholly into the antitypological mold. But in general the turn to population thinking and the defeat of essentialism or typology seemed, and still seems to many, to be the ''essential'' step in the grand advance of evolutionary biology. Anyone who doubts *this* ''central dogma'' is either a typologist or ''a neo-vitalist'' (Ruse, 1988).

Is *pure* population thinking necessary or even possible? Even the slight, almost (to us) imperceptible variations that selection has to work on must be variations *of*—some character, some property, some behavior—in short of the referent of some *predicate*. But predicates are *sortals:* they tell us about kinds, not particulars. Mayr appears to believe this is a philosopher's paradox: ''we''—that is, biologists (the good guys), he insists, have always admitted common properties across instances (Mayr, 1987). Of course they have; the question is, on their own premises, had they any right to do so? If there *are* only particulars aggregating to populations and ''characterized'' (whatever that may mean in ''population thinking'' terms) by very tiny dissimilarities among their constituents, what on earth can ''properties in common'' mean? Sober (1984), for example, who takes population thinking to an extreme, alleging that all of science up to Darwin (not only biology) exemplified its Aristotelian, ''essentialist'' opposite, nevertheless wants to stress the *properties* of organisms (or even groups of organisms) *for* which selection selects (cf. Waddington, 1957). But, again, properties are not and cannot be particulars; they cannot be assimilated to the particularism of population thinking. There must be something or other *of* which selection selects the somewhat better adapted variants.

Recurrently, biologists (not primarily philosophers; such prominent philosophers of evolution as Hull, Sober, and Ruse unanimously celebrate the populationist, and particularist, approach), uneasily aware of this difficulty, demand a return to an emphasis on form. In fact, as I see it, there has been since 1859 (if not earlier) a kind of see-saw in biological thinking between a stress on form (or ''unity of type''—which also includes, of course, diversity of type: in general, a respect for and attention to morphology) and, in contrast, the emphasis on gradually changing, multifarious ''conditions of life'' which produce, through contingent historical circumstances, what Waddington (1957) called ''the appearance of form.'' There is the flow of evolution: the only ''essences'' allowed in biological theory are relations of descent. Or there is the recognition of the primacy of the patterns that selection has to modify: patterns, whether of taxonomic characters in their hierarchical interrelations or of developmental channels that constrain evolutionary change.

In the literature here under discussion, the most explicit discussion of the primacy of form is Webster's (1984) essay on ''The relations of natural forms'' [but see also Goodwin (1988)].* Its first section, on ''the problem of sameness,'' and in part the second, on ''the problem of difference,'' set out eloquently and relatively clearly the limitations of a Darwinian (or ultra-Darwinian) position on these questions. Reverting to Geoffroy's ''Principle of Connections'' and to

*A paper by Webster and Goodwin (1982), which proposes a ''structuralist'' approach to evolutionary theory, seems to me no more satisfactory than Webster's or Goodwin's proposals in the collections we are now considering. ''Structuralism'' was a recognizable movement in linguistics and in anthropology, but as a philosophical epithet applied beyond those particular disciplines it seems to me not much more than a buzz word.

some of William Bateson's (1894) remarks in *Materials for the Study of Variation*. Webster proclaims the need to "abandon empiricist notions of the ontological (and epistemological) primacy of observable objects, qualities and relations and the concomitant equation of natural kinds with empirical regularities" (Webster, 1984, p. 206). Intrinsic necessities of form and of development need to be recognized as at least equally fundamental with (or perhaps more fundamental than?) the contingencies of history. He quotes Bateson's conclusion:

> For the crude belief that living beings are plastic conglomerates of miscellaneous attributes, and that order of form or symmetry have been impressed upon this medley by selection alone; and that by variation any of these attributes may be subtracted or any other attribute added in indefinite proportion, is a fancy which the study of variation does not support. (Bateson, 1984; Webster, 1984, p. 209.)

Those are fighting words: loyal synthesists will point out that the historical situation in the 1890s was very different indeed from what it is today, when Mendelism, biochemistry, and molecular biology have in turn supported rather than undercut a fundamentally Darwinian metatheory of evolutionary, and therefore, "ultimately," of all biological events. They will be right, moreover, in so doing; yet there is also something to be said for Webster's complaint. The trouble is that when it comes to a philosophical resolution of his own problem, he falls back on a position of "transcendental realism" that is, to this writer and probably to most philosophers, wholly unintelligible. Over and over, it seems, when someone calls for a return to taking form seriously, the positive alternative looks pretty unsatisfactory.

Nevertheless, the problem persists—perhaps, indeed, it is because its resolutions have been less than satisfactory that it persists! To an uneasy philosopher (and what else are philosophers for?), the roots of the problem are indeed located, as Webster argues, in an overempiricist and overparticularizing cast of thought that has all along characterized our Darwinian heritage. In the building of his "one long argument" it seems clear that Darwin was deeply influenced by Herschel, who was a "scientist" (as, following Whewell's usage, we could call him), rather than a philosopher, but a scientist reflecting (in his *Preliminary Discourse*) very much in the Baconian–Newtonian tradition on the practice and aims of science (Hodge, 1977; Ruse, 1979). And although Mill's major methodological work postdates the formation of Darwin's theory, Mill's hyperempiricism certainly appealed to Darwin; he was delighted by any favorable reaction to his views by that savant of contemporary philosophy (see, e.g., Hull, 1973, p. 276). In fact, when we think of its philosophical foundations, Darwinism appears to have been forged in the smithy of British empiricism, that edifice first outlined (if in much ambiguity and confusion) by John Locke (1690) and brought to completion—and therewith to self-contradiction—in the work of David Hume (1739). Here indeed all is particulars. Anything apparently general is an abstraction generated by contiguity and resemblance and the resultant "gentle force of association." Within this conceptual scheme (if "concepts" or "schemes" be indeed conceivable in its terms—as Hume, undoubtedly one of the great geniuses of philosophy in our tradition, certainly in our tongue, valiantly tried to show they could be)—within this conceptual scheme there are resemblances between particular, atomizable impressions and ideas, and among those "resembling" independently sortable items, also, of course, the very minute differences that Darwinism needs for its material. (In Hume it is a question of subjective particulars; for Darwinian purposes these have to be moved to the real world; but that is a side issue for my present argument.) What is lacking, however, in this model, is precisely the qualitative identity, or sameness, needed for the grasp of a given *form* constant over varying instances. Hull keeps accusing "essentialists" of demanding that things be similar, when they are not so (e.g., Hull, 1965, 1988). But it is the *same,* not the similar, that those who resist Darwinian flux are after. Consider homologue or homology: in Owen's terms (to quote the devil himself): "the same organ in different animals under every variety of form and function" (quoted in Boyden, 1943). In Darwin's limb list, how very *dis*similar are the organs mentioned! Indeed, it is the population thinker who relies, poor soul, on similarities, on Humean resemblances: "characters" (whatever, if anything, that term can mean to a thorough empiricist) that vary ever so slightly, so that selection may favor the slightly more advantageous: such characters must be, it seems, almost wholly similar, but only a little different. It is the empiricist who needs resemblances, and so blind is he (or she, though I hope not) to anything else, that the very nature of form, the very nature of natures, escapes his (her) conceptual net. It is impossible, given an empiricist cast of mind, to understand the problem: that is one reason, though not, alas, the only one, why attempts to formulate it as a problem, let alone to solve it, so often run aground.

At the same time, if we can only throw off the blinkers of empiricist thinking, there is no fundamental reason, it seems to me, why the two seemingly opposed positions in this case should not be reconciled. Perhaps in the 21st century evolutionists will overcome this conflict, much as the opposition between the various strains of "Darwinism" and of "Mendelism" that plagued the first decades of this century were overcome by the synthesis now seemingly under challenge.

Perhaps, indeed, this may happen sooner. The best evidence for such an optimistic prognosis comes, so far as my knowledge goes, from the work of Stuart Kauffman, exemplied in our present collections in an essay in Depew and Weber called "Self-organization, selective adaptation and its limits" (Kauffman, 1985). Oyama's contribution to Ho and Fox (Oyama, 1988; see Oyama, 1985) also suggests a starting point for a "new new synthesis," though from a different direction: starting from, and modifying, Lewontin's distinction between variational and transformational theories of evolution, and stressing the role of the organism as agent in evolution (cf. Lewontin, 1983). But for the present let me stay with Kauffman as evidence for my prediction, especially since at the moment I am dealing with the problem of form rather than development.

As I noted earlier, Kauffman is dealing in the Depew and Weber piece chiefly with regulatory genetic networks (Kauffman, 1985), but as his forthcoming book (Kauffman, 1990) will show on a much broader canvas, his work has foundations in, and implications for, the study of morphogenesis, the origin of life, and, in fact, for the structure of the theory of selection itself. I shall return to Kauffman's argument in connection with some of my remaining topics, but let me for now quote the conclusion of this paper. He writes:

> Evolutionary theory has grown almost without insight into the powerful self-organizing properties of complex systems, whose features are just beginning to be understood. I have carried out this discussion largely in terms of the highly ordered connectivity properties of well-scrambled genetic regulatory networks. An initial motivation for these inquiries lay in the discovery that such scrambled networks also exhibit extraordinarily coordinated patterns of gene expression paralleling many features of contemporary cell differentiation . . . , including the implication that each cell type differentiates directly into only a few neighboring cell types. Other powerful self-organizing properties appear to arise with respect to rhythmic phenomena, and the onset of spatial structuring in reaction-diffusion, or viscoelastic models of morphogenesis. . . .
>
> The fundamental problems are not only to understand such self-organizing properties but to clarify their interrelation to selective forces, design principles, and historical accidents in evolution. Inclusion in this task of systems with strongly self-organizing properties raises new conceptual problems and opportunities. These properties are themselves generators of order, relieving selection of the mantle which it has borne. On the other hand, selection either makes use of those self-organizing properties or opposes them. We have as yet little understanding of what it might mean for such internal self-organizing features of organisms to interact with selective forces, how to characterize such interactions theoretically, and how to assess them experimentally. (Kauffman, 1985, pp. 202–203.)

Such are the tasks for the future—although progress is made on them in some of the more recent work of Kauffman and others, which is summed up in his book. What is important to recognize here is that while on the one hand he is proposing theoretical perspectives that "relieve selection of the mantle it has borne," and thus seems to be joining the more extreme "challengers," on the other he is hoping to show how selection either "makes use of such self-organizing properties or opposes them." It is not only the limits of selection, but its opportunities and probable paths that he is studying. Indeed, his general aim is to restore to our attention D'Arcy Thompson's emphasis on form (Thompson, 1917), while at the same time accepting *fully* the achievements of the synthesis from the molecular to the macroevolutionary level, and attempting, through a variety of modeling techniques, to show more carefully and accurately how selection probably works. He sometimes speaks of the search for "ahistorical universals.'' Again, this looks at first sight like a contradiction of the Darwinian stress on historical contingency. Yet even if we accept Jacob's eloquent description of evolution as a ''tinkerer'' (Jacob, 1982), we have to admit that tinkerers must have something already there to tinker with, and it is possible that the study of such materials and processes will provide, not only further insight into constraints on selection, but new knowledge of what brings those materials and processes—and hence the course of selection itself—into the shapes and paths they have taken. Kauffman's research project, though many-dimensional, seeks to build, in particular biological contexts, and using research in a number of fields in physics, chemistry, and mathematics, a new synthesis, or a range of new directions for a new synthesis, that will deny no particle of the knowledge gained in the last 60 years of evolutionary research. Is this evolution at a crossroads? Once more, my answer is ''yes and no,'' but here again, not because one is turning aside from a defunct or defeated program, but because new vistas are coming into view, which enrich, rather than deny, the old perspectives.

EMBRYOLOGY

So, via form, to embryology. This field was not of course as prominent in the literature of the synthesis as was genetics, systematics, or paleontology. Dobzhansky wrote a foreword for the English version of Schmalhausen's *Factors of Evolution* (Schmalhausen, 1949), but this seems to have been rather a handwaving in the direction of the new harmonizing of disciplines than a true absorption in terms of the experimental support of evolutionary theory or the evolutionary explanation of advancing knowledge in a given area. Although all generalizations in the history of science are likely to be false, one can without too much risk hazard the statement that experimental embryology has developed in this century without much intimate interaction with major research in the areas dominated by the synthetic theory (e.g., Hamburger, 1980; Churchill, 1980). Modifier genes, regulatory genes, or rate genes all could help to suggest an easy transition from genotype to phenotype without undue concern for the study of development itself. Moreover, since selection works on available variation present in a given generation to bias the structure of the next generation's population, ''developmental constraints'' as well as historical constraints of every conceivable variety can be readily assimilated to a purely Darwinian conceptual context (Maynard Smith *et al.*, 1985). Still, not only in the case of a great embryologist like Lillie (1927), but recurrently and persistently, the practice of embryology has resisted assimilation to the synthesis, and a number of recent ''challengers'' ground their complaints in the alleged incompatibility of the ''facts'' of ontogeny with the ''dogmas'' of Darwinism (e.g., Løvtrup, 1984;

Webster, 1984; Goodwin, 1984*a,b,* 1988; Oyama, 1988). What about this? Must evolutionary theory take a radically new direction in the light of these objections?

Frankly, my problem with the more extreme positions taken in this context is that, with the best will in the world, I cannot understand how they are to replace, rather than to supplement, a Darwinian approach to evolutionary problems. A "critique" like that of Løvtrup (1984) is too wildly inaccurate to persuade; his ontogenetic reports, therefore, are just that; they seem to provide no viable alternative in terms of evolutionary theory. Goodwin's "generative paradigm," which is, presumably, the new deal hoped for also by Webster, is also difficult to assimilate (Goodwin, 1984 *a,b,* 1988; Webster, 1984). Goodwin (1984*a*) tells us, for example:

> what is required is the deduction of the correct relational order which generates the observed phenomena, and this order or organization is not directly observable, though it is real. This logical relational order is what defines the distinctive organizational properties of living organisms, and the proposition of this essay is that the appropriate mathematical description is provided by field equations. (Goodwin, 1984*a,* p. 239.)

Fine; in many sciences, progress is made by inference to mathematical relations that could not be directly "observed." All evolutionary discourse, in fact, has to make inferences to something presumably real but nonobservable, and more and more frequently such inferences issue in mathematical formulations. Take, e.g., the work of Kingsolver and Koehl (1985) on the changing function of insect wings, cited by Brandon in a forthcoming work on adaptive evolution (Brandon, 1990). Here physical modeling supports the hypothesis that insect wings served, up a given, mathematically calculated limit, for thermoregulation and then switched to the function of flight. And incidentally, if this is all Webster means by his "transcendental realism" (Webster, 1984), so be it. But how is this kind of mathematicization to *replace* the synthesis? The work of Oster, Alberch, and others, to which Goodwin, as well as Kauffman, refers, does indeed suggest, again, a "change of emphasis" in evolutionary theory (Alberch, 1982; Oster *et al.,* 1980; cf. Maynard Smith *et al.,* 1985). Goodwin (1984*b,* p. 117) quotes Oster *et al.* (1980, p. 231) as saying, "we view the evolutionary diversification of form as the product of regulation of developmental dynamics . . . In order to understand evolution we must understand developmental mechanics." Granted: we have in the findings of Alberch and of Oster *et al.* a necessary condition for a fuller theory of evolution that the synthesis (lacking the results of more recent embryological research) necessarily omitted. So far, this seems to present a situation parallel to that of evolutionary theory vis-à-vis molecular biology. G. L. Stebbins (in the Cornell lecture referred to earlier) uses the example of the brown-barred goose flying easily far above the head of Hilary atop Mt. Everest: as he tells the story, a single gene that controls the storage of oxygen, a gene which neither Hilary nor his contemporary evolutionists could have known about, enables the bird to perform without difficulty this, to us, impossible feat. Does such an advance in knowledge, which provides us with insight into the means for the evolution of this extraordinary adaptation, defeat the Darwinian theory? No more than Mendelism did. It gives it yet another experimental underpinning.

In his Pollard paper, however (''Changing from an evolutionary to a generative paradigm''), Goodwin (1984*b*) does shed more light on the fundamental reform he has in mind. His subject matter, like that of Alberch, Oster, and others, is the development (and evolution) of vertebrate limbs, and he calls for a radical rereading of this standard case of homology. From his point of view, it makes nonsense to ask of a four-digit animal, ''which digit did it lose?'' Tetrapod limbs, he holds, are not united in their structure:

> by virtue of descent from a common ancestor, with functional variants showing loss or gain of identifiable elements in the ancestral limb form. They are all members of a logical* class of structure united by common generative principles. What actually happens in the historical lineage of these forms tells us something about adaptation to external contingencies but nothing about internal organizational principles. It is now necessary to recognize that biological process conforms not only to extrinsic functional stability criteria of the type expressed in the concept of fitness, but also to intrinsic principles of order or organization arising, in the case of morphology, from the spatial ordering of generative fields. (Goodwin, 1984*b,* p. 113.)

So far, this sounds very similar to one (the morphogenetic) aspect of Kauffman's program, a program which is explicitly presented, in its totality, as supplementing rather than replacing the synthesis. The conclusion of the paragraph I have just been quoting, however, does state clearly the sense in which Goodwin would disagree with this assessment. He continues:

> Organisms are not aggregates of elements, whether molecules, cells, organs, skeletal or other components, whose random variation results in an unconstrained variety of forms. They are self-organizing wholes governed by laws describing spatial and temporal organization such that processes of biological change involve constrained transformation, whether ontogenetic or phylogenetic. *Evolution and development then emerge as aspects of this generative process over different time-scales and constrained by different categories of parametric change.* (Goodwin, 1984*b,* p. 113, my italics.)

If we ignore the fact that [except perhaps for Dawkins (1987), whose beloved computers are indeed aggregates of components, not organisms] no working biologist accepts the first statement, here presumably being ascribed to all Darwinian evolutionists, what does this trio of propositions tell us? It asserts, in its conclusion, that both evolution and development are aspects of a single ''generative process.'' If that is so, then indeed it will be the new ''theory of generative

*The term ''logical'' used here in a sense difficult for a philosopher to understand; the statement would be clearer without that disturbing adjective.

process'' that will embrace both ontogeny and phylogeny, development and evolution—and evolutionary biology, in particular the metatheory of natural selection, will no longer appear as the all-embracing foundation of the biological sciences. Indeed, Goodwin (1988) argues for the *replacement* of the historical approach of Darwinism by a *rational* biology: that is, a biology grounded squarely on the principles of form, on a science of form, that must precede and underlie history. Kauffman wants to use the very same kind of embryological results, among other data from a number of fields, to provide yet another underpinning for and expansion of the synthesis; Goodwin sees this ''generative process'' as the all-encompassing ''rational'' foundation for both development and evolution—a nondevelopmental replacement for both kinds of history. Who is right? Perhaps it is only a rhetorical question. The writers of the ''hardened'' synthesis displayed a certain smugness that made one want to resist their message (quite apart from the genuine conceptual problems it raised). In the case of the present would-be challengers there is an irritated tone of voice that is just as offputting. Even discounting that factor, however—or attempting to do so—, considering the history of the Darwinian tradition, the challenges it has overcome and assimilated in the past, I would predict that in this case again it is probably a Kauffmanlike expansion rather than a Goodwinlike replacement that will occur.

A recent, and, in my view, impressive piece of evidence supporting my prediction is to be found in Buss's (1987) *The Evolution of Individuality*. He announces in his preface, ''In this text I advocate a modification of the synthetic theory of evolution which I believe holds the potential for specific evolutionary predictions regarding the natural history of development, cell structure, and genomic organization'' (Buss, 1987, pp. vii–viii). Note: ''modification,'' not replacement. His argument is presented in four essays. In the first he seeks (with some plausibility) to explain how ''Weismann's Legacy'' has prevented evolutionists from assimilating the multifarious data on development in the vast range of living taxa. The fact that the classic work supporting the synthesis was done on organisms already existing as individuals (in the traditional biological sense of ''individual'') kept interest away from the question of how individuality had developed in the first place. The core of his positive thesis, supported by ample experimental evidence, is contained in the second and third essays, on the evolution of development and on life cycle evolution, respectively, where he argues (to put it summarily and crudely) that the history of life should be read as a history of competing units of selection. This leads him, in the concluding essay, to argue in favor of a hierarchical theory of evolution, as opposed to genic selectionism. His concern has been chiefly with levels of selection prior (both analytically, in terms of comparative biology, and phylogenetically, in terms of evolutionary history) to the establishment of the individuality characteristic of organisms like *Drosophila melanogaster* or *Homo sapiens;* but he is convinced that selection can also proceed at higher levels. Unfortunately, I am skeptical of his confidence that the "evolution of ideas" can also be easily assimilated to selection theory, but that disagreement is beside the point here. Let the evolutionary epistemologists rejoice for now; that fad, too, will pass away! What concerns me here is his firm assertion—which his whole exposition supports—that while "evolutionary theory is faced with new challenges in its capacity to explain the diversity of life"—and indeed, that is part of what makes it, in Lakatos's terms, "a progressive research program"—nevertheless, These new challenges can be accommodated without modification of the logical structure of the [sc. Darwinian] argument, by an expansion in scope of the units of selection to which this logical structure is applied" (Buss, 1987, p. 196). This is not so much, therefore—or better, it is not at all—a Goodwinian subsumption of evolution and development within a more comprehensive "generative" or "rational" theory. It resembles rather an attempt like Eldredge's (though, except for the emphasis on hierarchy, in a very different direction) to complete what was uncompleted, or like Kauffman's to assimilate new fields of research, both experimental and mathematical in their methodologies, in order to expand and support a theory unchanged in its basic foundations (Eldredge, 1985; Kauffman, 1985, 1990). The highway has been broadened; some may regret the loss of previous unploughed fields: mysteries now open to explanation—but such is the course of science. And it has been deepened, too. But still there is no impending crossroad in sight.

I have been glancing here at some work in embryology in its relation to the synthesis. If one broadens the reference of "development" to include the whole of a life history, rather than merely its early stages, one finds in such contributions as those of Bateson (1988), Gordon (1988, 1990), and Oyama (1988) a more general direction from which evolutionary theory may be, and is being, enriched: the study of behavior, of the *activity* of organisms, and their contribution to evolutionary processes. This approach, involving the role of ethology rather than embryology in evolution, deserves much more detailed consideration that I can give it here. But if one may venture to characterize the bent of these three essays, each very different in content and well worth study on its own, one may say that it is here the static and passive nature of the units of orthodox theory that is being questioned. As Gordon argues, for example, the rules of behavior (in her work the rules of behavior of harvester ant colonies) are themselves changing, and it is those changes that need systematic study. This means in the first instance changes in the methodology of research in invertebrate behavior, but it implies, ultimately, changes also in the investigator's perspective in evolutionary theory (Gordon, 1988, 1990). How radically, if at all, such developments will affect the physiognomy of Darwinism it is hard to predict. It is worth noting, however, that while Kauffman is concerned to introduce "ahistorical universals" into the study of evolution, and while Goodwin, going further, wants to *substitute* a "rational" for a "historical" foundation for biology, such authors as

Bateson, Goodwin, or Oyama propose to expand, and deepen, yet further the historical base of biological research. This presents for further reflection a paradoxical duality in the present challengers (or modifiers) of Darwinian orthodoxy: less history or more, which is it to be?*

BIOGEOGRAPHY

Geographical distribution was one of the classic supports for descent with modification, and, especially through Mayr's emphasis on allopatric speciation, geographical isolation has played a major role in the synthesis as well. At the same time, as I have already noticed, some cladists, notably Nelson and Platnick, rely on a particular biogeographical view to support their emphatic rejection of any form of Darwinism. So one could claim that while biogeography, with emphasis especially on dispersal to new areas, supported a Darwinian conception, panbiogeography, with emphasis on vicariance as against dispersal, opposes such a perspective. But birds do fly and so, *pace* Nelson, do fishes swim. Why *no* dispersal, ever? Among, for example, North and South American fauna or Hawaiian *drosophila* it seems pretty well established. On the other hand, of course, in the light of recent geological discoveries, vicariance must also be admitted as a ground for allopatry. But why not? There is nothing, in itself, "adaptive" and therefore selectionist about dispersal—true, it is a way of spreading varieties, but there is nothing as such adaptive about winds that blow where they list. Long ago, in an old-fashioned T. H. Huxley type of evolution course, we were taught about Wallace's line as one of the evidences for evolution; but there did not seem to be anything "adaptive" about it. It was an evidence for evolution, not, specially, for natural selection. Nor, on the other hand, does plate tectonics, for example, appear, as such, to undercut a Darwinian interpretation of organic evolution; it shows how habitats (sometimes) change, not how their inhabitants change in the makeup or structure of their populations. As I have already suggested, it may be that the panbiogeography-inclined cladists were looking for some supporting field to replace population genetics as the major foundation for their phylogenies. Or perhaps they just like to draw geographical cladograms to match their taxonomic ones. Whatever the explanation, I must confess, I find it difficult to turn into a major revolution work that takes seriously such widely speculative and overambitious tracts as Croizat's (1984) *Space, Time and Form* (Nelson and Platnick, 1984; Craw and Page, 1988). Philosophers may not be good for much, but we do have, I hope (most of us, at least) a nose for empty metaphysics when we smell it. Nevertheless,

*Ho's essay (1988) also intends to radicalize history; but unfortunately her attacks on Darwinism are so extreme and in part so poorly defended as to undercut the value of the positive point she presumably wants to make.

"transformed cladism" does constitute an anomaly in my attempt to minimize the significance of the crossroads metaphor as applied to contemporary evolutionary theory.

ECOLOGY

First, last, and always, natural selection has to do with changes in the relative adaptedness of members of populations (if at a number of levels) to their environments, in so far as those (heritable) differences issue in differential reproduction. The theory of natural selection is inconceivable apart from attention to organisms in relation to their environments. Yet ecology has developed, at least in part, and still goes on, I think one could say for the most part, in relative independence of evolutionary biology. Granted, there is now an "evolutionary ecology," but this is rather the assimilation (by some ecologists) of concepts and interests from evolutionary theory than the absorption of ecology into evolutionary theory—as the synthesis claimed for other disciplines. True, this is a very complicated story, as Kingsland's book (1985) and the papers of Collins (1986), Kimler (1986), Futuyma (1986), and others, from a 1985 conference, make clear. Nevertheless, one is inclined to ask why objections to the synthesis should not come from ecologists, whose work ought to be central to evolutionary biology, yet seems to remain peripheral. In the materials here under review, however, except for some of the problems raised by Gray (1988) no such thing happens. Eldredge does indeed distinguish between a genealogical and an ecological hierarchy; this seems to me a seminal distinction, and I hope it will be further developed in a number of directions (Eldredge, 1985; Grene, 1987). But Eldredge is a paleontologist trying to flesh out evolutionary theory, not an ecologist coming to the defense of his discipline and its place in evolutionary biology. In the Weber *et al.* (1988) collection, moreover, there are a couple of references to the work of Ulanowics, an ecologist who is interested in the new "thermodynamic" approach to evolution; but in general (as we shall see) the major motivation for that movement does not come from ecology either. Buss (1987) quotes Van Valen's remark that evolution is the control of development by ecology; he is trying to move development to center stage, but despite everyone's knowledge of the importance of environment to evolutionary processes, ecology, it seems, is still, in most cases, left waiting in the wings.

THE RELATION OF BIOLOGY TO THE "EXACT" SCIENCES

During the "biochemical revolution," such eminent evolutionists as Dobzhansky (1968) and Simpson (1965) hastened to the defense of their discipline against the (seeming) threat of reduction. A classic position on the uniqueness of biology is still presented, in powerful and persuasive form, by Mayr (1985). At the same time, however, *leitmotivs* from physics or chemistry are being introduced into biology, not, now, in the old reductionist spirit of a Crick or a Watson, but, allegedly, in an effort to enrich biological theory. These appear, in view of the evidence here under review, to be of three kinds: particular principles said to undermine Darwinism, global principles that claim to lay a new foundation for evolutionary theory, or particular principles, analogies, or models from other disciplines that are intended to supplement rather than to replace the principles of the synthesis.

Saunders and Ho present a thesis of the first kind in their "principle of minimum increase in complexity" (Saunders and Ho, 1984; see also Ho and Fox, 1988; Fox, 1988). The question that leaps from their pages is an obvious one: why do they think this principle resists a standard Darwinian explanation? Their first example is taken from a paper by Stebbins in the first volume of *Evolutionary Biology* (Stebbins, 1967), which uses minimal increase in complexity in the adaptive radiation of the flowering plants ("selection along the lines of least resistance") to *support* the theory of natural selection. Natural selection, Stebbins argues,

> will alter those characteristics in which significant changes are the most likely to occur by means of mutation and genetic recombination. This likelihood, in turn, depends upon the degree to which developmental pathways must be altered in order to produce the character favored by selection. (Stebbins, 1967, p. 131.)

The first case he presents to support this principle consists of two examples, the tulip and the magnolia, which have something in common but also differ significantly with respect to the probable path for selection to take:

> In a plant having a single large flower borne on a short stem or branch, such as a tulip or magnolia, production of a second flower of equal size would require a radical reorganization of the branching system and its developmental pattern, which could probably be carried out only by considerable changes at many gene loci. (Stebbins, 1967, p. 131.)

So far, so good. But the difference between the two is equally important. Stebbins continues:

> the difference between the structure of the gynecium of the tulip and the magnolia flower would alter the path of least resistance to selection for increased seed number in these two flowers. In the tulip, the number of carpels in the syncarpous ovary is three, as it is in most genera of the Liliaceae, and alteration of this number would require radical readjustments in the architecture of the entire flower. On the other hand, the number of ovules per carpel is already large, and this number could be increased by means of simple changes in the activity of the intercalary meristem responsible for carpel development. Consequently, in the tulip, increase in seed number is most likely to be accomplished by selection for an increased number of ovules per carpel. (Stebbins, 1967, p. 131.)

The case of the magnolia is different:

> In the magnolia, on the other hand, each carpel produces only one or two seeds, while the number of carpels per flower is very large. In such a flower, response to selection for increased seed number is most likely to occur via mutations which increase the amount of meristem from which carpels are produced and, consequently, the number of carpels per flower. (Stebbins, 1967, p. 131, 132.)

This is only a small part of a long and complex argument, all of which serves to assimilate adaptive radiation in the higher plants to the framework of the synthesis. That it should be cited as part of a *refutation* of the synthesis is astonishing.

Saunders and Ho's second example, of parallel evolution, seems too general to amount to much, and the third, mimicry, is of course one of the major cases easily susceptible to Darwinian analysis. In general, the situation here appears to be similar to that with respect to blind cave fishes and the like: such phenomena may seem at first to resist Darwinian interpretation, but the "efficiency" of natural selection in expending the least (or lesser) energy necessary to produce organisms somewhat better suited to their environments than their conspecifics, and hence at an advantage in terms of differential reproduction: this principle fits them easily into their theoretical niche within an overall Darwinian scheme.

More ambitious is the by now notorious Brooks and Wiley "entropic" approach, here exemplified also in Pollard (Brooks and Wiley, 1984). Here, of course, the second law is supposed to preside over evolution, not, now, as used to be (allegedly) the case in the 1950s, by allowing life to "feed on negentropy," but by the power of entropy itself. Only a competent physicist could judge the adequacy of this new program. As a philosopher looking on, I can only say I have my doubts—and not much more than doubts. The principle claims to do too much; and within its scope, evolution in any realistic detail has still to be written in at every stage on other evidence. Admittedly, the principle of natural selection is also global, and even, as Brandon has clearly shown (Brandon, 1981, 1990), devoid of concrete empirical content. But it is a principle that, in a metatheoretical role, is well suited to preside over, and has presided over, concrete experimental research in an indefinite number of empirical contexts. In the case of the entropy line, there seems to be no such likely input into actual research contexts. True, the volume edited by Weber *et al.* (1988) suggests that the turn to a thermodynamic perspective stems from biologists, physicists, and philosophers with a diversity of research interests and expertise, and not only from yet another pair of cladists impatient with the synthesis and looking for foundational principles elsewhere. Depew and Weber's (1988) thoughtful afterword may also temper one's scepticism a little, but, on the other hand, David Layzer's essay in this volume serves to strengthen it: Layzer argues convincingly (and it is not hard to do!) that the second law on its own is insufficient to support a theory of evolution (Layzer, 1988). His statement that "there is a single universal law governing processes that dissipate order, but order is generated by several hierarchically linked processes'' (Layzer, 1988, p. 39) seems to temper considerably the second law euphoria that characterizes much of the argument in this (at best) controversial area.*

Much more promising, in my view, is the judicious use of models or experimental results from many fields, including a number in the ''exact'' sciences, and also new mathematical and computer techniques, in order to enrich and expand the fundamental reasoning of Darwin and Darwinism. That seems to me to be the approach, for example, of Fitch and Upper (1988) in their study of the origin of the genetic code. On a more comprehensive scale, that appears also to be the approach taken by Kauffman, with no claim, so far as I can see, to any single overall principle that magically takes over everything. As I have also noted, only a small part of such a project is exemplified in Kauffman's (1985) essay in Depew and Weber, much more in his forthcoming book (Kauffman, 1990). Even the book is presented as work in progress, a collection of schemata, of lines for research, rather than as a set of set results. However, there is, in my view, very much more meat to it than is exhibited in the ''thermodynamic'' movement.

HISTORY AND PHILOSOPHY OF SCIENCE

In conclusion, let me touch very briefly on the lessons suggested by some of this material for our approach to the history and philosophy of science. First, a glaring paradox: Pollard celebrates the contribution by Sir Karl Popper to his volume, because Popperian philosophy has been so crucial in the development of ''post-Darwinian'' approaches to evolution (Pollard, 1984*b*). Meantime, however, of course, Popper himself has been converted to evolutionary epistemology of an utterly naive and ultra-Darwinian cast, so that his contribution to Pollard's collection wholly undercuts the supposed non-Darwinian methodology of the collection itself (Popper, 1984). How can one take at all seriously a position that so denies itself on principle? The point is perhaps too silly to be important. But in biology, in particular, for some reason obscure to me, Popper has been extremely influential in a number of quarters, and one is happy, therefore, to point out any folly that will support the slogan ''No Poppery!'' (Van Valen, 1988).

Another oddity in respect to the philosophy of science is presented by Hailman in the Ho and Fox collection (Hailman, 1988). That optimality theory

*I cannot help noting, in passing, that Layzer (1988) refers to Mayr's founder principle as an instance of the application of entropy to biological theory. In another context, that same principle was applied in the initiation of the theory of punctuated equilibrium. I suspect that the founder, in this case as in the other, will not be amused.

has needed, and may still need, critical analysis is evident (e.g., Dupré, 1987), but to base such a critique on so naive and outworn a philosophy as Bridgeman's "operationalism" is surprising indeed (Hailman, 1988). True, there are now more sophisticated forms of something resembling operationalism in fashion in philosophy: the "semantic theory," for example, is used with sympathy by both Lloyd (1987, 1988), and Beatty (1987). Be that as it may, I would argue that it is only in a realist interpretation that even this newer "operational" approach is of use in biology. If biologists are not talking, even at their most speculative, about what are or have been real events and entities in the real world which we, too, inhabit, then I cannot imagine what they are wasting their time (and our money) investigating.

More positively, both Burian's (1985) essay in Depew and Weber and Buss's (1987) book support a thesis that seems to me essential for an understanding of the Darwinian tradition. Burian is dealing, briefly, with the history of the concept of the gene, which he is treating (in collaboration with Doris Zallen) in a book now in progress. Central scientific concepts, he argues, may change their exact physiognomies while still functioning in the development of what is recognizably the same theory, even referring to the same reality. There *are* genes, and genetics, Mendelian, biochemical, molecular, stakes its claim on the *fact* of their existence. Nevertheless, through the progress of theoretical as well as experimental research (the latter, of course, mediated by advances in the technology available to experimenters), the *concept* of the gene may grow and alter in the course of these developments. Buss speaks, more generally, of the "flexibility of the Darwinian theory," but the point is the same.* One need not go as far as Hull (1985) and claim that Darwinism "has no essence"—or only the same "essence" as species: that is, genealogy and nothing but genealogy. But it is important to recognize that theories, let alone metatheories like the theory of natural selection, are congeries of human practices: actions, beliefs, and conformity to, as well as creation of, norms or values. Far from being fixed once for all, so that any anomaly suffices to defeat them, they are, when fruitful, capable of alteration and growth—and this in no epicyclic, but in a coherent, if by no means predictable, fashion. You and I have become the persons we are through our respective histories: not altogether "not in our genes," but not altogether in them either. Analogously, contemporary Darwinism, or the synthetic theory of organic evolution, is itself a historical entity that has developed astonishingly over the last decades, and in the face of the counterevidence presented in the volumes we have been looking at, still flourishes and bears in itself the potentiality for further amplification and development. Indeed, as we have seen, some of that potential, too, is evidenced in some of those same volumes. Resoundingly, more "No" than "Yes"!

*That is not to agree with Buss that science is "subjective," a thesis which he mistakenly elicits from Polanyi's (1958) *Personal Knowledge*. See, for example, Ziman (1967), *Public Knowledge*.

ACKNOWLEDGMENT

I would like to thank Dr. Richard Burian for his careful reading and criticism of an earlier version of this manuscript.

REFERENCES

Alberch, P., 1982, Developmental constraints in evolutionary processes, in: *Evolution and Development* (J. T. Bonner, ed.), pp. 313–332, Springer-Verlag, Berlin.

Ayala, F. J., 1985, Reduction in biology: A recent challenge, in: *Evolution at a Crossroads: The New Biology and the New Philosophy of Science* (D. J. Depew and B. H. Weber, eds), pp. 65–80, MIT Press, Cambridge, Massachusetts.

Bateson, P., 1988, The active role of behaviour in evolution, in: *Evolutionary Processes and Metaphors* (M. -W. Ho and S. F. Fox, eds.), pp. 191–208, Wiley, New York.

Bateson, W., 1894, *Materials for the Study of Variation,* MacMillan, London.

Beatty, J., 1987, On behalf of the semantic view, *Biol. Philos.* **2:**16–33.

Bock, W., 1979, The synthetic explanation of macroevolutionary change—a reductionistic approach, *Bull. Carnegie Mus. Nat. Hist.* **13:**20–69.

Boyden, A., 1943, Homology and analogy: A century after the definitions of "homologue" and "analogue" of Richard Owen, *Q. Rev. Biol.* **18:**228–241.

Brandon, R. N., 1981, A structural description of evolutionary theory, *Philosophy of Science Association* **1981:**427–439.

Brandon, R. N., 1985, Adaptation explanations: Are adaptations for the good of replicators or interactors? in: *Evolution at a Crossroads: The New Biology and the New Philosophy of Science* (D. J. Depew and B. H. Weber, eds.), pp. 81–96, MIT Press, Cambridge, Massachusetts.

Brandon, R. N., 1990, *Adaptation and Evolution,* Princeton University Press, Princeton, New Jersey.

Brandon, R. N., and Burian, R. M., eds., 1984, *Genes, Organisms, Populations: Controversies Over the Units of Selection,* MIT Press, Cambridge, Massachusetts.

Brent, L., Rayfield, L. S., Chandler, P., Fierz, P., Medawar, P. B., and Simpson, E., 1981, Supposed Lamarckian inheritance of immunological tolerance, *Nature* **290:**508–514.

Brent, L., Chandler, P., Fierz, W., Medawar, P. B., Rayfield, L. S., and Simpson, E., 1982, Supposed Lamarckian inheritance of immunological tolerance, *Nature* **295:**242–244.

Brooks, D. R., and Wiley, E. O., 1984, Evolution as an entropic phenomenon, in: *Evolutionary Theory: Paths into the Future* (J. W. Pollard, ed.), Wiley, New York.

Burian, R. M., 1985, On conceptual change in biology: The case of the gene, in: *Evolution at a Crossroads: The New Biology and the New Philosophy of Science* (D. J. Depew and B. H. Weber, eds.), pp. 21–42, MIT Press, Cambridge, Massachusetts.

Buss, L. W., 1987, *The Evolution of Individuality,* Princeton University Press, Princeton, New Jersey.

Campbell, J. H., 1985, An organizational interpretation of evolution, in: *Evolution at a Crossroads: The New Biology and the New Philosophy of Science* (D. J. Depew and B. H. Weber, eds.), pp. 133–167, MIT Press, Cambridge, Massachusetts.

Churchill, F. B., 1980, The modern evolutionary synthesis and the biogenetic law, in: *The Evolutionary Synthesis* (E. Mayr and W. B. Provine, eds.), pp. 112–122, Harvard University Press, Cambridge Massachusetts.

Collins, J. H., 1986, Evolutionary ecology and the use of natural selection in ecological theory, *J. Hist. Biol.* **19:**257–288.

Cracraft, J., 1989*a*, Introduction to the symposium, *Syst. Zool.* **37:**219–220.

Cracraft, J., 1989*b*, Deep-history biogeography: retrieving the historical pattern of evolving continental biotas, *Syst. Zool.* **37:**221–236.

Craw, R., and Page, R., 1988, Panbiogeography: Method and metaphor in the new biogeography, in: *Evolutionary Processes and Metaphors* (M. -W. Ho and S. W. Fox, eds.), pp. 163–190, Wiley, New York.

Croizat, L., 1964, *Space, Time, Form: The Biological Synthesis,* L. Croizat, Caracas, Venezuela.

Cullis, C. A., 1988, Control of variation in higher plants, in: *Evolutionary Processes and Metaphors* (M.-W. Ho and S. W. Fox, eds), pp. 49–62, Wiley, New York.

Darwin, C., 1859, *On the Origin of Species,* John Murray, London.

Dawkins, R., 1986, *The Blind Watchmaker,* Norton, New York.

Depew, D. J., and Weber, B. H., eds., 1985, *Evolution at a Crossroads: The New Biology and the New Philosophy of Science,* MIT Press, Cambridge, Massachusetts.

Depew, D. J., and Weber, B. H., 1988, Consequences of nonequilibrium thermodynamics for the Darwinian tradition, in: *Entropy, Information, and Evolution: New Perspectives on Physical and Biological Evolution*(B. H. Weber, D. J. Depew, and J. D. Smith, eds.), pp. 317–354, MIT Press, Cambridge, Massachusetts.

Dobzhansky, T., 1937, *Genetics and the Origin of Species,* Columbia University Press, New York.

Dobzhansky, T., 1968, On Cartesian and Darwinian aspects of biology, *Graduate J.* **8:**99–117.

Dobzhansky, T., Ayala, F. J., Stebbins, G. L., and Valentine, J. W., 1977, *Evolution,* Freeman, San Francisco.

Dupré, J., 1987, *The Latest on the Best,* MIT Press, Cambridge, Massachusetts.

Eldredge, N., 1971, The allopatric model and phylogeny in Paleozoic invertebrates, *Evolution,* **21:**156–167.

Eldredge, N., 1985, *Unfinished Synthesis,* Oxford University Press, Oxford.

Eldredge, N., and Cracraft, J., 1980, *Phylogenetic Patterns and the Evolutionary Process,* Columbia University Press, New York.

Eldredge, N., and Gould, S. J., 1972, Punctuated equilibrium: An alternative to phyletic gradualism, in: *Models in Paleobiology,* (T. J. M. Schopf, ed.), pp. 82–115, Freeman, San Francisco.

Fisher, R. A., 1930, *The Genetical Theory of Natural Selection,* Clarendon Press, Oxford.

Fitch, W. M., and Upper, K., 1988, The evolution of life—An overview of general problems and a specific study of the origin of the genetic code, in: *Evolutionary Processes and Metaphors* (M.-W. Ho and S. W. Fox, eds.), pp. 35–48, Wiley, New York.

Fox, S. W., 1984, Proteinoid experiments and evolutionary theory, in: *Beyond Neo-Darwinism: An Introduction to the New Evolutionary Paradigm* (M.-W. Ho and P. Saunders, eds.), pp. 15–60, Academic Press, London.

Fox, S. W., 1988, Evolution outward and forward, in: *Evolutionary Processes and Metaphors* (M.-W. Ho and S. W. Fox, eds.), pp. 17–33, Wiley, New York.

Futuyma, D. J., 1986, Reflection on reflections: Ecology and evolutionary biology, *J. Hist. Biol.* **19:**303–312.

Ghiselin, M. T., 1974, A radical solution to the species problem, *Syst. Zool.* **23:**536–544.

Gilinsky, N. L., 1986, Species selection as a causal process, in: *Evolutionary Biology, Vol.* 20 (M. K. Hecht and B. Wallace, eds.), pp. 249–273, Plenum Press, New York.

Goodwin, B. C., 1984*a,* A relational or field theory of reproduction and its evolutionary implications, pp. 219–241, Academic Press, London.

Goodwin, B. C., 1984*b,* Changing from an evolutionary to a generative paradigm in biology, in: *Evolutionary Theory: Paths into the Future* (J. W. Pollard, ed.), pp. 99–120, Wiley, New York.

Goodwin, B. C., 1988, Morphogenesis and heredity, in: *Evolutionary Processes and Metaphors* (M.-W. Ho and S. W. Fox, eds.), pp. 145–162, Wiley, New York.

Gordon, D. M., 1988, Behaviour changes—finding the rules, in: *Evolutionary Processes and Metaphors* (M.-W. Ho and S. W. Fox, eds.), pp. 243–254.

Gordon, D. M., 1990, Caste and change in social insects, *Oxf. Surv. Evol. Biol.* **6** (in press).

Gould, S. J., 1983, The hardening of the modern synthesis, in: *Dimensions of Darwinism,* (M. Grene, ed.), pp. 71–93, Cambridge University Press, Cambridge.

Gould, S. J., 1986, Evolution and the triumph of homology, or why history matters, *Am. Sci.* **74:**60–69.

Gray, R., 1988, Metaphors and methods: Behavioural ecology, panbiogeography and the evolving synthesis, in: *Evolutionary Processes and Metaphors* (M.-W. Ho and S. W. Fox, eds.), pp. 209–242, Wiley, New York.

Grene, M., 1987, Hierarchies in biology, *Am. Sci.* **75:**504–510.

Hailman, J. P., 1988, Operationalism, optimality and optimism: Suitabilities versus adaptations of organisms, in: *Evolutionary Processes and Metaphors* (M.-W. Ho and S. W. Fox, eds.), pp. 85–116, Wiley, New York.

Hamburger, V., 1980, Embryology and the modern synthesis in evolutionary theory, in: *The Evolutionary Synthesis* (E. Mayr and W. B. Provine, eds.), pp. 97–112, Harvard University Press, Cambridge, Massachusetts.

Hennig, W., 1950, *Grundzüge einer Theorie der phylogenetischen Systematik,* Deutscher Zentralverlag, Berlin.

Hennig, W., 1966, *Phylogenetic Systematics,* University of Illinois Press, Urbana, Illinois.

Ho, M.-W., 1988, On not holding nature still: Evolution by process, not by consequence, in: *Evolutionary Processes and Metaphors* (M.-W. Ho and S. W. Fox eds.), pp. 117–144, Wiley, New York.

Ho, M.-W., and Fox, S. W., eds., 1988*a*, *Evolutionary Processes and Metaphors,* Wiley, New York.

Ho, M.-W., and Fox, S. W., 1988*b,* Processes and metaphors in evolution, in: *Evolutionary Processes and Metaphors* (M.-W. Ho and S. W. Fox, eds.), pp. 1–16, Wiley, New York.

Ho, M.-W., and Saunders, P., eds., 1984, *Beyond Neo-Darwinism: An Introduction to the New Evolutionary Paradigm,* Academic Press, London.

Hodge, M. J. S., 1977, The structure and strategy of Darwin's "long argument," *Br. J. Hist. Sci.* **10:**237–245.

Howard, J. C., 1981, A tropical volute shell and the Icarus syndrome, *Nature* **290:**242–244.

Hull, D. L., 1965, The effect of essentialism on taxonomy. *Br. J. Philos. Sci.* **15:**314–326; **16:**1–18.

Hull, D. L., ed., 1973, *Darwin and His Critics,* University of Chicago Press, Chicago.

Hull, D. L., 1976, Are species really individuals?, *Syst. Zool.* **25:**174–191.

Hull, D. L., 1985, Darwinism as a historical entity: A historiographic proposal, in: *The Darwinian Heritage* (D. Kohn, ed.), pp. 773–812, Princeton University Press, Princeton, New Jersey.

Hull, D. L., 1988, *Science as a Process,* University of Chicago Press, Chicago.

Hume, D., 1739, *A Treatise of Human Nature,* John Noon, London [reprint Clarendon Press, Oxford, 1955].

Jacob, F., 1982, *The Possible and the Actual,* Pantheon, New York, Pantheon.

Janvier, P., 1984, Cladistics: Theory, purpose and evolutionary implications, in: *Evolutionary Theory: Paths into the Future* (J. W. Pollard, ed.), pp. 39–75, Wiley, New York.

Kauffman, S., 1985, Self-organization, selective adaptation, and its limits: A new pattern of inference in evolution and development, in: *Evolution at a Crossroads: The New Biology and the New Philosophy of Science* (D. J. Depew and B. H. Weber, eds.), pp. 169–207, MIT Press, Cambridge, Massachusetts.

Kauffman, S., 1990, *Origins of Order: Self-Organization and Selection in Evolution,* Oxford University Press, Oxford.

Kimler, W. C., 1983, Mimicry: Views of naturalists and ecologists before the modern synthesis, in: *Dimensions of Darwinism,* (M. Grene, ed.), pp. 97–127, Cambridge University Press, Cambridge.

Kimler, W. C., 1986, Advantage, adaptiveness, and evolutionary ecology, *J. Hist. Biol.* **19:**215–234.

Kingsland, S., 1985, *Modeling Nature: Episodes in the History of Population Ecology,* University of Chicago Press, Chicago.

Kingsolver, J. G., and Koehl, M. A. R., 1985, Aerodynamics, thermoregulation, and the evolution of insect wings: Differential scaling and evolutionary change, *Evolution* **39:**488–504.

Kluge, A., 1989, Parsimony in vicariance biogeography: a quantitative method and a Greater Antillean example, *Syst. Zool.* **37:**315–328.

Layzer, D., 1988, Growth of order in the universe, in: *Entropy, Information, and Evolution: New Perspectives on Physical and Biological Evolution* (B. H. Weber, D. J. Depew, and J. D. Smith, eds.), pp. 23–39, MIT Press, Cambridge, Massachusetts.

Lewontin, R. C., 1970, The units of selection, *Annu. Rev. Ecol. Syst.* **1:**1–18.

Lewontin, R. C., 1974, *The Genetic Basis of Evolutionary Change,* Columbia University Press, New York.

Lewontin, R. C., 1983, Gene, organism and environment, in: *Evolution from Molecules to Men* (D. S. Bendall, ed.), pp. 273–285, Cambridge University Press, Cambridge.

Lillie, R. S., 1927, The gene and the ontogenetic process, *Science* **64:**361–368.

Lloyd, E. A., 1987, Response to Sloep and Van der Steen, *Biol. Philos.* **2:**23–26.

Lloyd, E. A., 1988, *The Structure and Confirmation of Evolutionary Theory,* Greenwood Press, Westport, Connecticut.

Locke, J., 1690, *An Essay concerning Human Understanding,* Bassett, London [reprint of 4th ed. (1700), Clarendon Press, Oxford, 1975].

Løvtrup, S., 1984, Ontogeny and phylogeny, in: *Beyond Neo-Darwinism: An Introduction to the New Evolutionary Paradigm* (M.-W. Ho and P. Saunders, eds.), pp. 159–190, Academic Press, London.

Matsuno, K., 1984, Open systems and the origin of photoreproductive units, in: *Beyond Neo-Darwinism: An Introduction to the New Evolutionary Paradigm* (M.-W. Ho and P. Saunders, eds.) pp. 61–88, Academic Press. London.

Maynard Smith, J., 1982, Lamarckian inheritance and the puzzle of immunity, in: *Evolution Now* (J. Maynard Smith, ed.), pp. 91–93, Freeman, San Francisco.

Maynard Smith, J., Burian, R., Kauffman, S., Alberch, P., Campbell, J., Goodwin, B., Lande, R., Raup, D., and Wolpert, L., 1985, Developmental constraints and evolution, *Q. Rev. Biol.* **60:**265–287.

Mayr, E., 1942, *Systematics and the Origin of Species,* Columbia University Press, New York.

Mayr, E., 1974, Cladistic analysis or cladistic classification?, *Z. Zool. Syst. Evolutionsforsch.* **12:**94–128.

Mayr, E., 1982, *The Growth of Biological Thought,* Harvard University Press, Cambridge Massachusetts.

Mayr, E., 1983, How to carry out the adaptationist program, *Am. Nat.* **121:**324–334.

Mayr, E., 1985, How biology differs from the physical sciences, in: *Evolution at a Crossroads: The New Biology and the New Philosophy of Science* (D. J. Depew and B. H. Weber, eds.), pp. 43–63, MIT Press, Cambridge, Massachusetts.

Mayr, E., 1987, Answers to these comments, *Biol. Philos.* **2:**212–220.

Mayr, E. and Provine, W. B., eds, 1980, *The Evolutionary Synthesis,* Harvard University Press, Cambridge, Massachusetts.

McIntosh, R. P., 1985, *The Background of Ecology: Concept and Theory,* Cambridge University Press, New York.

Moorehead, P. S., and Kaplan, M. M., eds., 1967, *Mathematical Challenges to the Neo-Darwinian Interpretation of Evolution,* Wistar Institute Press, Philadelphia, Pennsylvania.

Nelson, G. and Platnick, N., 1984, Systematics and evolution, in: *Beyond Neo-Darwinism: An Introduction to the New Evolutionary Synthesis* (M.-W. Ho and P. Saunders, eds.), pp. 143–158, Academic Press, London.

Olson, E., 1960, in: *Evolution after Darwin* (S. Tax, ed.), Vol. I, pp. 523–545, University of Chicago Press, Chicago.

Oster, G. F., Odell, G., and Alberch, P., 1980, Mechanics, morphogenesis and evolution, *Lect. Math. Life Sci.* **13:**165–255.

Oyama, S., 1985, *The Ontogeny of Information,* Cambridge University Press, Cambridge.

Oyama, S., 1988, Stasis, development and heredity, in: *Evolutionary Processes and Metaphors* (M.-W. Ho and S. W. Fox, eds.), pp. 255–274, Wiley, New York.

Patterson, C., 1981, Significance of fossils in determining evolutionary relationships, *Annu. Rev. Ecol. Syst.* 12:195–223.

Patterson, C., 1982, Morphological characters and homology, in: *Problems of Phylogenetic Reconstruction* (K. A. Joysey and A. E. Friday, eds.), pp. 21–74, Academic Press, London.

Polanyi, M., 1958, *Personal Knowledge,* University of Chicago Press, Chicago.

Pollard, J. W., ed., 1984*a*, *Evolutionary Theory: Paths into the Future,* Wiley, New York.

Pollard, J. H., 1984*b*, Preface, *Evolutionary Theory: Paths into the Future* (J. W. Pollard, ed.), pp. xii–xiii, Wiley, New York.

Pollard, J. H., 1984*c*, Is Weissman's barrier absolute?, in: *Beyond Neo-Darwinism: An Introduction to the New Evolutionary Paradigm* (M.-W. Ho and P. Saunders, eds.), pp. 291–314, Academic Press, London.

Pollard, J. H., 1988, New genetic mechanisms and their implications for the formation of new species, in: *Evolutionary Processes and Metaphors* (M.-W. Ho and S. W. Fox, eds.), pp. 63–84, Wiley, New York.

Popper, K., 1984, Evolutionary epistemology, in: *Evolutionary Theory: Paths into the Future* (J. W. Pollard, ed.), pp. 239–255, Wiley, New York.

Provine, W. B., 1971, *The Origins of Theoretical Population Genetics,* University of Chicago Press, Chicago.

Rosen, D. E., 1984, Hierarchies and history, in: *Evolutionary Theory: Paths into the Future* (J. W. Pollard, ed.), pp. 77–97, Wiley, New York.

Ruse, M., 1979, *The Darwinian Revolution,* University of Chicago Press, Chicago.

Ruse, M., 1988, Booknotes, *Biol. Philos.* **3:**285–289.

Saunders, P. T., and Ho, M.-W., 1984, The complexity of organisms, in: *Evolutionary Theory: Paths into the Future* (J. W. Pollard, ed.), pp. 121–139, Wiley, New York.

Schmalhausen, I. I., 1949, *Factors of Evolution,* Blakiston, Philadelphia, Pennsylvania.

Simpson, G. G., 1944, *Tempo and Mode in Evolution,* Columbia University Press, New York.

Simpson, G. G., 1983, *Major Features of Evolution,* Columbia University press, New York.

Simpson, G. G., 1965, The crisis in biology, *Am. Scholar* **1965:**363–377.

Sober, E., 1984, *The Nature of Selection,* MIT Press, Cambridge, Massachusetts.

Sober, E., 1989, The conceptual relationship of cladistic phylogenetics and vicariance biogeography, *Syst. Zool.* **37:**245–253.

Stanley, S. M., 1975, A theory of evolution above the species level, *Proc. Natl. Acad. Sci. USA* **72:**646–650.

Stanley, S. M., 1979, *Macroevolution. Pattern and Process,* Freeman, San Francisco.

Stebbins, G. L., 1967, Adaptive radiation and trends of evolution in higher plants, in: *Evolutionary Biology,* Vol. 2 (T. Dobzhansky, M. K. Hecht, and W. C. Steere, eds.), pp. 101–142, Appleton-Century-Crofts, New York.

Steele, E. J., Gorczynski, R. M., and Pollard, J. W., 1984, The somatic selection of acquired characters, in: *Evolutionary Theory: Paths into the Future* (J. W. Pollard, ed.), pp. 217–237, Wiley, New York.

Temin, H. M., and Engels, R., 1984, Movable genetic elements and evolution, in: *Evolutionary Theory: Paths into the Future* (J. W. Pollard, ed.), pp. 173–201, Wiley, New York.

Thompson, W. D'A., 1917, *On Growth and Form,* Cambridge University Press, Cambridge.

Van Valen, L. M., 1988, Species, sets and the derivative nature of philosophy, *Biol. Philos.* **3**:49–66.

Vrba, E. S., 1984, Patterns in the fossil record and evolutionary processes, in: *Beyond Neo-Darwinism: An Introduction to the New Evolutionary Paradigm* (M.-W. Ho and P. Saunders, eds.), pp. 115–142, Academic Press, London.

Waddington, C. H., 1957, *The Strategy of the Genes,* Allen and Unwin, London.

Weber, B. H., Depew, D. J., and Smith, J. D., eds., 1988, *Entropy, Information, and Evolution: New Perspectives on Physical and Biological Evolution,* MIT Press, Cambridge, Massachusetts.

Webster, G., 1984, The relations of natural forms, in: *Beyond Neo-Darwinism: An Introduction to the New Evolutionary Paradigm* (M.-W. Ho and P. Saunders, eds.), pp. 193–217, Academic Press, London.

Webster, G., and Goodwin, B. C., 1982, The origin of species: A structuralist approach, *J. Soc. Biol. Struc.* **5**:15–42.

Williams, G. C., 1966, *Adaptation and Natural Selection,* Princeton University Press, Princeton, New Jersey.

Wimsatt, W. C., 1980, Reductionistic research strategies and their biases in the units of selection controversy, in: *Scientific Discovery: Case Studies* (T. Nickles, ed.), pp. 213–259, Reidel, Dordrecht.

Ziman, J., 1967, *Public Knowledge,* Cambridge University Press, Cambridge.

3

The Protein Molecular Clock

Time for a Reevaluation

SIEGFRIED SCHERER

FOSSILS, MOLECULES, AND EVOLUTION

Paleontology versus Molecular Biology?

The search for the origin of today's living organisms is based mainly on two different lines of evidence. The first is the investigation of the biological characteristics of species still alive today and thus available for in-depth studies. This field of origin research, commonly known as comparative biology in its widest sense, includes the comparison of macromolecular sequences. The second line of evidence relies on the fossil record. However, investigation of fossilized organisms usually is restricted by the availability of fossils.

One major question addressed by evolutionary theory concerns the time of divergence of living taxa. This time might be estimated by dating the strata housing the earliest fossil which is considered to represent the last common ancestor of two taxa. However, limitations are posed (1) by the availability of fossils, (2) by elucidating the taxonomic status of a particular fossil, and (3) by the uncertainty of whether a given stratigraphic appearance of an organism really represents its earliest presence in earth history.

The primary structure of proteins does not have these limitations and hence is considered by numerous researchers as constituting a very superior and more reliable source of phylogenetic information. The resulting rivalry between paleontologists and molecular biologists has been characterized somewhat ironically

Portions of the data reported in this chapter were presented at the First Congress of the European Society of Evolutionary Biology, Basel, Fall, 1987.

SIEGFRIED SCHERER • Faculty for Biology, University of Konstanz, D-7750 Konstanz, Federal Republic of Germany. At the time of writing, the author was on leave at the Department of Biochemistry, Virginia Polytechnic Institute and State University, Blacksburg, Virginia 24061.

by Jones and Rouhani (1986), stating that "the human fossil record is no exception to the general rule that the main lesson to be learned from paleontology is that evolution always takes place somewhere else." Bearing in mind that these authors are geneticists, it is not too surprising when they conclude that "unlike students of the human fossil record, molecular biologists know that the information they need to reconstruct human evolution is easily available and merely awaits their attention." Indeed, the study of the history of interpretation of, for example, *Ramapithecus* and related fossil remains reveals that the molecular clock concept, in favoring a recent split of the ape/human lineage, considerably influenced the development of paleanthropologic hypothesis (for review, see Hartwig-Scherer, 1989).

Comparing Protein Sequences

As was emphasized by Jones and Rouhani (1986), many biologists are disappointed by the seemingly inevitable subjectivity of "classical taxonomy" and the drastically limited data available from the fossil record and hope that molecular biology will serve as a source of indisputable, clearly-defined data. This would allow a more objective approach to taxonomy than was possible before sequences of biological macromolecules became available. For some proteins, comparison of sequences produces quantitative measurements of similarity. However, this is only possible in a rigorous sense if sequences show the same number of residues. Otherwise, deletions, insertions, duplications, or other mutational events must be postulated (e.g., Dayhoff, 1967–1978) which may or may not reflect the real history of the correspondent genes. In any case, a more or less unequivocal difference or similarity matrix is derived from a given set of sequences, which can be used for several purposes: (1) Classification of organisms by constructing a phylogenetic tree by using several algorithms [for review see, e.g., Felsenstein (1983)]. However, seemingly independent from the algorithm used, an increasing number of reports have been published, emphasizing a significant disagreement between molecular and classical taxonomy (Ambler *et al.*, 1979*a,b;* Bremer, 1988; Meyer *et al.*, 1986; Scherer and Binder, 1986; Scherer and Sontag, 1983, 1986; Scherer *et al.*, 1985). (2) Posing several assumptions, similarity matrices may serve to estimate the time of divergence of taxa when the fossil record does not provide sufficient information. For instance, microorganisms, plants, invertebrates, birds, and primates are in this category. The potential merits of this approach were expressed emphatically about 10 years ago by Doolittle:

> The geologic time scale is largely based on radiodating of characteristic strata and as such is subject to a certain amount of experimental error. Furthermore, the interpretation of fossil remains found in those strata is often arbitrary, and in many instances the

> fossil record is scanty. In such cases molecular clocks could, in theory, be very useful in fixing times of divergence. (Doolittle, 1979, p. 53.)

THE PROTEIN CLOCK: AN OVERVIEW

The concept that proteins can serve as molecular clocks of evolution was proposed more than 20 years ago by Zuckerkandl and Pauling (1965) [for review see Fitch (1976), Gillespie (1986*a*), Thorpe (1982), Wilson *et al.* (1976), Zuckerkandl (1987)]. Assuming that (1) many amino acid substitutions do not have major functional consequences, (2) the mutation rate is approximately constant over time for a given class of proteins, and (3) the DNA repair mechanisms are comparably efficient in the taxa under consideration, the degree of sequence differences between each two species considered should be found proportional to the time elapsed since the last common ancestor of these species existed. This presumed proportionality is known as the ''molecular clock.'' Each amino acid change which is fixed during evolution would represent a single ''tick'' of the clock, the constant of proportionality (which is expected to be different in different kinds of proteins), indicating the ticking rate of the clock. If the time of divergence between taxa is delivered by the fossil record, the sequences of these taxa can be used to calculate the average number of substitutions in that given time, usually expressed as minimum nucleotide replacements per 100 million years. This is calibrating the molecular clock. If, on the other hand, the time of divergence of taxa cannot be deduced by paleontology, the number of amino acid replacements may serve to estimate that time using a calibration derived from the fossil record. According to Ayala (1986), the rapid acceptance of this concept has ''revolutionized the reconstruction of evolutionary history and the timing of evolutionary events.''

The first assumption of Zuckerkandl and Pauling (1965), i.e., that many amino acid substitutions are selectively neutral, serves as basis of the neutral theory of molecular evolution, which was proposed later by Kimura [for review see Kimura (1983)]. If the neutral theory is correct, a ''statistical'' molecular clock would be predicted, because the underlying mutational events themselves are statistical processes. However, the overall probability of change would be constant, although some minor variations may occur. When this prediction of the well-formulated neutral theory is tested, the results reveal significant deviations from the theory. Thus, the molecular clock cannot be supported theoretically by the neutral theory [for details see Ayala (1986), Gillespie (1986*b*), and Hudson (1983), but see Kimura (1987)]. This failure holds not only for the neutral theory, but, according to Gillespie (1986*a*), also for all models suggested to describe the changes of macromolecular sequences in time. The situation has been characterized recently: "The forces responsible for molecular evolution remain a mystery" (Gillespie, 1986*a*). Although no theoretical basis for the constancy of the molecular clock seems to be available, this by no means indicates that the concept of a molecular clock is fallacious: It may be suggested that a theoretical basis can be constructed, but is not known yet.

THE PROTEIN CLOCK: AN EMPIRICAL TEST

Currently, the only possibility to test the molecular clock concept is by using empirical data (Ayala, 1986), that is, the given differences between protein sequences and the paleontological record delivering the estimates of divergence time. This empirical test can be performed with a variety of proteins (thousands of sequences are known) and part of the emerging evidence will be presented in this section.

Direct Comparison of Amino Acid Sequences

Superoxide Dismutase

This enzyme is essential in the defense of cells against the toxicity of oxygen. The sequences of Cu–Zn superoxide dismutase from 13 species have been published (Reinecke *et al.*, 1988; Steffens *et al.*, 1986) and the amino acid differences are given with regard to the geologic time scale in Fig. 1. It is obvious that at least three different rates of evolutionary change can be inferred from the data which differ by a factor of 60. It is, therefore, easily understandable that superoxide dismutase was considered a "very erratic" (Ayala, 1986) or "not acceptable" (Lee *et al.*, 1985) evolutionary clock.

Relaxin

This is a hormone consisting of two individual amino acid chains and, for example, is produced by pregnant mammals. It causes a widening of the symphysis pubis as well as softening of the cervix and vagina. Found also in sharks and a skate, where it is capable of widening the shark birth canal, it is considered as a hormone of viviparity. Interestingly, the shark relaxin also widens the pelvic bone of mice and guinea pigs (Schwabe, 1986). Seven sequences are available and the molecular clock model is tested using the percentage of amino acid difference (Fig. 2). Recently, Schwabe *et al.* (1989) reported that the relaxin of the mink whale (*Balaenoptera acutorostrata*) differ from that of the Bryde whale

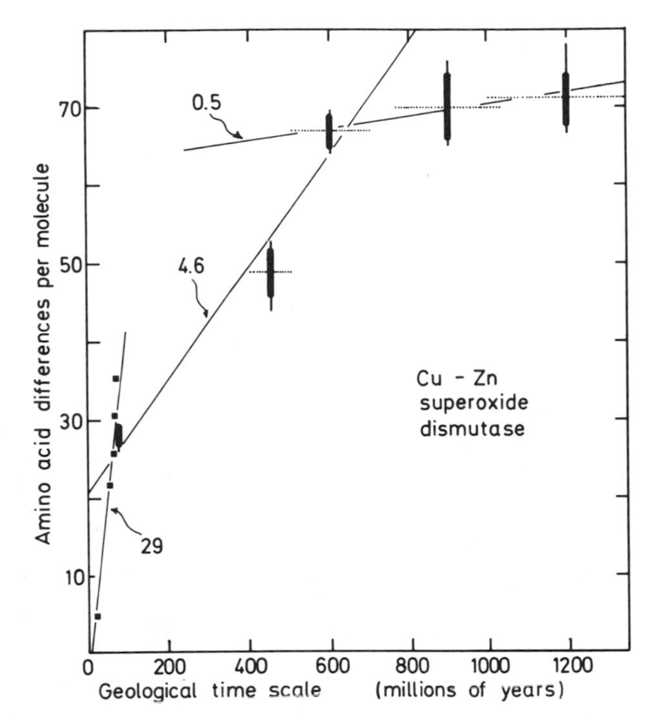

FIG. 1. Molecular clock model of Cu–Zn superoxide dismutase, based on 11 species from *Drosophila*, mammals, and plants. Sequences have been published in Reinecke *et al.* (1988) and Steffens *et al.* (1986). The numbers indicate the slope of the plot and give apparent evolutionary rates in amino acid exchanges per 100 million years and 100 residues.

(*B. edeni*) by three residues, whereas the latter differs from pig relaxin by only one residue. Although data are limited, it is evident that the molecular clock hypothesis is fallacious concerning the structure of relaxin.

Ferredoxins

These are ubiquitous iron–sulfur-containing electron transport proteins found in almost all organisms. In Fig. 3 plant-type ferredoxins, active in the terminal part of photosynthetic electron transport, are fitted to the molecular clock model. Cyanobacteria, eucaryotic algae, fern, moss, shave grass, monocotyledons, and dicotyledons have been used in the comparison. There is no correlation between amino acid differences and alleged time of divergence. It should be mentioned that the precambrian divergence times are not given unequivocally by the fossil record and therefore are subject to discussion. If, for

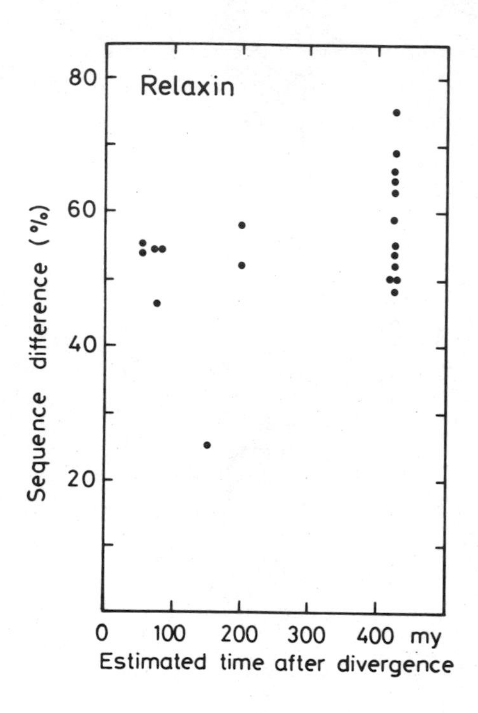

FIG. 2. Molecular clock model based on seven relaxin sequences. A difference matrix calculated by Schwabe (1986) was used to produce this figure.

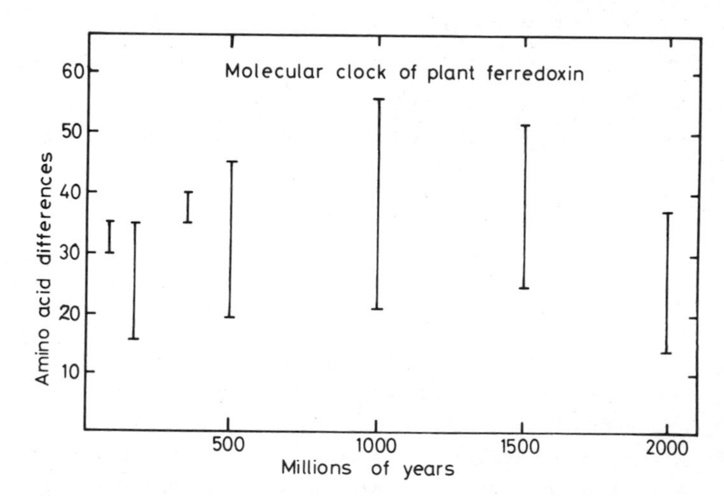

FIG. 3. Molecular clock model of plant-type ferredoxins based on 32 ferredoxin sequences from cyanobacteria, Rhodophyta, Chlorophyta, Equisetatae, Filicatae, and angiosperms. [Reprinted from Scherer and Binder (1986).]

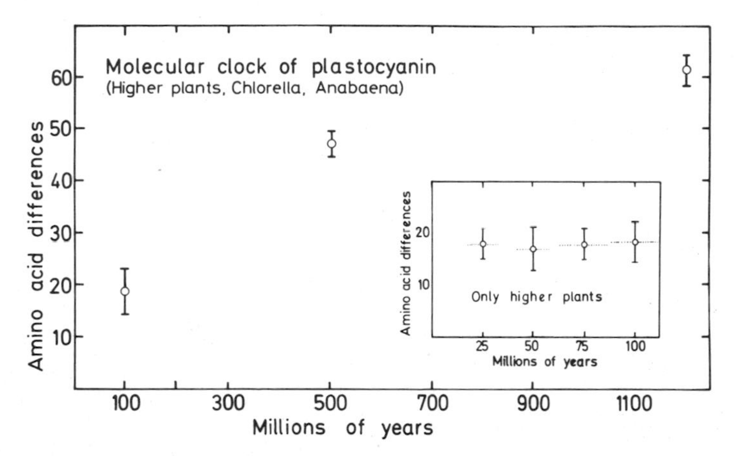

FIG. 4. Molecular clock model based on 12 plastocyanin sequences of a cyanobacterium, a green alga, and ten higher plants. [Reprinted from Scherer and Binder (1986).]

that reason, the test is restricted to divergence times less than 500 million years, it still yields no reliable correlation between alleged earth history and amino acid sequences. Although the fossil record of plants is poor, current paleontological models of plant evolution seem not to allow for such vast changes as to fit the ferredoxin data to the molecular clock model.

Plastocyanin

This copper-containing protein catalyzes the electron transport from the cytochrome *b/f* complex of chloroplasts or cyanobacteria to photosystem I and can be replaced in several algae by a soluble cytochrome *c*. Figure 4 shows plastocyanin to be an unreliable clock, since at an extended time scale the amino acid differences increase in a nonlinear manner (but note that this is based on limited data), whereas within higher plant chloroplasts a complete disagreement with the molecular clock concept emerges. Although the divergence times of higher plants are still a matter of discussion due to the poor fossil record, it seems not very probable that the sequence data will ever fit the clock model, although plastocyanin might be used at the family level as a valuable taxonomic marker (Boulter, 1978).

Fibrinopeptides

Nearly 30 fibrinopeptides have been sequenced and interpreted in terms of the molecular clock [for review see Doolittle (1979)]. Based on 24 sequences

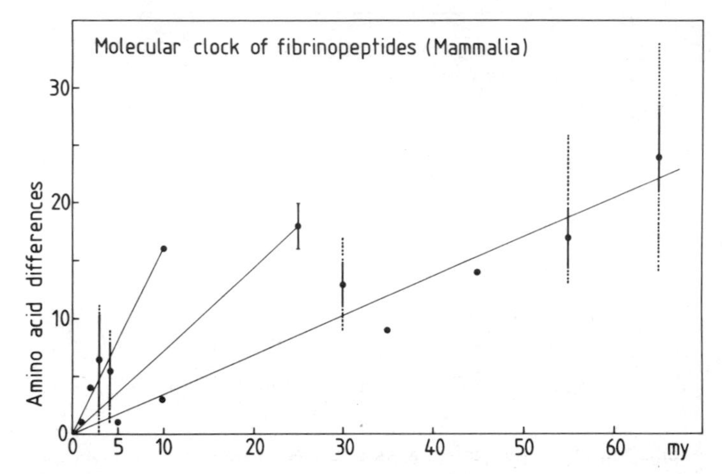

FIG. 5. Molecular clock model of fibrinopeptides, based on sequences from 26 species. The amino acid sequences were taken from Dayhoff (1967–1978); the sources for the divergence times are given in the legend of Fig. 9. Dotted lines indicate maximum and minimum values, thick lines the average value and standard deviation. Single points reflect only one pair of species being available for this divergence time.

from the mammalian orders Carnivora, Lagomorpha, Rodentia, Primates, Perissodactyla, and Artiodactyla, a clock model is shown in Fig. 5. It is hardly possible to evaluate the data in terms of a single linear slope. The obvious difference to textbook schemes is mainly due to the fact that in this diagram *all* values of the difference matrix have been included, whereas usually only selected comparisons were considered (e.g., Doolittle, 1979). This problem will be addressed in the discussion in more detail.

Comparing Minimum Nucleotide Reconstitutions

In the five examples discussed above the amino acid sequences have been compared directly, without taking into account that, due to the "degeneration" of the genetic code, in some cases more than one nucleotide reconstitution would have been necessary to yield an amino acid replacement. Several authors have therefore compared minimum nucleotide reconstitutions (sometimes called "minimum mutational distance"), which can be calculated from the amino acid differences, using the genetic code. The following examples deal with this type of data.

Hemoglobin and Myoglobin

The rates of nucleotide reconstitution for alpha- and beta-hemoglobins as well as for myoglobin have been estimated from a variety of protein sequences:

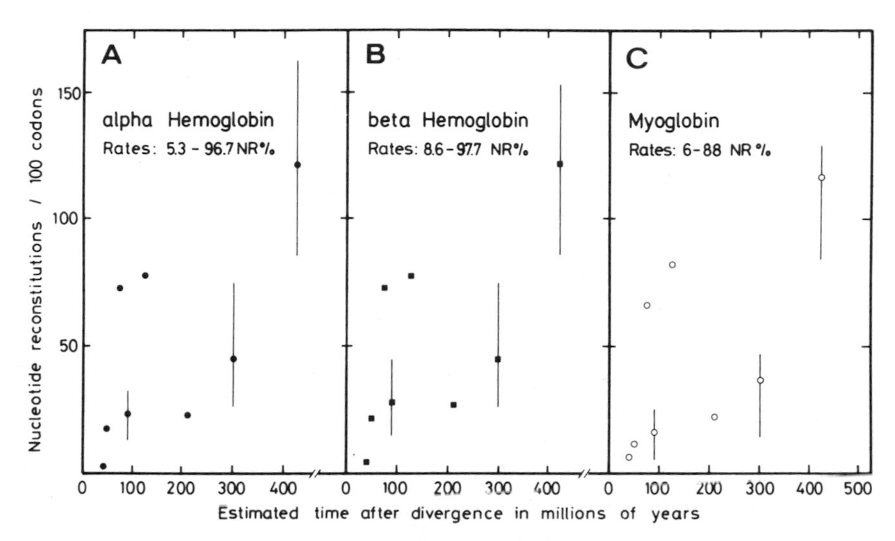

FIG. 6. Molecular clock model of (A) alpha-hemoglobin, (B) beta-hemoglobin, and (C) myoglobin, based on 294 individual sequences. Minimum nucleotide reconstitutions are taken from Baba *et al.* (1981) and Goodman *et al.* (1982). Note the similar pattern of these three proteins, which is probably due to the fact that the same species as well as the same fossil framework have been used.

According to the most parsimonous phylogenetic tree calculated from the sequences, minimum nucleotide reconstitutions between the nodes of the trees as well as between the nodes and the species living today were estimated (Baba *et al.*, 1981; Goodman *et al.*, 1982) and compared with times of divergence based on the fossil record (Fig. 6). There is a clear overall tendency of increasing nucleotide substitutions versus geologic time. On the other hand, it seems not advisable to use these molecules as a molecular clock, since the variability of nucleotide substitutions found for one divergence time seems unacceptably high. The "clock" is ticking with rates differing by a factor of 10–20. In other words, two species whose alpha-hemoglobins differ, for instance, by 25 nucleotide reconstitutions per 200 codons might have separated 50 or even 200 million years ago.

Cytochrome c

This protein has been served widely for both constructing phylogenetic trees and developing the molecular clock hypothesis. Commonly, data based on a limited number of sequences have been used in order to plot assumed times of divergence against similarity, seemingly yielding a pretty good linear relationship. However, during the last 15 years, several authors have raised doubts against a constant rate of evolutionary change of cytochrome *c* (Goodman *et al.*,

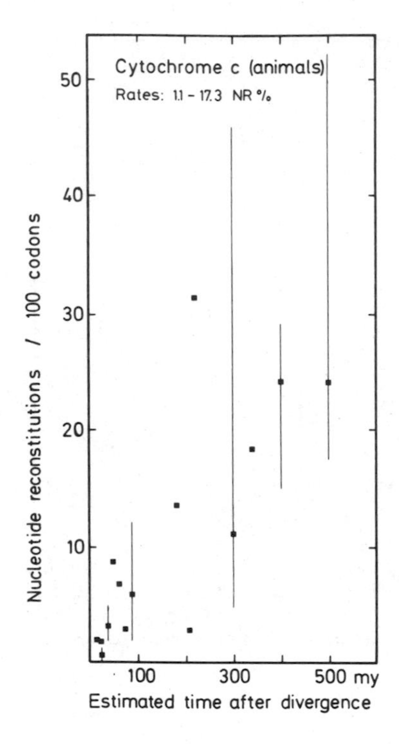

FIG. 7. Molecular clock model of mitochondrial cytochrome *c*, based on 87 individual sequences. Minimum nucleotide reconstitutions have been calculated by Baba *et al.* (1981). Lines indicate minimum and maximum values with the average. Single points represent only one comparison.

1982; Jukes and Holmquist, 1972; Penny, 1974). The data based on the comparison of 87 sequences have been compiled in Fig. 7 to demonstrate that, although there is an unequivocal tendency of increasing nucleotide reconstitutions versus geologic time, the data provide only a sloppy evolutionary clock, since the scattering of the data is tremendous. An assumed divergence time of 200 million years, for instance, is correlated to 3 as well as 31 nucleotide reconstitutions per 100 codons (See Fig. 6). Estimated rates of the cytochrome *c* clock have been reported as ranging between 1 and 42 nucleotide replacements per 100 codons and 100 million years (Baba *et al.*, 1981). The situation becomes even worse by comparing only cytochrome *c* of plant mitochondria, which, in contrast to cytochrome *c* of animals, do not allow for the construction of any reasonable phyogenetic tree (Scherer and Binder, 1986).

Alpha-Crystallin

This protein is found in eye lenses in high amounts, usually representing 25–50% of the total protein. Its length is around 170 residues and it has been sequenced from 41 species. Several phylogenetic trees have been constructed and compared by deJong and Goodman (1982), taking into consideration data from classical taxonomy as well as other protein sequences. Chicken and frog have been used as outgroups to aid the construction of the tree. From their lowest nucleotide replacement tree the molecular clock data of Fig. 8 has been calculated. As was stated also by these authors, dramatically different rates of molecular evolution show up if the clock model is applied. In Fig. 8 the evolutionary rates determined by the relative positions of the nodes of the tree have not been included. Baba *et al.* (1981), including the nodes, calculated average rates of nucleotide replacements (NR) and published figures ranging from 1.5 to 14.8% NR per 100 residues and 100 million years, the extremes being even lower and higher, respectively. Considering a long evolutionary period, i.e., including the outgroups, does not produce a linear graph (Fig. 8A). A similar nonclocklike behavior is seen if only the mammalia during a period of less than 100 million years are considered (Fig. 8B). This result will not be changed substantially by using any other solution provided by a different algorithm for tree construction (DeJong and Goodman, 1982). The standard deviations are evidence that the nonclocklike behavior of the crystallin sequences is highly significant.

Other Examples

Space does not allow the discussion of further protein classes in this chapter. However, insulin (Bajaj *et al.*, 1984; Schwabe and Warr, 1984), carbonic

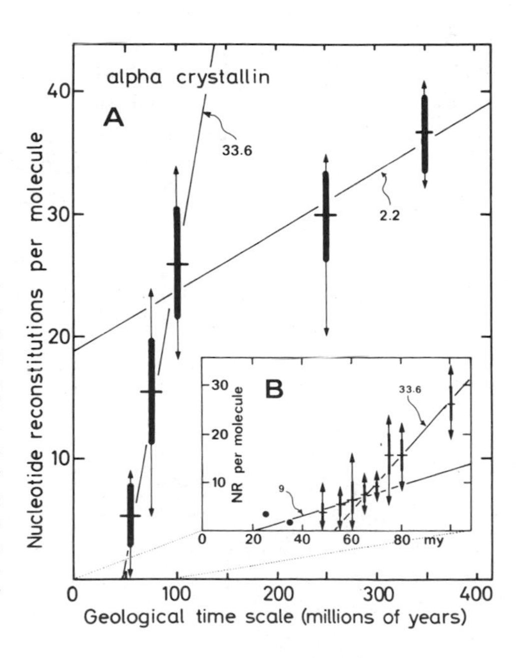

FIG. 8. Molecular clock model as applied to alpha-crystallin from 41 species. The numbers of minimum nucleotide replacements (NR) are taken from deJong and Goodman (1982). The arrows refer to maximum and minimum values, thick bars represent average and standard deviation. Single dots indicate that only one pair of species was available with that particular presumed divergence date. The numbers shown in the graph indicate apparent evolutionary rates, expressed as NR per 100 million years per 10 codons.

anhydrase (Goodman *et al.*, 1982; Tashian *et al.*, 1972), albumin (Gingerich, 1984), sea urchin proteins (Lessios, 1979), amylase (Hickey *et al.*, 1987), and other proteins have also been reported to supply very poor molecular clocks. In fact, with growing numbers of sequences available, it becomes increasingly difficult to demonstrate a protein class at all which may be considered as a reliable molecular clock.

DISCUSSION

Contradictory Results?

As is evidenced by the data presented in this chapter, the molecular clock model of protein evolution is fraught with potential error. A major question which the reader might have in mind at this point is the obvious difference between the data presented in this chapter and earlier publications on the validity of molecular clocks, which quickly were incorporated in textbooks. Three possible causes for this obvious contradiction can be given.

First, other studies have used a much more restricted data base. For instance, Lee and co-workers (1985) refer to cytochrome *c* as a rather exact molecular clock. However, they consider only the sequences of five species. Similarly, the usual evaluations of the cytochrome *c* clock are based on 15–20 species, whereas today more than 90 species have been investigated [compare Fig. 7 and Baba *et al.* (1981)]. Seemingly, with a growing data base the accuracy of the molecular clock is decreasing. Second, the more recent studies have started to consider not only extant organisms, but additionally hypothetical ancestral sequences of the nodes of most parsimonous phylogenetic trees (e.g., Goodman *et al.*, 1982). Since the apparent change of evolutionary rates revealed by this method is dramatic, the internal inconsistency of the molecular clock model is highlighted. Third, the calibration of the clock based on paleontological evidence is crucial. As an example, the molecular clock of the fibrinopeptides is considered. In Fig. 9A, a phylogenetic tree of the mammals with known fibrinopeptide sequences is given. It is this phylogenetic tree which has been used in calibrating the clock of Figs. 5 and 10A, yielding a rather poor clock both considering mammals in general or the order artiodactyla in particular. For comparison, the phylogenetic tree used by Doolittle (1979) in order to calibrate the fibrinopeptide clock is shown in Fig. 9B. Obviously, the calibration of Doolittle results in an enhanced linearity (Fig. 10B), using divergence dates of the artiodactyls, each one compared with cattle, of, e.g., 45 (pig), 40 (camel), 35 (giraffe), 30 (deer), 25 (pronghorn), and 20 million years (gazelle) (the symmetrical 5-million-year distances between the nodes of this particular phylogenetic tree probably need to be revised). Furthermore, independent of the

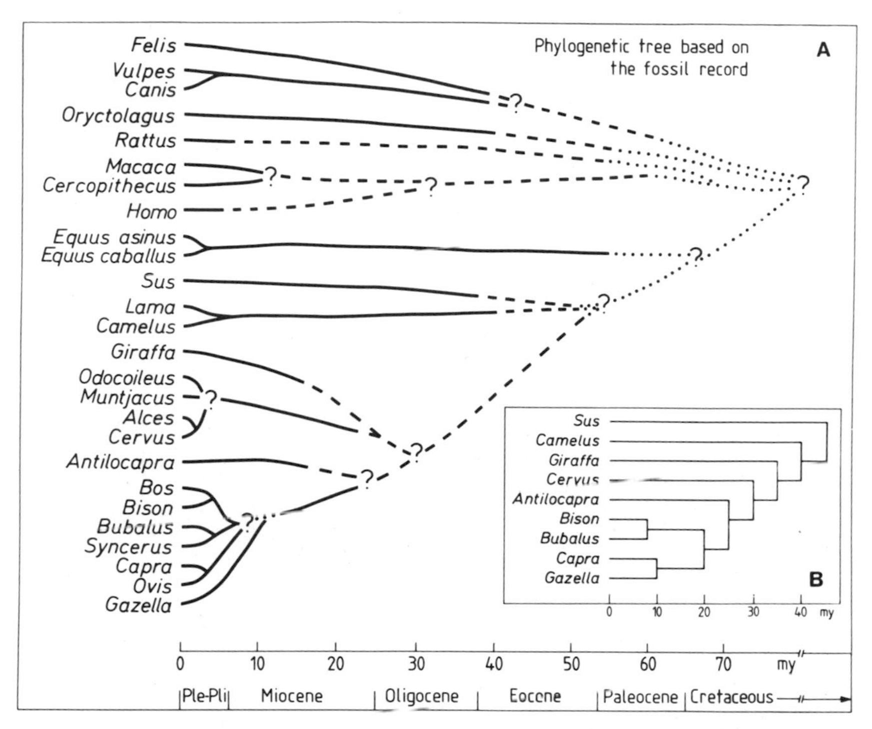

FIG. 9. (A) Evolutionary tree of the mammalian species with known fibrinopeptide sequences. Solid line: Family documented in the fossil record; dotted lines: Order documented in the fossil record; dashed lines: connections assumed. The question marks indicate that no "connecting link" has been discovered. This phylogenetic tree is based on data taken from Kuhn-Schnyder and Rieber (1984), Müller (1985), Romer (1966), and Thenius (1969, 1979). (B) Evolutionary tree of some artiodactyl genera, according to Doolittle (1979). Note the conflicting divergence dates compared with part A.

phylogenetic tree used for calibration, the standard deviation remains substantial. It is demonstrated by this example that even by choosing an appropriate phylogenetic tree, the linearity of the clock can be enhanced only to a limited extent.

Numerous suggestions have been made to account for possible "errors" of molecular clocks resulting in the above-mentioned nonlinearity. These will be discussed in the following sections.

Amino Acid Sequences or Minimum Mutational Distances?

The direct comparison of amino acid sequences (Figs. 1–5) as well as comparisons corrected due to the degeneration of the genetic code (Figs. 6–8,

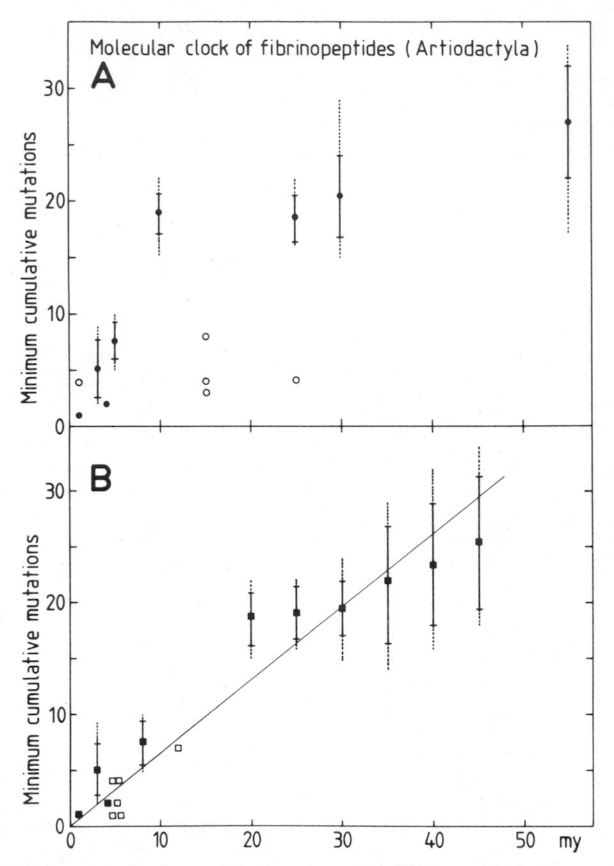

FIG. 10. (A) Molecular clock model of artiodactyl fibrinopeptides according to a phylogenetic tree constructed on the basis of amino acid sequences which have been converted beforehand to a minimum cumulative distance [Fig. 24 of Doolittle (1979)]. The divergence dates are taken from Fig. 9A. (B) Molecular clock model based on the same data set as in part A, but using the divergence data of Doolittle (1979, Figs. 24 and 26), including not only the cattle versus artiodactyl distances as given by Doolittle (1979), but all pairwise comparisons of the artiodactyls.

10) result in unsatisfactory molecular clocks. The comparison of Fig. 5 (molecular clock of fibrinopeptides based on the amino acid sequences) and Fig. 10A (minimum cumulative mutations) demonstrates that both methods fail to yield a linear relationship (i.e., a reliable molecular clock). This was also demonstrated by Ayala (1986) for the superoxide dismutase clock with a variety of correction factors applied to the distance matrix. In several cases it may be doubted whether correction factors of whatever kind would create a molecular clock with an acceptable predictive feature (Figs. 2–4).

The Apparent Saturation Effect

As has been mentioned (Dayhoff and Eick, 1968), the evolutionary history of a protein may involve reversions and multiple substitutions at a given residue. One should expect the number of these changes (which, unfortunately, cannot be detected directly) to increase with time. Therefore, Dayhoff *et al.* (1979) proposed a model which claimed to correct the molecular clock appropriately. It was reviewed recently by Wilbur (1985), who suggested that it is inconclusive. An extensive review of numerous correction formulas has been published by Gillespie (1986*a*), with the final statement that none is compatible with the data at hand. If Gillespie turns out to be right, then this is due to the "forces of molecular evolution remaining a mystery." However, Kimura pointed out that it simply becomes progressively difficult to recover hidden substitutions by ordinary statistical means as more and more amino acid changes occur, finally resulting in an "apparent saturation effect" (Kimura, 1987). In other words: The longer ago an evolutionary split occurred, the lesser the slope of the molecular clock plot will be, since the number of changes at a given site will be underestimated due to hidden substitutions and reversion. This could be accepted as an explanation of why the graphs of Figs. 1, 4, and 8 exhibit a pronounced saturation toward increased geological time. The saturation would be expected to start earlier considering "fast-evolving" proteins such as fibrinopeptides (Dickerson, 1971; Doolittle, 1979; Doolittle and Blomback, 1964; see also Perler *et al.*, 1980; Woese, 1987). As a result, one should predict that the molecular clock is most accurate for a medium level of divergence, since, on the other hand, at a very low level of divergence the statistical nature of the mutational process will greatly decrease the level of confidence. However, this prediction is not verified, since the data presented in this chapter show that the expected relationship (enhanced linearity for shorter times and increased confidence for medium divergence levels) generally is not found (see the figures).

Influence of Generation Time

The impact of generation time on the evolutionary rate of proteins has been the subject of controversy. By investigating rodent globin genes, Wu and Li (1985) suggested that the apparently faster evolutionary rate of mice and rat genes compared with human genes was due to the shorter cell cycle times of the rodents rather than to the 100fold greater generation time of humans. Although the theoretical study of Korey (1981) suggests that generation time should have a tremendous effect on the rate of protein evolution which would "undermine the main premise of the clock thesis, especially as it is applied to the dating of lineages not remotely separated,'' the controversy on the generation time effect is far from being settled (Esteal, 1985).

Horizontal Gene Transfer

The action of gene transfer is well known for bacteria, for instance, by the exchange of plasmids, but is increasingly found for plants and animals as well. Although the role of horizontal (or cross-species) gene transfer in perturbing the pattern of molecular evolution of bacteria has been questioned (e.g., Ambler *et al.*, 1979*a,b*), Syvannen (1987) suggested that, for instance, ''convergence'' in plant genes could reflect the operation of gene transfer by hybridization mechanisms that are much less active in animals. Indeed, the construction of reasonable phylogenetic trees of plants by means of protein sequences seems to be nearly impossible (e.g., Boulter, 1978; Bremer, 1988; Peacock and Boulter, 1975; Scherer and Sontag, 1983; Scherer *et al.*, 1985). In plants as well as in animals the action of retroviruses and transposable elements may contribute to a cross-species gene transfer not only between closely related organisms, since the viral host range can be quite variable: The human Epstein–Barr virus, for instance, cannot infect monkies, whereas the encephalitis virus, besides humans, also is found in monkeys, rodent, and even chicken. Quantitative data on horizontal gene transfer are not available, although single observations have been interpreted in these terms, such as a gene transfer of superoxide dismutase from *Photobacter leiognathi* to its host, the pony fish, *Leiognathus splendens* (Martin and Fridovich, 1981) (whether the failure of the superoxide dismutase clock can be explained by similar mechanisms remains purely speculative). Similar to the effect of generation or cell cycle time, the impact of horizontal gene transfer cannot yet be estimated.

Functional Constraints

The most important of all theoretical objections against the linearity of the molecular clock deals with the influence of natural selection on proteins. There can be no doubt of the action of selective forces on proteins: By comparing protein sequences it is evident that different parts exhibit different degrees of homology. Some parts, usually associated with active sites or recognition sites, are even nonvariable. The probability for each site to be converted due to a fixed mutation is certainly different. If the mutability of sites is strikingly different due to functional constraints, the modeling of all sites having the same mutability [as is, for instance, applied in the PAM model of Dayhoff *et al.* (1979)] is very likely to be fallacious.

The existence of functional constraints bearing on the variation of sequences is demonstrated nicely by the finding that noncoding regions exhibit a high variability compared with coding sequences (Wu and Li, 1985). Numerous workers have emphasized that functional constraints would lead to a distortion of the linearity of the molecular clock (e.g., Fitch and Langley, 1976; Gillespie, 1986*a,b;* Gillespie and Langley, 1979; Goodman *et al.*, 1982; Hudson, 1983; Langley and Fitch, 1974; Van Valen, 1974). Gillespie developed a statistical model which describes an "episodic molecular clock," which means that bursts of molecular evolution are followed by long periods of rather slow change, i.e., the evolutionary rate is changing *within* a lineage (Gillespie, 1984). A similar suggestion was published earlier by the group of Goodman, claiming that such "bursts" of molecular evolution are associated with periods of adaptive radiation of groups (Baba *et al.*, 1981; Goodman *et al.*, 1982, 1983). As a consequence, this group recently used "local molecular clocks" to date branching points in phylogenetic trees (Goodman *et al.*, 1988). This interesting theory, however, is not yet substantiated by data, since the selective forces which would result in such enhanced evolutionary rates of proteins remain to be demonstrated, even theoretically.

Other Perturbations of Molecular Clocks

The fixation rate of mutations may depend on the accuracy of DNA replication and the efficiency of DNA repair mechanisms. If these parameters have been subjected to evolutionary change, this could have happened in different lineages to a different extent, resulting in more or less severe perturbations of the clock. However, sufficient data to answer those questions are not at hand. Recently, the discussion on the impact of population size on the rate of molecular evolution has been dealt with in several papers. Avise *et al.* (1988) and DeSalle and Templeton (1988) reported significant effects of founder events ("bottlenecks") on the rate of molecular evolution. Since the historical biogeography of populations hardly can be traced back, such effects certainly contribute further to the difficulties in interpreting proteins as molecular clocks.

Saving the Protein Clock?

One of the founders of the protein molecular clock concept, Emile Zuckerkandl, recently reviewed the history and developments in the field. Considering the different types of molecular clocks, he concludes that "a return to the protein clock is conditionally recommended." Efforts to minimize obvious shortcomings of individual protein clocks have been reported: Since a single protein

constitutes a poor clock, the average of different proteins should give a more reliable measure. Two problems are associated with this approach: It is difficult to sequence five or more proteins of one species to obtain a reliable result. Since only a limited number of species have the same set of proteins sequenced, the approach is not easy to test. Unfortunately, in cases where such tests have been performed, the authors give no details on their approach, making an appropriate elucidation of the advanced linearity difficult (Ayala, 1986; Fitch, 1976; Fitch and Langley, 1976).

A molecular clock need not necessarily be represented by a linear relationship between geologic time and amino acid substitutions. Now and then nonlinear functions have been proposed to fit the data better than linear graphs, but, as Corruccini *et al.* (1980) pointed out, "vindication of their credibility does not seem imminent." Additionally, as illustrated by the data presented in this chapter, the variance of the number of amino acid substitutions associated with a particular time of divergence renders any function fitted to the data without usable predictive power.

The Relative Rate Test of the Molecular Clock

Since the Relative Rate Test of molecular evolution has been used frequently as an important tool in order to distinguish between "good" and "poor" molecular clocks (e.g., Avise *et al.*, 1988; DeSalle and Templeton, 1988; Easteal, 1988; Thorpe, 1982), it will be discussed here in some detail. It has been introduced by Sarich and Wilson (1973) and the principle is illustrated by Fig. 11. The test addresses the question of whether the evolutionary rates of the

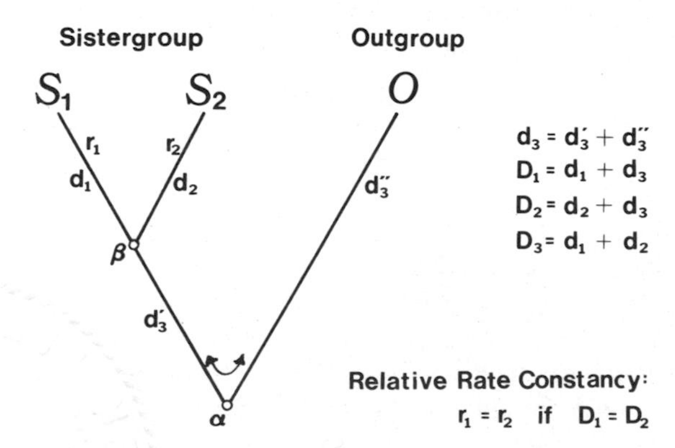

FIG. 11. The relative rate test of the molecular clock. r, rates of molecular evolution; d, theoretical mutational distances; D, observed amino acid differences; α and β, hypothetical ancestral sequences.

lineages leading to the two members, S_1 and S_2, of the sister group are identical. Since the hypothetical ancestral sequences α and β are not known, only relative rates with respect to an outgroup can be compared. As can be inferred from Fig. 11

$$d_1 = 0.5\ (D_1 + D_3 - D_2)$$

and

$$d_2 = 0.5\ (D_2 + D_3 - D_1)$$

Since D is the amino acid difference between a pair of species, the relative rates can be estimated directly from the difference matrix. The test is dependent only on the knowledge of the relative branching order of the organisms under consideration, but does not use estimates of geological divergence dates. Rate constancy is postulated, when d_1 equals d_2, which means D_1 is similar to D_2. It is not expected to be identical, since the statistical nature of the mutational process will produce slight deviations. Whether such deviations do warrant rejection of the rate constancy hypothesis can be tested by, e.g., the chi square test (Fitch, 1976).

The relative rate test, however, can not detect a change in evolutionary rates of different lineages when the rate is supposed to have changed independently, but by the same degree, in both lineages under consideration (Fitch, 1976). Although it would appear to be rather unlikely, it is, therefore, possible for a set of proteins to be classified as a ''good'' molecular clock, when it is actually a very ''poor'' clock. The plastocyanins are certainly a rather ''poor'' molecular clock (Fig. 4). However, when the relative rate test is applied to a test sample, the results are in favor of a ''good'' clock (Table I). The same holds for a sample of ferredoxins (compare Table I with Fig. 3). The Chi square values (calculated according to Fitch, 1976) do not allow rejection of the rate constancy hypothesis. For a final evaluation, however, the complete data set has to be analysed. This has been performed for superoxide dismutase (a ''poor'' clock as illustrated by Fig. 1), including a statistical evaluation of the results. As reported elsewhere (Scherer, 1989), the relative rate test, when applied to superoxide dismutase, clearly supports SOD as being a ''good'' molecular clock. Contradictions between the results of the relative rate test and the empirical test of the molecular clock by fitting the data to geological time have been reported recently for mitochondrial DNA (Hayasaka *et al.*, 1988).

These examples demonstrate that passing the relative rate test does not guarantee a uniform rate of molecular evolution. First, for theoretical reasons, the test cannot be used to detect simultaneous changes of the evolutionary rate in different lineages. Secondly, using real data sets as test cases, the relative rate test does not detect very poor molecular clocks. It is, therefore, inconclusive.

TABLE I. Relative Rate Test Applied to a Small Sample of Amino Acid Differences of Plastocyanins and Plant Type Ferredoxins[a]

Protein class	Member of sistergroup[b] S_1	S_2	Outgroup[b] (O)	Amino acid difference D_1	D_2	Chi square[c]
Plastocyanin	Potato	Spinach	*Anabaena*	56	58	0.24
	Potato	Spinach	*Chlorella*	46	47	0.06
	Spinach	*Chlorella*	*Anabaena*	58	50	1.36
Ferredoxin	Spinach	Alfalfa	*Nostoc*	34	33	0.05
	Spinach	Alfalfa	*Scenedesmus*	29	28	0.05
	Alfalfa	*Scenedesmus*	*Nostoc*	33	28	0.89

[a]For details of the statistical treatment see Scherer (1989).
[b]*Anabaena* and *Nostoc* are cyanobacteria, *Chlorella* and *Scenedesmus* are green algae.
[c]Chi square values have been calculated according to Fitch (1976). The critical Chi square value (one degree of freedom, 95% confidence limit) for the rejection of the rate constancy hypothesis is 3.84.

CONCLUSION

As has been outlined in some detail, I am not the first to raise doubts on the validity of the protein molecular clock. However, for the first time a comprehensive demonstration on the failure of the protein clock concept has been presented. Apart from this ''empirical test'' of the clock, numerous theoretical objections have been reviewed. Facing the data presented in the figures of this chapter, the issue narrows down to the question posed by Zuckerkandl (1977, p. 435): ''How sloppy a clock is one prepared to call a clock?.'' Considering the strong demands usually applied in experimental biology, it is hard to understand why the concept survived such a long period at all. It can neither be used as a tool for dating phylogenetic splits nor as reliable supportive evidence for any particular phylogenetic hypothesis.

SUMMARY

The protein molecular clock hypothesis (i.e., linearity of amino acid replacements compared with geologic time) has been tested empirically using ten different proteins, altogether representing more than 500 individual sequences from plants, animals, and prokaryotes. In no case a linearity within reasonable limits of confidence could be found as would be expected based on the clock concept: A reliable molecular clock with respect to protein sequences seems not to exist. This holds also for proteins such as cytochrome *c* or fibrinopeptides which usually have been considered as being reliable molecular clocks. It is shown that the relative rate test of the molecular clock is inconclusive. Thus, the prediction of divergence times based on protein structure is prone to error. Different approaches accounting for a nonconstancy of the rate of molecular evolution, questioning the molecular clock concept for theoretical reasons, are reviewed. It is concluded that the protcin molecular clock hypothesis should be rejected.

ACKNOWLEDGMENTS

I am grateful to Sigrid Hartwig-Scherer for stimulating discussions and to Carol Dellinger for improving the style of the manuscript.

REFERENCES

Ambler, R. P., Daniel, M., Hermoso, J., Meyer, T. E., Bartsch, R. G., and Kamen, M. D., 1979*a*, Cytochrome c_2 sequence variation among the recognised species of purple nonsulphur photosynthetic bacteria, *Nature* **278:**659–660.

Ambler, A. P., Meyer, T. E., and Kamen, M. D., 1979*b*, Anomalies in amino acid sequences of small cytochromes c and cytochromes c′ from two species of purple photosynthetic bacteria, *Nature* **278:**661–662.

Avise, J. C., Ball, R. M., and Arnold, J., 1988, Current versus historical population sizes in vertebrate species with high gene flow: A comparison based on mitochrondrial DNA lineages and inbreeding theory for neutral mutations, *Mol. Biol. Evol.* **5:**331–344.

Ayala, F. J., 1986, On the virtues and pitfalls of the molecular evolutionary clock, *J. Hered.* **77:** 226–235.

Baba, M. L., Darga, L. L., Goodman, M., and Czelusniak, J., 1981, Evolution of cytochrome c investigated by the maximum parsimony method, *J. Mol. Evol.* **17:**197–213.

Bajaj, M., Blundell, T., and Wood, S., 1984, Evolution in the insulin family: Molecular clocks that tell the wrong time, *Biochem. Soc. Symp.* **49:**45–54.

Boulter, D., 1978, Present status of the use of amino acid sequence data in plant phylogenetic studies, in: *Evolution of Protein molecules* (H. Matsubara and T. Tamanaka, eds.), pp. 243–250, Japan Scientic Society Press, Tokyo.

Bremer, K., 1988, The limits of amino acid sequence data in angiosperm phylogenetic reconstruction, *Evolution* **42:**795–803.

Corruccini, R. S., Baba, M., Goodman, M., Ciochon, R. L., and Cronin, J. E., 1980, Non-linear macromolecular evolution and the molecular clock, *Evolution* **34:**1216–1219.

Dayhoff, M. O., ed., 1967–1978, *Atlas of Protein Sequences and Structure,* Vols. 1–5, National Biomedical Research Foundation, Washington, D.C.

Dayhoff, M. O., and Eck, R. V., 1968, A model of evolutionary change in proteins, in: *Atlas of Protein Sequence and Structure 1967–1978* (M. O. Dayhoff, ed.), pp. 33–45, National Biomedical Research Foundation, Washington, D.C.

Dayhoff, M. O., Schwartz, R. M., and Orcutt, B. C., 1979, A model of evolutionary change in proteins, in: *Atlas of Protein Sequence and Structure,* Vol. 5, Supplement 3 (M. O. Dayhoff, ed.), pp. 345–352, National Biomedical Research Foundation, Washington, D.C.

DeJong, W. W., and Goodman, M., 1982, Mammalian phylogeny studied by sequence analysis of the eye lens protein crystallin, *Z. Säugetierkd.* **47:**257–276.

DeSalle, R., and Templeton, A. R., 1988, Founder effects and the rate of mitochondrial DNA evolution in Hawaiian *Drosophila, Evolution* **42:**1076–1084.

Dickerson, R. E., 1971, The structure of cytochrome c and the rates of molecular evolution, *J. Mol. Evol.* **1:**26–45.

Doolittle, R. F., 1979, Protein evolution, in: *The Proteins,* Vol. IV (H. Neurath, R. L. Hill, and C. L. Boeder, eds.), pp. 1–118, Academic Press, New York.

Doolittle, R. F., and Blomback, B., 1964, Amino acid sequence investigations of fibrinopeptides from various mammals—Evolutionary implications, *Nature* **202:**147–152.

Easteal, S., 1985, Generation time and the rate of molecular evolution, *Mol. Biol. Evol.* **2:**450–453.

Easteal, S., 1988, Rate constancy of globin gene evolution in placental mammals, *Proc. Natl. Acad. Sci. USA* **85:**7622–7626.

Felsenstein, J., 1983, Parsimony in systematics: Biological and statistical issues, *Annu. Rev. Ecol. Syst.* **14:**313–333.

Fitch, W. M., 1976, Molecular evolutionary clocks, in: *Molecular Evolution* (F. J. Ayala, ed.), pp. 160–178, Sinauer, Sunderland, Massachusetts.

Fitch, W. M., and Langley, C. H., 1976, Protein evolution and the molecular clock, *Fed. Proc.* **35:** 2092–2097.

Gillespie, J. H., 1984, The molecular clock may be an episodic clock, *Proc. Natl. Acad. Sci. USA* **81:**8009–8013.

Gillespie, J. H., 1986*a,* Rates of molecular evolution, *Annu. Rev. Ecol. Syst.* **17:**637–665.

Gillespie, J. H., 1986*b,* Natural selection and the molecular clock, *Mol. Biol. Evol.* **3:**138–155.

Gillespie, J. H., and Langley, C. H., 1979, Are evolutionary rates really variable?, *J. Mol. Evol.* **13:** 27–34.

Gingerich, P. D., 1984, Primate evolution: Evidence from the fossil record, comparative morphology, and molecular biology, *Yearb. Phys. Anthropol.* **27:**57–72.

Goodman, M., Weiss, M. L., and Czelusniak, J., 1982, Molecular evolution above the species level: Branching pattern, rates, and mechanisms, *Syst. Zool.* **31:**376–399.

Goodman, M., Braunitzer, G., Stangl, A., and Schrank, B., 1983, Evidence on human origins from hemoglobins of african apes, *Nature* **303:**546–548.

Goodman, M., Pedwaydon, J., Czelusniak, J., Suzuki, T., Gotoh, T., Moens, L., Shishikura, F., Walz, D., and Vinogradov, S., 1988, An evolutionary tree for invertebrate globin sequences, *J. Mol. Evol.* **27:**236–249.

Hartwig-Scherer, S., 1989, Ramapithecus—Vorfahr des Menschen? Zeitjournal, Berlin.

Hayasaka, K. T., Gojobori, T., and Horai, R., 1988, Molecular phylogeny and evolution of primate and mitochondrial DNA, *Mol. Biol. Evol.* **5:**626–644.

Hickey, D. A., Benkel, B. F., Boer, P. H., Genest, Y., Abukashawa, S., and Ben-David, G., 1987, Enzyme-coding genes as molecular clocks: The molecular evolution of animal alpha-amylases, *J. Mol. Evol.* **26:**252–256.

Hudson, R. R., 1983, Testing the constant-rate neutral allele model with protein sequence data, *Evolution* **37:**203–217.

Jones, J. S., and Rouhani, S., 1986, How small was the bottleneck?, *Nature* **319:**449–450.

Jukes, T. H., and Holmquist, R., 1972, Evolutionary clocks: Nonconstancy of rates in different species, *Science* **177:**530–537.

Kimura, M., 1983, *The Neutral Theory of Molecular Evolution,* Cambridge University Press, Cambridge.

Kimura, M., 1987, Molecular evolutionary clock and the neutral theory, *J. Mol. Evol.* **26:**24–33.

Korey, K. A., 1981, Species number, generation length and the molecular clock, *Evolution* **35:**139–147.

Kuhn-Schnyder, E., and Rieber, H., 1984, *Paläzoologie,* Thieme, Stuttgart.

Langley, C. H., and Fitch, W. M., 1974, An examination of the constancy of the rate of molecular evolution, *J. Mol. Evol.* **3:**166–177.

Lee, J. M., Friedman, D. J., and Ayala, F. J., 1985, Superoxide dismutase: An evolutionary puzzle, *Proc. Natl. Acad. Sci USA* **82:**824–828.

Lessios, H. A., 1979, Use of Panamanian sea urchins to test the molecular clock, *Nature* **280:**599–601.

Martin, J. P., and Fridovich, I., 1981, Evidence for a natural gene transfer from the ponyfish to its bioluminescent bacterial symbiont *Photobacter leiognathi, J. Biol. Chem.* **256:**6080–6089.

Meyer, T. E., Cusanovich, M. A., and Kamen, M. D., 1986, Evidence against use of bacterial amino acid sequence data for construction of all-inclusive phylogenetic trees, *Proc. Natl. Acad. Sci. USA* **83:**217–220.

Müller, A. H., 1985, *Lehrbuch der Paläozoologie,* Fischer, Jena.

Peacock, D., and Boulter, D., 1975, Use of amino acid sequence data in phylogeny and evaluation of methods using computer simulation, *J. Mol. Biol.* **95:**513–522.

Penny, D., 1974, Evolutionary clock: The rate of evolution of rattlesnake cytochrome c, *J. Mol. Evol.* **3:**179–188.

Perler, F., Efstratiadis, A., Lomedico, P., Gilbert, W., Kolodner, R., and Dodgson, J., 1980, The evolution of genes: The chicken preproinsulin gene, *Cell* **20:**555–566.

Reinecke, K., Wolf, B., Michelson, A. M., Puget, K., Steffens, G. J., and Flohe, L., 1988, The amino acid sequence of rabbit Cu–Zn superoxide dismutase, *Biol. Chem. Hoppe-Seyler* **369:** 715–725.

Romer, A. S., 1966, *Vertebrate Paleontology,* University of Chicago Press, Chicago.

Sarich, V. M., and Wilson, A. C., 1973, Generation time and genomic evolution in primates, *Science* **179:**1144–1147.

Scherer, S., 1989, The relative rate test of the molecular clock hypothesis: A note of caution, *Mol. Biol. Evol.* **6** (4)**:** 436–441.

Scherer, S., and Binder, H., 1986, Comparison of biological classification based on amino acid sequences and traditional taxonomy, in: *Studien zur Klassifikation 17* (P. O. Degens, H. J. Hermes, and O. Opitz, eds.), pp. 324–332, Index Verlag, Frankfurt.

Scherer, S., and Sontag, C., 1983, Amino acid sequences of 32 plant type ferredoxins—Endosymbiotic and evolutionary consequences, in: *Endocytobiology II* (H. E. A. Schenk and W. Schwemmler, eds.), pp. 863–870, Walter de Gruyter, Berlin.

Scherer, S., and Sontag, C., 1986, Zur molekularen Taxonomie und Evolution der Anatidae, *Z. Zool. Syst. Evolutionsforsch.* **24:**1–19.

Scherer, S., Binder, H., and Sontag, C., 1985, Occurrence and amino acid sequences of cytochrome c and plastocyanin of algae and plants: Endocytobiotic implications, *Endocytobiol. Cell Res.* **2:** 1–14.

Schwabe, C., 1986, On the validity of molecular evolution, *Trends Biochem. Sci.* **11:**280–283.

Schwabe, C., and Warr, G. W., 1984, A polyphyletic view of evolution: The genetic potential hypothesis, *Perspect. Biol. Med.* **27:**465–485.

Schwabe, C., Bullesbach, E. E., Heyn, H., and Yoshioka, M., 1979, Cetacean relaxin. Isolation and sequence of relaxins from *Balaenoptera acutorostrata* and *Balaenoptera edeni, J. Biol. Chem.* **264:**940–943.

Steffens, G. J., Michelson, A. M., Otting, F., Puget, K., Strassburger, W., and Flohe, L., 1986, Primary structure of Cu–Zn superoxide dismutase of *Brassica oleracea* proves homology with corresponding enzymes of animals, fungi and prokaryotes, *Biol. Chem. Hoppe-Seyler* **367:** 1007–1016.

Syvannen, M., 1987, Molecular clocks and evolutionary relationships: Possible distortions due to horizontal gene flow, *J. Mol. Evol.* **26:**16–23.

Tashian, R. E., Tanis, R. J., Ferrell, R. E., Stroup, S. K., and Goodman, M., 1972, Differential rates of evolution in the carbonic anhydrase isozymes of catarrhine primates, *J. Hum. Evol.* **1:** 545–552.

Thenius, E., 1969, *Handbuch der Zoologie,* Vol. 9, *Stammesgeschichte der Säugetiere,* Walter de Gruyter, Berlin.

Thenius, E., 1979, *Die Evolution der Säugetiere,* Thieme, Stuttgart New York.

Thorpe, J. P., 1982, The molecular clock hypothesis: Biochemical evolution, genetic differentiation and systematics, *Annu. Rev. Ecol. Syst.* **13;**139–168.

Van Valen, L., 1974, Molecular evolution as predicted by natural selection, *J. Mol. Evol.* **3:**89–101.

Wilbur, W. J., 1985, On the PAM matrix model of molecular evolution, *Mol. Biol. Evol.* **2:**434–447.

Wilson, A. C., Carlson, S. S., and White, T. J., 1976, Biochemical evolution, *Annu. Rev. Biochem.* **46:**573–639.

Woese, C. W., 1987, Bacterial evolution, *Microbiol. Rev.* **51:**221–271.

Wu, C. I., and Li, W. H., 1985, Evidence for higher rates of nucleotide substitution in rodents than in man, *Proc. Natl. Acad. Sci. USA* **82:**1741–745.

Zuckerkandl. E., 1977, Programs of gene action and progressive evolution, in: *Molecular Anthropology* (M. Goodman and R. E. Tashian, eds.), pp. 387–447, Plenum Press, New York.

Zuckerkandl, E., 1987, On the molecular evolutionary clock, *J. Mol. Evol.* **26:**34–46.

Zuckerkandl, E., and Pauling, L., 1965, Evolutionary divergence and convergence in proteins, in: *Evolving Genes and Proteins* (V. Bryson and H. J. Vogel, eds.), pp. 97–165, Academic Press, New York.

4

Molecular Evolution of the Alcohol Dehydrogenase Genes in the Genus *Drosophila*

DAVID T. SULLIVAN, PETER W. ATKINSON, and WILLIAM T. STARMER

INTRODUCTION

The development and now widespread utilization of efficient gene isolation and sequencing techniques has had a major if not revolutionary impact on the study of evolution. While it is clear that the rate of acquisition of nucleotide sequence information is likely to increase substantially in the next several years, there has already been sufficient information generated to improve markedly our understanding of the genetic events which occur over evolutionary time frames. Our understanding of the genetic changes which occur is a central problem that has existed for some time and has historically been approached using more indirect methods. DNA sequence comparisons tell us what changes have occurred, and persisted, but of course by themselves do not distinguish which events are of phenotypic consequence, which are remnants of history, or which are of subsequent evolutionary relevance. Nonetheless, the nucleotide sequence of the gene is the basic level of informational organization, and higher levels of organization must build upon it. In this light there has been a very large number of comparisons made among the nucleotide sequences of a given gene or gene sets in various lineages. Clearly the most extensively studied genes are the globin genes of vertebrates. However, in recent years the gene which encodes alcohol dehydrogenase (ADH) in *Drosophila* has also been analyzed at the nucleotide sequence level in a number of species. These data have not been summarized or

DAVID T. SULLIVAN, PETER W. ATKINSON, and WILLIAM T. STARMER • Department of Biology, Syracuse University, Syracuse, New York 13244.

systematically analyzed, hence the purpose of this review. In addition, it has become clear that a number of complex events have occurred at the *Adh* locus which are relevant to its evolution in the genus *Drosophila*. These include a duplication (Oakeshott *et al.*, 1982*a*; Batterham *et al.*, 1982), generation of a pseudogene (Fischer and Maniatis, 1985), and several changes in developmental regulation in various species groups (Dickinson, 1980; Dickinson *et al.*, 1984; Mills *et al.*, 1986).

The *Adh* gene is clearly the most extensively studied gene in the genus *Drosophila*. In *D. melanogaster*, it is currently the object of detailed molecular and genetic analyses of its structure and regulation. A review of this work has recently been published (Sofer and Martin, 1987) and only the essentials of these studies will be summarized here. The *Adh* gene encodes a protein of 254 or 256 amino acids, depending on the species. The active enzyme is a dimer of two products of the *Adh* locus. It is an especially abundant protein in *Drosophila* and has been estimated to represent on the order of 1–2% of the soluble protein of *D. melanogaster*. The ADH enzyme of *Drosophila* is clearly a different enzyme than the analogous enzyme of yeast, higher plants, or animals. *Drosophila* ADH differs from these other ADH molecules with respect to several properties, including molecular weight, number of subunits, and the absence of zinc. In contrast, ADH from these other groups are similar to each other. Furthermore, there is no appreciable amino acid sequence similarity between *Drosophila* ADH and the ADH of other organisms, but other ADH molecules do share stretches of similar amino acids with each other. These points have been expanded on by Place *et al.* (1986) and Jornval *et al.* (1977).

The principal function of ADH is to catabolize alcohols generated by microbial fermentation taking place in the habitat of larvae and/or adults. Consistent with this function, ADH is localized in the fat bodies, intestines, and Malpighian tubules of both larvae and adults of most species. However, appreciable variation of tissue localization of ADH exists in different species and this variation is proving a rich source of information for studies aimed at identifying tissue-specific control sequences (see below).

The gene encoding ADH is organized into three exons which are separated by introns of 60 to 70 base pairs (bp). The introns interrupt the coding region at amino acids 30 (or 32) and 166 (or 168), respectively. An especially interesting feature of the *Adh* gene in most, but not all, species is that transcription can initiate from either of two sites (Benyajati *et al.*, 1983). There is developmental specificity associated with this pattern of promoter utilization. The two sites are separated by about 700 nucleotides and splicing of a 5′ intron generates identical coding potential regardless of which promoter is used. The transcription start site nearest the coding sequence of the gene is referred to as the proximal or larval promoter, since it is used primarily in early larvae. The other site, the distal or adult promoter, is used primarily in adults, but also in some late third-instar larval cells. A detailed analysis of promoter utilization has been conducted by

Savakis *et al.* (1986). Interestingly, in species of the *mulleri* subgroup which have a duplication of the *Adh* gene, each gene has only one transcription start site, which is located close to the 5′ end of the gene (Fischer and Maniatis, 1985). However, the expression pattern of each gene, *Adh*-1 and *Adh*-2, is formally similar in a developmental context to the pattern of transcription directed from either the proximal or distal promoter, respectively (Batterham *et al.*, 1984). Several laboratories have conducted analyses, using various species, aimed at defining those sequence elements that are important for *Adh* expression. This topic will be addressed extensively in the section on the evolution of *Adh* regulation below.

The study of *Drosophila Adh* genes has also occupied a prominent position in evolutionary studies. The introduction of electrophoretic techniques to study allele frequencies in populations resulted in the discovery of extensive allozymic variation in natural populations. This became part of the now infamous selectionist–neutralist debate as to the interpretation of the meaning of that variation. ADH in *D. melanogaster* populations is generally dimorphic, having two principal alleles which encode proteins with different electrophoretic mobilities, which are named ADH fast (ADH^F) and ADH slow (ADH^S). Other, less frequent alleles have also been described (Thatcher and Sawyer, 1980). The ADH gene–enzyme system has been the object of intensive study by a number of investigators. ADH of *D. melanogaster* seems to offer an apparently ideal system to shed light on the problem of the role of allozymes and the forces which maintain the variation. The rationale for this was based on the following features of the system. Because the enzyme is a very abundant protein, it is likely to be physiologically important. Its function in catabolizing alcohols that are present in the habitat of all life stages suggested the possibility of important and measurable fitness differences for individuals of different genotypes. Finally, the polymorphism in *D. melanogaster* could be studied using a variety of powerful genetic manipulations and by employing the appropriate genetic backgrounds where needed to answer relevant questions.

Two general goals existed. The first goal is to show that different fitness values were associated with different *Adh* genotypes and the second was to make a connection between the biochemical properties of the allozymes and a specific physiological property of ADH that would be relevant to the viability or fecundity of the organisms. These goals have not been fully realized. Some of the specific problems which have been encountered in interpreting the results of these studies have been summarized (Gibson and Oakeshott, 1982). Nonetheless, there are compelling reasons to believe that the *Adh* polymorphism is maintained by selection. There is a global cline in *Adh* allele frequencies, with Adh^S increasing in frequency toward the equator in both hemispheres, which is difficult to interpret without invoking selective pressure (Oakeshott *et al.*, 1982*b*).

The property of either *Adh* allele that responds to selection has not been identified. *Adh*^{F} alleles are known to result in animals having higher enzyme activity, which at least in part is due to increased steady-state concentration of the enzyme. Recent studies (Laurie-Ahlberg and Stam, 1987; Laurie and Stam, 1988) have convincingly demonstrated that the difference in the activity between animals homozygous for *Adh*^{F} and *Adh*^{S} alleles is due to a difference in the concentration of the *Adh*^{F} protein. This is due to a property of the enzyme, and is not due to variation in the mRNA level. A recent review of the *Adh* polymorphism which summarizes aspects of this problem which are beyond our scope has been prepared by VanDelden (1982).

The stimulus for our review has been the relatively recent accumulation of nucleotide sequence information of 17 *Adh* genes from various species of the genus *Drosophila*. These include sequences from sets of closely related species at extremes of lineages within the genus and two outlying species. Five *Adh* genes from species of the melanogaster subgroup of the subgenus Sophophora (Kreitman, 1983; Cohn *et al.*, 1984; Coyne and Kreitman, 1986; Bodmer and Ashburner, 1984; Cohn and Moore, 1988) and one gene from a species of the obscura group (Schaeffer and Aquadro, 1987), which is another major division of the subgenus Sophophora, have been sequenced. From the subgenus Drosophila, the other major division of the genus, we have access to one sequence from the Hawaiian picture-winged group (Rowan and Dickinson, 1988) and ten genes from the repleta group (Fischer and Maniatis, 1985; Atkinson *et al.*, 1988; M. Menotti-Raymond and D. Sullivan, in preparation; J. Weaver and D. Sullivan, unpublished). This last set includes the products from at least two gene duplication events and three pseudogenes.

This review will be subdivided into several sections based on the analysis of these nucleotide sequences. We will use these data sets to compare nucleotide sequences and thus to derive a phylogeny of the *Adh* genes which can then be used to place other phylogenetic studies on the genus into perspective. Second, the amino acid translation of these DNA sequences will be used to compare aspects of protein sequence, structure, and function. Third, we will examine in some detail the origins of the *Adh* duplication and pseudogenes found in the repleta group. Fourth, we will examine the regulatory information flanking the *Adh* genes. We wish to determine if particular sequences are conserved and whether these conserved sequences are the ones shown by functional tests to be relevant to the control of *Adh* expression. Our goal is to identify candidate sequences which differ between species and which could be relevant to species differences in the developmental regulation of *Adh*. Finally, we will utilize specific examples of sequence comparisons and variations of the *Adh* genes which have relevance to general strategies and the assumptions often used for the calculation of evolutionary rates and other parameters under study in molecular evolution. Some of these examples point out the need to consider the full biological context of a particular gene. It should become evident that careful consideration of each gene, its functional context in the genome, and specific evolutionary history can be relevant to the assumptions that are used to interpret and derive conclusions from nucleotide sequence comparisons.

COMPARISON OF *Adh* NUCLEOTIDE SEQUENCES: CODING REGIONS

In this section we will present comparisons of nucleotide sequences which cover the coding portion of 17 *Adh* genes. The principal goal of this set of comparisons is to further our understanding of the phylogeny within the genus. We have chosen to utilize nucleotides from the ADH coding regions rather than introns or flanking regions. Nucleotides in flanking regions and introns are apparently subject to high rates of substitutions, so that comparisons across the breadth of the genus would present difficult or impossible alignment problems. In addition, some of the assumptions necessary to make sequence comparisons of noncoding sequences are difficult to rely on with certainty. Some examples of sequence properties which confound comparisons are presented below.

A tree diagram for the 17 sequences (Fig. 1) was generated by the parsimony method for inferring phylogenies from DNA sequences (DNAPARS from the phylogeny inference package, PHYLIP, provided by J. Felsenstein). This diagram shows the number of nucleotide substitutions inferred for each lineage and provides information about the origins of the duplicate genes and their pseudogenes in the species of the mulleri complex and *D. hydei*. The number of changes inferred for each branch of the tree was arbitrarily assigned when there were multiple equally parsimonious assignments for states. This was especially problematic for branches of longer lengths. It is thus obvious that the given tree is an approximation of the true history of *Adh* gene evolution as well as *Drosophila* species evolution. Problems associated with knowledge of ancient polymorphisms (Coyne and Kreitman, 1986), arbitrary state assignment, and the validity of assumptions underlying the parsimony methods make the detailed (i.e., branch lengths) reconstruction of the genetic history suspect. However, the overall typological features are probably useful, especially the sequences of linkage for different species and genes. The species in the subgenus Sophophora (*D. pseudoobscura* and above in the diagram) are clearly separated from the species in the subgenus Drosophila (*D. affinidisjuncta* and below in the diagram). This distinction is also apparent in the dendrograms based on DNA sequence similarity distance (Table I, Fig. 3) and substitutions per synonomous site [$K(s)$] distance (Table I, Fig. 2). However, the dichotomy of Sophophora and Drosophila species is slightly different when using $K(s)$ distances, because *D. affinidisjuncta* joins outside of the two main clusters that form the two subgenera.

Within the Sophophora subgenus the linkage of species is as expected; the

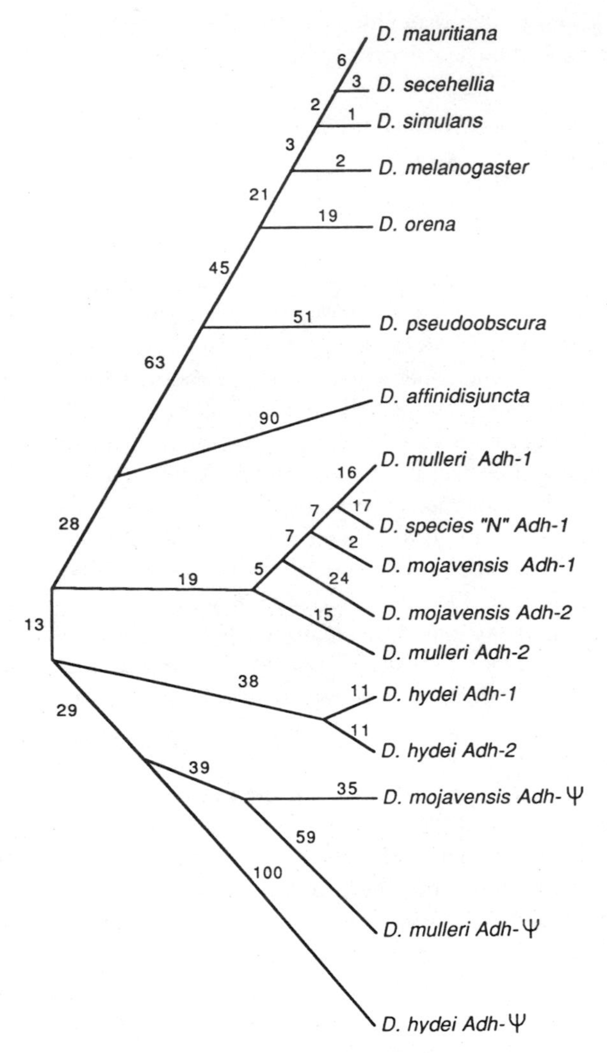

FIG. 1. Evolutionary tree of the 17 *Adh* sequences. The tree was constructed using the maximum-parsimony method, using the DNAPARS program of PHYLIP. Numbers on the branches represent the estimated number of nucleotide substitutions that were necessary to obtain the observed sequences.

closely related melanogaster group species *D. simulans, D. sechellia, D. mauritiana,* and *D. melanogaster* are linked together before being joined by *D. orena* and *D. pseudoobscura* (obscura group), in that order. The detailed relationship within the melanogaster group species is in basic agreement with earlier publications which used DNA sequences of *Adh* genes to reconstruct genetic history

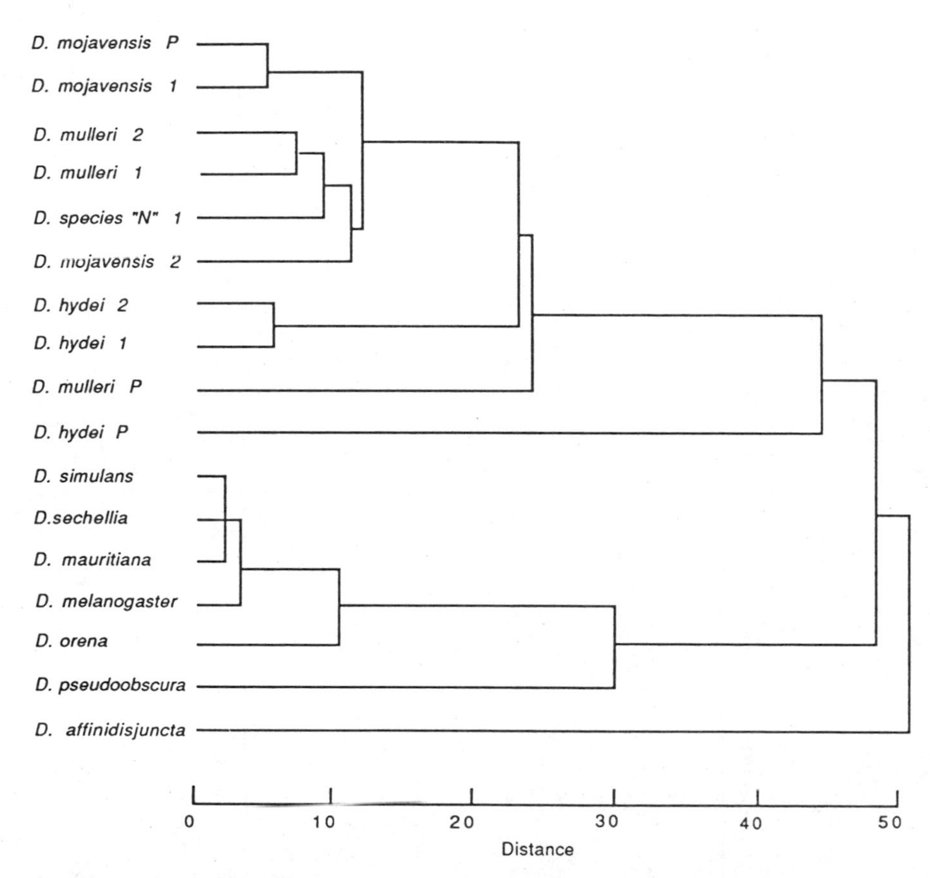

FIG. 2. Dendrogram of the 17 *Adh* sequences based on estimated $K(s)$ values (Table I). The dendrogram was constructed by the Fitch–Margoliash method, using the FITCH program of PHYLIP. The distances given in the dendrogram are arbitrary.

(Bodmer and Ashburner, 1984; Ashburner, *et al.*, 1984; Stephens and Nei, 1985; Coyne and Kreitman, 1986; Cohn and Moore, 1988). The minor differences among the resulting phylogenetic reconstructions at this level are most likely due to the methods used to infer genetic history, and will be resolved by the addition of more sequence information within species [polymorphisms; e.g., Kreitman (1983)], and the addition of more sequence information on other species. The differences in detailed relationship due to methods can be seen by comparing the tree diagram constructed with the maximum parsimony method with the dendrograms produced using the method of Fitch and Margoliash (1967), where the data are represented as distance matrices of either $K(s)$ or sequence distance information. In any event, the differences are small and probably unresolvable with the information from the coding region of one gene.

TABLE I. Percentage Similarity and $K(s)$ Values for *Adh* Comparisons among Species[a]

Drosophila species gene	D.ml	D.sc	D.ma	D.si	D.or	D.pu	D.af	D.hy1	D.hy2	D.hyP	D.mu1	D.mu2	D.muP	D.mj1	D.mj2	D.mjP	D.nl
D. melanogaster	—	98.3	97.9	98.8	94.8	85.5	75.7	76.9	77.1	69.6	79.0	80.0	72.8	77.9	77.5	74.3	80.3
D. sechellia	0.06	—	98.8	99.2	95.2	85.2	76.2	77.5	77.6	69.9	79.3	80.7	73.6	77.8	77.8	74.9	80.4
D. mauritiana	0.06	0.03	—	98.8	94.6	84.8	75.7	77.4	77.5	69.6	79.0	80.4	73.4	77.8	77.6	74.7	80.4
D. simulans	0.04	0.03	0.03	—	95.2	85.0	75.9	77.3	77.4	69.5	79.0	80.3	73.3	77.8	77.6	74.6	80.4
D. orena	0.17	0.17	0.20	0.17	—	86.0	76.5	77.4	77.1	69.7	78.8	80.4	73.3	77.4	77.6	73.9	80.1
D. pseudoobscura	0.61	0.61	0.64	0.63	0.53	—	76.7	77.4	78.0	72.1	80.1	80.3	74.1	77.8	78.6	75.4	80.0
D. affinidisjuncta	1.15	1.11	1.19	1.15	1.01	1.06	—	82.0	83.1	76.0	82.2	82.7	75.9	80.8	81.7	76.1	82.6
D. hydei Adh-1	1.18	1.11	1.11	1.15	1.09	1.11	1.03	—	97.3	81.1	88.9	89.4	81.0	88.4	88.9	82.8	89.0
D. hydei Adh-2	1.14	1.10	1.10	1.13	1.15	1.07	0.88	0.10	—	81.1	89.3	89.7	81.8	88.2	89.0	83.5	89.4
D. hydei Adh-P	2.09	1.96	2.03	2.14	1.94	1.50	1.34	0.78	0.79	—	80.0	80.2	79.0	79.3	79.8	80.1	80.1
D. mulleri Adh-1	0.92	0.87	0.91	0.92	0.91	0.78	0.93	0.47	0.42	0.94	—	95.8	82.4	94.5	93.6	84.9	95.7
D. mulleri Adh-2	0.86	0.81	0.83	0.86	0.81	0.83	0.94	0.47	0.45	0.89	0.13	—	83.2	93.7	94.4	84.9	95.2
D. mulleri Adh-P	1.14	1.04	1.05	1.08	1.04	1.02	1.21	0.67	0.63	1.09	0.56	0.53	—	81.2	81.8	91.7	82.4
D. mojavensis Adh-1	1.00	1.03	1.01	1.02	1.01	1.02	1.01	0.48	0.47	0.92	0.18	0.20	0.63	—	93.3	84.3	94.2
D. mojavensis Adh-2	1.04	1.04	1.04	1.05	0.98	0.92	0.95	0.44	0.42	0.88	0.21	0.21	0.64	0.20	—	84.5	94.1
D. mojavensis Adh-P	1.10	0.98	0.99	1.02	1.05	0.94	1.25	0.58	0.56	1.17	0.43	0.44	0.30	0.44	0.44	—	84.1
D. species "N" *Adh-1*	0.82	0.86	0.83	0.84	0.84	0.83	0.93	0.50	0.45	0.92	0.16	0.20	0.61	0.20	0.22	0.51	—

[a]Percent similarity values are above the diagonal, while $K(s)$ values are beneath the diagonal. Species listed in the columns are abbreviations of and in the same order as those listed as rows. *Adh*-P = *Adh*-pseudogene.

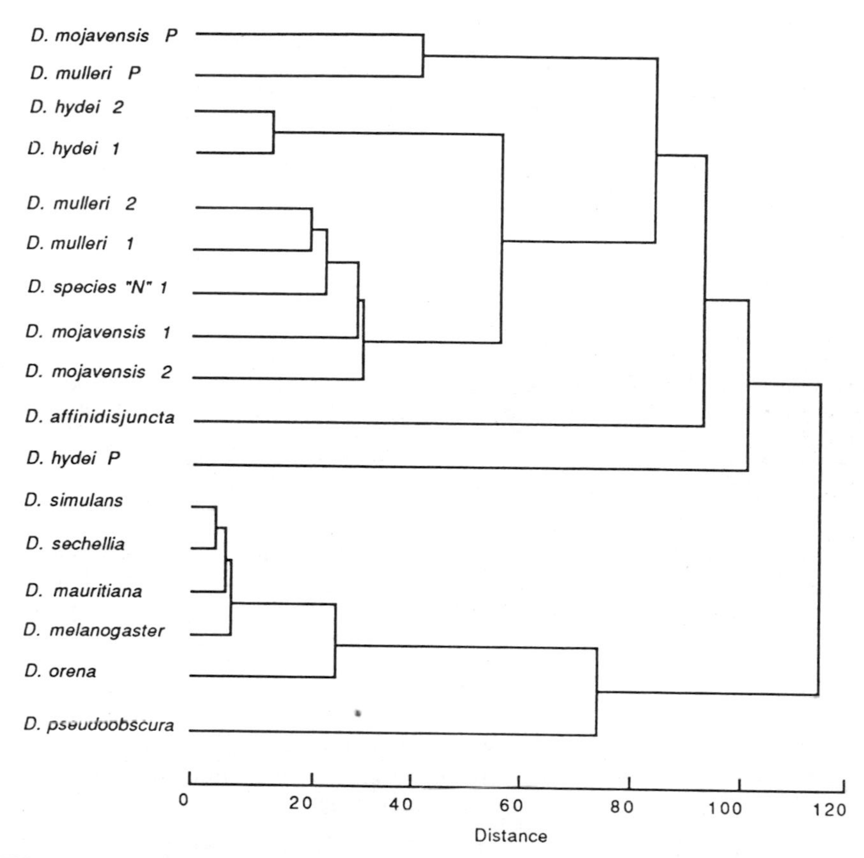

FIG. 3. Dendrogram of the 17 *Adh* sequences based on the DNA sequence distance values (1% similarity/100 of Table I). The dendrogram was constructed by the Fitch–Margoliash method, using the FITCH program of PHYLIP. The distances given in the dendrogram are arbitrary.

Within the cluster of species representing the subgenus Drosophila, all species (except *D. affinidisjuncta*) carry three copies of an *Adh* gene (only *Adh*-1 of species "N" is sequenced) and evaluation of species relationships can be made after recognizing the likely origin of the duplicate genes (see below). The mulleri subgroup species *D. mulleri, D. mojavensis,* and species "N" are clustered together as expected before linkage with *D. hydei* (hydei subgroup). This is true for *Adh*-1, Adh-2, and *Adh*-P (the pseudogene) lineages in the tree diagram (Fig. 1) as well as in the dendrograms (Figs. 2 and 3). The details of linkage order and distances do differ slightly for the mulleri subgroup species, depending on the method used. A more exact picture is probably only possible by including information on variation within species and more sequences from closely related species from within the mulleri subgroup.

TABLE II. Taxonomic Designation and Systematic Placement of the *Drosophila* Species

Species	Subgenus	Group	Subgroup	Complex
Drosophila melanogaster	Sophophora	melanogaster	melanogaster	melanogaster
D. mauritiana	Sophophora	melanogaster	melanogaster	melanogaster
D. simulans	Sophophora	melanogaster	melanogaster	melanogaster
D. sechellia	Sophophora	melanogaster	melanogaster	melanogaster
D. orena	Sophophora	melanogaster	melanogaster	melanogaster
D. psuedoobscura	Sophophora	obscura	obscura	obscura
D. affinidisjuncta	Drosophila	Hawaiian picture-wing		
D. mulleri	Drosophila	repleta	mulleri	mulleri
D. species "N"[a]	Drosophila	repleta	mulleri	mulleri
D. majavensis	Drosophila	repleta	mulleri	mulleri
D. hydei	Drosophila	repleta	hydei	

[a]*D.* species "N" is an undescribed species from Navojoa, Sonora, Mexico.

Overall the phylogenetic history of the genus as revealed by the *Adh* sequences is congruent with what has been postulated by Throckmorton (1975). A comparison of the taxonomic designations listed for each species in Table II with the tree diagram and dendrograms illustrates this congruence. This is also consistent with the work on immunological distances determined by micro-complement fixation of larval hemolymph protein for various species of higher diptera (Beverly and Wilson, 1982). However, we do not have information on species in the subgenus Scaptodrosophila and thus cannot comment on the placement of that group of species in relationship to the other subgenera.

The repleta group species for which sequence information on *Adh*-1, *Adh*-2, and *Adh*-P is available (i.e., *D. mulleri, D. mojavensis,* and *D. hydei*) can be used to infer the likely sequence of events that led to the origin of these genes. Atkinson *et al.* (1988) concluded that there are two duplication events that led to the formation of the three genes in *D. mulleri* and *D. mojavensis* and that both events preceded the speciation of *D. mojavensis* and *D. mulleri*. The lineage leading to the present pseudogenes in the two species was a product of the first duplication, while the divergence leading to *Adh*-1 and *Adh*-2 was more recent. This notion is supported by the overall analysis presented here and will be developed in more detail below.

COMPARISON OF ADH PROTEIN SEQUENCES

The conceptual translation of the 14 ADH coding sequences is shown in Fig. 4. Since our interest is in comparing the protein structure, we have not attempted to include a theoretical translation of the pseudogenes. Positions of amino acid sequence identity are shown by underlining. It is immediately apparent that there is a high degree of primary sequence conservation in *Drosophila* ADH, with 170 out of 254 positions having identity in all species. Of note is a difference of two amino acids near the N-terminal end. Species of the melanogaster species complex have two additional amino acids, which we have positioned at amino acids 3 and 4. These positions have not been considered in calculating the protein similarities. The average percent identity of amino acids through the ADH molecule is 67.0%. When this is broken down into the three exons, the percentage of amino acids which are identical in all species in exons 1, 2, and 3 are 56.7%, 67.9%, and 69.0%, respectively. The number of amino acids in each exon was used as the basis for the expected number in a χ^2 test. This test showed that the relative number of identical amino acids in an exon is similar. Therefore, it appears that, on the average, no appreciable differences in amino acid replacement rate occurs in different exons. In a comparison of *D. pseudoobscura* ADH and the ADH of species of the melanogaster species complex, Schaeffer and Aquadro (1987) pointed out that exon 2 showed significantly less divergence than exon 3. Although our comparisons do not indicate this difference as a general property of all *Adh* genes, our analysis in no way precludes the possibility that there is differential mutability of DNA sequences of different exons or that selection operates on a particular exon differently in different lineages. Consequently, we do not view our results as in conflict with those of Schaeffer and Aquadro (1987). Rather, the results of the *D. pseudoobscura* comparison when put in contrast to the average of all species may point out some interesting properties of the *D. pseudoobscura* ADH protein coding sequence.

The comparisons of the ADH proteins serve to reinforce the general conclusions of others that strong purifying selection is operating on the *Adh* locus. Visual inspection of Fig. 4 reveals that at many nonconserved positions the amino acids which are replaced are very conservative with respect to protein secondary structure. This conversation of protein secondary structure is visualized in Fig. 5 and 6. We have compared two sequences, ADH-1 from *D. mojavensis* and ADH from *D. melanogaster,* with respect to two aspects of secondary structure, probability of helicity (Fig. 5) and hydrophobicity (Fig. 6). These amino acid sequences are from species on divergent ends of the genus and as expected represent a case of extreme difference in terms of the number of amino acid replacements. It is evident that the secondary structure, as best as can be determined given the recognized uncertainty in these predictions, is strongly conserved. The examination of amino acid replacements which represent changes to amino acids with different chemical properties and might have been thought to be nonconservative appear to be conservative when considered in the context of secondary structure predictions. This is especially true for replace-

```
ADHDMU1 MA--IANKNIIFVAGLGGIGFDTSREIVKSGPKNLVILDRIENPAAIAELKALNPNVTVTFYPYDVTVPVAETTKLLQKIFDQLKT
ADHDMJ1 MA--IANKNIIFVAGLGGIGFDTSREIVKKGPKNLVILDRIENPAAIAELKALNPKVTVTFYLYDVTVSVAESTKLLQKIFDQLKT
ADHDNV1 MA--IANKNIIFVAGLGGIGFDTSREIVKSGPKNLVILDRIENPAAIAELKALNPKVTVTFYPYDVTVPVAETTKLLKTIFDKLKT
ADHDMU2 MV--IANKNIIFVAGLGGIGFDTSREIVKSGPKNLVILDRIENPAAIAELKALNPKVTVTFYPYDVTVSVAETTKLLKTIFDKLKT
ADHDMJ2 MA--IANKNIIFVAGLGGIGFDTSREIVKSGPKNLVILDRIENPAAIAELKALNPKVTVTFYPYDVTVSVAETTKLLKTIFDKLKT
ADHDHY1 MA--IANKNIIFVAGLGGIGLDTSREIVKSGPKNLVILDRIDNPAAIAELKAINPKVTVTFYPYDVTVSVAESTKLLKVIFDKLKT
ADHDHY2 MA--IANKNIIFVAGLGGIGLDTSREIVKSGPKNLVILDRIDNPAAIAELKAINPKVTITFYPYDVTVSVAESTKLLKVIFDKLKT
ADHDAF  MV--IANSNVIFVAGLGGIGLDTSREIVKSGPKNLVVLDRVDNPAAIAELKALNPKVTVTFYPYDVTVPLAETKKLLKTIFDKLKT
ADHDPU  MS--LTNKNVVFVAGLGGIGLDTSRELVKRNLKNLVILDRIDNPAAIAELKAINPKVTITFYPYDVTVPVAETTKLVKTIFAQVKT
ADHDOR  MAFTLTNKNVIFVAGLGGIGLDTSKELVKRDLKNLVILDRIENPAAIAELKAINPKVTVTFYPYDVTVPIAETTKLLKTIFAKLTT
ADHDMA  MAFTLTNKNVIFVAGLGGIGLDTSKELVKRDLKNLVILDRIENPAAIAELQAINPKVTVTFYPYDVTVPIAETTKLLKTIFAKLKT
ADHDML  MSFTLTNKNVIFVAGLGGIGLDTSKELLKRDLKNLVILDRIENPAAIAELKAINPKVTVTFYPYDVTVPIAETTKLLKTIFAQLKT
ADHDSC  MAFTLTNKNVIFVAGLGGIGLDTSKELLKRDLKNLVILDRIENPAAIAELKAINPKVTVTFYPYDVTVPIAETTKLLKTIFAKLKT
ADHDSM  MAFTLTNKNVIFVAGLGGIGLDTSKELLKRDLKNLVILDRIENPAAIAELKAINPKVTVTFYPYDVTVPIAETTKLLKTIFAKLKT
        |                                                                                    |
        1                                                                                   86

ADHDMU1 VDLLINGAGILDDYQIERTIAVNFTGTVNTTTAIMSFWDKRKGGPGGVIANICSVTGFNAIYQVPVYSASKAAALSSTNSLAKLAP
ADHDMJ1 VDLLINGAGILDDHQIERTIAVNFTGTVNTITAIMSFWDKRKGGPGGVIANVCSVTGFNAIYQVPVYSASKAAALSFTNSLAKLAP
ADHDNV1 VDLLINGAGILDDYQIERTIAVNFTGTVNTTTAIMSFWDKRKGGPGGVIANICSVTGFNAIYQVPVYSASKAAALSFTNSLAKLAP
ADHDMU2 VDLLINGAGILDDYQIERTIAVNFTGTVNTTTAIMSFWDKRKGGPGGIIANICSVTGFNAIYQVPVYSASKAAALSFTNSLAKLAP
ADHDMJ2 VDLLINGTGILDDHQIERTIAVNFTGTLNTTTAIMSFWDKRKGGPGGVIANICSVTGFNAILPVPVYSASKAAALSFTNSLARLAP
ADHDHY1 VDLLINGAGILDDYQIERTIAVNFAGTVNTTTAIMAFWDKRKGGPGGVIANICSVTGFNAIYQVPVYSASKAAALSFTNSLAKLAP
ADHDHY2 VDLLINGAGILDDYQIERTIAVNFAGTVNTTTAIMAFWDKRKGGPGGVIANICSVTGFNAIYQVPVYSASKAAALSFTNSLAKLAP
ADHDAF  VDLLINGAGILDDNQIERTIAVNFTGTVNTTTAIMDFWDKRKGGPGGVVANICSVTGFNSIYQVPVYSASKAAALSFTTSIAKLAH
ADHDPU  IDVLINGAGILDDHQIERTIAVNYTGLVNTTTAILDFWDKRKGGPGGIICNIGSVTGFNAIYQVPVYSGSKAAVVNFTSSLAKLAP
ADHDOR  VDVLINGAGILDDYQIERTIAVNYTGLVNTTTAILDFWDKRKGGPGGIICNIGSVTGFNAIYQVPVYSGTKAAVVNFTSSLAKLAP
ADHDMA  VDVLINGAGILDDHQIERTIAVNYTGLVNTTTAILDFWDKRKGGPGGIICNIGSVTGFNAIYQVPVYSGTKAAVVNFTSSLAKLAP
ADHDML  VDVLINGAGILDDHQIERTIAVNYTGLVNTTTAILDFWDKRKGGPGGIICNIGSVTGFNAIYQVPVYSGTKAAVVNFTSSLAKLAP
ADHDSC  VDVLINGAGILDDHQIERTIAVNYTGLVNTTTAILDFWDKRKGGPGGIICNIGSVTGFNAIYQVPVYSGTKAAVVNFTSSLAKLAP
ADHDSM  VDVLINGAGILDDHQIERTIAVNYTGLVNTTTAILDFWDKRKGGPGGIICNIGSVTGFNAIYQVPVYSGTKAAVVNFTSSLAKLAP
        |                                                                                    |
       87                                                                                  172
```

```
ADHDMU1 ITGVTAYSINPGITKTPLVHKFNSWLDVEPRVAELLLEHPTQTSLQCAQNFVKAIEANQNGAIWKLDLGTLEAIEWTKHWDSHI
ADHDMJ1 ITGVTAYSINPGITKTTLVHKFNSWLDVEPRVGELLLEHPTQTSLECAQNFVKAIEANQNGAIWKLDLGTLEAIEWTKHWDSHI
ADHDNV1 ITGVTAYSINPGITKTTLVHKFNSWLDVEPRVAELLLEHPTQTSLECAQNFVKAIEANQNGAIWKLDLGTLEAIEWTKHWDSHI
ADHDMU2 ITGVTAYSINPGITKTTLVHKFNSWLDVEPRVAELLLEHPTQTTLQCAQNFVKAIEANQNGAIWKLDLGTLEAIEWTKHWDSHI
ADHDMJ2 ITGVTAYSINPGITKTTLVHKFNSWLDVEPRVAELLLEHPTQTTLQCAQNFVKAIQANQNGAIWKLDLGTLEAIEWTKHWDSHI
ADHDHY1 ITGVTAYSINPGITKTTLVHKFNSWLDVEPRVAELLLEHPTQTSLQCAQNFVKAIEANQNGAIWKLDLGRLEAIEWTKHWDSGI
ADHDHY2 ITGVTAYSINPGITKTTLVHKFNSWLDVEPRVAELLLEHPTQTTLQCAQNFVKAIEANKNGAIWKLDLGRLDAIEWTKHWDSGI
ADHDAF  ITGVTAYSINPGITKTVLVHKFNSWLSVEPRVAELLLEHPTQTTLQCAQNFVKAIEANQNGAIWKLDLGRLDAIEWTKHWDSGI
ADHDPU  ITGVTAYTVNPGITKTTLVHKFNSWLDVEPRVAEKLLEHPTQTSQQCAETFVKAIELNKNGAIWKLDLGTLEPITWTQHWDSGI
ADHDOR  ITGVTAYTVNPGITRTTLVHKFNSWLDVEPQVAEKLLAHPIQSSLACAENFVKAIELNENGAIWKLDLGTLEAIKWSKHWDSGI
ADHDMA  ITGVTAYTVNPGITRTTLVHKFNSWLDVEPQVAEKLLAHPTQPSLACAENFVKAIELNQNGAIWKLDLSTLEAIQWTKHWDSGI
ADHDML  ITGVTAYTVNPGITRTTLVHKFNSWLDVEPQVAEKLLAHPTQPSLACAENFVKAIELNQNGAIWKLDLGTLEAIQWTKHWDSGI
ADHDSC  ITGVTAYTVNPGITRTTLVHKFNSWLDVEPQVAEKLLAHPTQPSLACAENFVKAIELNQNGAIWKLDLGTLEAIQWTKHWDSGI
ADHDSM  ITGVTAYTVNPGITRTTLVHKFNSWLDVEPQVAEKLLAHPTQPSLACAENFVKAIELNQNGAIWKLDLGTLEAIQWTKHWDSGI
        |                                                                                  |
       173                                                                                256
```

FIG. 4. The amino acid sequences (International Union of Biochemistry single-letter codes) for the 14 genes that encode ADH. The line beneath the sequences depicts regions that are invariant.

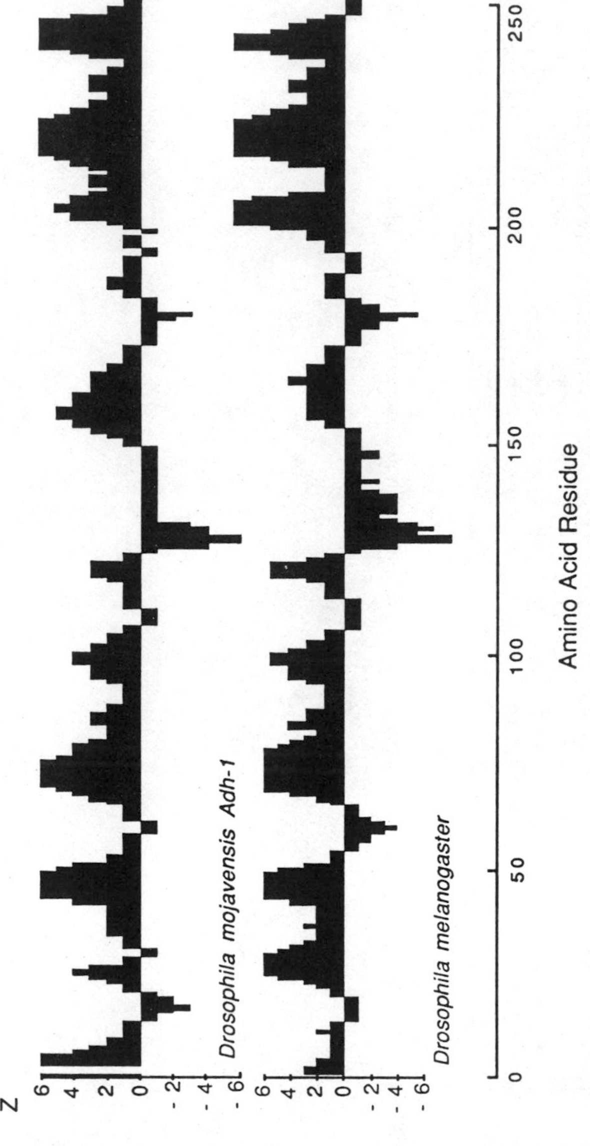

FIG. 5. A comparison of the likelihood of helicity across regions of ADH for *D. mojavensis* ADH-1 and *D. melanogaster* ADH. *Z* is the value used to predict helix formation.

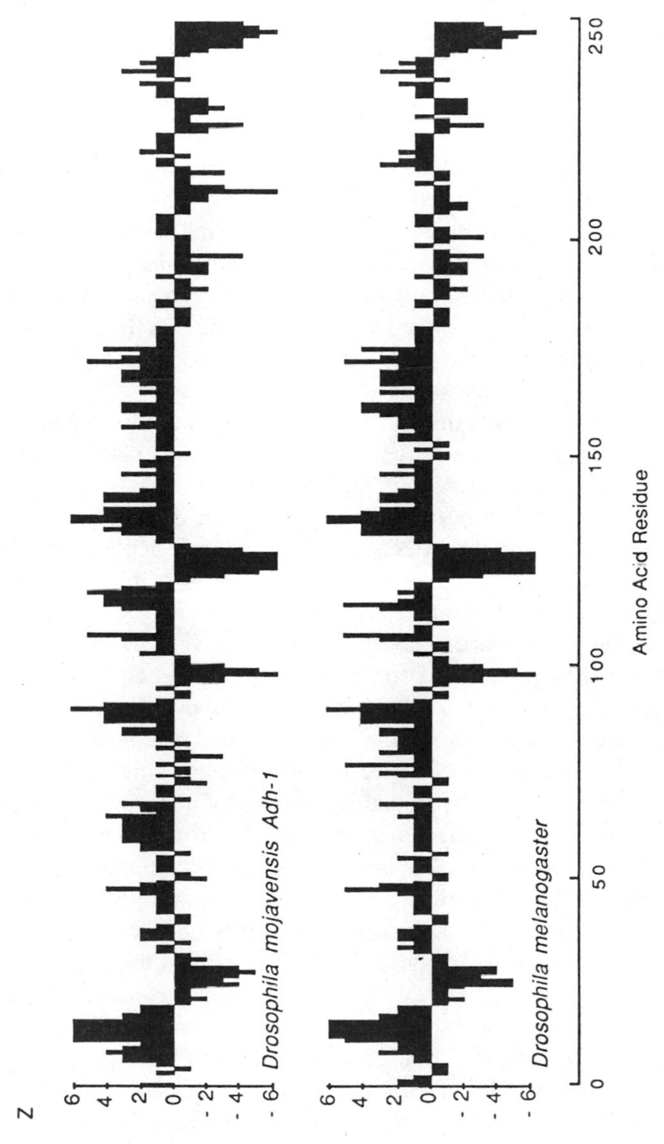

FIG. 6. A comparison of the likelihood of hydrophobicity across regions of ADH for *D. mojavensis* ADH-1 and *D. melanogaster* ADH. *Z* is the value used to predict hydrophobicity.

ments affecting amino acids in regions with high probability of being helical. We have probed this issue further by attempting to demonstrate which aspect of secondary structure is most conserved. The significance of amino acid replacements in the ADH polypeptides was evaluated by comparing secondary structure parameters used for predictions of hydropathy (*H*), alpha-helix formation (*A*), and beta-sheet formation (*B*) for different sequences. These comparisons were made by calculating the statistic $X_{ij} = \Sigma_{k}|(Z_{ik}-Z_{jk})|/N$, where Z_{ik} is the value used to predict the structural feature at amino acid position *k* for polypeptide *i*, Z_{jk} is the similar statistic for polypeptide *j*, and *N* is the number of amino acid replacements that differ between the two sequences. The expectations for *Z* values were calculated by averaging the hydopathic index (Kyte and Doolitle, 1982), the helix conformational parameter (Chou and Fasman, 1978, Table 3), and the beta-sheet conformational parameter (Chou and Fasman, 1978, Table 3) over an amino acid span of 7. In order to evaluate the significance of any one coparison, we calculated X_{ir} for 1000 randomized sequences of *N* amino acid replacements, where *i* is the *D. melanogaster* sequence and $r = 1$ to 1000 is the randomized sequence. A randomized sequence was produced by randomly substituting nucleotides in the DNA sequence (*D. melanogaster*) until *N* amino acid differences between the sequence of *D. melanogaster* and sequence *r* were obtained. By repeating this process for *N* ranging from 2 to 52, we constructed 95% confidence bands for the effects of *N* amino acid replacements on the secondary structure. This band would therefore include X_{ij} values that one would expect on the basis of random nucleotide substitutions that result in the given number *n* of amino acid replacements. It should be emphasized that the calculations that compared two sequences for hydropathy, alpha-helix, and beta-sheet formation were not predictions on the occurrence of these features in the ADH molecule, but were the differences in the conformational parameters that are used to make decisions about the occurrence of those features. Results (Figs. 7–9) show that the differences in hydropathic features of the ADH molecule are significantly outside of the range of that expected from random amino acid replacements, indicating that hydrophobicity has been conserved in the molecule. Both features, alpha-helix and beta-sheet formation, appear to fall within the 95% confidence band (although near the lower boundary), and thus amino acid replacements are less likely conserved with respect to these features of the ADH molecule.

There are a number of positions where multiple amino acid replacements in the ADH lineage have occurred. Using the phylogeny of these genes (Fig. 1), it is possible to determine that for at least ten positions, replacement resulting in the same amino acid has occurred independently. These are positions 42, 52, 69, 73, 83, 90, 216, 231, 242, and 244. With the exception of the replacements at position 90 and possibly 231, all of these amino acids are in regions predicted to have a probability of being alpha-helical. Each of these replacements could be

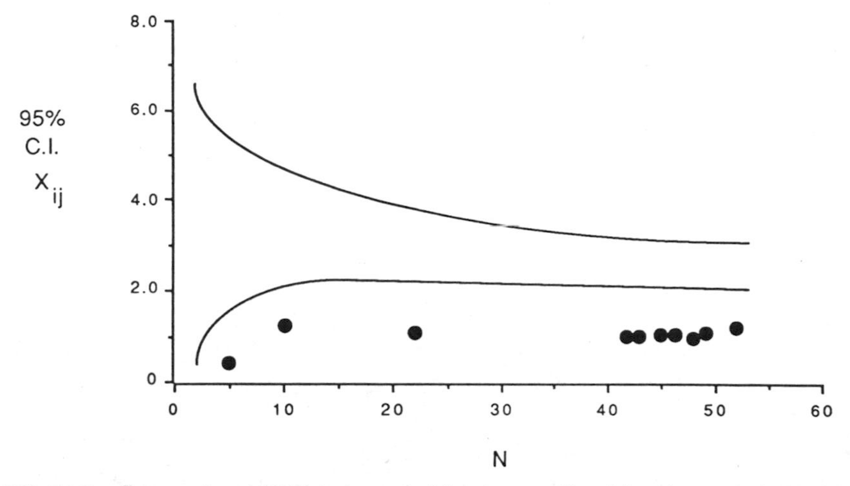

FIG. 7. Confidence band (95% is bounded by the two lines) for the statistic X which measures the average deviation between the value used for predicting hydropathy in *D. melanogaster* ADH and 1000 randomized sequences. The random sequences have 2–52 amino acid replacements (abscissa). The solid circles represent actual X values for ADH of *D. melanogaster* compared to the other ADH molecules.

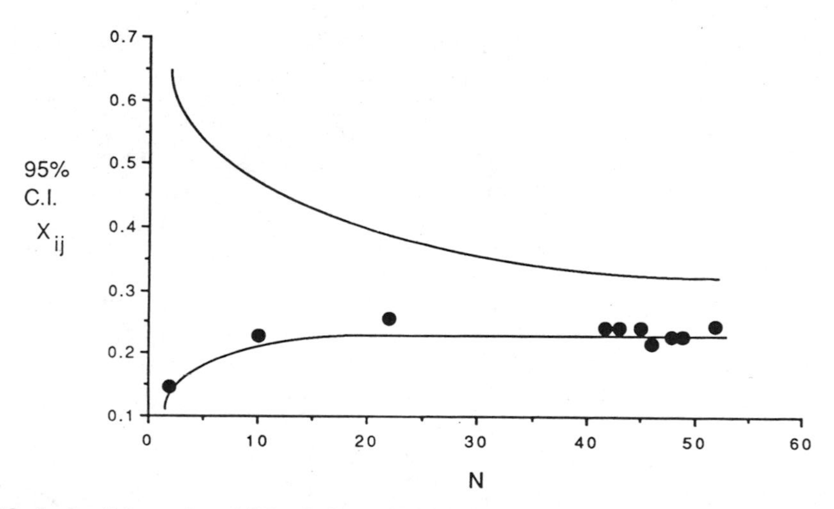

FIG. 8. Confidence band (95% is bounded by the two lines) for the statistic X which measures the average deviation between the value used for predicting helicity in *D. melanogaster* ADH and 1000 randomized sequences. The random sequences have 2–52 amino acid replacements (abscissa). The solid circles represent actual X values for ADH of *D. melanogaster* compared to other ADH molecules.

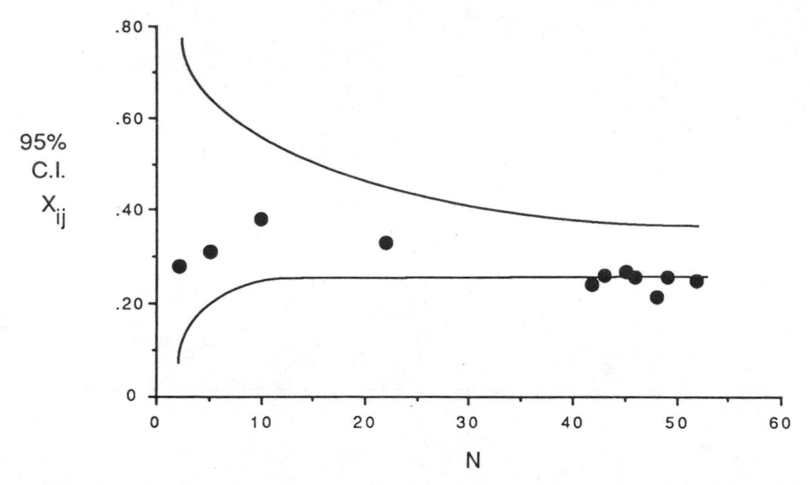

FIG. 9. Confidence band (95% is bounded by the two lines) for the statistic *X* which measures the average deviation between the value used for predicting beta-sheet structure in *D. melanogaster* ADH and 1000 randomized sequences. The random sequences have 2–52 amino acid replacements (abscissa). The solid circles represent actual *X* values for ADH of *D. melanogaster* compared to the other ADH molecules.

accomplished with a single nucleotide substitution. It seems possible that if neutral amino acid replacements have occurred and remained in the ADH molecule, these are the most likely candidates. However, in the absence of positive evidence from studies of protein function, this remains a weakly supported position.

DUPLICATIONS AT THE *Adh* LOCUS

During the evolution of the genus *Drosophila* there have been at least two separate duplications of the *Adh* locus. The earliest, which probably occurred very early during the evolution of the genus or conceivably in an earlier lineage, is now evidenced by the existence of a gene which is apparently functional and has been identified very close to the 3′ end of *Adh* in species of the melanogaster and obscura subgroups (Schaeffer and Aquadro, 1987; Cohn and Moore, 1988). Schaeffer and Aquadro (1987) have analyzed the structure of this gene and its relationship to the *Adh* gene in *D. pseudoobscura*. This 3′ gene has been judged to be a product of a duplication of the *Adh* gene based on its close proximity to *Adh* (approximately 400 bp separates the two coding regions), its similarity of nucleotide sequence in exon regions, (58%), and the strong similarity of the 3′ gene and *Adh* in the exon–intron spacing pattern. The protein encoded by the 3′

gene is similar to ADH with respect to hydrophobic domain organization and has a 38% amino acid sequence similarity. The function of this presumed protein is not known. It has yet to be determined whether the gene 3′ to *Adh* is also present in species of the subgenus *Drosophila*. We have inspected the available sequences from *D. hydei* and *D. mojavensis* which extend about 900 and 700 bp, respectively, beyond *Adh*-1, the most 3′ *Adh* gene. No evidence of another gene similar to *Adh* is present, but of course this gene could be only slightly displaced relative to its position in *D. pseudoobscura* and may have been missed in this comparison.

Several years ago it was observed (Oakeshott *et al.*, 1982*a*; Batterham *et al.*, 1982) that several species of the mulleri and hydei subgroups contain isozymes of ADH which are the result of a gene duplication. An extensive analysis of the ADH isozyme content of larvae and adults from a number of species from throughout the genus indicated that the gene duplication had occurred during the origins of the repleta group (Batterham *et al.*, 1984). The repleta group is composed of four major subgroups, repleta, mercatorum, mulleri, and hydei (Wasserman 1982) and only the mulleri and hydei groups contain species with duplicate *Adh* genes. The entire *Adh* region from three species has been sequenced. These are *D. mulleri* (Fischer and Maniatis, 1985), *D. mojavensis* (Atkinson *et al.*, 1988), and *D. hydei* (M. Menotti-Raymond and D. Sullivan, in preparation). In addition, the *Adh*-1 gene of *D.* species "N" has been sequenced (J. Weaver and D. Sullivan, unpublished).

Comparison of the expression patterns of *Adh*-1 and *Adh*-2 with the differential promoter utilization of *D. melanogaster* and other species having one gene with two promoters reveals an interesting correlation (Batterham *et al.*, 1983). *Adh*-1 is expressed in cells, including larval fat body, gut, and Malpighian tubules, which in species with a single gene initiate *Adh* transcription at the proximal promoter. Similarly, cells of *D. mulleri* and *D. mojavensis* which contain ADH-2 initiate transcription at the distal promoter in species with one *Adh* gene. In *D. hydei* there is a different but related pattern of *Adh*-2 expression which resembles the expression pattern of a gene which has two promoters. However, there is no evidence of the usually highly conserved sequence of a distal promoter near the *Adh*-2 gene.

These data and intraspecific comparison of the sequences of the *Adh* genes were used by Atkinson *et al.* (1988) to derive a model for the evolution of the locus structure in the mulleri subgroup. This model assumes that ADH is required at all stages of larval and adult life and that a single-gene *Adh* structure, typified by *D. melanogaster,* is ancestral to the duplicate gene structure. For several reasons having to do with the origins of the two promoters of the *Adh* gene, it had first appeared attractive to suggest that the duplicate locus structure was ancestral. One could then propose that the two promoters arose by a simple deletion of an intervening structural gene. This proposal fails on several counts.

First, the phylogeny presented in Fig. 1 demonstrates the close relationship of *Adh*-1 and *Adh*-2. The degree of sequence similarity reflects a relatively recent divergence, not an ancient origin of the two genes. Second, since species with duplicate genes are all clustered in one phylogenetically coherent position of the tree, a model with the duplicate gene structure being ancestral to the two-promoter, one-gene structure would require several independent equivalent events in generating species with one *Adh* gene. This is unlikely.

We envision a series of steps which have converted a *D. melanogaster*-like *Adh* locus into a locus with three genes, as is found in *D. mulleri, D. hydei,* and *D. mojavensis.* This is depicted in Fig. 10. The loci in the mulleri and hydei subgroups contain three genes with a pseudogene as the most 5′ gene. The first event would have been a duplication to generate a locus with two genes. Inspection of the sequences around each of the genes of *D. mulleri, D. mojavensis,* and *D. hydei* reveals that only the pseudogenes have remnants of a distal promoter. This suggests that the first duplication eventually resulted in a downstream gene without a distal promoter either because the duplication did not include sequences which extend 5′ or because of a subsequent deletion that removed the sequences around the distal promoter. There is at present no basis upon which to choose between these alternatives. The second event was a duplication of the downstream gene leading to *Adh*-1 and *Adh*-2. This conclusion relies on comparisons of nucleotide sequence similarity of the three *Adh* genes within the three species. In each case *Adh*-1 and *Adh*-2 are more similar to each other than either is to the pseudogene. This is true for intron sequences and synonymous coding

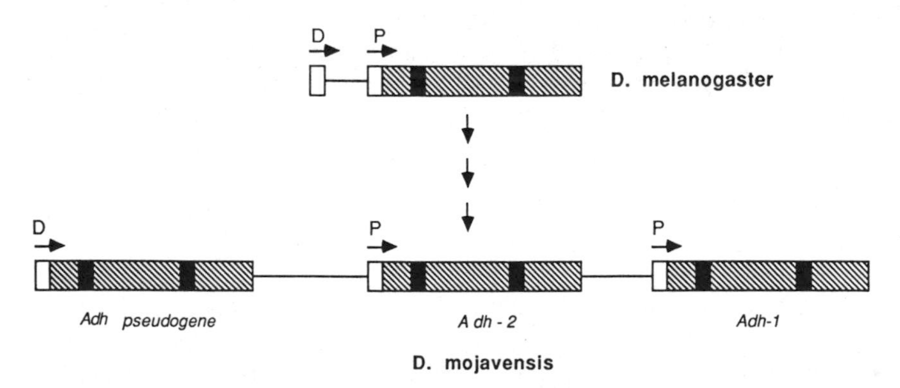

FIG. 10. Structures of the single *Adh* gene region represented by *D. melanogaster* and the three-gene region presented by *D. mojavensis.* Hatched boxes represent exons and filled boxes, introns. Promoter regions are labeled P or D based on sequence similarity and position with respect to the coding region. D is a distal-like promoter, P is a proximal-like promoter. Horizontal arrows indicate direction of transcription. Identical arrows represent a series of intermediate *Adh* locus structures.

nucleotides as well as the non-synonymous nucleotides. Therefore, we are confident that this conclusion is not affected by a comparison between a coding gene and pseudogene.

A third event which occurred during the evolution of the *Adh* locus in the repleta group was the translational inactivation of the 5′ gene resulting in what we now refer to as the *Adh* pseudogene. Calculations of the time of pseudogene inactivation using the method of Li *et al.* (1981) have substantial variance and this method has not proved useful in attempting to estimate the time when this occurred. In the model for *Adh* evolution in the repleta group developed by Atkinson *et al.* (1988) the inactivation of the pseudogene is proposed to occur following the first duplication. This model utilizes both interspecific and intraspecific comparisons of $K(s)$ and substitutions per nonsynonomous site [$K(a)$] of the *D. mulleri* and *D. mojavensis Adh* genes. These comparisons suggest strongly that the pseudogene was a functional gene for a fraction of its evolutionary history following the first duplication event. It is less clear if this might have occurred before or after the second duplication. However, an extensive search within the mulleri subgroup has failed to identify a species with two or three genes in which the ancestor of the pseudogene has coding capabilities. This is consistent with a relatively early inactivation of the gene that is now the pseudogene. Inspection of the pseudogenic regions of *D. hydei, D. mojavensis,* and *D. mulleri* reveals that the three species have in common several specific alterations which are probably relevant to the lack of expression of this *Adh* gene. The most striking of these is a deletion of about 870 nucleotides whose 5′ end is just proximal to the distal promoter sequence and extends to a point about 36 nucleotides upstream of what would be an initiation codon. The second event of significance is a common frameshift mutation (one base deletion) in the second codon. Therefore, we would argue that the pseudogene was inactivated sometime after the first duplication event, but before the divergence of the mulleri and hydei subgroups.

We have considered the possibility of gene conversion being responsible for the greater similarity of *Adh*-1 and *Adh*-2 as compared to the pseudogene (Atkinson *et al.*, 1988) and determined that this is an unlikely explanation. Intergenic gene conversion events are to be expected whenever tandemly repeated sequences occur in a genome. It is likely that two conversion events have been identified in the *Adh* region. The most certain of these has been identified in the *Adh*-1 gene of *D. mulleri* by Fischer and Maniatis (1985). They found a segment of nucleotide sequence of *Adh*-1 and *Adh*-2 to be identical. This includes exon 1 and extends through intron 1 into exon 2. They interpreted the cause of this sequence identity to be gene conversion.

Another possible conversion event may have occurred in *D. hydei.* The exons and introns of *Adh*-1 and *Adh*-2 are 97.3% and 93.2% similar, respectively. These genes are significantly more similar to each other than the *Adh*-1 and *Adh*-2 genes of *D. mojavensis* or *D. mulleri* are to each other. However, this high similarity of *D. hydei Adh*-1 and *Adh*-2 does not extend into the 3′ untranslated region. Consequently, we suggest that a conversion event between *Adh*-1 and *Adh*-2 may have occurred. The limits of converted sequence seem to extend from exon 1 though exon 3. However, there is an alternative explanation. There could have been a more recent and separate second duplication in the *D. hydei* lineage. This is not a parsimonious hypothesis since it would require additional independent duplications in either or both of the anceps and meridiana complex lineages (these are also species complexes in the mulleri subgroup). Species of these complexes share specific aspects of the pattern of *Adh* expression with *D. hydei,* and a common origin appears likely, but is not yet firmly established. Consequently, the most likely hypothesis we can offer is that the mulleri and hydei subgroups diverged following the second duplication event and that the high degree of sequence similarity between *Adh*-1 and *Adh*-2 of *D. hydei* is due to gene conversion.

EVOLUTION OF REGULATORY INFORMATION

In recent years there has been a developing interest in evaluating the role that changes in regulatory information have had in microevolutionary events. The hypothesis has been put forth that changes in patterns of gene regulation are relatively more important in speciation than are genetic changes which result in changes in the structure of proteins (Wilson *et al.*, 1977; Hedrick and McDonald, 1980; MacIntyre, 1982). This hypothesis is based on the observation that closely related species often show more differences in the timing, amount, or localization of gene products than in their primary structure. Clearly the formal distinction between "structural" and "regulatory" information can become blurred, but that does not affect the basic thrust of the hypothesis. The regulation of *Adh* in *Drosophila* is a useful system for evaluation of this hypothesis. We will focus on those genetic changes which fall into the class of *cis*-acting elements. These can be properly thought of as part of the *Adh* region and the available sequences can be compared in noncoding regions for the existence of conserved regions. These may be able to be shown to have regulatory function with functional tests using transformation experiments.

An initial question worth addressing is whether interspecific phenotypic variation exists and whether this can be legitimately classified as having a basis in regulatory information. The answer is clearly yes. A number of studies have identified interspecific differences in tissue-specific *Adh* expression. The most extensive studies are those of Dickinson and colleagues on *Adh* expression in the Hawaiian picture-winged *Drosophila*. Dickinson and Carson (1979) reported that ADH is found in the fat body, carcass, midgut, and Malpighian tubules of *D. grimshawii,* but only in the fat body of *D. orthofascia.* Hybrids between these species reveal ADH with the electrophoretic mobility of *D. grimshawii* in the carcass, midgut, and tubules and hybrid dimers in the fat body. These data provided the initial suggestion that *cis*-acting elements were responsible for the tissue expression differences between species. Similar results were reported when the expression of ADH in the midgut of *D. differens* and *D. heteroneura* was compared. In *D. heteroneura* no ADH is found in the midgut, while *D. differens* does have midgut ADH. In species hybrids between *D. differens* and *D. heteroneura* the midgut was found to contain *D. differens* ADH, while the fat body contains ADH of both parental species and most significantly heterodimeric molecules. In a subsequent study, which included the analysis of several enzymes in 27 species of Hawaiian *Drosophila,* Dickinson reported that species differences in tissue enzyme content were quite common (Dickinson, 1980). In later studies, Rabinow and Dickinson (1981) reported that the ADH mRNA is found in fat bodies of *D. grimshawii* and *D. orthofascia* and the midgut of *D. grimshawii,* but not the midgut of *D. orthofascia.* This further suggests a difference in *cis*-acting sequences between the two species which function in the control of transcription. More extensive analysis (Rabinow and Dickinson, 1986), involving four species, *D. affinidisjuncta, D. grimshawii, D. formella,* and *D. hawaiiensis,* revealed consistent correlations of ADH activity and *Adh* mRNA content in tissues that showed *Adh* expression differences. Finally, Rowan and Dickinson (1986) have explored *Adh* promoter usage during development of 12 species of Hawaiian *Drosophila.* The basic structure of the *Adh* genes in the Hawaiian species is similar to the *Adh* locus of *D. melanogaster,* having both distal and proximal promoters (Rowan *et al.,* 1986). Two species, *D. formella* and *D. prostopalpis,* have a pattern of promoter usage which varies significantly from the pattern common to the other species. In *D. formella,* transcripts from the proximal and distal promoters are abundant in adults. Both transcripts are found in fat-body, Malpighian tubules, head, and gut. In *D. prostopalpis* distal transcripts are found to be predominant in the carcass of third-instar larvae, both transcripts are about equally abundant in fat body and tubules, and only proximal transcripts are present in midgut. This represents a relative increase in larval distal transcript as compared to other species.

To summarize the data from the Hawaiian *Drosophila:* they indicate substantial variation in *Adh* expression pattern among species, this variation is demonstrable at the transcript level, it involves specific aspects of promoter usage in development, and all species hybrid data are consistent with species variation being the result of variation in *cis*-acting genetic elements.

There are also examples of interspecific tissue expression in other groups of *Drosophila.* In the mulleri subgroup, two species, *D. mojavensis* and *D. arizonae,* have high levels of *Adh*-1 expression in the cytoplasm of nurse cells of the ovary. Other species have either no or low amounts of ADH in the ovary and no other adult *Adh*-1 expression. Hybrids between species with *Adh*-1 expression differences have only ADH-1 from the ovary-expressing parental species (Mills *et al.*, 1986; Batterham *et al.*, 1984), again suggesting difference in *cis*-acting elements.

In all species of the mulleri species complex, *Adh*-2 expression follows a developmental pattern quite similar to the pattern of distal promoter usage in the *Drosophila* species having one *Adh* gene. However, the studies of sequence divergence (Atkinson *et al.*, 1988) revealed that the upstream region of *Adh*-2 was derived from a proximal promoter. Therefore, this presents an interesting case of the control region of *Adh*-2 being homologous to a proximal-type promoter but analogous to a distal-type promoter. Species classified within the mulleri subgroup, but not in the mulleri species complex (e.g., meridana complex and anceps complex), and species in the hydei subgroup express *Adh*-2 with a different developmental pattern. In these species *Adh*-2 expression begins early, either in embryos or first-instar larvae, and continues for the duration of the larval stages and through adult life. This pattern of expression is what would be expected of an *Adh* gene having two promoters. However, analysis of the *D. hydei* *Adh*-2 gene has revealed that it has a single proximal promoter (M. Menotti-Raymond and D. Sullivan, in preparation). The *D. hydei* locus structure is essentially similar to the locus in *D. mulleri* (Fischer and Maniatis, 1985) and *D. mojavensis* (Atkinson *et al.*, 1988). Therefore, it is evident that a series of events has occurred during the evolution of *Adh*-2 in the mulleri and hydei subgroups. The gene originated from a duplication event that as far as can be determined included only sequences from a proximal promoter control region which usually directs larval expression. There are two known expression patterns of *Adh*-2 genes. One, represented by species of the mulleri species complex, is reminiscent of distal promoter-directed expression. The other, represented by *D. hydei* and species of an anceps and meridiana species complexes, is similar to the expression pattern of an *Adh* gene with both promoters. The basis for these interesting changes of *Adh*-2 expression is not fully understood, but may in part be related to sequences 5′ to the *Adh* pseudogene which are known to be necessary for *Adh*-2 expression (Fischer and Maniatis, 1986; C. Bayer and D. Sullivan, in preparation). These are described below.

It has been recognized for some time that the specific activity of ADH is very different in different *Drosophila* species (McDonald and Avise, 1976; Batterham *et al.*, 1984). These differences could be due to either different catalytic properties of the enzyme (i.e., structural differences) or regulatory differences which control the amount of enzyme. It is of note that most species have a higher specific activity in larvae than in adults. However, two species, *D. hydei* and *D. melanogaster,* have much higher activity in adults than in larvae. In addition, there are differences between these two species in pattern of *Adh* expression. *D.*

melanogaster has a larval specific activity in a range typical of many species, but has an extremely high activity level in adults. *D. hydei,* on the other hand, has a typical adult activity, but very low larval activity.

There has been an analysis of one interspecific activity difference, between *D. simulans* and *D. melanogaster* (Dickinson *et al.*, 1984). The ADH activity of these species is only slightly different in larvae, but *D. melanogaster* has about a fourfold higher adult activity. Species hybrids have a preferential expression of the *D. melanogaster* gene in adults. The elevated activity in *D. melanogaster* is correlated with an increase in ADH protein content and is partially accounted for by higher ADH mRNA content. While the mechanisms controlling ADH levels in the two species are not fully understood, it seems clear that there are differences in the pattern of *Adh* regulation in the two species.

The analysis of regulatory information near the *Adh* gene has been directly studied through use of P-element-mediated transformation (Rubin and Spradling, 1982). The initial experiments involved the introduction of a *D. melanogaster* gene into *D. melanogaster* ADH null mutants (Goldberg *et al.*, 1983). In these experiments the transfected gene was expressed at approximately typical levels and with the correct stage- and tissue-specific pattern. These initial transformants contained 5.5 kilobases (kb) 5′ and 4.5 kb 3′ of the coding region. Subsequent experiments (Posakony *et al.*, 1985), using smaller DNA fragments, identified three regions to be important in *Adh* expression. It was ascertained that an element more than 2 kb upstream of the proximal promoter is required for full larval expression. Another region located within the sequences -69 to -660 of the distal promoter is required for correct developmental expression from the distal promoter, and a third region within 386 nucleotides of the proximal promoter is required for correct expression from the proximal promoter.

A transient assay has been developed which has greatly facilitated the dissection of the *Adh* control regions (Martin *et al.*, 1986). In this assay, *Adh* containing test plasmid DNA is injected into embryos of ADH null mutants and these animals are then analyzed for ADH content as third-instar larvae. The ADH distribution in tissues is mosaic because not all cells contain plasmids, but this does not interfere with determining the overall *Adh* expression pattern. This approach has been utilized by Shen *et al.* (1989) to dissect the proximal control region of *D. melanogaster Adh.* Starting with a cloned gene that contained 450 nucleotides 5′ to the translation start site and which is expressed in a typical manner, a series of internal and terminal deletions were constructed and assayed for their ability to support ADH expression. A terminal deletion that removed sequences 5′ to position −126 relative to the translation start did not support *Adh* expression. Two regions of importance were identified in this upstream region. A 10-bp region represented by deletion ns1, between −410 and −400 relative to the translation start, was found to be essential for *Adh* expression. The second region of critical importance, represented by deletions ns6 and ns7, contains the

TATA box (deletion ns7) and about 80 bp immediately upstream of the TATA box (deletion ns6). What may be a very important outcome of this analysis is the discovery of an interaction between control sequences. The effects of deletions ns1 and ns6, but not ns7, are not seen in constructs containing two *Adh* genes. One gene has the deletion ns1 or ns6 in the control region and is otherwise a complete *Adh* gene. The other gene is an *Adh* gene with 459 bp of the 5′ end intact. When these two genes are injected as two plasmids, *Adh* expression from only the intact gene occurs. However, when the two genes are fused into a single plasmid and injected into embryos, expression from both genes is seen. Therefore it appears that the control region can be divided into three parts. These parts are a region around the TATA box which is essential and two other regions, one close to the TATA box (i.e., within 50 bp), and a second region about 350 bp upstream. Each of the two latter regions acts, in *cis,* but can be compensated for with a distant functionally equivalent region. This may be equivalent to saying that they are enhancers which function independent of position. If so, there may be some important ideas about how enhancers work forthcoming from these studies.

The use of P-element transformation in analyzing species-specific tissue expression has been greatly facilitated by the observation that genes cloned from distantly related species of *Drosophila* will retain their expression specificities following transformation into *D. melanogaster.* This provides a functional test with which to analyze the role that specific sequences might play in tissue-specific gene expression. Studies of this type have been conducted using genes from *D. mulleri* (Fischer and Maniatis, 1986, 1988) and Hawaiian *Drosophila* (Brennan and Dickinson, 1988; Brennan *et al.,* 1988). The most extensive analysis of this type has been conducted by Fischer and Maniatis (1986, 1988) using the *Adh* genes of *D. mulleri.* Transformation of the *Adh*-1 or *Adh*-2 genes of *D. mulleri* into *D. melanogaster* results in flies which express *Adh* with the expected *D. mulleri* pattern, which, although similar to that of *D. melanogaster,* has identifiable differences. Generally the expression pattern of *Adh*-1 is related to that of the proximal promoter. *Adh*-2 has a pattern of expression similar to that of a distal promoter. Analysis of the upstream regions by the use of *in vitro*-constructed deletions reveals that all sequences necessary for *Adh*-1 expression are 3′ to position −350 (positions are relative to the transcription start). Deletion to −170 abolishes *Adh*-1 transcription. The analysis of *Adh*-2 expression reveals a more complex pattern for the arrangement of control sequences. Sequences which convey the tissue-specific expression pattern of *Adh*-2 are located within 1200 bp of the transcription start site. However, transformants having only this upstream region express at 5–10% of the level of transformants which also contain additional upstream DNA sequences. These are located between 400 and 2500 bp 5′ to the pseudogene and 2400–4500 bp upstream of *Adh*-2. In addition, a sequence located in excess of 600 bp 3′ to *Adh*-2 seem to be required for expression in Malpighian tubules.

In a subsequent study Fischer and Maniatis (1988) used the linker substitution method to analyze the detailed organization of the *Adh*-1 control region. In the 350-bp region upstream from the transcription start, two important regions which are needed for *Adh*-1 expression in addition to the TATA box have been identified. One, referred to as box B, is located from −181 to −269 with respect to the transcription start. Deletion or linker substitution within this region results in a depression of transcription of the *Adh*-1 gene. A second element is located between −91 to −60. Linker substitution in this region, named box A, also results in a depression of transcription. Since deletion of both box B and box A results in constructs with a low level of *Adh* expression, the region 3′ to the *Adh* gene was examined for information important to *Adh* expression. It was discovered that a region, not precisely delimited, in the 3′ end can fulfill a role similar to box B. These regions can substitute for one another, but neither can replace box A. These two regions seem to have the properties of an enhancer and furthermore seem particularly important for the high level of expression in the larval fat body. Box A seems important for expression of *Adh* in tissues other than the fat body. It seems evident that correct *Adh* expression is the result of an interaction, either direct or indirect, of a number of sequence elements.

Set against this background, which demonstrates that interspecific transformation generally results in expression patterns appropriate for the gene of the donor species and that deletion or substitution experiments identify *cis*-acting regions which are important for expression, we have conducted a comparative analysis of sequences upstream of *Adh*. Figure 11 presents a comparison of the *Adh*-1 genes of *D. hydei* and *D. mojavensis* with the start of translation at position zero. The upstream region has a fairly erratic pattern of sequence conservation. Notably almost all of the regions are more similar than are introns 1 or 2 of these genes. One might conclude that the majority of nucleotides are important in *Adh* expression. However, the transformation experiments indicate that less than 350 nucleotides are needed for correct expression and not even the majority of nucleotides within this region are functional. It would appear that there is a degree of sequence conservation in the upstream regions that is not easily explained. Figure 12 presents these regions in two other comparisons, *D. melanogaster* and *D. orena* proximal promoter regions and *Adh*-1 of *D. mojavensis* and *D. mulleri*. Again there seem to be erratic stretches of sequence conservation to at least 500 nucleotides upstream.

We have conducted an extensive search of the upstream regions of the *Adh* genes in an attempt to find regions of strong sequence conservation which might, in the light of the transformation studies, be candidates for control regions. Two regions stand out. Figure 13 shows the sequence of the proximal promoter

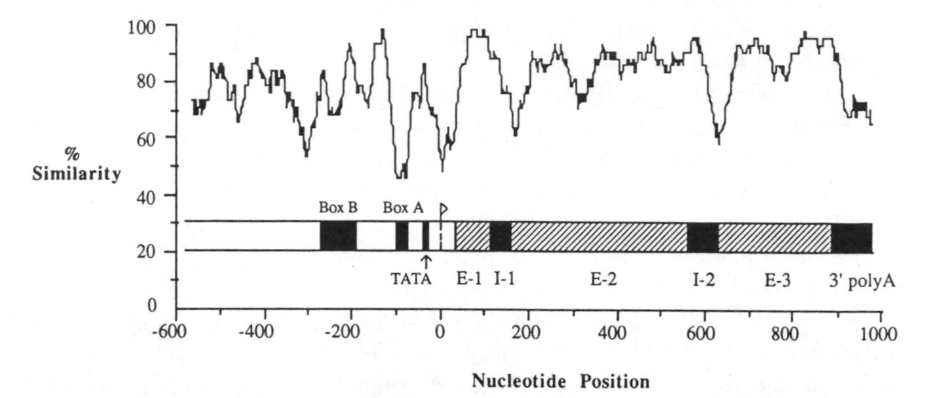

FIG. 11. Percent similarity of the *Adh-1* genes of *D. mojavensis* and *D. hydei* from a point 500 nucleotides upstream from the ATG initiation codon to the poly A sequence of *D. mojavensis.* Percent similarity was determined using an algorithm which calculated the percent similarity of sequences contained within successive, overlapping windows of 40 nucleotides. Sequences were aligned using the algorithm of Wilbur and Lipman (1983), after which nucleotides opposite gaps were removed before the similarities were calculated. The relative positions of the structural and functional features of these *Adh-1* genes are also shown. See text for details.

including the TATA box and upstream sequences through the box A region as defined by Fischer and Maniatis (1988). In this box A region there are clear blocks of sequence conservation (underlined). These are most evident in the repleta group, but are also identifiable in the more distantly related species. The region of the TATA box is perfectly conserved in these genes. The extent of one the deletions, ns6, is indicated. This is the deletion demonstrated by Shen *et al.* (1989) to be essential for expression of the *Adh* proximal promoter. The equivalent regions of the *Adh*-2 genes of the mulleri subgroup species are also included in this analysis even though they have a different expression program. This serves to indicate their evolutionary derivations from a proximal promoter-containing gene. It may also be relevant that there is a region of substantial divergence between the TATA and box A regions in these genes.

The other region of striking sequence similarity is found upstream of the distal promoter of *D. melanogaster* and *D. affinidisjuncta* and upstream of the pseudogene of *D. mojavensis, D. mulleri, and D. hydei* (Fig. 14). This region was previously identified by Rowan and Dickinson (1989) in their comparison of the nucleotide sequences of the *Adh* genes of *D. affinidisjuncta* and *D. melanogaster*. Its presence upstream of the pseudogenes adds weight to the argument that the pseudogenes are derived from a gene once having had a distal promoter. It is of particular note that this region of sequence conservation is located in a region known to be important for adult expression of *Adh*-2 (Fischer and Maniatis, 1986; C. Bayer and D. Sullivan, in preparation).

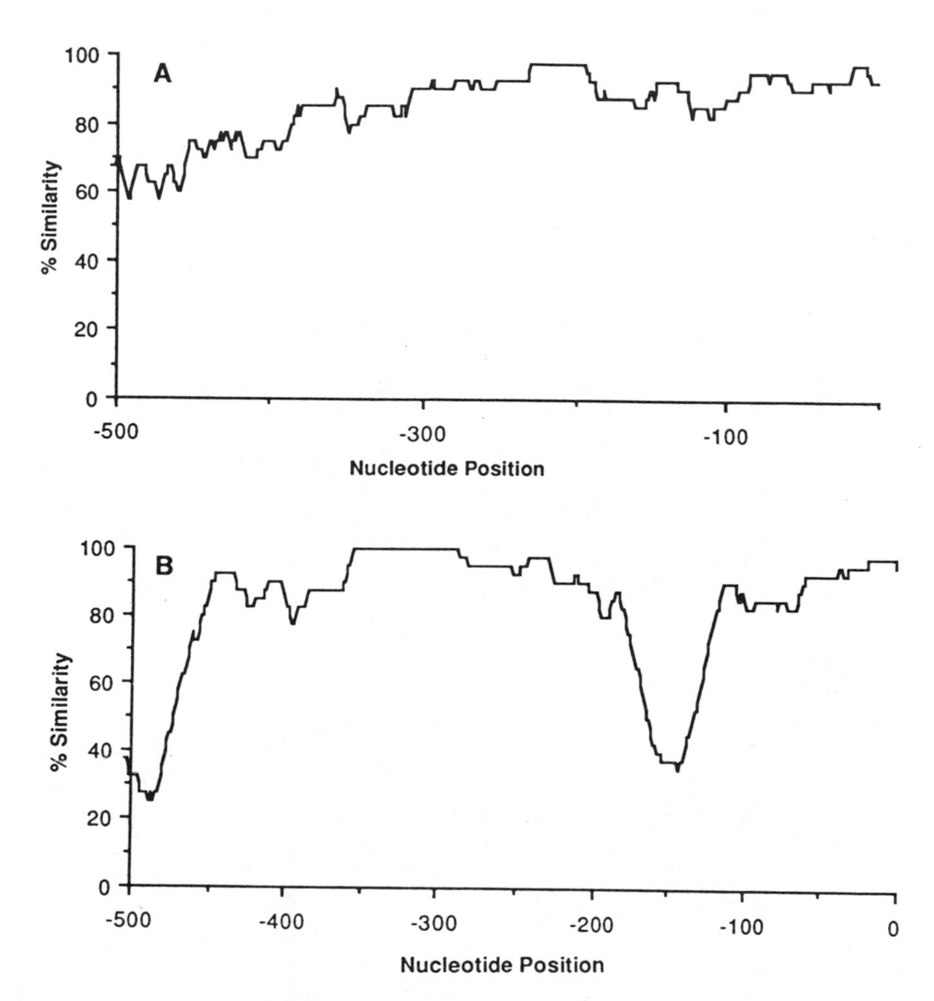

FIG. 12. Percent similarity of upstream regions of the Adh-1 genes of (A) *D. mojavensis* and *D. mulleri;* (B) proximal upstream regions of *D. melanogaster* and *D. orena* DNA. Sequences 500 nucleotides upstream from the respective ATG initiation codons were compared using an algorithm which calculated the percent similarity of sequences contained within successive, overlapping windows of 40 nucleotides. Sequences which were obviously highly homologous were aligned by eye, while sequences which were not were aligned using the algorithm of Wilbur and Lipman (1983). Nucleotides opposite gaps were then removed before the similarities were calculated.

```
MOJ1        GCCGACTGCG--GCCTTCGTTATTGATAAGCCA                    TGTGCTGCGT GGGCAG     TGGCCGTGTGGCTATAAATA
MUL1        GCCGACCGCGCGGCCAGTGGTATTGATAAGACA                    TGTGCTGCGT GGGCA   GCGTGGCCGTGTAGCTATAAATA
N1          GCCGACTGCGCGGCCTGTGATGTTGATAAGCCA                    TGTGCTGCGT GGGCA   GCGTGGCCCTGCGGCTATAAATA
MOJ2      GTCAGCTTCGCG  GTCAGCGTTATTGATAAGCCAACTGAGA GATTTGCTTGCAATACGACCGGCAGGCATGGCACTGCAGTGCGAGTTATAAATA
MUL2      GTCAGCTTCGC   GTCAGC TTATTGATAAGCCAACTGAGA GATTTGCTTGCAATATGACCGGC              AGTGC AGTTATAAATA
HYD2  GTCTACGTCTAGGTCGCTGCCAGCGTTATTGATAAGCCAACTGAGA GATTTGCTGGCAATACGAC CG   GCA   GCGTGCCACTGCGGTTATAAATA
HYD1           GAATTCGCCGCGAGCGGTATTGATAAGC-AACAAAGA GATTCGCCGTCAATGGCAA CGTGGGCA   GCGTGGCCCTACGGCTATAAATA
PSE                GTAATGCGAGA      GATAAG AAACAGAAAAGCTCGACGTGAAAGAAGCTTCTGGCGGAAGCTCTTGCCGAACTCCCTATAAATA
ORE   GTGAATAGCCGAGAGATCGCGTAATGATA GATAAA      GAAA GCTCTACGTAAGCGAAGCTTCTGGGGGA        TAGATCTTCCTATAAATA
AFF   GAGCAGTGCTGATGATCCGCCAACATTACTGATAAA    CAGCTGCGAG GAAAT ACTCAAACTT    AGCAGACTGTTCGCC       TATAAATA
MEL   CGGCTGAGCGCCGAGATCGCGTAACGGTA GATAAT      GAAAAGCTCTACGTAACCGAAGCTTCT       GCTGT ACGGATCTTCCTATAAATA

ns6   CGGCTGAGC-------------------------------------------ACGTAACCGAAGCTTCT       GCTGT ACGGATCTTCCTATAAATA
```

FIG. 13. Alignment of sequences immediately upstream from the TATA sequences of the *Adh* genes: MOJ1—*D. mojavensis Adh-1;* MUL1—*D. mulleri Adh-1;* N1—species "N" *Adh-1;* MOJ2—*D. mojavensis Adh-2;* MUL2—*D. mulleri Adh-2;* HYD2—*D. hydei Adh-2;* HYD1—*D. hydei Adh-1;* PSE—*D. pseudoobscura* proximal promoter; ORE—*D. orena* proximal promoter; AFF—*D. affinidisjuncta* proximal promoter; MEL—*D. melanogaster* proximal promoter. The boxed sequence in *D. mulleri* depicts a sequence identified by Fischer and Maniatis (1988) as box A. The sequence ns6 depicts a deletion (broken line) described by Shen *et al.* (1989) which significantly reduces *Adh* expression in *D. melanogaster.* The sequences contained within box A and the deletion in ns6 are clearly present, to varying degrees, in the corresponding locations in all other proximal-type *Adh* upstream regions. The more conserved of these are underlined.

```
MEL-D:-564 TGCAAATTAAGCCGAAGTTCAAT T   GCGACCGCAGCAACAACACGATCTTTCTACACTTCTCCTTGCTAT GC TTGACATTCAC  AA GGT -478
AFF-D:-260 TGCAAAT       AA TGCAATCT   GCGAACATAACAACAGCAACAACAACAATATAAGCCGACTAATATTGCTTTGACATTCAGTGAA GGT -177
HYD-P:-441 TGC                TCAATTT   GCGACCACAACAAA      CAAGATTCTTGACTATGTC     TATTGCTTTGACATTCACTGAA GGT -373
MUL-P:-523 TGC                TCAATTTGCTGCGACCACAACAAAAATAACAAAGTTCTTGACTACGTC     GATTGCTTTGACATTCAGTGAAAGGT -447
MOJ-P:-493 TGC                TCAATTT   GCGACCACAACAAAAATAACAAAGTTCTTGACTACGTC     GATTGCTTTGACATTCAGTGAA GGT -420

CONSENSUS: TGC...............TCAATTT...GCGACCACAACAAAAACAACAA..TTCTT.ACT.C..C....TATTGCTTTGACATTCAGTGAA.GGT
                                C                   C GT C                    G                   C
```

FIG. 14. Alignment of a region upstream from the distal promoters of *Adh* genes of *D. melanogaster* (MEL-D) and *D. affinidisjuncta* (AFF-D) with a region upstream from the pseudogenes of *D. hydei* (HYD-P), *D. mulleri* (MUL-P), and *D. mojavensis* (MOJ-P). A consensus sequence is also shown for this region. A consensus sequence is constructed when five out of five or four out of five nucleotides are identical at a position. When three of five are identical, alternative nucleotides are shown. Distances shown are nucleotides upstream from the respective transcription start sites. This region of similarity between *D. melanogaster* and *D. affinidisjuncta* was previously identified by Rowan and Dickinson (1988). The region of similarity shown here was initially identified using the algorithm of Wilbur and Lipman (1983).

Adh AND ISSUES OF MOLECULAR EVOLUTION

Studies in molecular evolution, i.e., protein or DNA sequence comparisons generally have as their goal the establishment of phylogenetic relationships or the elucidation of the mechanisms of genetic changes which occur during evolution. It is often useful when performing sequence comparisons to focus upon individual sections of particular genes, since it is likely that these will reveal different patterns of change due to different evolutionary histories. Li *et al.* (1986) have made an extensive comparison of divergence in coding, intron, and surrounding regions using nucleotide sequence data from a large number of vertebrate genes. In this light we will summarize what the comparison of the *Drosophila Adh* genes reveals about the evolution of sequence classes with different functions. We are interested in whether regions vary with respect to the rate of nucleotide substitution, whether any particular region has properties which indicate little or no selective pressure on the nucleotide sequence, and whether there is a basis to select *a priori* a particular class of sequence for phylogenetic comparisons.

Using the convention of writing genes from left to right, we begin at the 5′ flanking region. A simple question is, What are the dimensions of the 5′ flanking region? Clearly the only way to answer this question is with functional tests, i.e., *in vitro* mutagenesis followed by transformation. We have presented information which suggests that the pattern of sequence divergence upstream of the *Adh* genes is complex and indicates some unexplained conservation. It may be easy to understand the conservation of stretches of sequence which are important in the control of a variety of genes. However, it is obvious that there will be substantial variation among different genes in the size and arrangement of multiple-element control elements. Furthermore, our comparisons above suggest sequence conservation in regions which as yet have not been identified as having a function with respect to *Adh* expression. A further complication is the difficulty posed by overlapping genes and the lack of available guidelines to suggest where one gene ends and another begins. Consequently and not surprisingly, sequence comparisons of the 5′ flanking regions will have limited value in molecular evolution beyond identifying regions of conservation which might then be subjected to functional tests.

Within the coding regions there are the two classes of nucleotide sites, synonymous and nonsynonymous. Synonymous sites are those nucleotide which are able to change without a resulting amino acid replacement. In any comparison between genes it is generally observed that the extent of nucleotide substitution at synonymous sites exceeds that at nonsynonymous sites. This provides clear evidence for the strong purifying selection which operates on coding regions. In a large comparison among mammalian genes, Li *et al.* (1986)

report that the rate of substitution at synonymous sites averages about five times that of nonsynonymous substitutions. Both nonsynonymous and synonymous substitution rates vary in different genes, although the magnitude of variation is much smaller for synonymous sites (about a factor of 10) as compared to nonsynonymous sites (a factor of 700). In Table III we present the $K(a)$ values and the ratio of $K(s)/K(a)$ for all comparisons of *Adh* genes. Generally the values for $K(s)/K(a)$ ratios of coding genes cluster in the range of 8–10. The three very high values, *D.ml/D.sc, D.ml/D.si,* and *D.mu2/D.nl*, each result from very small $K(a)$ values for the closely related species and may not be statistically significant. Interestingly, all comparisons involving *D. hydei* and *D. affindisjuncta* genes are in the higher range of this distribution due to higher $K(s)$ values. This could reflect some specific aspect of mutation in these species. The comparisons involving the *Adh* pseudogenes of *D. mojavensis* and *D. mulleri* typically are at the low range, which is consistent with their loss of coding capability.

Synonymous sites have frequently been utilized in sequence comparisons, since it is thought that the rate of nucleotide substitution of these sites may approximate the mutation rate. This depends on the assumption that there is little or no functional significance for these positions. It has long been recognized that organisms utilize different synonymous codons with different frequency. Therefore, it is commonly appreciated that the bias in codon usage will have the effect of reducing the rate of substitutions to a value below the mutation rate. The bases for codon bias has been reviewed by Li *et al.* (1986) and appears to be due primarily to tRNA availability and codon–anticodon interaction. Consequently, the effect of codon bias on sequence comparisons is thought to be modest and expected to be constant in related species. However, we recently have studied the codon utilization pattern in *Adh* genes reported here and found a clear difference in codon bias within the genus *Drosophila* (Starmer and Sullivan, 1989).

It has been pointed out by Bodmer and Ashburner (1984) and Ashburner *et al.* (1984) that codon use in *Drosophila Adh* is nonrandom, with a deficiency of A and a more modest deficiency of T in the third position. We have calculated the third-position usage of the four bases in 28 *Drosophila melanogaster* coding genes (including *Adh*) and found 10.5% A, 17.8% T, 30.7% G, and 41.0% C. *Drosophila melanogaster Adh* has 3.4% A, 14.5% T, 30.9% G, and 50.8% C. Therefore it appears that the codon bias in *Drosophila Adh* is more extreme than in an average *Drosophila* gene. We have also examined the codon utilization of the *Adh* gene of the other species considered here (Table IV). It is apparent that the codon utilization of the *Adh* genes in the melanogaster and obscura groups is different than in species of the subgenus Drosophila. Therefore the calculations $K(s)$ values between genes from species of the two subgenera could be influenced by differences in codon bias.

With the exception of nucleotides used as splice signals, the nucleotides of

TABLE III. *K*(*a*) Values and Ratio of *K*(*s*)/*K*(*a*) for Comparisons of *Adh* among Species[a]

Drosophila species gene	*D.ml*	*D.sc*	*D.ma*	*D.si*	*D.or*	*D.pu*	*D.af*	*D.hy1*	*D.hy2*	*D.hyP*	*D.mu1*	*D.mu2*	*D.muP*	*D.mj1*	*D.mj2*	*D.mjP*	*D.nl*
D. melanogaster	—	0.002	0.007	0.002	0.018	0.057	0.137	0.120	0.120	0.182	0.118	0.109	0.199	0.126	0.123	0.182	0.108
D. sechellia	36.94	—	0.005	0.000	0.016	0.060	0.132	0.116	0.116	0.180	0.119	0.105	0.197	0.126	0.119	0.182	0.104
D. mauritiana	9.10	6.80	—	0.005	0.017	0.062	0.134	0.118	0.118	0.182	0.120	0.107	0.195	0.128	0.121	0.183	0.106
D. simulans	22.94	—	6.70	—	0.016	0.060	0.132	0.116	0.116	0.180	0.119	0.105	0.193	0.126	0.119	0.182	0.104
D. orena	9.69	10.49	11.58	10.49	—	0.064	0.140	0.119	0.118	0.185	0.122	0.109	0.198	0.135	0.127	0.186	0.109
D. pseudoobscura	10.70	10.17	10.32	10.50	8.27	—	0.126	0.112	0.104	0.169	0.113	0.103	0.185	0.124	0.118	0.167	0.107
D. affinidisjuncta	8.39	8.41	8.88	8.71	7.24	8.41	—	0.056	0.054	0.119	0.063	0.053	0.130	0.076	0.067	0.123	0.057
D. hydei Adh-1	9.83	9.55	9.39	9.87	9.16	9.96	18.41	—	0.009	0.102	0.037	0.028	0.120	0.044	0.042	0.106	0.029
D. hydei Adh-2	9.50	9.48	9.32	9.75	9.75	10.29	16.51	11.25	—	0.100	0.040	0.030	0.113	0.047	0.044	0.099	0.033
D. hydei Adh-P	11.47	10.89	11.15	11.91	10.47	8.88	11.26	7.65	7.89	—	0.098	0.100	0.099	0.111	0.106	0.072	0.099
D. mulleri Adh-1	7.81	7.32	7.60	7.74	7.46	6.94	14.80	12.76	10.59	9.57	—	0.018	0.115	0.022	0.028	0.104	0.012
D. mulleri Adh-2	7.89	7.71	7.76	8.19	7.42	8.04	17.79	16.93	14.92	8.88	7.34	—	0.108	0.029	0.017	0.100	0.009
D. mulleri Adh-P	5.73	5.29	5.38	5.60	5.25	5.51	9.29	5.59	5.62	11.02	4.87	4.95	—	0.123	0.111	0.035	0.106
D. mojavensis Adh-1	7.94	8.21	7.88	8.10	7.48	8.29	13.28	11.05	9.96	8.25	8.38	6.90	5.10	—	0.035	0.112	0.021
D. mojavensis Adh-2	8.42	8.76	8.61	8.84	7.72	7.78	14.13	10.48	9.61	8.33	7.49	12.24	5.78	5.67	—	0.106	0.019
D. mojavensis Adh-P	6.04	5.40	5.40	5.59	5.65	5.63	10.13	5.46	5.65	16.24	4.10	4.43	8.59	3.94	4.16	—	0.099
D. species "N" *Adh-1*	7.60	8.27	7.83	8.06	7.71	7.77	16.47	16.90	13.56	9.29	13.65	23.80	5.74	9.93	11.47	5.16	—

[a] *K*(*a*) values are above the diagonal, while the ratios are beneath the diagonal. Species listed in the columns are abbreviations of and in the same order as those listed as rows. *Adh*-P = *Adh*-pseudogene.

TABLE IV. Codon Position III Base Utilization for *Adh* Alleles in Species of Two Subgenera of the Genus *Drosophila*[a]

	A	T	G	C	Total	G+C, %
Subgenus Sophophora						
D. melanogaster	10	37	79	130	256	81.64
D. sechellia	9	39	77	131	256	81.25
D. mauritiana	11	37	76	132	256	81.25
D. simulans	9	36	78	133	256	82.42
D. orena	9	39	80	128	256	81.25
D. pseudoobscura	8	52	78	116	254	76.38
Subgenus Drosophila						
D. affinidisjuncta	32	71	66	85	254	59.45
D. mulleri Adh-1	24	47	74	109	254	72.05
D. mulleri Adh-2	21	44	74	115	254	74.41
D. mojavensis Adh-1	26	53	72	103	254	68.90
D. mojavensis Adh-2	36	46	70	102	254	67.72
D. species "N" *Adh-1*	25	44	72	113	254	72.83
D. hydei Adh-1	30	61	64	99	254	64.17
D. hydei Adh-2	29	60	66	99	254	64.96
D. hydei Adh-pseudogene	55	67	57	74	253	51.78
D. mojavensis Adh-pseudogene	39	47	71	91	248	65.32
D. mulleri Adh-pseudogene	36	50	66	89	241	64.32

[a]Sources for sequences: *D. melanogaster* (Kreitman, 1983); *D. sechellia* (Coyne and Kreitman, 1985); *D. mauritiana, D. orena, D. simulans* (Bodmer and Ashburner, 1984); *D. pseudoobscura* (Schaeffer and Aquadro, 1987); *D. affinidisjuncta* (Rowan and Dickinson, 1988); *D. mulleri Adh-1, Adh-2, Adh*-pseudogene (Fischer and Maniatis, 1985); *D. mojavensis Adh-1, Adh-2, Adh*-pseudogene (Atkinson *et al.*, 1988), *D.* species "N" *Adh-1* (D. Sullivan, unpublished), *D. hydei Adh-1, Adh-2, Adh*-pseudogene (M. Menotti-Raymond and D. Sullivan, unpublished).

intron sequences are generally thought to have no function, although some exceptions to this are known. In the case of *Adh* it has been shown through the use of *in vitro* constructs of ''intronless'' genes that the *Adh* gene can function normally without the two introns which interrupt the coding sequence (Shen *et al.*, 1989). A small fraction of the nucleotides of the intron are required for proper transcript splicing, but most of the nucleotides are expected to be able to change with no selective constraints. Comparison of the extent of substitutions in the intron to the silent sites in the coding genes of *D. mojavensis* and *D. mulleri* (Atkinson *et al.*, 1988) generally supports this view, since the extent of substitutions in introns is greater than at silent sites. However, there are several anomalies in intron comparisons among the *Adh* genes that suggest caution in making assumptions about rates of substitutions in introns. Several of these involve observations which compare divergence of introns with silent sites. This is relevant because it is becoming apparent through observations on codon bias and nonrandom distribution of silent site substitutions along the *Adh* gene (Shaeffer and Aquadro, 1987) that the rate of silent site substitution is likely to be constrained by selection. Therefore, the argument can be made that if substitution rates in introns are equal to or less than silent site rates, some selective constraint might be operating on these sites as well. Shaeffer and Aquadro (1987) noted that the sequence divergence in intron 2 exceeds that of intron 1 and while these divergence values were similar to those obtained for the silent sites of exons 1 and 3, they greatly exceeded the divergence observed in silent sites in exon 2. Furthermore, comparisons of substitutions in silent sites to introns in species of the melanogaster complex (Kreitman, 1983; Bodmer and Ashburner, 1984; Cohn *et al.*, 1984) indicated that silent site substitutions are more frequent than intron substitutions. Observations in our laboratory have also revealed some puzzling comparisons. Intron 1 of *Adh*-1 of *D.* species "N" and intron 1 of *Adh*-1 of *D. mojavensis* are 96.4% similar, whereas the similarity of intron 2 in these genes is 57.7% which is in the range of expected values predicted from other intron sequence comparisons of *Adh* genes within the mulleri subgroup. We can offer no reason to explain this apparent sequence conservation in intron 1. We have also noted (Atkinson *et al.*, 1988) that an unexpectedly high degree of conservation is revealed in comparison of the introns of the pseudogenes of *D. mojavensis* and *D. mulleri*. However, we believe this is a result of sequence conservation over the entire pseudogene region and is not confined to the introns.

Comparisons of sequences of the 3′ end of genes suffer from a problem similar to that of 5′-end comparisons. These are twofold. First, there is little reason which enables one to make any assumptions with regard to function or lack thereof of sequences near the 3′ end of a gene. Second, one is also unable to assume anything with respect to the placement or orientation of an adjacent gene. *Adh* provides a particularly relevant example of this point. Kreitman (1983) pointed out a region of unexpectedly high sequence conservation downstream from *Adh* of *D. melanogaster*. Subsequent observations by Cohn and Moore (1988) and detailed analysis by Schaeffer and Aquadro (1987) explained this apparent anomaly as simply being due to another gene whose location is quite close to the 3′ end of *Adh*. Consequently, only in very well characterized genomic regions will comparison outside of the immediate coding regions be particularly useful to sequence comparison studies of molecular evolution.

Pseudogenes have been suggested to be the "paradigm of neutral evolution" (Li *et al.*, 1981). Since these are sequences with no known function, the expectation is that they should diverge free of selective constraints and therefore provide a useful baseline whose nucleotide substitution rate would approximate the mutation rate. This may well be the case in the majority of cases. Again, however, *Adh* pseudogenes reveal a degree of sequence conservation which is currently unexplainable by any known function. Atkinson *et al.* (1988) compared the sequences of the *Adh* pseudogenes which are at the 5′ ends of the *Adh* regions of *D. mojavensis* and *D. mulleri* and found an unexpectedly high degree of sequence conservation. This is reflected in the introns and flanking regions of the pseudogenes having a higher degree of similarity than the introns in either coding genes. The nonsynonymous substitution rates are similar to those in the coding genes in interspecific comparisons and the overall similarity of the pseudogenes to each other is greater than that of the coding genes of these species. The functional reasons for this unexpectedly high sequence similarity is unknown, but this should serve as a caution against a simple assumption of lack of function of these regions.

Finally, we comment briefly on the issue of rates of nucleotide substitutions in the genus *Drosophila*. If the rate of substitution can be determined, then the phylogenetic relationships determined through nucleotide sequence comparisons can be put into a time frame. In addition, a full understanding of the mechanisms of evolution will require knowledge of either average or maximal rates of nucleotide substitutions. In the genus *Drosophila* there is a minimal fossil record. Therefore it will remain difficult to determine absolute rates. Generally calculations of nucleotide substitution rates have used the estimate of 60 million years ago as the time of origin of the genus *Drosophila*. This is the estimate of Throckmorton (1975), which is based on biogeographic observations.

Several studies have used DNA–DNA reassociation studies to calculate the average rates of substitutions in single-copy DNA. Zwiebel *et al.* (1982) compared the thermal stabilities of *D. melanogaster* and six interspecific hybrid duplexes. They used a 40-million-year divergence time of *D. virilis* and *D. melanogaster,* assumed a 1% mismatch for 1°C lowering of the duplex melting temperature T_m, and estimated a rate of 0.66% of bases per million years. However, it is important to note that this is the rate for the most conserved sequences, since it was calculated using the 18.6% of the single-copy DNA which formed a stable duplex under the conditions used. The remaining 81.4% of the single-copy DNA did not form a duplex and presumably was diverging at a higher average rate. Powell *et al.* (1986) also used the change in T_m to estimate rates of divergence. They compared *D. melanogaster* with *D. yakuba* and argued that a change in melting temperature of 1°C should equal a 1.5–2.0% mismatch. The divergence time of *D. melanogaster* and *D. yakuba* is estimated from biogeographic observations to be between 2 and 6 million years ago. Therefore the estimates of rates of sequence divergence derived from the data of Powell *et al.* (1986) are at least twofold higher than the estimates of Zwiebel *et al.* (1982). In addition, Zwiebel *et al.* (1982) noticed that in DNA reassociation studies between *D. melanogaster* and its close relatives *D. simulans* and *D. mauritiana* a substantial fraction, 35%, of the single-copy DNA failed to hybridize. This is a surprisingly high value considering the close relationship of these species. Zwiebel *et al.* (1982) suggested that two different evolutionary processes were operating, one the normal accumulation of mutations, the other a rapid process leading to failure of DNA reassociation between DNA of these species. The data we have summarized here with respect to the *Adh* gene indicate that it is very difficult to identify a class of sequence that is not under some selective constraints. Furthermore, Cohn *et al.* (1984) have noted that the *Adh* gene (coding sequence plus intron regions plus some flanking region) is diverging at a rate approximately equivalent to the slowly diverging component of the single-copy DNA. Consequently, the rapidly diverging class of sequences (Zwiebel *et al.*, 1982) may represent sequences under minimal selective constraints having substitution rates which approach the neutral rate. The more slowly diverging class, e.g., *Adh,* may be those sequences whose divergence is slower due to natural selection. This suggests, as have other reports, that the neutral mutation rate in *Drosophila* may be relatively high.

In all likelihood, studies on the molecular evolution of *Adh* and other genes in *Drosophila* in the future will be of great significance for understanding the details of evolutionary events. There are two areas in particular that deserve attention. First, both we in this work and many other workers, have compared the sequence of a gene in a species and then deemed that sequence as representative of that species. This is of course an oversimplification. The value of the studies of Kreitman (1983) cannot be overemphasized in this regard. He has sequenced 11 alleles from defined populations and thereby provided a far more detailed view of *D. melanogaster Adh* than we have for any other species. Second, it will be necessary to have comparative sequence data from several different genes and gene classes from sets of species from throughout the genus. If there are meaningful differences between genes, they must be put into the perspective of a set of genes. With the continued improvement of technology, particularly including polymerase chain reaction techniques and other rapid DNA sequencing techniques, it will be a manageable task to obtain this information.

Acknowledgment

Research from our laboratory reported here was supported by NIH grant GM 31857.

REFERENCES

Ashburner, M., Bodmer, M., and Lemeunier, J., 1984, On the evolutionary relationships of *D. melanogaster, Dev. Genet.* **4:**295–312.

Atkinson, P. W., Mills, L. E., Starmer, W. T., and Sullivan, D. T., 1988, Structure and evolution of the *Adh* genes of *Drosophila mojavensis, Genetics* **120:**713–723.

Batterham, P., Starmer, W. T., and Sullivan, D. T., 1982, Biochemical genetics of the alcohol longevity response of *Drosophila mojavensis*, in: *Ecological Genetics and Evolution* (J. S. F. Barker and W. T. Starmer eds.), pp. 307–321, Academic Press, New York.

Batterham, P., Lovett, J., Starmer, W. T., and Sullivan, D., 1983, Differential regulation of duplicate alcohol dehydrogenase genes in *D. mojavensis, Dev. Biol.* **96:**346–354.

Batterham, P., Chambers, G. K., Starmer, W. T., and Sullivan, D. T., 1984, Origin and expression of an alcohol dehydrogenase gene duplication in the genus *Drosophila, Evolution* **38:**644–657.

Benyajati, C., Spoerel, H., Haymerle, H., and Ashburner, M., 1983, The messenger RNA for *Adh* in *D. melanogaster* differs in its 5′ end in different developmental stages, *Cell* **33:**125–133.

Beverly, S. M., and Wilson, A. C., 1982, Molecular evolution in *Drosophila* and higher diptera I. Microcomplement fixation studies of a larval hemolymph protein, *J. Mol. Evol.* 18:251–264.

Bodmer, M., and Ashburner, M., 1984, Conservation and change in the DNA sequences coding for *Adh* in sibling species of *Drosophila, Nature* **304:**425–430.

Brennan, M. D., and Dickinson, W. J., 1988, Complex developmental regulation of the *Drosophila affinidisjuncta* alcohol dehydrogenase gene in *Drosophila melanogaster, Dev. Biol.* **125:**64–74.

Brennan, M. D., Wu, C. -Y, and Berry, A. S., 1988, Tissue-specific regulatory differences for the alcohol dehydrogenase genes of Hawaiian *Drosophila* are conserved in *Drosophila melanogaster* transformants, *Proc. Natl. Acad. Sci. USA* **85:**6866–6869.

Chou, P. V., and Fassman, G. D., 1978, Prediction of the secondary structure of proteins from their amino acid sequence, *Adv. Enzymol.* **47:**45–148.

Cohn, V. H., and Moore, G. P., 1988, Organization and evolution of the alcohol dehydrogenase gene in *Drosophila, Mol. Biol. Evol.* **5:**154–166.

Cohn, V., Thompson, M., and Moore, G., 1984, Nucleotide sequence comparison of the *Adh* gene in three drosophilids, *J. Mol. Evol.* **20:**31–37.

Coyne, J. A., and Kreitman, M., 1986, Evolutionary genetics of two sibling species, *Drosophila simulans* and *D. sechellia, Evolution* **40:**673–691.

Dickinson, W. J., 1980, Evolution of patterns of gene expression in Hawaiian picture-winged *Drosophila, J. Mol. Evol.* **16:**73–94.

Dickinson, W. J., and Carson, H. L., 1979, Regulation of the tissue specificity of an enzyme by a *cis*-acting genetic element: Evidence from *Drosophila* hybrids, *Proc. Natl. Acad. Sci. USA* **76:**4559–4562.

Dickinson, W. J., Rowan, R., and Brennan, M., 1984, Regulatory gene evolution: Adaptive differences in expression of alcohol dehydrogenase in *D. melanogaster* and *D. simulans, Heredity* **52:**215–225.

Fischer, J., and Maniatis, T., 1985, Structure and transcription of the *Drosophila mulleri* alcohol dehydrogenase genes, *Nucleic Acids Res.* **13:**6899–6917.

Fischer, J. A., and Maniatis, T., 1986, Regulatory elements involved in *Drosophila Adh* gene expression are conserved in different species and separate elements mediate expression in different tissues, *EMBO J.* **5:**1275–1289.

Fischer, J., and Maniatis, T., 1988, Drosophila *Adh:* A promoter element expands the tissue specificity of an enhancer, *Cell* **53:**451–461.

Fitch, W. M., and Margoliash, E., 1967, Construction of phylogenetic trees, *Science* **155:**279–284.

Gibson, J., and Oakeshott, J., 1982, Tests of the adaptive significance of the alcohol dehydrogenase polymorphism in *Drosophila melanogaster:* Paths, pitfalls and prospects, in: *Ecological Genetics and Evolution* (J. S. F. Barker and W. T. Starmer, eds.), pp. 291–306, Academic Press, New York.

Goldberg, D., Posakony, J., and Maniatis, T., 1983, Correct developmental expression of a cloned alcohol dehydrogenase gene transduced into the *Drosophila* germ line, *Cell* **34:**59–73.

Hedrick, P. W., and McDonald, J. F., 1980, Regulatory gene adaptation: An evolutionary model, *Heredity* **45:**83–97.

Jornvall, H., 1977, Structural and functional changes in different alcohol dehydrogenase during evolution, in: *Alcohol and Aldehyde Metabolizing Systems Enzymology and Subcellular Organelles* (R. G. Thurman, J. R. Williamson, H. R. Drott, and B. Chance, eds.), pp. 145–155, Academic Press, New York.

Kreitman, M., 1983, Nucleotide polymorphism at the alcohol dehydrogenase locus of *D. melanogaster, Nature* **304:**412–417.

Kyte, J., and Doolittle, R. F., 1982, A simple method for displaying the hydropathic character of a protein, *J. Mol. Biol.* **157:**105–132.

Laurie, C. C., and Stam, L. F., 1988, Quantitative analysis of RNA produced by Slow and Fast alleles of *Adh* in *Drosophila melanogaster,* **85:**5161–5165.

Laurie-Ahlberg, C. C., and Stam, L. F., 1987, Use of P-element mediated transformation to identify the molecular basis of naturally occurring variants affecting *Adh* expression in *Drosophila melanogaster, Genetics* **115:**129–140.

Li, W.-H., Gojobori, T., and Nei, M., 1981, Pseudogenes as a paradigm of neutral evolution, *Nature* **292:**237–239.

Li, W.-H., Luo, C. -C., and Wu, C. -I., 1986, Evolution of DNA sequences, in: *Molecular Evolutionary Genetics* (R. J. MacIntyre, ed.), pp. 1–94, Plenum Press, New York.

MacIntyre, R. J., 1982, Regulatory genes and adaptation: Past, present, and future, *Evol. Biol.* **15:**247–285.

Martin, P., Martin, A., Osmani, A., and Sofer, W., 1986, A transient expression assay for tissue-specific gene expression of alcohol dehydrogenase in *Drosophila, Dev. Biol.* **117:**547–580.

McDonald, J., and Avise, J., 1976, Evidence for the adaptive significance of enzyme activity levels: Interspecific variation in alpha-GPDH and ADH in *Drosophila, Biochem. Genet.* **14:**347–355.

Mills, L. E., Batterham, P., Alegre, J., Starmer, W. T., and Sullivan, D. T., 1986, Molecular genetic characterization of a locus that contains duplicate *Adh* genes in *Drosophila mojavensis* and related species, *Genetics* **112:**295–310.

Oakeshott, J. G., Chambers, G., East, P., Gibson, J. B., and Barker, J., 1982*a,* Evidence for a genetic duplication involving *Adh* in *D. buzzatii* and related species, *Aust. J. Biol. Sci.* **35:**73–84.

Oakeshott, J. G., Gibson, J. B., Anderson, P. R., Knibb, W. R., Anderson, D. G., and Chambers, G. K., 1982*b,* Alcohol dehydrogenase and glycerol-3-phosphate dehydrogenase clines in *Drosophila melanogaster* on different continents, *Evolution* **36:**86–96.

Place, A. R., Anderson, S. M., and Sofer, W., 1986, Introns and domain-coding regions in the dehydrogenase genes, in: *Multidomain Proteins—Structure and Evolution* (D. G. Hardie and J. R. Coggins, eds.), pp. 175–194, Elsevier, Amsterdam.

Posakony, J. W., Fischer, J. A., and Maniatis, T., 1985, Identification of DNA sequences required for the regulation of *Drosophila* alcohol dehydrogenase gene expression, *Cold Spring Harbor Symp. Quant. Biol.* **50:**515–520.

Powell, J. R., Caccone, A., Amato, G. D., and Yoon, C., 1986, Rates of nucleotide substituion in *Drosophila* mitochondrial DNA and nuclear DNA are similar, *Proc. Natl. Acad. Sci. USA* **83:**9090–9093.

Rabinow, L., and Dickinson, W. J., 1981, A *cis*-acting regulator of enzyme tissue specificity in *Drosophila* is expressed at the RNA level, *Mol. Gen. Genet.* **183:**264–269.

Rabinow, L., and Dickinson, W. J., 1986, Complex *cis*-acting regulators and locus structure of *Drosophila* tissue ADH variants, *Genetics* **112:**523–537.

Rowan, R., and Dickinson, W. J., 1986, Two alternate transcripts coding for alcohol dehydrogenase accumulate with different development specificities in different species of picture-winged *Drosophila, Genetics* **114:**435–452.

Rowan, R. G., and Dickinson, W. J., 1988, Nucleotide sequence of the genomic region coding for alcohol dehydrogenase in *Drosophila affinidisjuncta, J. Mol. Evol.* **28:**43–54.

Rowan, R. G., Brennan, M. D., and Dickinson, W. J., 1986, Developmentally regulated RNA transcripts coding for alcohol dehydrogenase in *Drosophila affinidisjuncta, Genetics* **114:**405–434.

Rubin, G. M., and Spradling, A. C., 1982, Genetic transformation of *Drosophila* with transposable element vectors, *Science* **218:**348–353.

Savakis, C., Ashburner, M., and Willis, J. H., 1986, The expression of the gene coding for alcohol dehydrogenase during the development of *Drosophila melanogaster, Dev. Biol.* **114:**194–207.

Schaeffer, S. W., and Aquadro, C. F., 1987, Nucleotide sequence of the *Adh* gene regions of *Drosophila pseudoobscura:* Evolutionary change and evidence for an ancient gene duplication, *Genetics* **117:**61–73.

Shen, N. L. L., Subrahmanyam, G., Clark, W., Martin, P., and Sofer, W., 1989, Analysis of *Adh* gene regulation in *Drosophila:* Studies using somatic transformation, *Dev. Genet.,* In press.

Sofer, W., and Martin, P., 1987, Analysis of alcohol dehydrogenase gene expression in *Drosophila, Annu. Rev. Genet.* **21:**227–236.

Starmer, W. T., and Sullivan, D. T., 1989, A shift in the third-codon-position nucleotide frequency in alcohol dehydrogenase genes in the genus *Drosophila, Mol. Biol. Evol.,* In press.

Stephens, J. C., and Nei, M., 1985, Phylogenetic and analysis of polymorphic DNA sequences at the *Adh* locus in *Drosophila melanogaster* and its sibling species, *J. Mol. Evol.* **221:**289–300.

Thatcher, D. R., and Sawyer, L., 1980, The complete amino-acid sequence of 3 alcohol dehydrogenase EC-1.1.1.1 allele enzymes (Adh^{n11}, Adh^{S} and Adh^{F}) from the fruit fly *Drosophila melanogaster, Biochem. J.* **187:**875–886.

Throckmorton, L., 1975, The phylogeny, ecology, and geography of *Drosophila,* in: *Handbook of Genetics* (R. C. King, ed.), Vol. 3, pp. 421–469, Plenum, New York.

VanDelden, W., 1982, The alcohol dehydrogenase polymorphism in *Drosophila melanogaster:* Selection at an enzyme locus, *Evol. Biol.* **15:**187–222.

Wasserman, M., 1982, Evolution of the repleta group, in: *Genetics and Biology of Drosophila,* Vol. 3b (M. Ashburner, H. L. Carson, and J. N. Thompson, ed.), pp. 61–139, Academic Press, New York.

Wilbur, W. J., and Lipman, D. J., 1983, Rapid similarity searches of nucleic acid and protein data banks, *Proc. Natl. Acad. Sci. USA* **80:**726–730.

Wilson, A. C., Carlson, S. S., and White, T. J., 1977, Biochemical evolution, *Annu. Rev. Biochem.* **46:**573–649.

Zwiebel, L., Cohn, V., Wright, D., and Moore, G., 1982, Evolution of single-copy DNA and the *Adh* gene in seven Drosophilids, *J. Mol. Evol.* **19:**62–71.

5

Population Genetics of the Polymorphic Meadow Spittlebug, *Philaenus spumarius* (L.)

OLLI HALKKA and LIISA HALKKA

INTRODUCTION

The number of abundant and widespread insect species is small. Onc of the most ubiquitous of such insects is the meadow spittlebug or cuckoo spit insect, *Philaenus spumarius* (L.) (Homoptera, Aphrophoridae). This insect is found throughout Europe and occurs in Asia from the Urals to Japan betwccn parallels 45° N and 65° N. In North America, it has two areas of distribution, one from British Columbia to California, and a larger one from the easternmost prairies to the Atlantic coast (Weaver and King, 1954). It has also been found in North Africa and South America. In 1960 it was recorded from New Zealand for the first time (Archibald *et al.*, 1979; Lees, 1988) and appears to be spreading there. From the most arctic, alpine, and arid parts of Eurasia and North America, *Philaenus* has not been found.

Philaenus has been recorded from a height of over 3000 m in the mountains of Afghanistan (Dlabola, 1957). It is found in northern Europe as far north as 69° 50′ N (Halkka *et al.*, 1974*b*). These two extremes show that cold climate, if moist enough, does not deter *Philaenus*. But in Finnish Lapland the nymph stage may last up to mid-August, long enough for nymphs to be killed in the first autumn frosts (M. Raatikainen, personal communication). It seems that the sum

---

OLLI HALKKA and LIISA HALKKA • Department of Genetics and Tvärminne Zoological Station, University of Helsinki, SF-00100 Helsinki, Finland.

of 700–800 daydegrees (above +5°C) during the growth season sets the limit in the Arctic and at high elevations.

Weaver and King (1954, p. 61) stated that "Humidity is perhaps the most important factor determining survival of the spittlebug in all stages of its existence." *Philaenus* is able to live only in a climate in which evaporation from the spittle masses does not exceed the capacity of the nymphs to produce spittle components.

Correspondingly, in North America the prairies and other arid central parts are free of *Philaenus,* both in the United States and Canada. Similarly, in the southeastern Soviet Union and in the western Asiatic part of the Soviet Union, the distribution limit of *Philaenus* rims the northern limit of the steppe area.

In moderate or highly humid parts of Europe, Asia, and North America, the spittlebug often occurs at very high densities. Evans (1975) noted that a dense *Philaenus* population distorted the pyramid of numbers in an abandoned field in Michigan.

In North America, from alfalfa fields nymph densities as high as 2550 nymphs/m^2 are on record (Wiegert, 1964). Prior to the introduction of effective insecticides and resistant alfalfa cultivars (Nielson and Lehman, 1980) *Philaenus* was a pest of considerable economic importance (Weaver and King, 1954). In Europe, fourth-instar nymph densities generally remain under 1000 nymphs/m^2 even in the best of the meadows. On strawberries, *Philaenus* still is a yield pest in Northern Europe and North America (Zajac and Wilson, 1984).

The extensiveness of the distribution area in three continents, the obviously rapid spreading in New Zealand after introduction, and the high densities in the most suitable habitats in Europe and the United States are characteristics of a successful species. Not very many species of insect about 6–7 mm long can beat *Philaenus* in the production of total global biomass.

Common and abundant species often show greater versatility and flexibility in filling available niches than species which are rare and sparse. The present chapter attempts to discuss causal connections, if any, between the *color polymorphism* globally present in spittlebug populations and the success of the populations.

MATERIAL BASIS AND METHODS OF *PHILAENUS* RESEARCH

Color morph frequencies in *Philaenus* populations are on record from ten European countries, from some parts of Asia, particularly Siberia, from the United States and Canada, and from New Zealand. The total number of adult individuals investigated is well over one-half million. From Tvärminne, Finland, alone, about 260,000 individuals have been analyzed in the years 1956 and 1960–88 with regard to sex and color morph.

In Tvärminne the field methods used in *Philaenus* research are such that interference with the normal life of the populations is minimized. All the research with isolated island populations was done in such a way that the individuals were captured, their sex and color morph recorded in the field, and then all the individuals released upon the plants.

In taking sweep net samples from isolated island populations in Tvärminne it is possible to capture more than 50% of the total population from all but the largest of the populations. From a population consisting of about 7000 individuals, more than 2000 individuals (about 32%) can easily be captured.

In some of the Tvärminne islands, miniature cages were used for isolating nymphs or groups of nymphs within the spittle masses. The largest diameter of these circular, flat-bottomed, flat-topped boxes is only about 50–60 mm. The plant stem on both sides of the spittle mass was pressed between the two halves of the cage with the help of string clips (Halkka *et al.*, 1967*a*, 1970). The cages were opened after emergence into adults of the nymphs. After sex and color morph were recorded, the adults were released into the meadow.

From other populations than those living in islands in the Tvärminne area, samples of killed individuals were taken to the laboratory for sex and color morph identification.

The miniature cages were used in crossing experiments. Isolated virgin females and males known to be homozygous p^t/p^t individuals (see the section "Genetics of the dorsal color polymorphism") were used in first parental crosses. The parents were put on red clover enclosed in a nylon-voilé cage. The eggs were laid on red clover, overwintered in the cages in natural conditions, and developed next summer on the same or a new plant to nymphs and adults. The emerging specimens were daily collected from the cage, determined for their sex and morph, and isolated in miniature cages. Specimens with known phenotypes and genotypes were then used for further controlled crosses. A continuous series of crossing experiments was performed in the years 1969–1982. About 80% of the crossings were successful, and produced from 1 to 96 offspring from each mated pair, the mean being 22.

ECOLOGY

Life Cycle and Natural Enemies

Philaenus appears to have univoltine life cycle in all parts of its area of distribution. In northern Europe, the phenology of this hemimetabolic insect seldom deviates from the following timetable:

Egg	Late August to mid-June
Nymph	Late May to late July
Adult	Late June to October

The spittlebug thus invariably overwinters at the egg stage, resides at the nymph stage within the protective spittle mass over many weeks, and at the adult stage has a preoviposition period longer than 1 month. In Central Europe and over most of North America, the non-egg part of the life cycle is 1–3 months longer than in northern Europe.

Witsack (1973) noted 178 eggs per female in greenhouse rearings in Germany. In Finland, the maximal progeny from a pair reared on red clover in a greenhouse numbered 96 adult individuals. In natural conditions, egg progenies probably are not very different from this, but adult progenies from a single female are much smaller.

Philaenus, being a xylem feeder, may in high densities cause considerable losses in the primary production of field layer vegetation (Wiegert, 1964). In normal weather a nymph may ingest 200 times its own weight of plant sap in 24 hr (Horsfield, 1978). In natural meadows the supporting capacity of the habitat is not strained in normal years, but dry periods causing some withering of the plants may change the situation dramatically. *Philaenus* nymphs try hopelessly to find enough xylem for forming the protective spittle mass and die while still attached to the plants. Survival of the nymphs is strongly positively correlated with air and soil moisture (Halkka and Halkka, 1979; A. Morikka and O. Halkka, in preparation).

An egg just about to hatch and newly hatched nymphs are probably the most precarious phases of the life cycle of *Philaenus*. Dry weather, spring night frosts, and probably numerous site-specific adverse conditions may kill terminal-stage eggs and first- and second-instar nymphs. Total mortality from oviposition to the adult stage is about 80–98% in stable *K*-selected populations.

Some of the enemies of the spittlebug are able to pick up nymphs from the spittle mass, and nymphs changing sucking site on the plant are very susceptible to predation. Invertebrate enemies of *Philaenus* are on record from the Arachnida, Coleoptera, Hymenoptera, and Heteroptera, and in some populations, parasitizing Pipunculidae (Diptera) may cause considerable losses (Harper and Whittaker, 1976; Whittaker, 1969). The vertebrate enemies range from frogs to shrews. In the United States, some Passerine birds are important as predators (Evans, 1964; Wiegert, 1964). In Europe, the prime bird predators have yet to be found (Halkka and Kohila, 1976). The color morphs TYP and TRI appear to be cryptic against some types of background (Hutchinson, 1963; Thompson, 1973), but the distinctly black-and-whitish morphs may show warning coloration (Thompson, 1973; Gibson, 1980) (see Fig. 1).

Food Plants

Thousands of species of dicotyledons and a number of monocotyledons have been noted as suitable as food plants for the nymphs in Europe and the United States (Halkka *et al.*, 1967*a;* Weaver and King, 1954). From Tvärminne, Finland, 46,952 nymph–food plant associations are on record from the years 1969–1978. Nymphs were observed on 71 different food plants, and the following 7 species were used most frequently. *Filipendula ulmaria* (FU) (28.6%), *Lysimachia vulgaris* (LV) (18.6%), *Solidago virgaurea* (SV) (15.8%), *Lythrum salicaria* (LS) (15.6%), *Potentilla palustris* (PP) (4.9%), *Chrysanthemum vulgare* (CV) (5.2%), and *Tripleurospermum maritimum* (TM) (3.6%). The remaining 7.7% was shared by 64 different species of plants. But *Philaenus* is not an indiscriminating polyphag: *Veronica longifolia* and *Vicia cracca* grow on many of the Tvärminne islands, but practically never are used as food plants.

The different food plants are used to a greatly variable extent in different years. For instance, 40.3% of the nymphs lived on FU in 1972, but only 8.6% in 1976; only 2.9% of them on SV in 1972, but 30.5% in 1976. During the period 1969–1978, there was a negative correlation between nymph frequencies on FU and SV. In years of high nymph survival (into adults), the percentage of nymphs was high on FU but low on SV, and vice versa in years of low nymph survival. The years of high nymph survival coincided with normal or high humidity of the air and soil in June.

As mininiches or ''grains'' of the environment, the food plants are very dissimilar, and every one of the food plant species depends heavily upon soil moisture in its ability to produce xylem for nymphs. In the years 1969–1978 the survival percentage of nymphs varied between dry and humid years from 8 to 97% on *Solidago virgaurea,* from 20 to 95% on *Filipendula ulmaria,* and from 55 to 98% on *Lysimachia vulgaris. Lysimachia* grows in those parts of the meadows which during droughts are relatively well able to retain soil moisture (Halkka and Mikkola, 1977).

There are thus large weather-dependent differences among food plants and consequently among the different parts of the meadows in carrying capacity for *K*-selection. The different color morphs (Fig. 1) are not all neutral with regard to this variability. The morph LAT appears to live disproportionately often on *Filipendula ulmaria,* 34% of the 407 LAT adults from 1969–1973 having been recorded from this plant species. On the other hand, only 10% of the LAT individuals were found on *Lysimachia vulgaris* (Halkka and Mikkola, 1977). Conversely, 9% of the 212 TRI individuals were recorded from *Filipendula,* but 26% of them from *Lysimachia.* The color morph TYP appears neutral with regard to food plants, and the morph MAR nearly so, but FLA, LCE, and LOP, in addition to LAT and TRI, ''favor'' some of the plants and ''disfavor'' others.

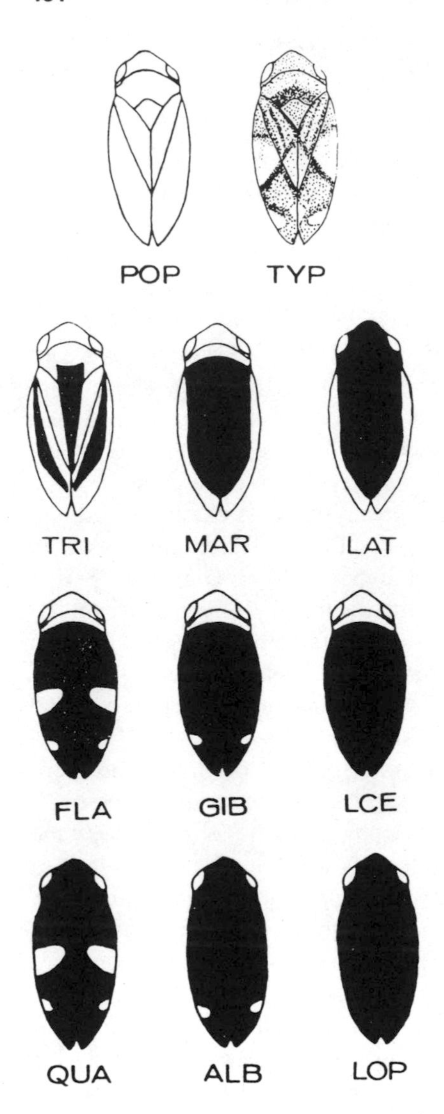

FIG. 1. The north European color morphs of *Philaenus spumarius*. The full names of the morphs are given in the text.

Over periods lasting 3–4 years, significant χ^2 values in the sense of extensive or restricted use of a food plant species by a color morph were obtained for many of the color morph–food plant combinations in 2 × 2 settings. In each 2 × 2 setting, the number of individuals of color morph *x* on food plant *y* and on all other food plants is compared with the number of all other color morphs on *y* and on all other food plants. With irregular intervals, an exceptionally hot and dry or cool and wet year often ensues, increases nymph mortality differentially on different food plants, and breaks down the significance of the χ^2 values in many of the meadow habitats.

The uneven occurrence of the different color morphs on the most important food plants may be due to plant physiological, general edaphic, or microclimatic variability. A color morph may live disproportionately often on a certain food plant because the plant grows in a microclimatic or edaphic environment suitable for the morph. Whatever the reason for the recorded uneven distribution, such a distribution may give a key for understanding dissimilar morph frequencies in different types of meadows. The very complex, spatially and temporally variable relationship between color morphs and their food plants will be treated in another context (A. Morikka and O. Halkka, in preparation).

Coexistence with Other Aphrophoridae Species

All over the world, *Philaenus* occupies habitats which are suitable for one or more other members of the family Aphrophoridae. Rules of coexistence between the different species have been investigated in Finland (Halkka *et al.*, 1977) and the United States (McEvoy, 1986). There is niche overlap, to the extent that not very uncommonly *Philaenus* is found in the spittle mass together with one, two, or three other species. This is one form of protocooperation, each one of the two or three species saving energy at the expense of another species in producing the protective spittle mass.

In the vast majority of the spittle masses, nymphs of a single species are found. The average number of *Philaenus spumarius* (PS) nymphs in one spittle clump is about 1.7, but on *Angelica archangelica,* up to 70 nymphs may live within the same spittle clump. PS shows a tendency to settle higher on the food plant than the commonest companion species, *Lepyronia coleoptrata* (LC). Calculated with the help of the Pianka (1974) formula, niche overlap in vertical distribution in 1978 on *Filipendula ulmaria* between PS and LC was 0.19, on *Galium palustre* 0.91, and on *Potentilla anserina* 1.00. PS also is much less often found on monocotyledons than is LC. These tendencies reduce niche overlap between PS and LC, which varied between the limits 0.60–0.73 with regard to food plant use in 1975–1978. Another common coexisting species, *Neophilaenus lineatus* (NL), lives almost exclusively on monocots, and thus there was only a slight food plant niche overlap, 0.01–0.05 in the years 1975–1978 between PS an NL. It seems that the Aphrophoridae have during their evolution diverged and specialized in using partly different species of food plants and different sections of the stems of the same food plants. During the growth of the nymphs, the food plants grow taller, and probably there is practically no competition between the species, in spite of considerable phenological overlap of adults: between any two of the three species 0.81–0.99 in 1977, 0.90–0.98 in

1978, and 0.93–0.97 in 1979. Although phenological overlap between nymphs was not studied, it cannot be very much less than that of the adults. Mainly due to different modes of polyphagy, resource utilization is maximized and niche overlap minimized between the coexisting species (Halkka *et al.*, 1977).

Philaenus appears to be the most effective colonizer among the Aphrophoridae. In Tvärminne archipelago at least, in 90 islands or tiny islets with often rather dense *Philaenus* populations, the other Aphrophoridae are very sparse or absent.

VISUAL POLYMORPHISM

Color Polymorphisms of *Philaenus spumarius*

As late as the 19th century, the color morphs of *Philaenus* were often considered each to have the rank of a species, but Haupt (1917) and some earlier students established their true status. Several authors published lists of morph frequencies in the 1930s to 1950s (Halkka, 1962*b*). These papers often dealt mainly with the noxiousness of *Philaenus* as a pest on alfalfa and other plants.

In 1962, color polymorphic equilibria were described from Finnish (Halkka, 1962*a*) and North American (Owen and Wiegert, 1962) populations. Hutchinson (1963) compared samples taken in 1920 and 1963 from the same meadow in Cherryhinton, England. In these three papers it was indicated that natural selection was probably responsible for the polymorphic balances observed.

The most intensively investigated of three visual polymorphisms on record from *Philaenus* has been the *dorsal color polymorphism*. In addition to the variability of the dorsal side pigmentation, the spittlebug is also variable with regard to *coloration of abdominal pleurites* (white, black, carrot red) (unpublished observations; Beregovoy, 1970) and *figuration of frontoclypeus* (Svala and Halkka, 1974).

In Europe, at least 15 different dorsal color morphs have been described. The morph *typicus* (TYP) is the most common both in females and males. The very pale morph *populi* (POP) represents a weak (underdeveloped) expression of TYP pattern. Another morph found in both sexes is the *trilineatus* (TRI). The rest of the melanic phenotypes are regularly found only in females (occasionally in males); *marginellus* (MAR), *lateralis* (LAT), *flavicollis* (FLA), *gibbus* (GIB), *leucocephalus* (LCE), *quadrimaculatus* (QUA), *albomaculatus* (ALB), and *leucophthalmus* (LOP). These 11 morphs (Fig. 1) found in mainland Finland are determined by seven different pigmentation alleles and by a small group of modifiers (Halkka *et al.*, 1973, 1975*a;* see below). In addition to these, the morphs *praeustus* (PRA), *vittatus* (VIT), and *ustulatus* (UST) are found in more southern parts of Europe and sporadically in North America. The three-letter codes of color morphs given above are used throughout the text.

Some of the color morphs of *Philaenus* are endemic to a small geographic region. The morphs *melanocephalus* and *hexamaculatus* are known only from Tatkul in the southern Ural region of the USSR (Beregovoy, 1970, 1972).

Genetics of the Dorsal Color Polymorphism

Direct evidence for genetic determination of dorsal pigment distribution comes from crosses started with parentals taken from Finnish populations and continued in some lines to the F11 generation. Crosses between and within color morphs have produced about 9800 individuals from about 440 controlled pairs. The results show that the genes determining the dorsal pigmentation are alleles of the pigmentation locus *p*. The main results of our crossing experiments are as follows (Halkka *et al.*, 1973, 1975*a;* L. Halkka, in preparation) (see also Figs. 1 and 2):

1. The allele p^T for the phenotype *trilineatus* (TRI) is top dominant in both sexes.

2. The other alleles for melanic pigmentation are the p^M for *marginellus* (MAR), p^L for *lateralis* (LAT), p^F for *flavicollis* (FLA), p^C for *leucocephalus* (LCE), and p^O for *leucophthalmus* (LOP). In females, these alleles constitute a dominance hierarchy. The alleles producing white areas in head and/or elytral margins are dominant over the alleles producing black pigment in corresponding areas. When the head/elytral white patterns are brought together by different alleles, these alleles are codominant.

The phenotype MAR can be produced by the true gene p^M or by the codominant combination of alleles p^L and p^C in the heterozygote p^L/p^C. The rare phenotype MAR–FLA is expressed by the heterozygotes p^M/p^F and p^L/p^F.

3. None of the alleles has been shown to be lethal when homozygous, but some p^O/p^O homozygotes may be subvital.

4. The allele p^t for *typicus* (TYP) has a specific position among the *p* locus alleles. In the females, p^t is the bottom recessive out of the seven pigmentation alleles.

It has been found that in females the young melanic morphs are first green, then develop a faint TYP pattern, and then, in about 1 hr, *take another color pattern*. This is true even in case of double melanic heterozygotes (e.g., p^C/p^L).

This can be understood only by taking the true TYP pattern as determined by genes in another locus or other loci outside the pigmentation locus *p*. In females, the presence of one or more melanic alleles in locus *p* suppresses the TYP pattern. Only the homozygous condition of the bottom recessive "zero allele" p^t/p^t allows the TYP phenotype to become visible.

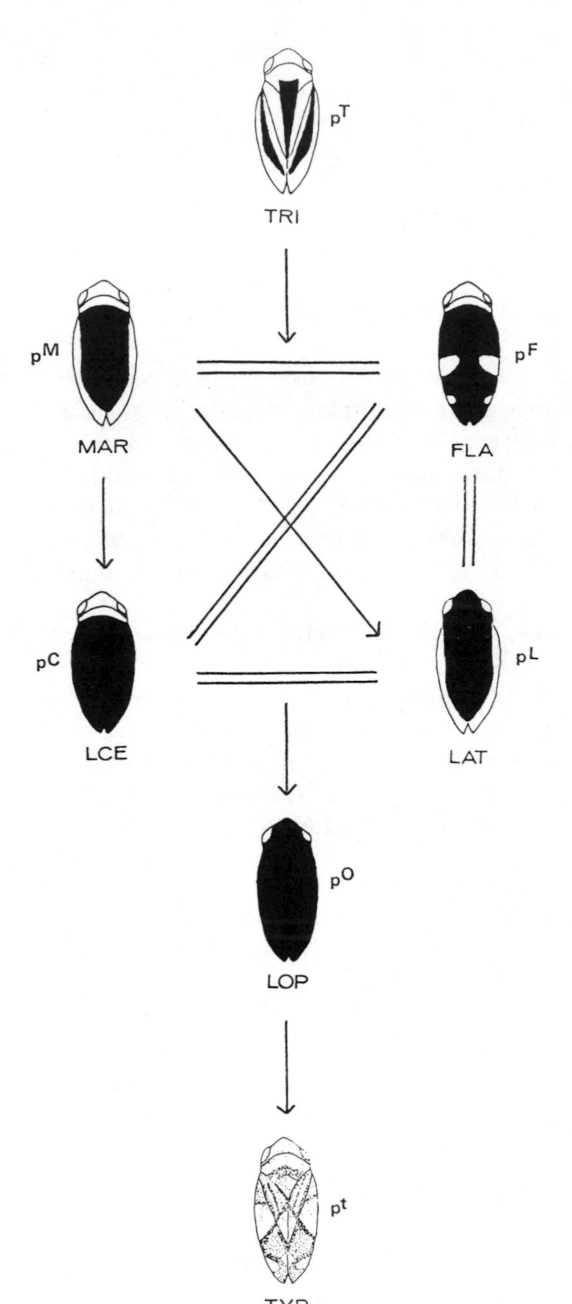

FIG. 2. A scheme of the dominance and codominance relationships in the color allelomorph system. Then letter "p" denotes "pigmentation."

5. In males p^t is dominant over all other alleles but the p^T. Even double melanic heterozygotes (e.g., p^C/p^L) have TYP phenotype. In males some kind of sex-bound switch mechanism makes the TYP-pattern genes epistatic over the pigmentation locus *p* alleles (except for its allele p^T). The homozygotes of the melanic alleles p^C or p^O, however, are often (but not always) black in males also.

6. Phenotypes from a very pale morph *populi* (POP) through a usual *typicus* to a quite dark form *xantocephalus* ("almost FLA") seem to belong in the same (quantitative) variation range of TYP pattern.

7. The number of white dots (zero, two, or four dots) in black elytreal margins depends most probably on modifier genes lying outside the *p* locus. Thus, the melanic allele p^C can be expressed as *leucocephalus* (LCE, black wings), *gibbus* (GIB, with two dots in wings), or *flavicollis* (FLA, four dots in wings). Correspondingly, the allele p^O can be expressed as *leucophthalmus* (LOP, black wings), *albomaculatus* (ALB, two dots) or *quadrimaculatus* (QUA, four dots). The melanic phenotype may stay in the four- or two-dot stage or darken through a series of four–two–zero dots in a few hours to a few days.

Recently, Stewart and Lees (1987, 1988) crossed individuals from Cynon Valley, Wales. In this heavily pigmented population with industrial melanism, many dark morphs present almost exclusively in the female in Northern Europe and North America occur rather frequently in males as well. Crossing experiments have shown that dominance hierarchy rules and morph frequencies are nearly the same in males as in females. It is not known if the same is true with individuals with less dark pigmentation from Wales, England, or Scotland.

In Tvärminne, Finland, the sex-bound expression of melanic phenotypes is "leaky" in some island populations. In a pair of populations rich in p^L and p^Oin Stora Västra Långgrundet, and in the population Grisselgrundet, atypical LAT and ALB males have been recorded, but at a very low frequency (unpublished data). These males may represent melanic homozygotes or double heterozygotes of melanic genes. In sweep net samples taken from northeastern Finland, LCE and LAT males were found [0.2% LCE in males and 10% in females in Kuusamo (Halkka *et al.*, 1967*b*, 1974*b*). These morphs were also found in the Ukraine (Halkka *et al.*, 1980) and in Italy together with MAR males (Raatikainen, 1971). In North America, MAR and LAT phenotypes were absent among well over 10,000 males collected by Thompson (1984) from three Canadian provinces and 23 American states, but a number of LCE-like males were found in this material.

In all probability, TYP is the original color morph of *Philaenus*. The basic TYP pattern is found in many other Aphrophorid spittlebugs as well (e.g., in the genera *Lepyronia* and *Aphrophora*).

The buildup of the gene complexes regulating the pigmentation of the dorsal side almost certainly dates back to times preceding the birth of the species *Philaenus spumarius*. This can be indirectly concluded from the fact that exactly the same color phenotypes are found in *P. signatus* Melichar, a Mediterranean species slightly larger than *P. spumarius* (Halkka and Lallukka, 1969). There are 11 morphs common to both species, a parallelism hardly resulting from convergent evolution, but probably due to the presence of all the pigmentation alleles in the ancient species which radiated into *P. spumarius* and *P. signatus*. This and similar instances from the snail genus *Cepaea* show that changes, even those leading to speciation, may take place in "strongly held polymorphisms" and thus these polymorphisms are not necessarily "end results of evolution," as supposed by Wright (1982).

The melanic morph LOP may be one of the oldest phenotypes in addition to TYP. Melanic mutants are known in many insects without complex polymorphism. Perhaps once in the history of *Philaenus* the populations consisted of no other morphs than TYP and LOP. The mutant gene, the gene p^O for LOP, is second from the bottom in the dominance hierarchy system.

It may be further imagined that in some body parts weak pigmentation failures developed as a result of mutations. These mutations gave clearer expression against the dark LOP pigmentation background than against the light TYP background. These pigmentation failures confused the searching image of predators, afforded apostatic protection for the species, and were gradually established as distinct color morphs such as LAT and LCE. Some morphs might have developed by fixing different melanic alleles in same supergene (MAR by combination of p^L and p^C and FLA by combination of p^C and a four-dot pattern).

It seems possible that color alleles high in the hierarchy developed later than those close to the bottom. This situation would accord nicely with the hypothesis presented by Sheppard (1958) that dominance in polymorphism evolves gradually, being intensified by disruptive or frequency-dependent selection.

In many North European populations the frequency of the TYP phenotype and thus of p^t/p^t genotype is 0.8 or even higher. The other color morphs are mostly heterozygotes carrying a melanic allele p^X and a TYP allele p^t. The probability of melanic double heterozygotes or homozygotes is very low in most natural populations. Taking the square root of the frequency of the phenotype TYP and halving the frequencies of other morphs give a fairly accurate measure of actual *p*-locus allele frequencies in such populations.

The fact that many natural populations contain FLA and MAR phenotypes with two different genetic backgrounds causes slight indeterminacy in the calculation of allele frequencies. The value of this source of error is calculable and of importance only in populations in which p^L and p^C are common. A third type of uncertainty is due to the recessiveness of the other melanic alleles in the presence of p^T, but this is also calculable.

In grouping *Philaenus* morphs, the *typicus* (TYP) group of morphs includes also the *populi* (POP); and the *leucocephalus* (LCE) group includes also the modified phenotypes *gibbus* (GIB) and *flavicollis* (FLA). The *leucophthalmus* (LOP) group includes also the modified phenotypes *albomaculatus* (ALB) and *quadrimaculatus* (QUA). The *trilineatus* (TRI), *marginellus* (MAR), and *lateralis* (LAT) generally do not show appreciable variability in phenotype expression.

Geographical Clines (South to North and West to East)

None of the pigmentation alleles except the allele for TYP, p^t, is present in all parts of the distribution area of *Philaenus*. TYP, sometimes accompanied by LAT, is the only morph in the easternmost populations of Siberia (Vilbaste, 1968; Whittaker, 1972). Among 761 specimens collected from many localities in Kamtschatka, Vilbaste (1980) found no other morphs than TYP.

The color morph TRI was found to be absent from practically all of the northernmost populations of Finnish, Swedish, and Norwegian Lapland (Halkka *et al.*, 1974*b*). In the apparently very newly established populations of New Zealand, no other morphs than TYP and LOP have been found (Archibald *et al.*, 1979; Thompson, 1984; Lees, 1988).

The absence of other morphs than two from New Zealand can be explained with the help of the founder principle. But the rarity of non-TYP morphs in easternmost Siberia and absence of TRI from northernmost Lapland represent the extreme ends of frequency decrease clines.

A large material, 37,660 individuals, was collected from Finland, Norway, and Sweden between the parallel 64° N and the Arctic Sea (Halkka *et al.*, 1974*b*). Grouping these populations into those north of the Arctic Circle (group N, 36 populations) and south of it (group S, 19 populations) results in the following morph frequency distribution (Table I).

The morph TYP is more frequent in group N than in S, but TRI, MAR, and LOP all are clearly more frequent in S than in N. The morphs LAT and LCE appear neutral with regard to latitude.

TABLE I. The Frequencies of the Color Morphs TYP, TRI, MAR, LAT, LCE, and LOP in Females and Males North (N) and South (S) of the Arctic Circle[a]

		TYP	TRI	MAR	LAT	LCE	LOP	Total sample size
♀	N	91.4	0.5	0.9	3.9	2.2	1.1	4887
	S	84.7	2.5	2.4	4.1	2.5	3.8	7675
♂	N	99.2	0.5	—	—	0.1	0.2	5096
	S	95.5	2.7	—	—	0.2	1.6	7749

[a]TYP is more, and TRI, MAR, and LOP, less common in the N than in the S populations.

The same rare morphs are found at low frequencies throughout northern Lapland (Halkka *et al.*, 1974*b*). Their rarity thus can only to a minor extent be due to the founder principle. The low frequency of alleles p^T, p^M, and p^O north of the Arctic Circle must be due to ecophysiological properties of heterozygotes for these three alleles. The frequency of the allele p^t is 0.96 in N and 0.92 in S, and thus the contribution of other heterozygotes than p^T/p^t, p^M/p^t, and p^O/p^t is negligible. The three genotypes appear to behave as if they were *biotypes*. Possibly there is tight linkage of these color alleles with alleles of a gene or group of genes (supergene) with ecophysiological effects.

Nothing is known about the possible genetic basis of such morphs as *vittatus* (VIT), *ustulatus* (UST), and *praeustus* (PRA). PRA occurs in Europe mainly south of latitude 60° N, VIT has an even more southern distribution, and UST is practically absent from countries north of West Germany, East Germany, and Poland. Possibly these morphs are due to alleles such that the heterozygotes in question are successful only in climatically and edaphically more favorable regions than Finland or northern Sweden. Starting with the morph least able to tolerate the harsh climate of northern Europe and ending with the most tolerant morphs, the order is: UST → VIT → PRA → TRI → MAR → other morphs → TYP.

In eastern Central Europe, the frequency of the morph LOP increases from northeastern Italy through Hungary and Czechoslovakia to Lithuania, Latvia, and Estonia. Along the same transect, the frequency of the morph MAR decreases clinally (Halkka *et al.*, 1975*d;* see also Raatikainen, 1971).

From the Estonian SSR through the Russian SSR, the Carelian Isthmus, and into southeastern Finland, a distinct decrease in the frequencies of the color morphs LCE and LOP takes place (Halkka *et al.*, 1976).

A clear west–east cline is found in the increase of LCE frequency from western parts of Swedish Lapland toward the easternmost parts of northern Finland (Halkka *et al.*, 1974*b*).

Thompson (1984) made an extensive survey of morph frequencies in North America. He noted that four of the melanic morphs were more common in northern than in the southern populations. Similar observations were made by Whittaker (1972) in Russia and central Siberia.

From the material collected by the six American or Canadian authors cited by Halkka (1962*b*), and more recently by Farish and Scudder (1967), Thompson and Halkka (1973), Boucelham *et al.* (1988), and Thompson (1984), it appears that the morphs LAT, LCE, and LOP are extremely rare in the United States east of the longitude 85° W. The morph MAR is low in frequency in Maine, New Hampshire, and Massachusetts, and gets more common toward the south, reaching frequencies of 5–10% in West Virginia and Maryland.

The clinal changes in color morph frequencies thus span much larger distances in the North American continent than in Europe. This is probably due to the fact that in North America continuous land masses are not interrupted by such isolating barriers as the Baltic Sea, the Carpathians, or the Alps.

Lowland to Highland Clines

Halkka *et al.* (1980) noted an increase in *leucophthalmus* frequency toward high altitudes in the Carpathian Mountains. In three Scottish glens, Berry and Willmer (1986) found the frequencies of the melanic morphs to be positively correlated with altitude, but there the same was true of the morph TYP as well. Honek (1984) says that ''in Czechoslovakia the melanism has obviously little to do with altitude.'' But taking the altitude 600 m and the frequency 15% of melanic morphs as borderlines dividing the 62 populations into four groups, the following is true. Beneath altitude 600 m, 11 populations have less and 13 populations more than 15% melanics. Above that altitude, 12 populations have less and 26 populations more than 15% melanics. A closer analysis is not easy, because Honek studied males and females separately in some but together in other populations.

The frequency of the color morph TRI correlates negatively with altitude in many lowland–highland transects. Berry and Willmer (1986) noted a statistically significant negative correlation of TRI frequency with altitude in Scotland. They were able to show that the reflectance of light from the dorsal surface of *trilineatus* was larger, but its ability to gain heat from light smaller, than light reflectance or heat gain of the melanics.

Honek (1984) omitted attempts to correlate TRI frequency with environmental variability in the 62 samples taken from various parts of Czechoslovakia. Fortunately, Honek recorded the altitudes of the habitats, which were found to range from 130 to 1200 m above sea level. Grouping the total material in the four altitude categories (1) 0–300 m, (2) 301–600 m, (3) 601–900 m, and (4) 901–1200 m makes it possible to compare TRI frequencies from different altitudes. Negative correlation of *trilineatus* frequency with altitude is evident: $\chi^2 = 137.8$, d.f. $= 9$, $p < 0.001$.

Boucelham and Raatikainen (1987) collected 33 samples along two transects from Finland over northern central Sweden into central Norway, crossing twice the Scandinavian range of mountains. The total material of 12,190 individuals was divided into altitude groups 0–100 m, 101–250 m, and over 250 m. The percentages of TRI individuals in the three groups were, respectively, 1.2%, 1.4%, and 0.6%. Again, TRI was least frequent at high altitudes.

Boucelham *et al.* (1988) obtained 8602 specimens from 19 populations along a transect from the Finger Lakes Region in New York to the Great Smoky Mountains in North Carolina. They recorded within the Great Smoky Mountains a statistically significant negative correlation of TRI frequency with altitude ($r = -0.813$, $p < 0.05$).

TRI frequencies thus correlate negatively with altitude in Scotland, Czechoslovakia, the Scandinavian mountain range, and the Great Smoky Mountains of the United States. Moreover, TRI frequencies correlate negatively with northern latitudes in Sweden, Norway, and Finland. The color morph TRI obviously is a biotype with a low fitness at high altitudes or high northern latitudes (Berry and Willmer, 1986).

In addition to the four lowland–highland TRI clines described, a fifth cline probably is present in the northern Caucasus mountains. Among the many good *Philaenus* samples taken by Beregovoy (1972) from the Urals and northern Caucasus, the large sample ($n = 1869$) from the mountain population Guzerpil was the only one which was completely devoid of TRI.

The evidence for good tolerance of LOP of harsh mountain climate is weaker than the convincing evidence for bad tolerance of TRI. However, Halkka *et al.* (1967*b*) in Norway and Halkka *et al.* (1980) in the Carpathian mountains noted high frequencies of LOP at high altitudes.

Area Effects

In 1967, two papers were published describing area effects in color polymorphism in *Philaenus* populations from two continents. Farish and Scudder (1967) noted that in British Columbia, populations in dry areas differ significantly from those in wet areas. Halkka *et al.* (1967*b*) noted six different types of area effects from Finland, Norway, Sweden, and the Baltic states of the USSR. Additional samples were taken from these countries and from the adjoining areas in Russia in 1967–1986 (Halkka *et al.*, 1976; Boucelham and Raatikainen, 1984, 1987) and revealed seventh and eighth area effects.

The following eight area effects are shown in Fig. 3:

1. The Østfold–Ångermanland group (ØÅ, Norway to Sweden) of populations is characterized by very low LCE, rather low TRI, MAR, and LOP frequencies, and relatively high LAT frequencies.

2. The Västmanland–Uppland group (VU, Sweden) is high in LAT and LOP, relatively high in LCE, and low in TRI and MAR.

3. The Åland group (Å, Finland) is low in TRI, MAR, LAT, LCE, and LOP, with frequencies higher than 2% being in general reached only by LAT and LOP.

4. The southwest Finland group (SWF) is moderately rich in LAT (about 6–7%), with TRI, MAR, LCE, or LOP practically always present in frequencies lower than 6%.

5. The northeast Finland group (NEF) is rich in LOP, and one of the populations is moderately rich in TRI as well.

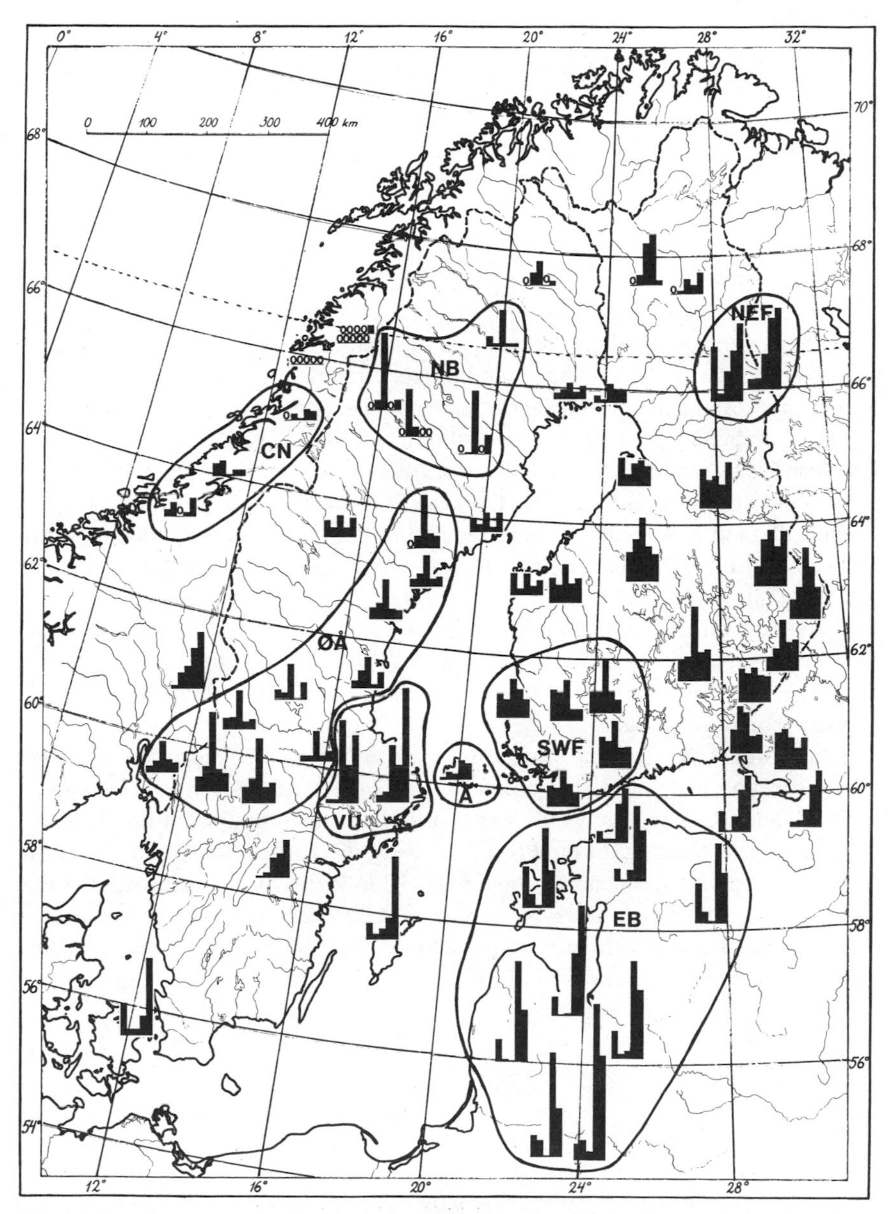

FIG. 3. The eight area effect regions in northern Europe: central Norway (CN), Norrbotten (NB), northeastern Finland (NEF), Østfold–Ångermanland (ØÅ), Västmanland–Uppland (VU), Åland (Å), southwestern Finland (SWF), and east Baltic (EB). Note the cline around the Gulf of Finland (between SWF and EB) running from the easternmost populations of EB to the easternmost populations of SWF. Descriptions of the area effects are given in the text. The morph TYP is not included and the bars denote, from left to right, TRI, MAR, LAT, LCE, and LOP.

6. The east Baltic group (EB, Estonia, Latvia, Lithuania, USSR) is high in LOP, low in LAT and MAR, never high in LCE, but occasionally high in TRI.

7. The Norrbotten group (NB) is rich in LAT (occasionally MAR) and poor or devoid of TRI and LCE.

8. The central Norway group (CN) is very poor in LAT, and poor in TRI, MAR, LCE, and LOP. Toward the northeast other morphs than TYP become extremely rare or are absent from populations.

These eight different modes of polymorphic balance exhibit a remarkable stability. Normally, samples taken from the same site with 10 or more years intervening do not differ significantly from each other.

The constancy of allele frequencies in each of the eight regions indicates that selective forces extending their influence over thousands of square kilometers are responsible for balance maintenance. The Västmanland–Uppland region is drier and warmer than the adjoining Østfold–Ångermanland region. The Åland isles are characterized by a maritime climate and habitually have a dry and cooler weather period in May to June. The southwestern part of Finland is climatically and edaphically harsher than both the Åland and the Baltic regions. Northeastern Finland characteristically has a cool and relatively humid climate with very snow-rich winters.

The area effect regions, in fact, overlap and largely coincide with the plant geographic regions delineated by Ahti *et al.* (1968). It is perhaps not surprising that allele frequencies of the xylem-sucking spittlebug coincide with plant geographic regions.

In North America area effects may extend over huge distances (Fig. 4). Thompson and Halkka (1973) noted that *Philaenus* samples from California, Oregon, and Washington were poorer in MAR than samples from Illinois, Wisconsin, and Quebec. In the Pacific states of the United States and in British Columbia, the TRI frequencies range from 10 to 20%. In the states west of the Great Lakes, TRI frequencies between 5 and 15% are found. The ecological factors causing these differences may depend on climatic rather than edaphic dissimilarities. Thompson (1984, 1988), in an extensive study on North American populations, attributes differences in morph frequencies to thermal properties of the morphs.

Industrial Melanism

The urban suspended particulate matter pollution in the Chicago industrial area was not intensive enough to cause increases in frequencies of the dark melanic morphs (Thompson and Halkka, 1973), but particulate pollution, sulfur dioxide, and other gases from five chimneys of a phurnacite plant in Cynon Valley, Wales, induced industrial melanism in *Philaenus* populations (Lees and

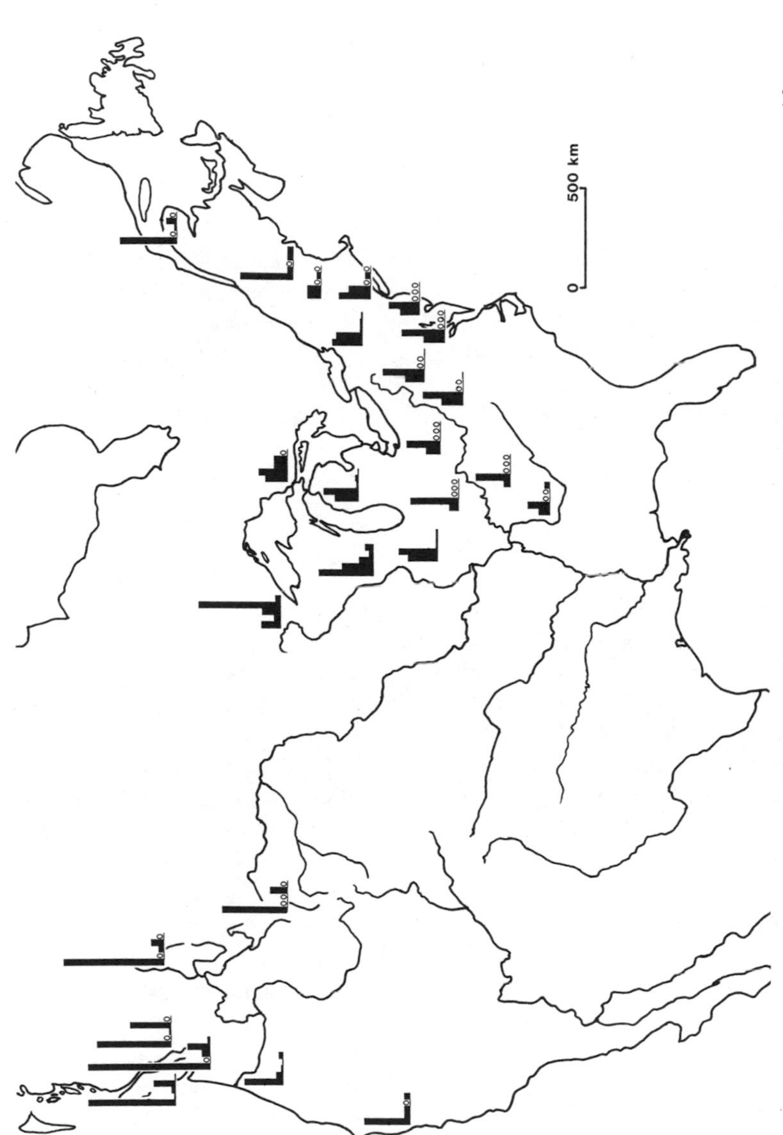

FIG. 4. Area effects in North America cover huge distances. Each of the histograms shows the averaged frequencies of all the samples from an American state or a Canadian province. The only exception is British Columbia, which is represented by three samples. The morph TYP is not included and the bars denote, from left to right, TRI, MAR, LAT, LCE, and LOP. Western populations are rich in TRI and occasionally relatively rich also in LCE. Populations in the Appalachian region and east of it are devoid of LAT and poor in LCE and LOP.

Dent, 1983). Maximum melanic frequencies near the phurnacite plant were found to be far higher than any reported from elsewhere in the species' range through Europe, Asia, and North America. A very significant feature of the Cynon Valley spittlebug population was breakdown of inhibition of expression of the melanic phenotypes in males. In contrast to most of the populations studied outside the British Isles, frequencies of LCE and LOP in males were almost as high as in the females.

Lees and Dent (1983) consider it unlikely that selective predation is responsible for the very high melanic frequencies in the Cynon Valley. The prevalence of LCE and LOP is attributed to ecophysiological selective factors acting through unknown physiological mechanisms.

Lees *et al.* (1983) observed some indication of urban influence on melanic frequency in other parts of Wales and in certain large regions in England. These effects were, however, much less pronounced than the intense industrial melanism in Cynon Valley. In the Cardiff docks, however, frequencies of melanic morphs in both females and males were almost as high as the corresponding frequencies in Cynon Valley (Lees and Stewart, 1988).

MIGRATION

Philaenus Is a Good Hopper and a Weak Flier

The leptokurtic, or skewed, mode of dispersal typical of insects (Davis, 1980) is also shown in *Philaenus* (Halkka *et al.*, 1967*a*). When forced to fly, the spittlebug usually descends after some 10–20 m of flight.

Weaver and King (1954) cite a report by the Canadian Department of Agriculture from 1945: "A spectacular concentration of *Philaenus*—appeared in early July along the Lake Ontario shore of the Niagara peninsula—migrating in a northeasterly direction at all levels—as they struck the tree tops they sounded like rain—Beneath the trees the excreta could be felt falling as a mist—."

From this description, it appears that in the New World, *Philaenus* occasionally makes mass migrations. Such migratory ability may explain the circumpolar distribution of the species and its occurrence in isolated habitats surrounded by unsuitable terrain. But much of the dispersal must have occurred passively in the transported hay parcels of migrant humans. The arrival of *Philaenus* recently in New Zealand (Archibald *et al.*, 1979) was probably anthropochoric.

The three modes of dispersal (1) leptokurtic, (2) mass migration, and (3) anthropochoric explain the present distribution of *Philaenus spumarius*. One subtype of the leptokurtic, skewed mode of dispersal is the anemohydrochoric migration to island habitats.

Anemohydrochoric Dispersal in Archipelagoes

In the archipelago of the westernmost parts of the Gulf of Finland, an area about 20 km long in the west to east direction and 6–10 km broad in the south to north direction was investigated with regard to the presence or absence of *Philaenus* on the islands. The area lies between longitudes 23° 5′ E and 23° 35′ E and south of latitude 60° N, and is facing open sea in the south. The northwestern part of the area is shown in Fig. 5. In this archipelago, 262 islands, islets, and rocks were investigated with regard to presence or absence of the spittlebug in June to early August 1970. Stands of vascular plants appearing adequate to serve as habitats for *Philaenus* were found on 135 of the islands, and spittlebugs were found on 91 such islands. On some very small rocky islets, spittlebugs were observed in plant stands the dicotyledonous plants of which would hardly fill a medium-sized flower vase. Obviously, *Philaenus* is able to immigrate into almost all the suitable habitats an archipelago can afford (Halkka *et al.*, 1971).

In archipelagoes, *Philaenus* dispersal probably is mainly anemohydrochoric. Specimens which in one way or other, e.g., by wind, are detached from the plants and get onto the surface of the sea are able to float on the waves. Long-distance floating appears to be uncommon, for *Philaenus* is not mentioned in a comprehensive review of literature on passive transportation of insects to the shores of the Baltic (Palmén, 1944).

In the best experiment on the floating power of *Philaenus* performed by our group, 1510 specimens marked by basic fuchsin powder were released onto the sea surface. About 95 min later, 41 of them were found landing on an island 1.1 km from the point of release, and 10 were seen floating alive. Most of the remaining 1459 had probably drowned.

This experiment made it clear that *Philaenus,* after possibly having been carried by wind some distance, is able to float on the sea surface (salinity in Tvärminne about 0.6%) for some time and still is able to land on an island. This anemohydrochoric mode of dispersal probably has enabled *Philaenus* to immigrate to many of the 91 islands in the Tvärminne and Jussarö archipelago now populated by the species.

FOUNDER PRINCIPLE

The Pigmentation Locus, *p*

Some of the most remote islands of the Tvärminne and Jussarö archipelagoes in the Gulf of Finland have constant *Philaenus* populations. The islands Segelskär, Myggan, Klovaskär 2, and Klovaskär 3 about 10 km from the main-

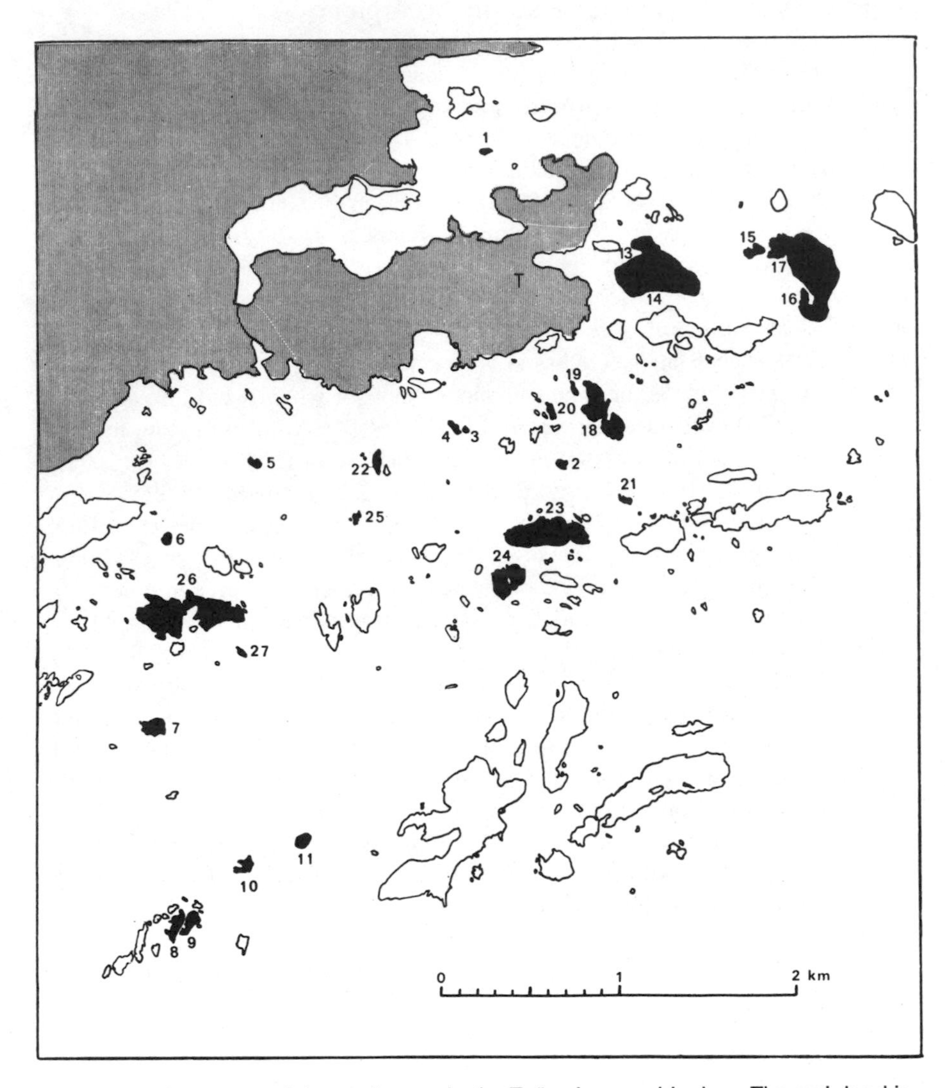

FIG. 5. The inner part of the study area in the Tvärminne archipelago.The mainland is shown in grey, the islands investigated by the *Philaenus* study group in black. The letter T refers to the population Tvärminne A. The following islands (populations) are mentioned in the text: Allgrundet (2), Porskobben (6), Östra Mellanspiken (8), Jofskär (13, 14), Gulkobben (15), Brännskär (16, 17), Rovholmen (18), Rovholmsgrundet (19), Norra Grisselgrundet (20), and Vindskär (26). The islands Segelskär, Klovaskär 2 and 3, and Myggan lie 10–12 km southeast from Tvärminne A.

land all have well-established spittlebug populations. But none of these island populations constantly possesses a complete set of pigmentation alleles (Halkka *et al.*, 1974*a*).

In Segelskär, the population remained practically monomorphic in 1969–1988. In addition to TYP, the prevalent morph of Segelskär, morphs LAT, LOP, and TRI have been observed sporadically. Small boat traffic to Segelskär increased substantially in the 1970s, and possibly some of the non-p^t pigmentation alleles have been brought to Segelskär in these boats. The *founder effect* possibly has prevented the non-p^t alleles from increasing their frequency.

In the island Myggan, the alleles for TYP, TRI, MAR, and LAT have been present at nearly constant frequencies in 1969–1988. Myggan is very rich in TRI, about 25% of the individuals belonging to this morph. The polymorphic balance maintains the frequency of TYP at about 65–70%, MAR at about 3%, and LAT at 5%. Both LCE and LOP are absent from Myggan (Fig. 6).

The islands Klovaskär 2 and Klovaskär 3 are separated by a narrow, about 50-m-broad strait. The heavy winds of the outer archipelago may easily throw individuals from Klovaskär 2 to 3 and vice versa. The islands serve as *stepping stones* for each other. It is thus not surprising that the same two alleles, p^t (TYP) and p^L (LAT), are common in both islands. In addition to these, p^O (LOP) occurs at low frequencies on both Klovaskärs. The allele p^M (MAR) occurs on Klovaskär 2 constantly at extremely low frequencies. The alleles p^C (LCE) and p^T (TRI) occur sporadically and apparently have failed to establish themselves on the Klovaskärs.

Segelskär, Myggan, and the two Klovaskärs thus have only one, four, and four established alleles, respectively, in their populations. Absence of two or more of the alleles from these remote islands probably is due primarily to the founder principle and secondarily to the disappearance of rare alleles through the founder effect (random genetic drift). There are some indications, however, that some *founder selection* is involved in establishment and maintenance of pigmentation alleles in remote outer island populations (Halkka *et al.*, 1974*a*).

Enzyme Gene Loci

A number of populations from the Tvärminne mainland and from six of the inner and outer islands of the Tvärminne and Jussarö archipelagoes were investigated with regard to enzyme gene heterozygosity (Saura *et al.*, 1973). The enzyme gene loci investigated were: *G-3-pdh, Ald, 6-P-gdh, Aph, Xdh-2, Est-2, Idh-2, Adk, Pgm-2, Tpi, Ao, Est-3, Est-5, Idh-1, Me, Lap, To-2, α-Gpdh-1*, and *Est-4*.

The populations were found to have different average degrees of heterozygosity, the most isolated population, Segelskär, being the least polymorphic.

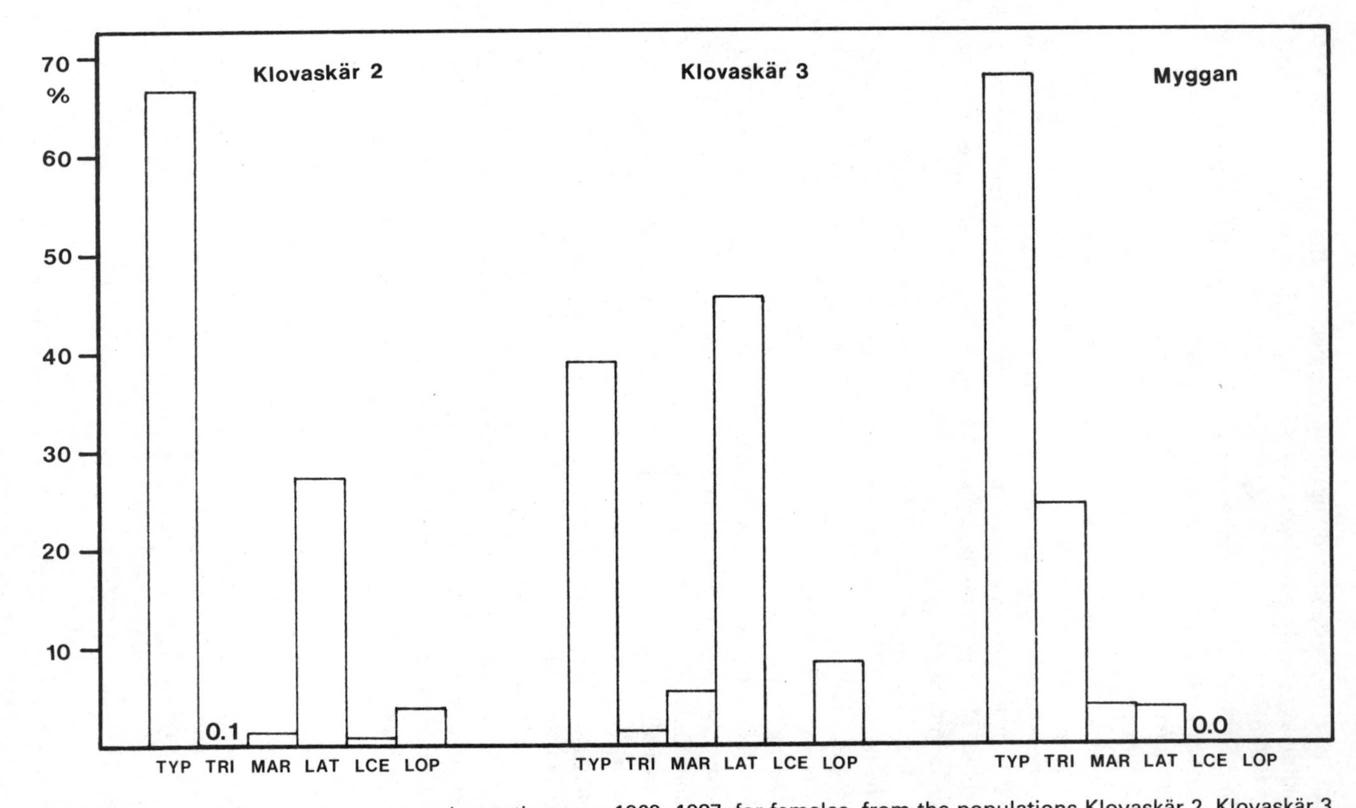

FIG. 6. Color morph frequencies averaged over the years 1969–1987, for females, from the populations Klovaskär 2, Klovaskär 3, and Myggan. Note the total absence of LCE from Klovaskär 3 and of LOP from Myggan.

Segelskär was, in fact, polymorphic with regard to four loci only. The Tvärminne mainland population, on the other hand, was polymorphic with regard to 16 loci. The five islands between the mainland and Segelskär were intermediate between the extremes, with 7–11 heterozygous loci. As a whole, heterozygosity thus correlated negatively with remoteness and isolatedness of the populations. This variability parallels nicely the variability in the degree of heterozygosity of the pigmentation locus *p*, which similarly correlates negatively with isolatedness.

The most effectively isolated islands have dissimilar sets of enzyme gene alleles. The arrival of alleles to the remote islands probably is almost exclusively dictated by chance. Establishment of a newly immigrated allele, however, is not necessarily governed solely by random genetic drift.

MULTINICHE SELECTION

Constant Meadows, Constant Equilibria

No meadow is maintained at a fully constant state, and no polymorphic equilibrium is absolutely constant, but in some cases the amplitude of change is very narrow. Hutchinson (1963) reported considerable constancy of British *Philaenus* populations over 43 years, and Farish and Scudder (1967) the same in Canadian populations over 16 years. The Finnish mainland population Tvärminne A retained the same polymorphic balance from 1956 to 1988. The samples of 1986 and 1987 deviated from the southwestern Finland balance, being slightly "too rich" in LCE and LOP.

Meadows in tiny rocky islets of the Tvärminne archipelago are much less well buffered against extreme climatic changes than the mainland meadows. Shallow soil and small watershed surface make them prone to desiccation. In slightly larger, wooded islands the conditions come closer to those prevailing in the mainland. In Tvärminne, the islands Gulkobben and Rovholmen have constant enough water economy for the meadows to retain almost stable coverages of the most important food plants of *Philaenus*. Morph frequency records from Gulkobben and Rovholmen from 1969 to 1988 (20 years) were grouped in pairs 1969–1970, 1971–1972, etc., which produced ten packages of two consecutive years from both islands. Chi squares were then calculated to compare all the adjoining pairs with each other. None of the pairs except one yielded a significant *P* value (Fig. 7). But between pairs with 4–18 years intervening, statistically significant differences were observed. This result indicates that the morph frequencies tend to be retained over many generations, but that gradual changes may take place over a time span of a decade or two.

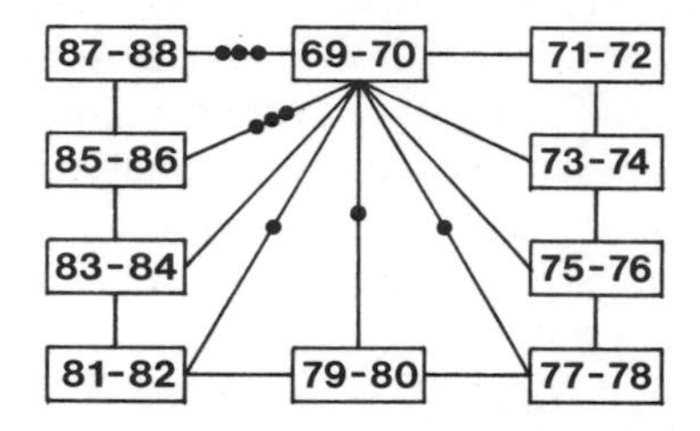

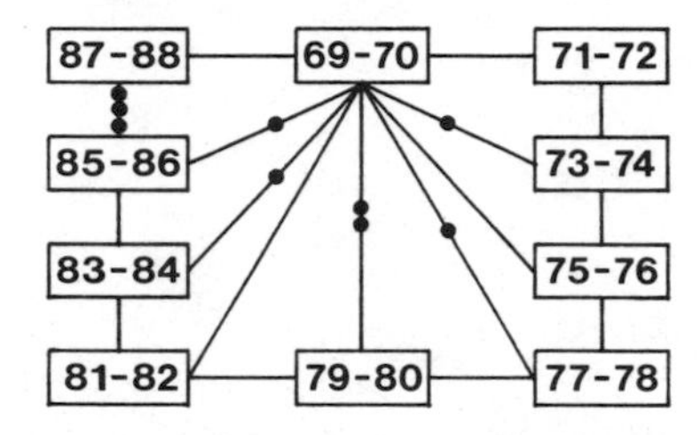

FIG. 7. The χ^2 "families" of Gulkobben and Rovholmen; d.f. 5, and significance levels 0.05, 0.01, and 0.001 are shown by one, two, and three dots, respectively.

The morph frequency equilibria in Gulkobben (Fig. 8) deviate only slightly from the southwestern Finland equilibrium maintained by, e.g., the mainland populations Tvärminne A and Nurmijärvi A (Figs. 9 and 10). Three of the other wooded islands of the Tvärminne archipelago, Vindskär, Jofskär, and Brännskär, have populations with typical southwestern Finland equilibria (Halkka *et al.*, 1974*a*). It seems that relatively old, established populations in islands *converge* toward this equilibrium.

The situation is different in small, rocky, treeless islands. Each such island maintains an equilibrium which deviates clearly from the southwestern Finland equilibrium. One example, allele frequencies from the population in the islet Norra Grisselgrundet, is given in Fig. 11. Temporal variability in morph frequencies of the population living in this treeless, tiny, rocky islet is much more extensive than in Tvärminne A, Nurmijärvi, or Gulkobben. The island-specific mode of morph frequencies is evident, however. The unique, island-specific equilibria may partly result from founder principle and founder effect during the initial stages of colonizing a new island. The remote islands Segelskär, Klovaskär 2, Klovaskär 3, and Myggan, lacking or practically lacking 2–5 of the alleles, still bear signs of stochastic events which took place during the founding phases. The incomplete allele sets are now, however, maintained at specific

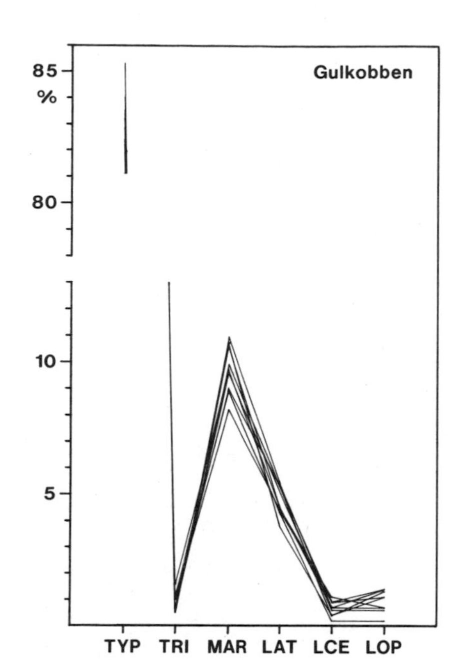

FIG. 8. Color morph frequency diagrams for pairs of years, 1969–1970, 1971–1972, etc., up to 1987 for females in the population on the island of Gulkobben. Abscissa: morph group; ordinate: percentage.

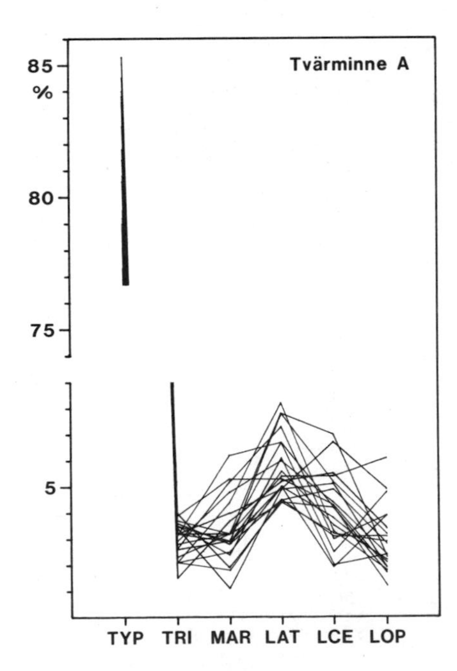

FIG. 9. Color morph frequency diagrams for females, for the years 1956, 1960–1963, 1965, 1972, 1974–1981, and 1984–1988 from the population of Tvärminne A.

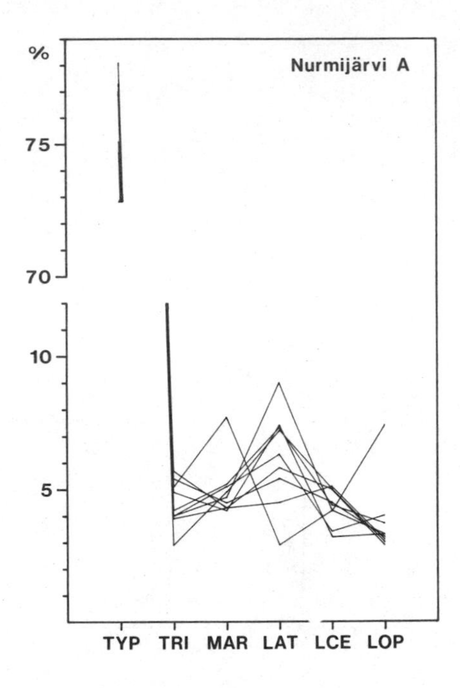

FIG. 10. Color morph frequency diagrams for females, for the years 1960–1967 and 1975 in the population of Nurmijärvi A, about 110 km northeast of Tvärminne.

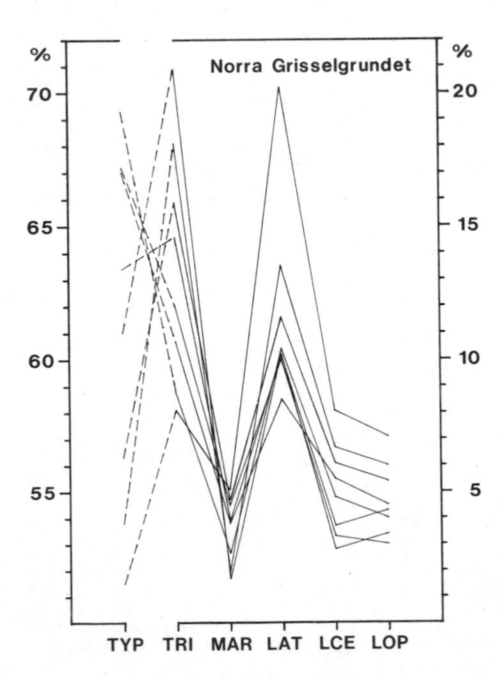

FIG. 11. Color morph frequency diagrams for pairs of years 1969–1987 for females from the population of Norra Grisselgrundet. No sample was taken in 1981 and 1982.

equilibria by balancing selection. From the population Myggan, for instance, the following morph frequencies (percentages) were recorded in 1972 and 1986, respectively: TYP 68.1–67.9, TRI 23.6–21.8, MAR 3.4–5.0, LAT, 4.9–3.3. The sample sizes were (females) 471 for 1972 and 636 for 1986.

The populations of Rovholmsgrundet and Norra Grisselgrundet (Halkka *et al.*, 1974*a*) diverge from each other and from other Tvärminne archipelago populations. They are in a *divergent phase* of allele frequency minievolution. It will take perhaps 50 or more generations until any one of these populations will reach the *convergent phase* exemplified by Gulkobben and Rovholmen.

Why do *Philaenus* populations experience a divergence phase followed by a convergence phase? The *multiniche selection theory* (e.g., Levene, 1953; Arnold and Anderson, 1983) may explain the divergence–convergence phenomenon. Selection possibly acts on spittlebug populations both indirectly and directly through variable floral diversity. The meadows in which the divergent *Philaenus* populations live are in what is called the ''early stage'' of their development. They are all poor in floral diversity, but individual meadows lack different plant species, which means that the meadows diverge from each other.

The situation is quite different on the wooded islands, on which meadows tend to be like each other and like those in the mainland. The meadows show convergence in the structure of their plant communities and can be said to be at a ''late stage'' of development. The *Philaenus* populations inhabiting such meadows exhibit in their allele frequencies convergence toward the southwestern Finland allele equilibrium. Part of this convergence is, in some instances, due to slight lessening of isolation. But convergence probably is mainly caused by increased uniformity of the environments of the different islands. Thanks to the phenomenon of land upheaval, we can be positive that the divergent and convergent phases are successive. The land upheaval is an isostatic lift of the earth crust that slowly compensates the sinkage caused by the continental ice sheet of the last Ice Age. In southwestern Finland, this land upheaval measures about 35 cm/100 hundred years and lifts meadows on low, rocky islands slowly out of the action of the waves.

Changing Meadows, Changing Equilibria

Shallowness of the soil layer of meadows in treeless, rocky islets renders the meadows vulnerable to climatic and biotic catastrophes. The climatic catastrophes are of two types: flooding and drought. There is no tide in the Baltic Sea archipelagoes, and the normal weather-dependent fluctuation in sea level has an amplitude of about +60 to −40 cm. Occasionally, extremely high water levels accompany heavy storms and cause flooding and soil impoverishment of the most low-lying meadows. On January 9, 1975, a heavy storm raised the water level up to 110 cm above normal level. The vegetation of some meadows changed drastically through changes in soil nutrition and direct detachment of turfs by the high waves. The meadow Östra Mellanspiken A lost most of its dicot stands, which were replaced by monocots in the next 2 years. The *Philaenus* population almost disappeared, and only six individuals were captured in 1976 (Halkka, 1978). In the succeeding years, dicots came back, and the spittlebug population gradually assumed again the pre-1975 allele frequency equilibrium. The percentages for 1974 (sample of females 618) and 1979 (sample size 819) were, respectively: TYP 65.2–69.5, TRI 1.1–0.0, MAR 16.3–10.5, LAT 12.5–15.9, LCE 4.7–3.1, and LOP 0.3–1.1

A corresponding change took place in the island Allgrundet. About one-tenth of the meadow was completely abolished, and about one-third of the rest lost some of the most nutritive layer of its soil.

The most severe biotic threat to island populations of *Philaenus* lies in the ability of the field vole, *Microtus agrestis,* to develop extremely high population densities. On treeless islands, *Microtus* may eat almost all the food plants of the spittlebug. The effects of high densities of *Microtus* in meadows A and B in the island Östra Mellanspiken were described by Halkka *et al.* (1975*c*). Here it may suffice to say that the damage caused by the voles to the meadows was accompanied by population size reduction, allele frequency changes, and subsequent restoration of original allele frequencies in the *Philaenus* populations.

Flooding and voles may alter meadow ecology profoundly, but these disasters strike sporadically and not upon all types of meadows. The only recurrent type of catastrophe in the Baltic Sea archipelago is drought, especially in the form of low June precipitation. Monthly June precipitation of only 5–10 mm, combined with generally warm, sunny weather, parches the meadows and kills large numbers of *Philaenus* nymphs. In the island Allgrundet, the numbers of nymphs isolated and adults emerged alive were the following in the four successive years 1971–1974:

Year	Nymphs isolated	Adults emerging	Survival (%)
1971	859	468	57.0
1972	5103	4661	91.3
1973	5084	708	13.9
1974	859	465	54.2

The low survival in 1973 was due to drought. The numbers reveal two features important for understanding the peculiar mode of *Philaenus* multiniche polymorphism. Given optimal conditions, the spittlebug has the capacity of increasing the size of the mating pool about tenfold from generation n to $n + 1$. A comparison of nymph numbers of 1972 and 1973 indicates that the maximal carrying capacity of a restricted habitat may perhaps be reached (and lost) in a single generation.

Droughts seem to be able to affect allele frequencies only on the condition that coverages of dicots change appreciably. This does not necessarily happen after a single dry growth season. Consequently, changes in allele frequencies were hardly discernible after the dry summer 1973 (Halkka, 1978, Figs. 3.2 and 3.3).

EXPERIMENTAL PERTURBATION OF ALLELE FREQUENCIES

Principles

Whenever there are indications of the presence of multiniche polymorphism, the associations between environmental parameters and genotypes can be studied in two different ways (Hedrick, 1986). *Environmental perturbations* interfere with the environment in such a way that parameters considered important are deliberately changed. This is not easy with natural populations, and studies of this type have indeed been done with, e.g., *Drosophila* or *Tribolium* populations in captivity. Fortunately, isolated *Philaenus* populations studied by the Finnish research team were subjected to natural environmental perturbations by flooding, voles, or drought. The consequences of these environmental perturbations are described in the preceding section, and in more detail in the publications cited therein. The lesson taught by following up the consequences of these environmental perturbations was straightforward. Allele frequencies changed with changing plant coverages and were restored with the recovery of dicot coverages toward the original structure of the plant community.

Genetic perturbations (Hedrick, 1986) are possible with isolated natural populations and much more easily done than the artificial environmental perturbations. Isolated Baltic Sea island populations of *Philaenus* are ideal subjects for such experiments.

The Exchange Experiment

An exchange experiment involving the populations of Allgrundet and Porskobben was started in 1970. Results spanning six generations (1969–1974) were published in 1975 (Halkka *et al.*, 1975*b*). Here the followup is extended to 1987 (the 19th generation).

The results are given in Figs. 12 and 13. In 1969 and 1970 it was noted that the population of Allgrundet was rich in LCE, but probably devoid of LOP. In contrast, the population of Porskobben was poor in LCE, but rich in LOP. If this allele frequency balance somehow follows from island-specific environmental

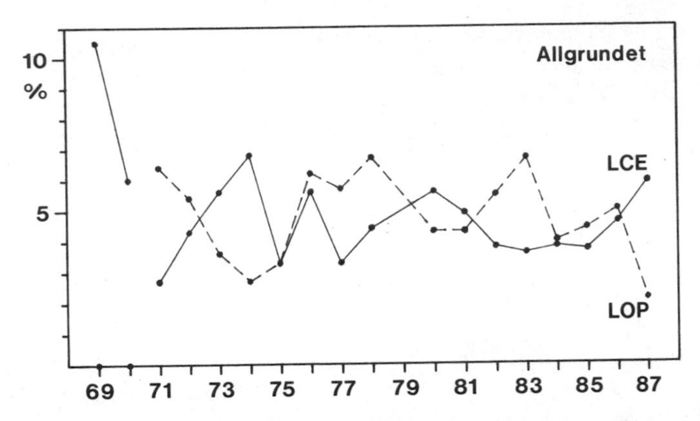

FIG. 12. Frequencies of the color morphs LCE and LOP from 1969 to 1987 in the population of Allgrundet. The diagrams left of the break give morph frequencies in an undisturbed population, those right of the break give morph frequencies after exchange of individuals between Allgrundet and Porskobben. Black dots on the abscissa denote zero frequency of LOP in 1969 and 1970. Note that restitution of preexchange frequencies took place before the flooding catastrophe in 1985.

conditions, genetic perturbation perhaps cannot cause a lasting change of frequencies.

In July 1970, 497 adult individuals were transferred from Allgrundet to Porskobben and 443 individuals from Porskobben to Allgrundet. The samples transferred probably comprised at least two-thirds of the populations from which they originated. After the transfer both populations consisted of about one-third original and two-thirds introduced individuals.

In Allgrundet, the increase of LCE and the decrease of LOP frequency, i.e., restoration of the original island-specific balance, took place in 1972–1974. In January 1975, a heavy storm which destroyed large parts of the meadow complex in Allgrundet probably removed thousands of *Philaenus* eggs. Monocots, mainly grasses, conquered the lower part of the meadow. In the late 1970s and in the 1980s the original dominating dicots, *Filipendula* and *Lysimachia,* gradually gained back the lost ground. It is perhaps not surprising that allele frequencies of *Philaenus* fluctuated irregularly in 1975–1985, or during a period of drastic changes in the coverages of the most important food plants. The equilibrium of the 1987 generations does not deviate very much from that of 1970 (Fig. 12).

Porskobben lies between large islands and was affected by the January 1975 storm much less than Allgrundet. In Porskobben, restoration of the original situation, low LCE and high LOP frequency, was evident in 1972–1987 (Fig. 13).

In both Allgrundet and Porskobben, the heavy introduction of alien genes in 1970 was reflected by the allele frequencies of 1971. In both populations, the

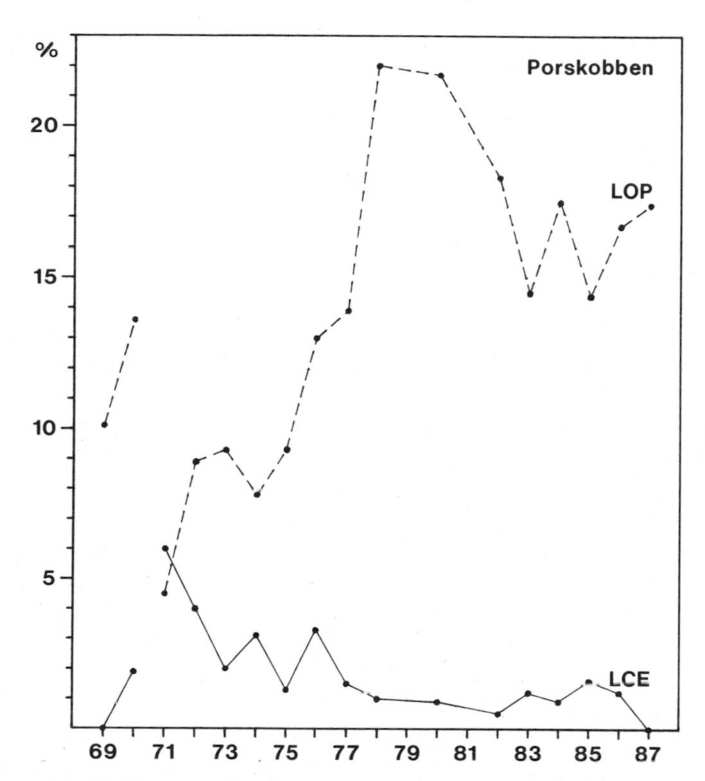

FIG. 13. Frequencies of the color morphs LCE and LOP from 1969 to 1987 in the population of Porskobben. The diagrams left of the break give morph frequencies in an undisturbed population, those right of the break give morph frequencies after exchange of individuals between Porskobben and Allgrundet.

frequencies moved toward pretransfer values in 1972–1974. After 1974, the ecology of the Allgrundet meadow was no longer the same as in 1969–1974. There was thus no reason to expect that the started restoration could go on in 1975 and in the following years. Although some changes in food plant coverages took place in Porskobben, they were not extensive enough to inhibit steady and clear restoration of original island-specific allele frequencies. This restoration took place in a gene pool which after genetic perturbation contained about 60–70% introduced, alien genes. How can the selection pressures be stronger than the very high migration pressures caused artificially?

The key to this enigma obviously is the high capacity to reproduce shown by *Philaenus*. The number of females in Allgrundet and Porskobben obviously was at least 300 immediately after the transfer in 1970. These females produced on both islands about 15,000–30,000 eggs. This means that during the egg and nymph stages, some 14,500–29,500 individuals were killed, in all probability partly selectively. The law of large numbers together with a high selection coefficient "explains" the rapid allele frequency changes under multiniche selection.

THE SELECTION REGIME OF *PHILAENUS SPUMARIUS*

From what is known about *Philaenus* populations in Europe, north America, and Asia, it appears that the selection regime of the spittlebug comprises at least three main components: *visual* (apostatic), *climatic, and multiniche* selection. The dorsal color polymorphism, in most parts of the world expressed fully only in the females, probably evolved via di- and trimorphism into polymorphism. Each of these steps made it more difficult for potential predators to build up a prey-specific searching image. The overwhelmingly most common morph *typicus* now is encountered by predators more frequently than the other morphs. The predators probably learn to follow a TYP-shaped searching image as a result of these encounters. According to the principle of *apostatic selection* (Clarke, 1962, 1969), color pattern variability causes confusion in the choice of suitable prey in a hunting predator. The males, unnecessary for the population after copulation, and often expressing the TYP phenotype (about 95% in many populations), are eaten and protect in a passive way "altruistically" their offspring. The offspring of a male resides in the form of diploid eggs in the abdomen of a female or in several females. The females, fully polymorphic, need protection during the preoviposition period, which lasts at least 4–5 weeks (Witsack, 1973).

For apostatic selection to be effective in population, the presence of the full array of morphs probably is more important than the frequencies of individual morphs. The frequencies show extensive clinal and area effect variability in a geographic scale. The clines appear to follow humidity or humidity/temperature gradients. The area effects coincide nicely with climatic or climatic/edaphic plant geographic regions. A large body of indirect evidence and certain experimental results indicate that the color morphs are dissimilar ecophysiologically. They are *biotypes,* in addition to being differentially colored and patterned. The biotype status of *trilineatus* is experimentally well documented. *Typicus* is the most sturdy and versatile biotype and the best colonizer, and *lateralis* appears to be the second-best generalist and colonizer.

In the selection regime of *Philaenus, climatic or climatic/edaphic selection* dictates the frequencies of the biotypes. The color morphs are not climatically neutral as such. Variable darkness of pigmentation means variable reflectance and differential heat gain in relatively low sunlight intensities (Berry and

Willmer, 1986). However, almost certainly there are tightly linked ecophysiologically reacting genes without any pigmentation functions in the immediate neighborhood of the pigmentation locus *p*. A large part of the clinal and area effect geographic variability may be due to such genes. If this is true, then the color alleles act as *markers* revealing the distribution of the "ecophysiology alleles."

The third component of the selection regime of *Philaenus* is *multiniche selection* (Levene, 1953; Arnold and Anderson, 1983). Relatively small populations in tiny, isolated islands exist and have proven to be ideal subjects for research directed toward multiniche selection in action. Three sources of evidence from such populations indicate that multiniche selection is operating in *P. spumarius* populations. First, meadows retaining constant food plant coverages maintain allele frequencies within a narrow amplitude over a decade or two. Second, catastrophes affecting the ecological balance of a meadow tend to change allele frequencies of the resident *Philaenus* population. Restoration of the original plant coverages is accompanied by at least partial restoration of *Philaenus p*-locus allele frequencies. Third, attempts to change existing allele frequencies through massive introduction of alien genotypes change the frequencies only temporarily. A partial or complete restoration of the original frequencies may take place in 2–4 years (generations).

The three components, apostatic, climatic, and multiniche selection, probably constitute a large part of the total selection regime of *Philaenus spumarius*. The three components interact with each other and with possible additional, unidentified components of the selection regime.

DISCUSSION

In the selection regime of *Philaenus,* multiniche selection perhaps is the most important component. The rules governing this type of polymorphism are known, and the collected theoretical knowledge and the empirical evidence both show how flexible and modifiable multiniche polymorphism can be (Hedrick *et al.*, 1976; Hedrick, 1986). Multiniche polymorphism is so effectively buffered against migration pressure that practically always the composite selection coefficient s vastly exceeds any imaginable immigration m [see Felsenstein (1976) and Slatkin (1987) for migration versus selection]. Genetic load, in the form of dysmetric load, is an integral part of the evolutionary force regulating allele frequencies. Hard and soft selection intermingle in such a situation (Wallace, 1968; Christiansen, 1975).

Populations maintaining several alleles in a multiniche environment are in need of a high reproductive capacity r. *Philaenus spumarius* populations may often experience losses of large numbers of tiny instar nymphs. Losses from the adult population are much smaller and involve "sacrifice" of males and minimal mortality, i.e., protection of the zygote-bearing females.

It seems that the most important factor behind the success of *Philaenus* as a widespread and abundant insect is the composite nature of the species. The interbreeding color morphs appear to be *biotypes* showing variable degrees of ecological specialization (Diehl and Bush, 1984). Rare morphs occupy small niches and in this respect are similar to rare independent species. The broader the niche available, the more frequent the morph specializing in this fraction of the habitat. A generalist, *typicus,* mates randomly with the specialists, which similarly mate randomly among themselves (Halkka *et al.*, 1967*a;* Farish and Scudder, 1967). Biotype frequencies are adjusted according to availability of different types of niches throughout the distribution area of the species. The whole array of biotypes is not a *conditio sine qua non* for success, however. In New Zealand, *Philaenus* is rapidly spreading, but only a very limited assemblance of color morphs is present in New Zealand populations.

The *Philaenus* biotype situation is not the only one in the Homoptera. Müller (1987, 1988) observed vitality and fertility differences between different larval color morphs of the leafhopper *Mocydia crocea,* a polyphag on gramineous plants. Each one of the six morphs showed a unique mode of vitality on the 22 food plant species in the experiments. For instance, morph "=r" was superior to "||g" on *Lolium perenne,* but inferior to it on *Festuca rubra*. Neither in *Philaenus* nor in *Mocydia* do the morphs show any signs of forming independent mating pools and becoming sibling species.

Templeton (1980) thinks that habitat-specific divergence is a rare mode of speciation. It may be true that without spatial isolation, some other change such as a mutation in the genes determining the mating signal or a mutation leading to assortative mating is necessary for such sympatric speciation. Some instances among the Homoptera Auchenorrhyncha are indicative of such a development. Guttman *et al.* (1981) believe that the host races of the Membracid *Enchenopa binotata* are on the verge of becoming biological species. Similarly, Claridge (1987) has found that the Macropsid *Oncopsis flavicollis* is a composite of a number of distinct, but as yet unidentified sibling species.

Multiniche selection and apostatic selection interact in *Philaenus* populations. Rules governing their interaction, and the role of intrahabitat migration, are schematically presented in Fig. 14, which is a modification of a scheme presented by Arnold and Anderson (1983). The flexibility characterizing this modification of the original mode perhaps explains the tenacity of *Philaenus* populations in avoiding extinction under adverse conditions. All the models based on the two homozygotes, one heterozygote principle in a protected polymorphism are subject to criticism in lacking robustness in a changing environment (Hoekstra *et al.*, 1985). The skewed type of polymorphism found in

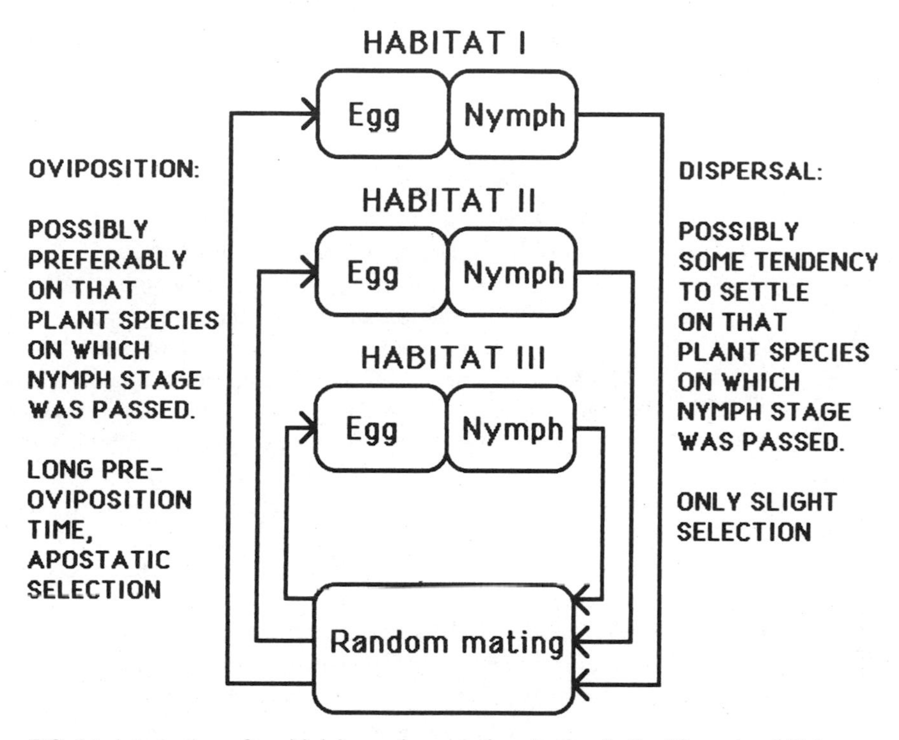

FIG. 14. Interaction of multiniche and apostatic selection in the life cycle of *Philaenus.*

Philaenus may escape such criticism in having one of the several homozygotes (p^t/p^t) ecologically dominant over most of the habitat. The other homozygotes are in most populations too rare to have a significant role in any theoretical model. The phenotypes, or color morphs, coexisting with p^t/p^t at appreciable frequencies are all heterozygotes of a type p^t/p^x, the symbol x denoting any one of the rare alleles. Provided that the color morphs really are biotypes, it will suffice to adjust frequencies of the rare alleles according to the p^t/p^x-specific carrying capacities in each one of the habitats in Fig. 13. No additional adjusting of p^x/p^x frequencies is necessary for maintaining multiniche polymorphic balance and minimizing the dysmetric load.

Temporal environmental variability may often result in random fluctuations in the direction of selection and may mimic random genetic drift. But in cool, temperate regions such as northern Europe, the climatic and the climate-dependent edaphic fluctuations follow a certain, the more extreme, the more seldom rule and are thus statistically predictable. Because of the existing spatial variability, the predictable temporal variability changes the carrying capacities of the different parts of the mosaic multiniche environment differentially. Any poikilothermous organisms, plants or animals, will survive and show moderate to high fitness only in the condition that their genetic heterogeneity does contain buffering capacity for even the most extreme types of variability. The setup of conditions contains a limited number of parameters important for each species living in this predictably variable environment. This means that a small number of di- or multiallelic loci may be enough for sufficient buffering capacity. It would, even with a limited number of environmental setups, take a very large genetic load if very many loci would be participating in responding to environmental variability. Combining, perhaps through transpositions, two or more loci in supergenes or in tight linkage positions would make the buildup of allele combinations easier.

Some of the loci crucial for fitness maintenance certainly contain genes responsible for ecophysiological tolerance of dry periods. In populations living in small islands without ponds or other water reservoirs, correlation between general humidity and nymph or adult survival often attains values $r \geq 0.8$ ($P = 0.01–0.001$) (A. Morikka and O. Halkka, in preparation).

The Nordic environment might be called cyclic-stochastically recurrent (CSR). In the northern European countries, this situation prevails as long as the CO_2 concentration of the atmosphere does not change radically and as long as there are no drastic changes in the volume, speed, and direction of the Gulf Stream.

Climatic situations deviating far from the average recur several times in a century and are experienced differentially by individuals living in microclimatically different environments. In a mosaic space-and-time setup, some parts of the habitats occupied by a univoltine poikilothermous animal or plant carry the heaviest selection pressures. In other words, small compartments of the total habitat suffer more than the average during a subextreme climatic period. Subpopulations living in these compartments constitute reservoirs for the alleles responding to climatic extremes.

There are observations indicating that a limited number of years under a new selection pressure is enough to change allele frequencies appreciably [see this chapter, section "Changing meadows, changing equilibria"; and also Lees *et al.*, (1983)]. The loci responsible for the climatic buffering capacity of *Philaenus* respond by allele frequency alteration effectively to any CSR variability.

It may well be that the number of significant heterozygous loci is less than 50 in the populations in question (Hedrick, 1973). In general, it seems possible that it is crucial to be able to identify the relatively few key genes of an organism showing high fitness in a defined setup of environmental conditions. Species-specific fitness may depend on a small number of genes not only during the speciation process (Templeton, 1980; Bush, 1975), but also in niche partitioning and in regional differentiation.

If the number of the ecophysiological loci in poikilothermous organism is limited to, say, 40, with possibly some supergenes included, and if the egg or other propagule productivity is high enough (some 50–200 per female), very strong selection pressures are possible in the CSR environment. Indeed, as indicated by the results of the massive transfer experiment, the selection pressures in a case such as *Philaenus* may amount to values appreciably exceeding 0.1.

ACKNOWLEDGMENTS

Our heartfelt thanks are due to other senior members of the *Philaenus* group: the late J. Vilbaste, M. and T. Raatikainen, R. Hovinen, and A. Vasarainen. The devoted help in the field work of V., A., E., and S. Halkka and K. Roukka is acknowledged with warm thanks. Many students participating in the course in ecological genetics in Tvärminne were very helpful in assisting in field work in seven island or mainland populations.

A. Morikka participated effectively in many phases of this work, statistical tests in particular. J. Mänttäri drew the figures and devotedly assisted in shaping the manuscript into its final form. We thank them both cordially.

REFERENCES

Ahti, T., Hämet-Ahti, L., and Jalas, J., 1968, Vegetation zones and their sections in the northwestern Europe, *Ann. Bot. Fenn.* **5**:169–211.

Archibald, R. D., Cox, J. M., and Deitz, L. L., 1979, New records of plant pests in New Zealand. III. Six species of Homoptera, *N. Z. J. Agric. Res.* **22**:201–207.

Arnold J., and Anderson, W. W., 1983, Density-regulated selection in a heterogeneous environment, *Am. Nat.* **121**:656–668.

Beregovoy, V. E., 1970, Variation in color differences in male and female spittlebug (*Philaenus spumarius* L.), *Dokl. Akad. Nauk SSSR* **191**:1156–1159 [in Russian].

Beregovoy, V. E., 1972, A study of polymorphism and quantitative evaluation of variability in populations, *Philaenus spumarius* (L.) taken as an example, *Zh. Obshch. Biol.* **33**:740–750 [in Russian, with English summary].

Berry, A.J., and Willmer, P. G., 1986, Temperature and the colour polymorphism of *Philaenus spumarius* (Homoptera: Aphrophoridae), *Ecol. Entomol.* **11**:251–259.

Boucelham, M., and Raatikainen, M., 1984, The colour polymorphism of *Philaenus spumarius* (L.) (Homoptera, Cercopidae) in different habitats, *Ann. Entomol. Fenn.* **50**:43–46.

Boucelham, M., and Raatikainen, M., 1987, The effect of altitude, habitat and geographical location on the polymorphism of *Philaenus spumarius* (L.) (Hom. Cercopidae), *Ann. Entomol. Fenn.* **53**:80–86.

Boucelham, M., Hokkanen, H., and Raatikainen, M., 1988, Polymorphism of *Philaenus spumarius* (L.) (Hom. Cercopidae) at different latitudes, altitudes and habitats in the USA, *Ann. Entomol. Fenn.* **54**:49–54.

Bush, G. L., 1975, Modes of animal speciation, *Annu. Rev. Ecol. Syst.* **6**:339–364.

Christiansen, F. B., 1975, Hard and soft selection in a subdivided population, *Am. Nat.* **109**:11–16.

Claridge, M. F., 1987, Insect assemblages–Diversity, organization, and evolution, in: *Organization of Communities, Past and Present* (27th British Ecological Society Symposium), pp. 141–162.

Clarke, B., 1962, Natural selection in a mixed population of two polymorphic snails, *Heredity* **17**: 319–345.

Clarke, B., 1969, The evidence for apostatic selection, *Heredity* **24**:347–352.

Davis, M. A., 1980, Why are most insects short fliers?, *Evol. Theory* **5**:103–111.

Diehl, S. R., and Bush, G. L., 1984, An evolutionary and applied perspective on insect biotypes, *Annu. Rev. Entomol.* **29**:471–504.

Dlabola, J., 1957, Die Zikaden Afghanistans (Homopt. Auchenorrhyncha), *Mitt. Münch. Entomol. Ges.* **48**:265–303.

Evans, F. C., 1964, The food of Vesper, Field and Chipping Sparrows nesting in an abandoned field in southeastern Michigan, *Am. Midl. Nat.* **72**:57–75.

Evans, F. C., 1975, The natural history of a Michigan field, in: *Prairie: A Multiple View* (M. K. Wali, ed.), pp. 27–51, University of North Dakota Press, Grand Forks, North Dakota.

Farish, D. J., and Scudder, G. G. E., 1967, The polymorphism of *Philaenus spumarius* (L.) (Hemiptera, Cercopidae) in British Columbia, *J. Entomol. Soc. Br. Columbia* **64**:45–51.

Felsenstein, J., 1976, The theoretical population genetics of variable selection and migration, *Annu. Rev. Genet.* **10**:253–280.

Gibson, D. O., 1980, The role of escape in mimicry and polymorphism: I. The response of captive birds to artificial prey, *Biol. J. Linn. Soc.* **14**:201–214.

Guttman, S. I., Wood, T. K., and Karlin, A. A., 1981, Genetic differentiation along host plant lines in the sympatric *Enchenopa binotata* Say complex (Homoptera: Membracidae), *Evolution* **35**: 205–217.

Halkka, O., 1962*a*, Equilibrium populations of *Philaenus spumarius* L., *Nature* **193**:93–94.

Halkka, O., 1962*b*, Polymorphism in populations of *Philaenus spumarius* close to equilibrium, *Ann. Acad. Sci. Fenn. Ser. AIV Biol.* **59**:1–22.

Halkka, O., 1978, Influence of spatial and host-plant isolation on polymorphism in *Philaenus spumarius,* in: *Diversity of Insect Faunas* (L. A. Mound and N. Waloff, eds.), pp. 41–55, Royal Entomological Society of London, London.

Halkka, O., and Halkka, L., 1979, Climatic selection in *Philaenus spumarius, Hereditas* **91**:301–302.

Halkka, O., and Kohila, T., 1976, Persistence of visual polymorphism, despite low rate of predation, in *Philaenus spumarius* (L.) (Homoptera, Aphrophoridae), *Ann. Zool. Fenn.* **13**:185–188.

Halkka, O., and Lallukka, R., 1969, The origin of balanced polymorphism in the spittlebugs (*Philaenus,* Homoptera), *Ann. Zool. Fenn.* **6**:431–434.

Halkka, O., and Mikkola, E., 1977, The selection regime of *Philaenus spumarius* (L.) (Homoptera), in: *Measuring Selection in Natural Populations* (F. B. Christiansen and T. M. Fenchel, eds.), pp. 445–463, Springer-Verlag, New York.

Halkka, O., Raatikainen, M., Vasarainen, A., and Heinonen, L., 1967*a*, Ecology and ecological genetics of *Philaenus spumarius* (L.) (Homoptera), *Ann. Zool. Fenn.* **4:**1–18.

Halkka, O., Raatikainen, M., and Vilbaste, J., 1967*b*, Modes of balance in the polymorphism of *Philaenus spumarius* (L.) (Homoptera), *Ann. Acad. Sci. Fenn. Ser. AIV Biol.* **107:**1–16.

Halkka, O., Raatikainen, M., Halkka, L., and Lallukka, R., 1970, The founder principle, genetic drift and selection in isolated populations of *Philaenus spumarius* (L.) (Homoptera), *Ann. Zool. Fenn.* **7:**221–238.

Halkka, O., Raatikainen, M., Halkka, L., and Lokki, J., 1971, Factors determining the size and composition of island populations of *Philaenus spumarius, Acta Entomol. Fenn.* **28:**83–100.

Halkka, O., Halkka, L., Raatikainen, M., and Hovinen, R., 1973, The genetic basis of balanced polymorphism in *Philaenus* (Homoptera), *Hereditas* **74:**69–80.

Halkka, O., Raatikainen, M., and Halkka, L., 1974*a*, The founder principle, founder selection, and evolutionary divergence and convergence in natural populations of *Philaenus, Hereditas* **78:**73–84.

Halkka, O., Raatikainen, M., and Halkka, L., 1974*b*, Radial and peripheral clines in northern polymorphic populations of *Philaenus spumarius, Hereditas* **78:**85–96.

Halkka, O., Halkka, L., Hovinen, R., Raatikainen, M., and Vasarainen, A., 1975*a*, Genetics of *Philaenus* colour polymorphism: The 28 genotypes, *Hereditas* **79:**308–310.

Halkka, O., Halkka, L., and Raatikainen, M., 1975*b*, Transfer of individuals as a means of investigating natural selection in operation, *Hereditas* **80:**27–34.

Halkka, O., Raatikainen, M., Halkka, L., and Hovinen, R., 1975*c*, The genetic composition of *Philaenus spumarius* populations in island habitats variably affected by voles, *Evolution* **29:** 700–706.

Halkka, O., Raatikainen, M., and Vilbaste, J., 1975*d*, Clines in the colour polymorphism of *Philaenus spumarius* in eastern Central Europe, *Heredity* **35:**303–309.

Halkka, O., Raatikainen, M., and Vilbaste, J., 1976, Transition zone between two clines in *Philaenus spumarius* L. (Hom., Aphrophoridae), *Ann. Entomol. Fenn.* **42:**105–111.

Halkka, O., Raatikainen, M., Halkka, L., and Raatikainen, T., 1977, Coexistence of four species of spittle-producing Homoptera, *Ann. Zool. Fenn.* **14:**228–231.

Halkka, O., Vilbaste, J., and Raatikainen, M., 1980, Colour gene allele frequencies correlated with altitude of habitat in *Philaenus* populations, *Hereditas* **92:**243–246.

Harper, G., and Whittaker, J. B., 1976, The role of natural enemies in the colour polymorphism of *Philaenus spumarius* (L.), *J. Anim. Ecol.* **45:**91–104.

Haupt, H., 1917, Die Varietäten von *Philaenus graminis* Degeer, *Stettiner Entomol. Z.* **78:**173–185.

Hedrick, P. W., 1973, Genetic variation in a heterogeneous environment. I. Temporal heterogeneity and the absolute dominance model, *Genetics* **78:**757–770.

Hedrick, P. W., 1986, Genetic polymorphism in heterogeneous environments: A decade later, *Annu. Rev. Ecol. Syst.* **17:**535–566.

Hedrick, P. W., Ginevan, M. E., and Ewing, E. E., 1976, Genetic polymorphism in heterogeneous environments, *Annu. Rev. Ecol. Syst.* **7:**1–32.

Hoekstra, R. F., Bijlsma, R., and Dolman, A. J., 1985, Polymorphism from environmental heterogeneity: Models are only robust if the heterozygote is close in fitness to the favoured homozygote in each environment, *Genet. Res.* **45:**299–314.

Honek, A., 1984, Melanism in populations of *Philaenus spumarius* (Homoptera, Aphrophoridae) in Czechoslovakia, *Vestn. Cesk. Spol. Zool.* **48:**241–247.

Horsfield, D., 1978, Evidence for xylem feeding by *Philaenus spumarius* (L.) (Homoptera: Cercopida), *Entomol. Exp. Appl.* **24:**95–99.

Hutchinson, G. E., 1963, A note on the polymorphism of *Philaenus spumarius* (L.) (Homopt., Cercopidae) in Britain, *Entomol. Mon. Mag.* **99:**175–178.

Lees, D. R., 1988, Studies on *Philaenus spumarius* in a novel environment: New Zealand, *Tymbal Auchenorrh. Newsl.* **11:**9.

Lees, D. R., and Dent, C. S., 1983, Industrial melanism in the spittlebug *Philaenus spumarius* (L.) (Homoptera: Aphrophoridae), *Biol. J. Linn. Soc.* **19:**115–129.

Lees, D. R., and Stewart, A. J. A., 1988, Localized industrial melanism in the spittlebug *Philaenus spumarius* (L.) (Homoptera: Aphrophoridae) in Cardiff docks, South Wales, *Biol. J. Linn. Soc.* **31:**333–345.

Lees, D. R., Dent, C. S., and Gait, P. L., 1983, Geographic variation in the colour/pattern polymorphism of British *Philaenus spumarius* (L.) (Homoptera: Aphrophoridae) populations, *Biol. J. Linn. Soc.* **19:**99–114.

Levene, H., 1953, Genetic equilibrium when more than one ecological niche is available, *Am. Nat.* **87:**331–333.

McEvoy, P. B., 1986, Niche partitioning in spittlebugs (Homoptera: Cercopidae) sharing shelters on host plants, *Ecology* **67:**465–478.

Müller, H. J., 1987, Über die Vitalität der Larvenfortmen der Jasside *Mocydia crocea* (H.-S.) (Homoptera Auchenorrhyncha) und ihre ökologische Bedeutung, *Zool. Jahrb. Syst.* **114:**105–129.

Müller, H. J., 1988, Die Vitalität der aus verschiedenen Larvenformen hervorgegangenen Adulten von *Mocydia crocea* (H.-S.) (Homoptera Auchenorrhyncha: Cicadellidae) bei der Überwinterung, *Zool. Jahrb. Syst.* **115:**117–127.

Nielson, M. W., and Lehman, W. F., 1980, Breeding approaches in alfalfa, in: *Breeding Plants Resistant to Insects* (F. G. Maxwell and P. R. Jennings, eds.), pp. 277–312, Wiley, New York.

Owen, D. F., and Wiegert, R. G., 1962, Balanced polymorphism in the meadow spittlebug, *Philaenus spumarius, Am. Nat.* **96:**353–359.

Palmén, E., 1944, Die anemohydrochore Ausbreitung der Inskten als zoogeographischer Faktor. Mit besonderer Berücksichtigung der Baltischen Einwanderungsrichtung als Ankunftsweg der Fennoskandischen Käferfauna, *Ann. Zool. Soc. Zool.-Bot. Fenn. 'Vanamo'* **10:**1–262.

Pianka, E. R., 1974, Niche overlap and diffuse competition, *Proc. Natl. Acad. Sci. USA* **71:**2141–2145.

Raatikainen, M., 1971, The polymorphism of *Philaenus spumarius* (L.) (Homoptera) in Northern Italy, *Ann. Entomol. Fenn.* **37:**72–79.

Saura, A., Halkka, O., and Lokki, J., 1973, Enzyme gene heterozygosity in small island populations of *Philaenus spumarius* (L.) (Homoptera), *Genetica* **44:**459–473.

Sheppard, P. M., 1958, *Natural Selection and Heredity,* Hutchinson, London.

Slatkin, M., 1987, Gene flow and geographic structure of natural populations, *Science* **236:**787–792.

Stewart, A. J. A., and Lees, D. R., 1987, Genetic control of colour polymorphism in spittlebugs (*Philaenus spumarius*) differs between isolated populations, *Heredity* **59:**445–448.

Stewart, A. J. A., and Lees, D. R., 1988, Genetic control of colour/pattern polymorphism in British populations of the spittlebug *Philaenus spumarius* (L.) (Homoptera: Aphrophoridae), *Biol. J. Linn. Soc.* **34:**57–79.

Svala, E., and Halkka, O., 1974, Geographical variation of frontoclypeal colour polymorphism in *Philaenus spumarius* (L.) (Homoptera), *Ann. Zool. Fenn.* **11:**283–287.

Templeton, A. R., 1980, Modes of speciation and inferences based on genetic distances, *Evolution* **34:**719–729.

Thompson, V., 1973, Spittlebug polymorphism for varning coloration, *Nature* **242:**126–128.

Thompson, V., 1984, Distributional evidence for thermal melanic colour morphs in *Philaenus spumarius,* the polymorphic spittlebug, *Am. Midl. Nat.* **89:**348–359.

Thompson, V., 1988, Parallel colour form distributions in European and North American populations of the spittlebug *Philaenus spumarius* (L.), *J. Biogeogr.* **15:**507–512.

Thompson, V., and Halkka, O., 1973, Colour polymorphism in some North American *Philaenus spumarius* (Homoptera: Aphrophoridae) populations, *Am. Midl. Nat.* **89:**348–359.

Vilbaste, J., 1968, *Über die Zikadenfauna des Primorje Gebietes,* Estonian SSR Academy of Sciences [in Russian, with German summary].

Vilbaste, J., 1980, On the Homoptera-Cicadina of Kamtchatka, *Polska Akad. Nauk Inst. Zool. Ann. Zool. 35* **24:**367–418.

Wallace, B., 1968, Polymorphism, population size and genetic load, in: *Population Biology and Evolution* (R. Lewontin, ed.), pp. 87–108, Syracuse University Press, Syracuse, New York.

Weaver, C. R., and King, D. R., 1954, Meadow spittlebug *Philaenus leucophthalmus* (L.), *Ohio Agric. Exp. Stn. Res. Bull.* **741:**1–99.

Whittaker, J. B., 1969, The biology of Pipunculidae (Diptera) parasitising some British Cercopidae (Homoptera), *Proc. R. Entomol. Soc. Lond. Ser. A Gen. Entomol.* **44:**17–24.

Whittaker, J. B., 1972, Polymorphism in *Philaenus spumarius* (Homoptera) in USSR, *Oikos* **23:** 366–369.

Wiegert, R. G., 1964, Population energetics of meadow spittlebugs (*Philaenus spumarius* L.) as affected by migration and habitat, *Ecol. Monogr.* **34:**217–241.

Witsack, W., 1973, Experimentell-ökologische Untersuchungen über Dormanz-Formen von Zikaden (Homoptera Auchenorrhyncha). 2. Zur Ovarial-Parapause und oblikatorische Embryonal-Diapause von *Philaenus spumarius* (L.) (Aphrophoridae), *Zool. Jahrb. Syst.* **100:**517–562.

Wright, S., 1982, The shifting balance theory and macroevolution, *Annu. Rev. Genet.* **16:**1–19.

Zajac, M. A., and Wilson, M. C., 1984, The effects of nymphal feeding by the meadow spittlebug, *Philaenus spumarius* (L.) on strawberry yield and quality, *Crop Prot.* **3:**167–175.

6

Pollination of Terrestrial Orchids of Southern Australia and the Mediterranean Region

Systematic, Ecological, and Evolutionary Implications

A. DAFNI and P. BERNHARDT

INTRODUCTION

Most pollination biologists studying the Orchidaceae would agree that Darwin (1877) remains seminal to the development of the discipline. However, the work of Darwin emphasized temperate, terrestrial species, often those populations adjacent to his country home. Consequently, his examination of the adaptive floral morphology of tropical epiphytes is best regarded as secondary. Darwin completed no fieldwork on any neotropical orchid taxon, so his conclusions are based on the reports of correspondents and extrapolation following the examination/dissection of greenhouse specimens. His work did provide direct stimulus to contemporary naturalists to attempt field studies in the tropics (Allan, 1977; Bernhardt, 1986).

Ironically, as we conclude the latter half of the twentieth century, the original preference for fieldwork has completely shifted. Most modern reviews of orchid pollination now emphasize contributions made on tropical epiphytic/lithophytic genera (Dodson, 1962; van der Pijl and Dodson, 1966; Dressler, 1981; Ackerman, 1986). Furthermore, most orchid pollination studies in the neotropics have tended to concentrate, with some notable exceptions, on the coevolution of orchids and euglossine bees (Dressler, 1968; Williams, 1982).

A. DAFNI • Institute of Evolution, Haifa University, Haifa 31999, Israel. P. BERNHARDT • Department of Biology, St. Louis University, St. Louis, Missouri 63103.

Such emphases are appropriate, as the Epidendroideae *sensu lato* remains the largest subfamily in the Orchidaceae. Furthermore, the Orchidaceae shows its greatest diversity within the montane tropics and research must progress logically from these centers of radiation.

We offer one critical argument to this commendable program of research. There are four subfamilies of monandrous orchids (*sensu* Burns-Balogh and Funk, 1986) which may be distinguished by column structure, and the gross morphology of this organ directly influences construction and dispersal of pollinia. The epidendroid orchids reflect the most specialized modifications to the column and pollinarium (Burns-Balogh and Bernhardt, 1985). Furthermore, adaptive radiation of many neotropical epidendroids depends on the exploitation of comparatively few euglossine bee taxa and the euglossines are entirely restricted to the neotropics (Michener, 1979). Therefore, while concentrating on the pollination biology of epidendroids will undoubtedly explain trends in floral evolution within Epidendroideae, it does not appear sufficient to understand floral evolution with the family, Orchidaceae.

To redress this unintentional imbalance, we respond with a review of nontropical taxa. There are two excellent reasons for combining studies in southern Australia (south of the Tropic of Capricorn) and the Mediterranean (i.e., Mediterranean Europe and the Middle East). First, neither region is dominated by epiphytic Epidendroideae. This permits an examination of the species in the neglected subfamilies Neottioideae, Orchidoideae, and Spiranthoideae (*sensu* Burns-Balogh and Funk, 1986). Second, parallelisms in environmental pressures permit us to compare levels of convergencies in two orchid floras that show a high degree of endemism and discrete speciation.

BIOGEOGRAPHY AND PHENOLOGY

Geography and Diversity

The Australian continent is regarded as remarkably depauperate in orchid taxa in proportion to its size. There are only 85 genera recognized, representing less than 500 species (*sensu* Clements, 1982). The epiphytic species are confined, primarily, to relictual rain forest on the north coast and to the moist-temperate jungle gullies as far southeast as Gippsland (eastern Victoria). None of these epiphytic genera is regarded as endemic, nor is the one representative of the relictual subfamily Apostasioideae, *Apostasia* (Dockrill, 1969). *Apostasia,* all epiphytes, and certain terrestrial but tropical genera (e.g., *Habenaria*) are regarded as disjuncts of the paleotropical flora isolated following the breakup of the Indo-Malaysian landbridge (Barlow, 1981). These taxa of moist-tropical zones make up 49% of the native orchid flora. The remainder of the Australian orchid taxa are terrestrial and monandrous, but cannot be classified exclusively within a broad epidendroid–vandoid context. These terrestrial genera show their greatest diversity along the southern coast of the continent and about 29 genera have been interpreted traditionally as autocthonous elements or at least endemic to temperate Australasia (i.e., Australia, Tasmania, and New Zealand) (Lavarack, 1981). Approximately 23% of the terrestrial genera are monotypic. The distribution of orchids on the Australian continent and Tasmania appears restricted due primarily to climatic changes stressing aridity and the trend toward sclerophyllic vegetation (Dockrill, 1969; Barlow, 1981). Only two geophytic orchid species are distributed within central Australia (Weber, 1981) and this arid region comprises over one-third of the continent. For this reason, Australian Orchidaceae are often described as coastal in distribution, although it is fallacious to suggest that the diversity of terrestrial species is highest within a few kilometers of the sea. Although these terrestrial orchids have been considered unique to the Australian flora, they are uncommon in the relicts of Gondwanian climax forest (Barlow, 1981), and these Tasmanian forests are neither centers of diversity for the terrestrial orchids nor ''living museums'' of the most primitive orchid genera. Rather, the diversity of southern Australian terrestrial species is highest in open woodlands and acidic heaths dominated by dwarf shrubs in the Epacridaceae and Myrtaceae. The diversity of the Australian terrestrials peaks in the Mediterrancan climate of the southwest. A second, smaller region of diversity is found in the temperate southeast. The terrestrial orchid flora competes unsuccessfully with the sclerophyllous, woody dicots. Density of orchids appears highest in those regions subject to cyclical summer fires (Gill, 1981).

The European–Mediterranean orchid flora comprises about 250 species (Baumann and Kuenkele, 1982) and all of these are terrestrial. Therefore, specific diversity of geophytic orchids in the Mediterranean closely parallels that of southern Australia. However, the main Mediterranean genera are *Ophrys* (about 20–50 species), *Orchis* (25–40), *Dactylorhiza* (5–28), and *Cephalanthera* (2–10). Taxonomic parameters remain unsettled, but 75% of the orchid genera in the Mediterranean appear to be monotypic. Unlike the monotypic genera of southern Australia, all Euro-Mediterranean monotypes tend to share a low growth habit, small, often dull-colored flowers, and functional nectaries (in direct contrast to the tall, large-flowered, often nectarless and dominant genera, *Orchis* and *Dactylorhiza*). Furthermore, the Mediterranean monotypes show a capacity to colonize montane habitats and may be less common in Mediterranean lowlands.

Regions receiving true Mediterranean climates through Eurasia (Aschmann, 1973) prove to be the real centers of diversity for Orchidaceae. While some of these species are distributed as far north as Scandinavia, orchid diversity declines as one moves further north (e.g., 20 species in Norway versus 100 in Greece) and toward arid zones of southwest Asia (e.g., Turkey, 86 species; Syria, 40; Israel, 29; Egypt, 1). As a tropical, Laurasian origin for the family Orchidaceae has never been challenged (Raven and Axelrod, 1974), it is presumed that the Mediterranean genera derive from eastern ancestors and show a secondary invasion of the northern area of the African plate.

The habitats in the Mediterranean supporting the most diverse orchid floras, as in southern Australia, are the exposed, warm, dwarf-shrub communities (phrygana) growing on calcareous in lieu of acidic sites. As many as 50 orchid species have been found in a single phrygana. Once again, geophytic orchids become less common in subalpine ecosystems of the Mediterranean.

Comparative Floral Phenology

The orchid floras of both regions exhibit their greatest diversity from early to mid spring, with few species in flower from late summer to early winter. In southern Australia, for example, geophytic orchids bloom throughout the year, although February–July (late summer–mid winter) remains extremely depauperate in flowering taxa. In southern Australia, however, seasonal and intraseasonal variation in taxonomic diversity are also marked by general shifts in pollination syndromes that occur in three overlapping pulses.

The first pulse peaks from late winter until early spring and is dominated by genera that exploit mycetophilid gnats, e.g., *Acianthus, Corybas,* and *Pterostylis*. These genera flower while both Mediterranean-type scrublands and temperate woodlands receive winter rains and the sporocarps of fungi emerge to complete the life cycles of the fungus gnats. Consequently, these orchid flowers employ biochemical and morphological modifications (see pp. 220, 212, 218) that exploit the feeding, mating, and ovipositional habits of the gnats (Vogel, 1978; Beardsell and Bernhardt, 1982). The diversity of the fungus gnat orchids are in decline by late spring, overlapping with the flowering comprising the second pulse.

Orchid taxa comprising this second pulse flower from early spring until early autumn, although phenological diversity is greatest in mid spring. These flowers offer no nectar (p. 210) and are pollinated primarily by bees and wasps. These flowers are either food mimics or sexual mimics (p. 213). Sexual mimics are dependent primarily on male wasps representing three families. Orchid taxa exploiting sexual mimesis are almost absent from the southern Australian flora by mid summer. However, putative food mimics may show limited visibility through early autumn (e.g., *Eriochilus* species) and probably play a minor role in the late phenology of subalpine zones [e.g., *Thelymitra venosa*(Burns-Balogh and Bernhardt, 1988)].

The third pulse peaks from mid spring until early summer and concludes by

TABLE I. Nectariferous Orchids—Mediterranean Area

Genus (subfamily)[a]	Number of species (number studied)	Pollinator[b]	Reference
Platanthera[c] (O)	6–7 (1) (2)	HW, BT (2)	Nilsson (1978); Hammersted (1980; cited in Nilsson, 1983*b*)
Anacamptis (O)	1	BT	Darwin (1877)
Orchis[d] (O)	3 (2) (3)	SB (2), HB, CB, SW, BT (1)	Voeth (1975, 1984); Peisl and Forster (1975); Dafni and Ivri (1979)
Epipactis (N)	(7)	SW (5), SY (3), CB (2) FL, AN, SB, SA (1)	Compiled by Burns-Balogh *et al.* (1987)
Neottia (N)	—	FL	Compiled by Burns-Balogh *et al.* (1987)
Limodorum (N)	2 (1)	SB, CB	Compiled by Burns-Balogh *et al.* (1987)
Listera (N)	2 (1)	IE, SB, IC	Nilsson (1981) (and sources cited there)
Herminium (O)	(1)	PD	Nilsson (1981) (and sources cited there)
Himantoglossum (O)	(3)	SB (2), HB (1)	Teschner (1975)
Coeloglossum (O)	(1)	SW, WA, BE	van der Pijl and Dodson (1966)

[a]Subfamilies (*sensu* Burns-Balogh and Funk, 1986): S, Spiranthoideae; N, Neottiodeae; O, Orchidoideae; E, Epidendroideae.
[b]Pollinator: AN, ants; BE, beetles; BT, butterflies; CB, carpenter bees; EB, eusocial bees; FG, fungus gnats; FL, flies; HB, honey bees; HW, Hawkmoths; IC, ichneumonid wasps; PD, parasitic dipterans; SA, sawflies; SB, solitary bees; SY, syrphids; WA, Wasps.
[c]The genus includes about 50 species, but only six or seven are present in the region.
[d]The other species in the genus are nectarless.

mid summer (*Spiranthes sinensis*) (Cady and Rotherham, 1970). It is comprised of small-flowered, nectariferous flowers usually arranged in tight racemes which are pollinated by insects with short mouthparts (p. 200). Pollinia vectors are recorded in the orders Coleoptera, Diptera, Hymenoptera, and Lepidoptera.

Of the three pulses described above, the second pulse shows the greatest degree of taxonomic diversity as well as the longest duration. Exceptions to these flowering pulses appear to be dependent on latitude and climate. Western Australian, Mediterranean-type populations appear to flower earlier than eastern, temperate populations. Northern populations appear to flower earlier than southern populations. In western Australia, *Spiculea ciliata* is a tiphiid wasp sexual mimic common to granite outcrops, yet it flowers at the height of summer (late

TABLE II. Nectariferous Orchids—Australia[a]

Genus (subfamily)	Number of species (number studied)	Pollinator	Reference
Acianthus (E)	6 (1)	FG (1)	Jones (1974)
Habenaria[b] (O)	17 (2)	BT (2)	Dixon (1986)
Microtis (N)	10 (2)	AN (1), WA (1)	Jones (1975), Bates (1981)
Peristeranthus[b] (E)	1 (1)	BE (1)	Wallace (1980)
Prasophyllum (S)	83 (9)	FL (4), BE (2), SY, SY (1), SB (1), WA (2)	Bates (1984*b*), Beardsell and Bernhardt (1982), Bernhardt and Burns-Balogh (1986*b*), Coleman (1933*a*), Jones (1972)
Spiranthes (S)	1 (1)	BT? (1)	Cady and Rotherham (1970)

[a]See Table I for subfamily and pollinator abbreviations.
[b]Tropical distribution.

November–January) as water and nutrients may be stored in the upper peduncle following rapid disintegration of the root system with the cyclical summer drought (Stoutamire, 1986).

The Mediterranean regions of Europe and the Middle East are distinguished by a divisive dry versus wet season (Aschmann, 1973). Furthermore, orchids of these regions are distributed primarily in areas receiving at least 400 mm of annual rainfall, although population density declines where summers are wet (e.g., *Ophrys*) (Kullenberg *et al.*, 1984*a*). Unlike most Australian orchids, many Mediterranean species (e.g., *Ophrys* and *Orchis* species) are common to disturbed habitats created by human activity (pastures, road sides, excavations, etc.). Such regions are also rich in solitary bees (Michener, 1974), and these insects remain the dominant pollinators of most orchid flowers (Tables I and II).

In the southern Mediterranean region, orchids flower as early as late November (late autumn) and cease by July, with the peak flowering season from March to April. Toward the northern limits of their range the orchid floral phenology extends from June to August, probably due to the prevalence of lower temperatures. The distinctive pollination pulses are not found in the Mediterranean taxa, but floral phenology may determine how the bees are rewarded. In general, deceptive flowers, often based on the exploitation of recently emerged (naive) bees, appear earlier in the flowering season, while the species offering edible rewards flower later in the season. This parallels the bee-pollinated orchids of southern Australia comprising the second and third pulses. The proportion of orchid species pollinated by some type of deceit syndrome remains higher through the southern Mediterranean compared to the north, as this area remains the center of diversity for the genera *Cephalanthera, Dactylorhiza, Ophrys,* and *Orchis* (pp. 195, 196). The density and diversity of solitary bees and wasps appear to coincide with the peak orchid flowering periods for each region of the Mediterranean.

VEGETATIVE REPRODUCTION

The vast majority of geophytic Orchidaceae bear some sort of fleshy, subterranean storage organs (Dressler, 1981). These storage organs may commonly control the extent of asexual increase via ramets. This mode of reproduction has been recorded often in spiranthoid and neottioid taxa in southern Australia, as adventitious roots will produce new root-stem tuberoids. The root-stem tuberoid is virtually the only portion of the orchid that survives the extended period of dormancy (Pate and Dixon, 1981). Extensive clonal colonies of *Chiloglottis, Caladenia,* and *Thelymitra* have been observed by P. Bernhardt (unpublished). In fact, this mode of asexual reproduction is so successful that cultivated stocks have been increased by the establishment of a Tuber Bank by the Australian Native Orchid Society. *Gastrodia, Dipodium,* and *Epipogium* are epidendroid genera, but they also increase by fleshy storage organs in Australia. In these genera, however, the storage organ is a rhizome or a stem-tuber consisting of cylindrical swellings of lateral shoots. In *Gastrodia,* there are lateral additions to the original stem-tuber each season, and each new stem tuber may live several years (Pate and Dixon, 1981).

Vegetative reproduction of rhizomatous orchids remains a rare phenomenon in the Mediterranean region, although it has been recorded in *Cephalanthera, Epipactis,* and *Limodorum.* These rhizomatous orchids sometimes create dense colonies when they inhabit forests. The same taxa occur in lower densities in open habitats. The remaining genera bear root-stem tuberoids (e.g., *Ophrys, Orchis, Dactylorhiza*), but tend only to produce a solitary storage organ each year to replace that of the previous season.

POLLINATION MECHANISMS

Reward Systems

The most frequent floral rewards are nectar and pollen (Faegri and van der Pijl, 1979), but other sources have been documented (e.g., edible oils, food bodies, floral fragrances), (Simpson and Neff, 1981). The pollen of monandrous orchids is usually organized as compact pollinia and rarely consumed by the pollinator (p. 204). The most common edible reward produced by geophytic orchids in Australia and the Mediterranean remains nectar and/or stigmatic secretions. Comparatively few orchid genera in southern Australia bear functional nectaries compared to genera of the Mediterranean. The presentation of nectar differs in both regions due to differences in the taxonomy and/or feeding morphology of the dominant pollinators.

Nectariferous Flowers

Taxonomic limitations of nectar secretion in the Australian orchids seem paradoxical. Only six genera are known to bear flowers with functional nectaries (Table II): *Acianthus* (Epidendroideae), *Habenaria* (Orchidoideae), *Microtis* (Neottioideae), *Peristeranthus* (Epidendroideae), *Prasophyllum* (Spiranthoideae), and *Spiranthes* (Spiranthoideae). Limited nectar secretion may occur in *Corybas* (Epidendroideae) (Jones, 1970) and possibly in *Genoplesium sensu stricto* (Neottioideae). *Peristeranthus* is monotypic and tropical. *Habenaria,* while almost pandemic, is also confined to the Australian tropics. Furthermore, only *Spiranthes sinensis* is found in Australia and that same species is distributed from southern Australia north to Japan! At first glance, it would seem that nectariferous species are extremely rare in southern Australia. However, *Prasophyllum sensu lato* represents the second largest genus in Australia (Clements, 1982) and shows peak diversity in the south.

Nectar concealed within long floral tubes (cuniculi) or at the tip of hollow spurs are almost absent in the orchids of south Australia. Only the perianth of *S. sinensis* offers a narrow, constricted tube (Fig. 1), but this tube appears considerably shorter compared to many outcrossing *Spiranthes* species on other continents (Luer, 1975). In all other southern Australian genera, regardless of subfamily, nectar always collects at the base of the labellum and either the whole labellum or the hypochile functions as a very shallow cup (Beardsell and Bernhardt, 1982).

Convex droplets of stigmatic fluid are commonly observed in mature flowers of the Australian orchids. The exudate has been referred to as a ''nectar'' by some authorities, as insects have been observed consuming the stigmatic secretions of *Thelymitra, Diuris,* and *Corybas* (Cady and Rotherham, 1970; Jones, 1970; Coleman, 1933*b*, 1932). Whether such droplets serve as a consistent ''nectar substitute,'' however, remains debatable. In general, the Neottioideae tend to produce a copious, stigmatic exudate, as they bear a liquid to semiliquid viscidium. Furthermore, the convex droplet is typical in pollen/stigma interactions requiring the erosion of friable/sectile pollinia (Burns-Balogh and Bernhardt, 1985). Where insects have been observed drinking stigmatic fluid, it is important to note that there are few records of pollinia becoming attached to the mouthparts or to ventral positions on the head. The consumption

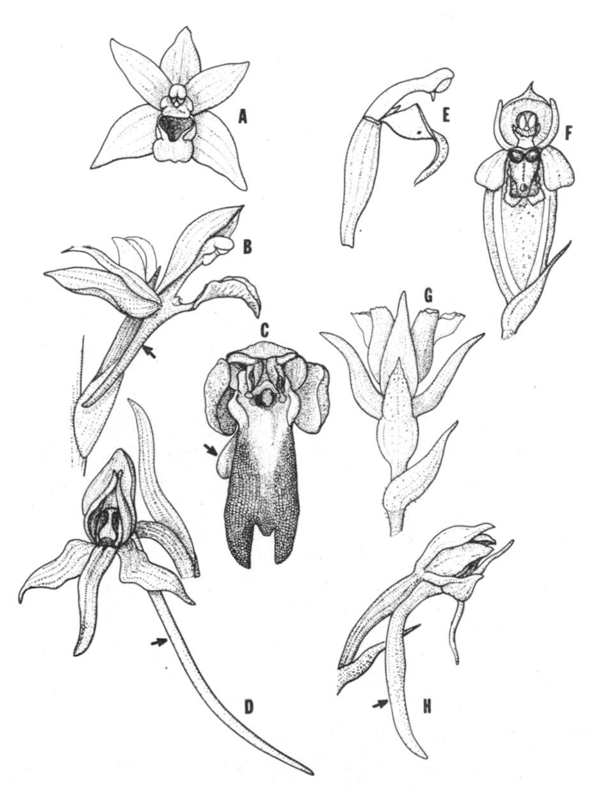

FIG. 1. Nectariferous orchid genera of (A–D) the Mediterannean region versus (E-H) Australia. Arrows indicate nectar-filled spurs or pouches. (A) *Epipactis purpurata* (×4); (B) *Limodorum abortivum* (×4); (C) *Coeologlossum viride* (×21); (D) *Platanthera chlorantha* (×5); (E) column-labellum orientation in *Acianthus exsertus* (sepals and lateral sepals removed) (×7.5); (F) *Microtis biloba* (×19); (G) dorsal view of *Spiranthes sinensis* (showing campanulate perianth) (×11); (F) *Habenaria fernandii* (tropical) (×10).

of stigmatic fluid by syrphid flies or halictid bees appears to be a form of robbery. Microdipterans often die after they attempt to feed on stigmatic sections, becoming glued to the lateral lobes of some *Thelymitra* species (P. Bernhardt, personal observation). The most important exception to this rule has been found in *Thelymitra antennifera*. Ventral depositions of soft pollinia are found on the head, mouthparts, and thorax of naturalized *Syrphus* species and native perugid wasps that have been observed attempting to drink stigmatic fluid (Dafni and Calder, 1987). However, the actual mechanism of pollinia dispersal was not observed in this orchid, and pollination probably involves mimetic ornamentation on the anther that directs the foraging insect toward the underlying stigma. A combination of reward and deception, as has been described in the

Mediterranean orchids [e.g., *Dactylorhiza fuchsii* (Dafni and Woodell, 1986)] may occur in some Australia genera (p. 204).

The shallow, exposed presentation of floral nectar can be attributed to the pollinator spectrum in southern Australia. The majority of Australian orchids are pollinated by short-tongued colletid and halictid bees as well as short-tongued syrphid flies, wasps, and beetles. Although long-tongued bee families (e.g., Megachilidae, Apidae, and Anthorphoridae) are found in Australia, they are not so diverse as the Halictidae and Colletidae. Furthermore, the Syrphidae and wasp families, like the Tiphiidae, have undergone intense radiation in Australia (Armstrong, 1979). Consequently, floral presentation in the nectariferous orchids of the south represents a trend toward domination by short-tongued foragers. As usual, the pollinia of the nectariferous orchids are deposited on the head and dorsally on the thorax. In *Peristeranthus* species (Wallace, 1980) and a few *Prasophyllum* species (Coleman, 1933*a*) beetle pollination may result in some dorsally deposited pollinia on the abdomen. In contrast, the tropical *Habenaria* species bear the same long spur indicative of their congeners in the Mediterranean and in most of the rest of the world. Unlike the nectariferous genera of the south, these tropical *Habenaria* species appear to be pollinated by long-tongued, tropical Lepidoptera (Dixon, 1986).

It must be noted, though, that the actual degree of specialization between nectariferous orchids of the south and the mouthparts of their insects remains unclear. While *Microtis parviflora* is known to be pollinated by worker ants (Jones, 1975), the pollinator spectrum of *Microtis unifolia* also includes wasps and beetles (Bates, 1981; Beardsell and Bernhardt, 1982). The pollination system of *Prasophyllum odoratum* emphasizes syrphid flies, but male colletid bees are of secondary importance (Bernhardt and Burns-Balogh, 1986*b*). In *Prasophyllum elatum* and its allies, the dark, mildly scented flowers are pollinated almost exclusively by male tiphiid wasps. The female wasps are wingless and arrive at the flowers only *en copula* (see below, Fig. 7) (Bates, 1984*b;* Bernhardt, 1987).

While most of the monotypic genera of the Mediterranean region are nectariferous, only one large genus (*Epipactis*) offers nectar (Table I). The multispecific genera *Orchis, Dactylorrhiza,* and *Ophrys* are dominated by deceptive mechanisms, with nectariferous taxa in the minority.

Mediterranean species that offer nectariferous rewards are found in both the Neottieae and Orchidinae. With the exception of *Epipactis,* though, these taxa do not expose nectar on their labella (Fig. 1), but conceal the reward in short spurs (*Himantoglossum, Herminium*) or long spurs (*Platanthera, Anacamptis*). The presentation of stigmatic fluid as an edible reward is very rare (Dafni and Woodell, 1986). However, as in the Australian genera, most of the nectariferous orchids in the Mediterranean cluster many flowers along racemose inflorescences. *Platanthera* and *Habenaria* are often regarded as closely allied, ''sister genera'' (Luer, 1975). Dressler (1981) notes that Lepidopteran pollination is indicated for most of the Habenariinae. It is not surprising, then, to observe that butterflies and moths dominate the pollination systems of tropical Australian *Habenaria* and Mediterranean *Platanthera* species (Tables I and II).

Nectariferous orchids of the Mediterranean region do not show quite the same trend toward pollination by short-tongued insects and successful transfer of pollinia appears less generalist than in southern Australia. For example, although *Platanthera chlorantha* is pollinated by 28 species of Mediterranean insects, only 6 insect species are responsible for 97% of the total pollinations (Nilsson, 1978). The same narrow range of consistent successful pollinators appears true for *Epipactis palustris* (Brantjes, 1981; Nilsson, 1978), *Herminium monorchis* (Nilsson, 1979), *Orchis coriophora* (Dafni and Ivri, 1979), and *Dactylorhiza maculata* ssp. meyeri (Voeth, 1983). It is possible that the pollinators of nectariferous orchids in southern Australia also show this trend toward narrow polylectisism (*sensu* Michener, 1979), but documentation must wait for a complete insect taxonomy.

References to the pollination of nectariferous orchids by beetles (Tables I and II; *Coeloglossum, Peristeranthus, Prasophyllum*) should not be dismissed for either the Mediterranean or Australia. Classic concepts of cantharophily are changing to show that the digestive systems of certain beetles are modified for the consumption of pollen *and* nectar (Crowson, 1981) and the flower-visiting scarab genus *Amphicoma* has a well-developed color vision (Dafni *et al.*, in press). In Australia, buprestid, cerambycid, and cantharid species forage on the shallow nectaries of endemic Myrtaceae and Burseraceae and are regarded as true pollinators (Hawkeswood, 1981, 1987) of the shrubs and trees.

Shelter

Shelter flowers are unknown in Australia, but the flowers of Mediterranean *Serapias* species offer insects a shelter in unfavorable weather conditions and a nocturnal lodge (Godfrey, 1925, 1931; Vogel, 1975; Gumprecht, 1977; Voeth, 1980; Dafni *et al.*, 1981). Since the flowers act as a solar heater which can reach up to 3°C above the ambient temperature, the poikilothermic insects receive an energy gain (Dafni *et al.*, 1981). The suggestion is that the flowers mimic a sleeping hole for the male solitary bees (Fig. 2), which are the main pollinators of the orchid (Voeth, 1980; Dafni *et al.*, 1981). Richard (1982) also noted that insects sleep in the tunnel-shaped flowers of *Cephalanthera* species, which results in pollination.

Combined Systems of Reward and Deception

Deceptive mechanisms may appear concurrently with rewards in the same orchid flower. The ultimate result is an increase in capsule production. This combination of syndromes is found in taxa of both regions, but the extent of such

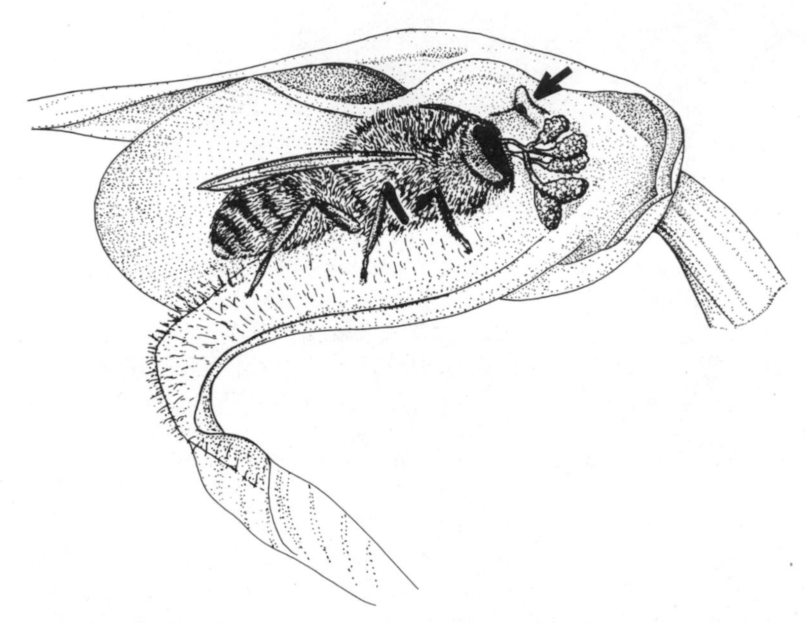

FIG. 2. Transverse section of the floral "sleep chamber" of *Serapias vomeracea* showing an inactive *Eucera* sp. (×15). The floral segments form a subcampanulate tube, which becomes more constricted at its base, so the bee must sleep with its head under the column (arrow), permitting the viscidium to make dorsal contact with the head. The presence of more than one pollinarium on the bee indicates it has entered more than one sleep chamber perhaps over a couple of days.

pollination systems is not fully understood in southern Australia. The nectar glands of Australian *Corybas* has been discussed, but it is clear that the flower adopts a brood-site deception syndrome (p. 212). Members of the genera *Diuris* and *Thelymitra* are regarded as food-deception flowers (p. 210), but they produce pollinia that are so friable that monads or tetrads may be scraped off the visicidia and added to the scopal load of bees. Presumably, female bees attempt to feed their larvae the pollen of *Diuris maculata* (Beardsell *et al.*, 1986) and *Thelymitra pauciflora* (J. Armstrong, personal communication). The consumption of stigmatic secretions produced by these flowers suggests overlapping syndromes. Although *Prasophyllum elatum* is nectariferous and wasp-pollinated, it is scentless (to humans) and the dark maroon/iodine pigmentation is reminiscent of the pseudocopulatory systems which attract male tiphiid wasps (p. 213).

A few Mediterranean genera present pollination syndromes consistently interpreted as a combination of deception and reward. *Dactylorhiza fuchsisi* offers stigmatic fluid as a reward which is used by honeybees, while bumblebees are deceived by the empty spurs (Dafni and Woodell, 1986). *Epipactis consimilis* offers nectar, but the labellum mimics aphids. Aphidophagous male and female hoverflies consume the nectar, but the females also attempt to oviposit on the prey models (Ivri and Dafni, 1977). *Limodorum abortivum* offers nectar which is consumed by large, solitary bees, while small bees are attracted to the flowers due to color convergence with *Cistus creticus* (Cistaceae) and these small bees attempt to collect the pollinia (Burns-Balogh *et al.*, 1987).

Food Deception Syndromes

It is selectively advantageous for an insect to associate a suite of attractants (floral shape, pigmentation, and scent) with edible rewards (nectar, pollen, oils). Although the response to a specific range of attractants appears to be innate in most insects, some insect taxa appear to prefer different suites within a habitat over time. Some plant taxa exploit the conditioning of foragers to certain floral models and/or exploit naive foragers, which are enticed by a wide variety of visual and olfactory cues prior to initial conditioning. If an orchid is a fraud flower, it may imitate the suite of attracting cues [e.g., floral mimicry in *Thelymitra* (Bernhardt and Burns-Balogh, 1987)] or it may imitate a portion of the model, such as the anther cone or staminal cluster [e.g., pseudanthery (Dafni, 1984)].

The food-mimic orchids of both southern Australia and the Mediterranean region detain floral foragers with dummy structures devoid of either nectar or pollen. A portion of the floral perianth will be modified to form a spur, pouch, or tubular throat which never fills with nectar. Of course, these nectar mimics share habitats with a number of coblooming taxa bearing nectar-filled cups, tubes, or spurs. Furthermore, some of the nectariferous orchids of the Mediterranean region bear nectar-filled organs (e.g., *Platanthera* and *Gymnadenia*). The nectar mimics appear to have been derived from nectariferous ancestors and often belong to genera which still retain one or more species bearing flowers with nectar-filled spurs [e.g., *Orchis* (Dafni, 1987)].

The empty spur or tube mode of presentation is almost absent in the terrestrial orchids of southern Australia, unless one includes *Dendrobium* species with short spurs or pouches (Calder *et al.*, 1982). However, this epidendroid genus is lithophytic, not geophytic, and is best regarded as a paleotropical taxon on the edge of its cool southern distribution (Lavarack, 1981). The dichotomy in floral form between the Mediterranean and southern Australian regions once again reflects the comparatively low selective pressure of long-tongued insects on the geophytic orchid communities of southern Australia.

Dummy anthers, on the other hand, are found on the flowers of terrestrial orchids in both regions (Figs. 3 and 4). The labellum often bears round, dark spots or a large, yellow, wrinkled callus and/or is ornamented by a series of stalked calli bearing globose-acuminate heads that are heavily cutinized,

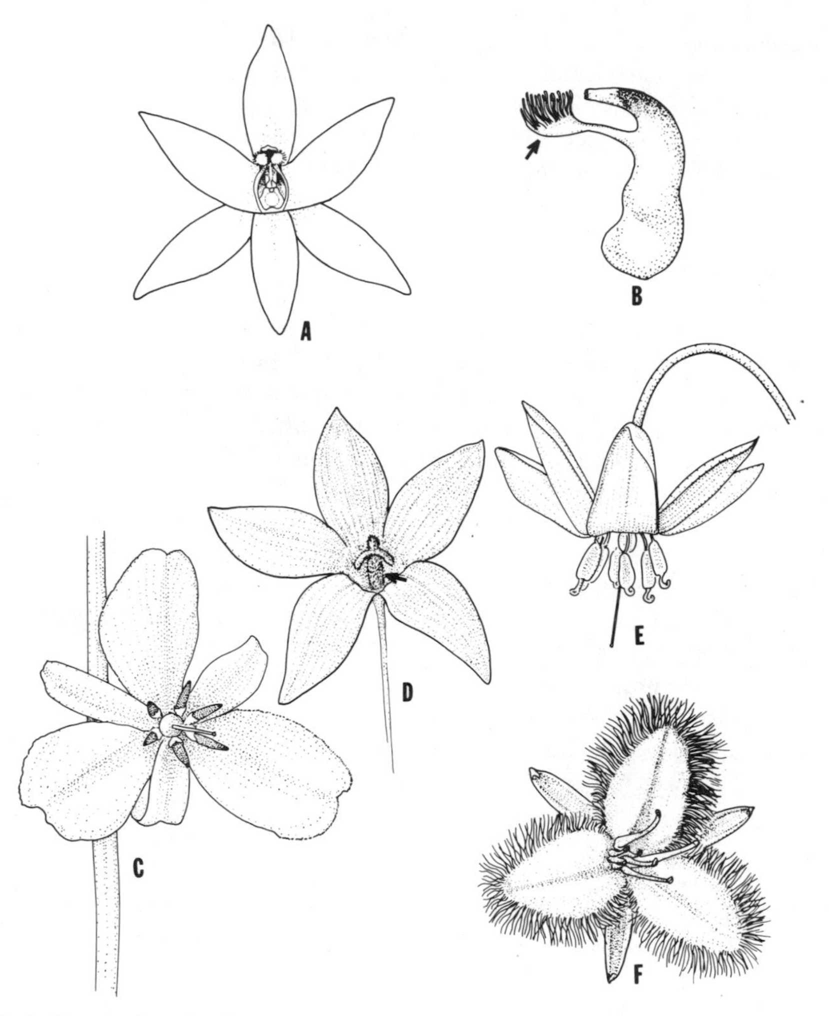

FIG. 3. Floral mimesis of the vernal guild of nectarless, blue-purple pollen flowers by *Thelymitra* spp. (A) Schematic drawing of a flower in the *Thelymitra nuda-pauciflora* complex (×3) (labellum indicated by arrow). (B) Side view of staminodal hood (×10) (trichome brush arm indicated by arrow). (C–F) Putative models. (C) *Dichopogon* sp. (×5.5); (D) *Wahlenbergia* sp. (×8) (hairy pollen presenter indicated by arrow); (E) *Stypandra* sp. (×5); (F) *Thysanotus* sp. (×4).

wrinkled, and often brightly pigmented. In the Australian genus *Thelymitra,* the column encourages pseudanthery via fusion of the two staminodes to the filament of the fertile anther and subsequent ornamentation-pigmentation of the column "hood" (Fig. 4) (Burns-Balogh and Bernhardt, 1988).

Free, granular pseudopollen is produced within a depression on the labellum

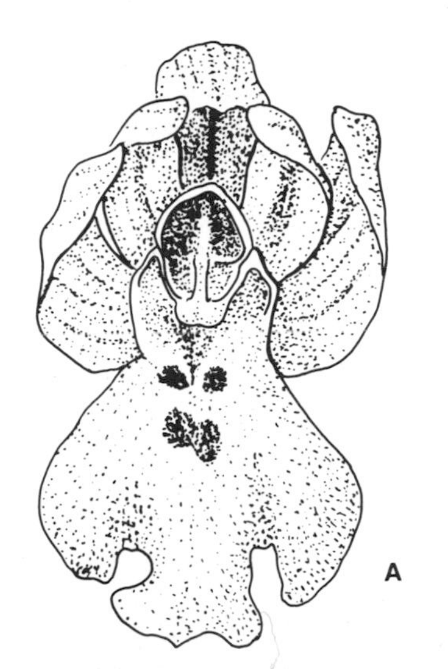

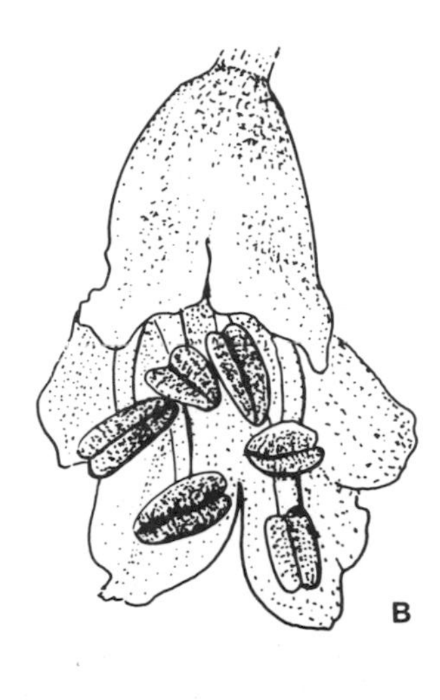

FIG. 4. Floral mimesis of coblooming herbs by *Orchis*. (A) *Orchis israelitica* (×4) (mimic); (B) *Bellevalia flexuosa* (×6) (Liliaceae, model). Dark spots on the labellum of the orchid against a light-colored background attract pollen-collecting bees that take pollen and nectar from a guild of geophytes which bear nodding, light-colored perianths and dark anthers.

of the Australian saprophytic epidendroid *Gastrodia sesamoides*. Selective staining (P. Bernhardt, unpublished) suggests that the pseudopollen is modified multilayered, epidermis containing starch granules and a few lipid droplets. This pseudopollen is removed by foraging bees and deposited in their scopae (Jones, 1981; Beardsell and Bernhardt, 1982).

The food mimicry syndromes in the terrestrial orchids of both regions tend to show a broad overlap. Four basic categories are described below, including Batesian, nonmimicry, integrated Muellerian–Batesian (guild mimicry), and integrated reward-deception systems (Dafni, 1986).

Batesian Mimicry

Batesian mimicry occurs when a low-density, nonrewarding species mimics the flowers of a model species offering copious rewards at a much higher density (Schemske, 1981). This system has also been referred to as Dodsonian mimicry (Pasteur, 1982) and deceptive floral mimicry (Little, 1980, 1983). This syndrome is characterized by two critical interlocking factors. First, the mimic must remain at a low frequency (Ducke, 1901; Macior, 1970, 1971; Yeo, 1972; Carlquist, 1979). Second, the abundant model species provides a constant edible

TABLE III. Batesian Floral Mimicry in Mediterranean Orchids

Model	Mimic	Pollinator[a]	Reference
Cistus salviifolius (Cistaceae)	*Cephalanthera longifolia*	SB	Dafni and Ivri (1981*b*)
Campanula trachelieum (Campanulaceae)	*Cephalanthera damasonium, C. longifolia*	?	Szlachetko, cited in Burns-Balogh *et al.* (1987)
Campanula spp. (Campanulaceae)	*Cephalanthera rubra*	SB	Nilsson (1983*b*)
Lathyrus vernus (Fabaceae)	*Orchis pallens*	SC	Voeth (1982)
Bellevalia flexuosa (Liliaceae)	*Orchis israelitica*	SB, SY	Dafni and Ivri (1981*a*)
Scabiosa columbaria (Dipsacaceae)	*Traunsteinera globosa*	SB?	Dafni (1987)
Knautia sylvatica (Dipsacaceae)	*Traunsteinera globosa*	SB?	S. Vogel, cited in Dafni (1987)

[a]See Table I for abbreviations.

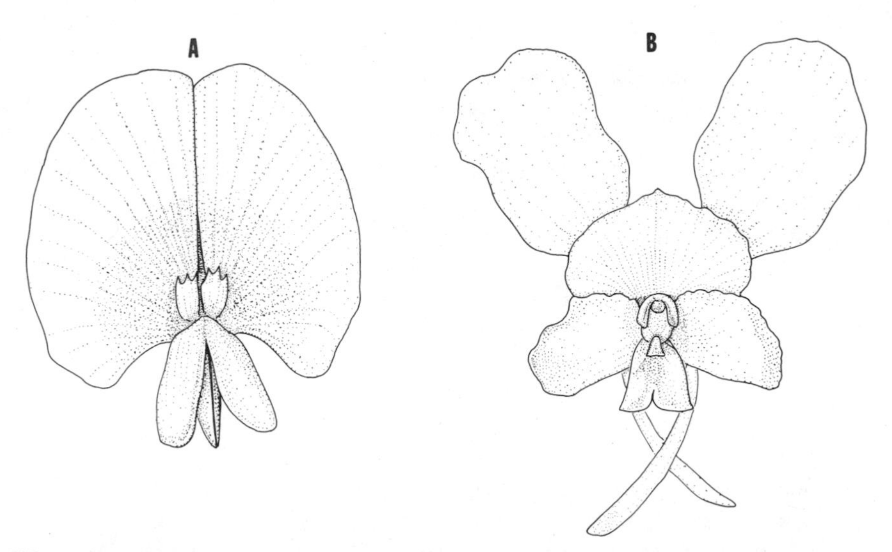

FIG. 5. Floral mimesis of coblooming papilionoid legumes by *Diuris* spp. (A) *Isotropis* sp. (×18) (model); *Diuris longifolia* (×8) (mimic). *Diuris* flowers tend to be larger or wider than the flowers of most model legume flowers. The "keel and wing" petals of the orchid flower are really a single trilobate labellum (wing petals = lateral labellum lobes; keel = central bifid lobe).

reward during the blooming period of the mimic species (Vogel, 1975; Heinrich, 1977; Voeth, 1982; Nilsson, 1983*a*).

The known examples of Batesian mimicry in the Mediterranean region are summarized in Table III. True Batesian mimicry is still unknown in the Australian terrestrial orchids. This may reflect the absence of fieldwork on large, key genera such as *Prasophyllum sensu stricto*. Plants of the genus *Diuris* undoubtedly show elements of Batesian mimicry (Fig. 5) (Beardsell and Bernhardt, 1982), as the flowers of these orchids undoubtedly mimic vernal, papilionoid legumes (Beardsell *et al.*, 1986) and the orchids are pollinated, at least in part, by female colletid bees that feed their offspring legume pollen (Armstrong, 1979). However, there is still no evidence that one *Diuris* species mimics one papilionoid species. Furthermore, field observations suggest that some *Diuris* species display peak flowering periods 1 or 2 weeks prior to the massed flowering of their models and deceive early, emergent, and naive foragers. Until there is more evidence to the contrary, *Diuris* species are best placed among the guild mimics (see Table V, below).

Nonmimicry Deception

In this syndrome the mimic species lacks any specific model. Little (1983) refers to this as "mimicry based on naiveté." The nonmodel mimic tends to flower during a period in which there are large numbers of inexperienced pollinators with polytrophic and/or polylectic foraging habits. In Mediterranean regions nonmimic deception species tend to flower early in spring and in large numbers (e.g., *Orchis* and *Dactylorhiza*). Examples of this mode of deception in the Mediterranean are summarized in Table IV.

Nonmimic deception is common from early to mid spring in southern Australia, with brightly colored neottioids offering dummy anthers, e.g., *Caladenia* Sect. Eucaladenia sensu lato (Bates, 1985*b*) and the satellite genera *Glossodia* and *Elythranthera*. From late spring through early summer there are the epidendroid *Dipodium punctatum* and *Gastrodia sesamoides*. Nonmimic deception

TABLE IV. Nonmimicry Deception

Plant	Pollinator[a]	Reference
Dactylorhiza sambucina	EB	Nilsson (1980)
Orchis mascula	SB	Patterson and Nilsson (1983)
Dactylorhiza sambucina	EB	Nilsson (1983*a*)
Orchis mascula ssp. *mascula*	SB	Nilsson (1983*a*)
Orchis morio	EB	Nilsson (1984), Dafni (1987)
Orchis dinsmorei	SB	Dafni (1987)

[a]See Table I for abbreviations.

in autumn may be attributed to *Eriochilus* species (Neottioideae). Some of the neottioid species form extensive ramet/genet populations within a habitat. However, fieldwork is still too fragmentary to totally rule out guild mimicry in any of the genera listed above, as all of them appear to be pollinated by polylectic/polytrophic insects (Armstrong, 1979) that may not show initial discrimination between the pseudanthery of the orchid labellum and similarly pigmented, sweetly scented flowers that comprise the local guild of flowers that bear no nectar but offer pollen as their only edible reward.

Guild Mimic Systems

Guild mimicry implies that each habitat will have a minimum of three coblooming species showing a convergence in floral attractants and presentation. At least two species mimic each other, but both offer some version of the standard edible reward. These two species represent a Muellerian trend and form the true guild. One or more species is a Batesian-type mimic of the guild members, the mimic's floral attractants and presentation converge with the Muellerian models but the mimic offers no reward. Guild mimicry appears far more common in southern Australia than in the southern Mediterranean.

Orchis caspia, in the Mediterranean, offers no reward, yet attracts about four species of solitary bees. The bee species are the pollinators of the orchid, but the bees forage upon, and legitimately pollinate, *Asphodelus aestivus* (= *A. microcarpus*), *Bellevalia flexuosa,* and *Salvia fruticosa* (two lilioid monocots and a member of the Labiatae). *Orchis caspia* supposedly exploits the bees which are unable to discriminate between the reward-offering, similarly attractive models and the rewardless orchid (Fig. 3). The bees are sustained by the rewards offered by the guild members despite an established frequency of error when the bees visit the orchid species (Dafni, 1983). This syndrome gives *O. caspia* a purportedly broad ecological amplitude, and it maintains relatively large populations in different plant communities.

In southern Australia, guild mimicry probably dominates most of the species in the genera *Diuris* and *Thelymitra* (Table V). *Diuris* species would be considered the more specific mimics, as the flowers of *Diuris* species appear to resemble three or four genera of vernal, papilionoid legumes in southeastern Australia (Beardsell and Bernhardt, 1982; Beardsell *et al.*, 1986; Bates, 1986). *Diuris maculata* is pollinated both by polytrophic halictid bees and wasps and by *Trichocolletes* species (Colletidae) which appear to feed their young primarily on pollen of papilionoids (Armstrong, 1979).

Thelymitra species with blue, pink-purple flowers and a column hood terminating in a pair of trichome brushes are mimics of nectarless vernal lilioids that present porose-poricidal anthers which require buzz pollination (*sensu* Buchmann, 1983) and also nectarless blue flowers in the monotypic family Wahlen-

TABLE V. Australian Orchid Food Mimics[a]

Genus (subfamily)	Number of species (number studied)	Pollinator	Remarks	References
Caladenia (N)	67 (5)	SB (3), BE, FL, WA (2)	Pseudostamens on the labellum (NM)	Bates (1985*b*), Stoutamire (1983)
Dipodium (E)	8 (1)	Bee (1)	Scentless, purple (NM)	Bernhardt and Burns-Balogh, (1983)
Diuris (S)	26 (4)	BW (4)	Pea-guild mimic	Beardsell *et al.* (1986), Coleman (1932, 1933*a*), Bates (1986)
Elythranthera (N)	2 (1)	Bees (1)	Labellum pseudostamens (GM)	Buchman (1983)
Eriochilus (N)	3 (1)	Bees (1)	Labellum pseudostamens (NM)	Erickson (1951)
Gastrodia (E)	2 (1)	Bees (1)	Scented labellum pseudopollen (NM)	Jones (1981)
Thelymitra (N)	34 (7)	Bees (5)	Buzz-pollination mimics (GM)	Burns-Balogh and Bernhardt (1988)
		SB (1)	General yellow model (GM)	Dafni and Calder (1987)

[a]See Table I for subfamily and pollinator abbreviations. GM, guild mimic; NM, no model.

bergiaceae (Fig. 4) (Bernhardt and Burns-Balogh, 1986*b;* J. D. Armstrong, personal communication). *Thelymitra antennifera,* on the other hand, is a vernal mimic of members of the yellow-flower guild including the nectarless, buzz-pollinated dicots (e.g., species of *Hibbertia,* Dilleniaceae) and nectariferous *Goodenia* species (Goodeniaceae) (Dafni and Calder, 1987).

Reproductive Deception

Unlike food deception, reproductive deception requires mimicry of stimuli intrinsic to a reproductive phase in the life cycle of the pollinator. Flowers pollinated by female insects tend to mimic standard oviposition sites such as carrion, dung, or the fruiting body of fungi. Consequently, one refers to sapromyophily and mycetophily in flowers pollinated by fleshflies or fungus gnats, respectively, and coprocantharophily in flowers pollinated by dung scarabs.

Flowers pollinated by male insects tend to imitate the scent and/or form of female insects, which may lead the male to attempt to copulate with the dummy female (pseudocopulation). In some tropical orchids (e.g., *Oncidium*) the flower presents a dummy of a male bee and the real bee attacks the blossom and pollinates while it attempts to drive the dummy away (pseudantagony) (Dressler, 1981). The latter mode of reproductive deception is unknown in either Mediterranean or south Australian regions.

Brood Site Deception

Both south Australia and the Mediterranean region have one genus, respectively, containing taxa which deceive ovipositing dipterans. However, these orchid taxa may combine the deceit syndrome with a nectar reward. In the Mediterranean region only *Epipactis consimilis* induces female syrphid flies to lay eggs on the flower, as labellum ornamentation probably mimics aphids (host oviposition site). *Epipactis consimilis* also offers nectar (Ivri and Dafni, 1977). The fly's eggs, of course, have no chance of survival upon hatching.

Jones (1970) interpreted *Corybas diemneicus* as an example of brood site deception, and this is currently the only example of the syndrome recognized in Australia. *Corybas* species in this latitude flower in late winter to early spring and appear to be confined to shady, wet gullies. Photographic evidence from New Zealand (Johns and Molloy, 1985) shows that the geoflorous *Corybas* species are often partially covered by fallen leaves and twigs at anthesis, so only long perianth segments are visible above the detritus. Like *Acianthus* and many *Pterostylis* species, *C. diemenicus* appears to depend on adult fungus gnats in the family Mycetophilidae for pollination (p. 196). However, in *C. diemenicus* the floral segments form a dark brown to iodine-colored chamber complete with a gill-like interior and a moist, clammy surface. This is interpreted as a mimic of the fruiting bodies of basidiomycetes that emerge at the end of the wet winter. Female gnats attempt to oviposit in this chamber. The pollinia are deposited dorsally on the gnat's thorax.

Sexual Deceit

Bergstroem (1978) suggests a trend toward the evolution of pseudocopulation in which the ''indiscriminant sexual selection'' by the male insect could be exploited by the plant. This system may represent an intermediate stage in pseudocopulatory systems toward true pseudocopulation. Sexual deceit is found in one orchid genus in both regions. In the Mediterranean region sexual deceit is exemplified by *Orchis galilea,* which is pollinated exclusively by male *Lasioglossum marginatum* (Bino *et al.*, 1982). The flowers emit a strong, musky odor which probably acts as a species-specific sexual attractant. The male bees appear to "pounce upon" the dark spots on the orchid's labellum, but the bee had not been observed actually attempting to copulate with the dark spot.

This intermediate state appears to have its parallel in south Australian *Caladenia patersonii*. The flowers offer the usual cluster of dark, short-stalked, acuminate, and shiny calli on the labellum as well as a trend toward linear filamentous sepals and lateral petals (p. 219). This mode of presentation is typically associated with pseudocopulatory syndromes in the genus *Caladenia* (Sec. *Calonema*) with pollination performed by male wasps in the family Tiphiidae. However, despite the usual suite of pigment and morphological modifications, the flowers are pollinated by tiphiid males and other insects of both sexes, including bees and syrphid flies searching for food (Stoutamire, 1983). Sexual deceit in *C. patersonii* appears to intergrade with nonmimicry deception (p. 209).

Pseudocopulation

Male insects attempt to copulate with the flower but receive depositions of pollinia. The concept was first suggested by Pouyanne (1917) and confirmed much later by the extensive studies of Kullenberg (1961). Pasteur (1982) has suggested the term "Pouyannian mimicry." However, in the Mediterranean region it is found only in the genus *Ophrys*. In southern Australia at least nine orchid genera present an attractant system that stimulates some degree of pseudocopulation (Table VI).

In general, those flowers with a pseudocopulatory syndrome have a brown, brick, or maroon labellum resembling the dark body of the female insect (Kullenberg, 1961; Wallace, 1978; Stoutamire, 1974, 1981; see color cover photographs in Peakall, 1986). Some dummy females incorporate a shiny, raised speculum (Fig. 6) (e.g., *Ophrys* and *Calochilus*) which may show ultraviolet reflection (Kullenberg, 1961). Barth (1985) compares the speculum of some Mediterranean *Ophrys* species pollinated by male wasps to the shiny spot that appears on the back of a female wasp when she crosses her wings. This speculum is absent in the majority of pseudocopulatory orchids of southern Australia pollinated by wasp genera in the Tiphiidae. Tiphiid wasps of Australia have wingless females (Fig. 7). However, many female tiphiids have rather glossy bodies (see cover photographs in Peakall, 1986) and the tiphiid pseudocopulatory orchids of southern Australia tend to bear labella with glossy, dummy females (Fig. 8).

Labellum vestiture will tend to correspond to that of the body of the female insect imitated by the flower. For example, the females of some ichneumonid wasps tend to be rather smooth, as is the labellum of *Cryptostylis ovata* species (Fig. 9), which appear to be pollinated almost exclusively by male *Lissopimpla semipunctata* (Beardsell and Bernhardt, 1982). In contrast, female wasps in the families Scoliidae and Sphecidae and certain species of Apoideae have rather

TABLE VI. Australian Orchids with Sexual Mimicry[a]

Genus (subfamily)	Number of species (number studied)	Pollinator	Remarks	Reference
Anthrochilus (N)	41 (1)	WA (1)	Labellum hinged	Rotherham (1968), Stoutamire (1986)
Caladenia (N)	67 (5)	WA (11), bee? (2)	Hammer glands (osmophores?) on filamentous perianth segments, calli dark, congested, compressed; flower color maroon-iodine, labellum classed passive movement	Rogers (1931), Stoutamire (1983)
Calaena (N)	1 (1)	SA (1?)	Articulated labellum, passive movement	Cady (1965)
Calochilus (N)	7 (2)	WA (2)	Labellum without passive movement, but highly ornate with hundreds of appendages	Fordham (1946), Jones and Gray (1974), Bernhardt and Burns-Balogh (1987)
Chiloglottis (N)	5 (4)	WA (4)	Flowers iodine-maroon, labellum clawed, passive movement, central appendage (dummy females)	Dockrill (1956), Stoutamire (1975)
Cryptostylis (S)	5 (3)	WA (3)	Labellum rigid; one wasp species pollinates all species; no hybrids	Coleman (1927, 1928, 1929), Leonard (1970)
Drakea (N)	3 (2)	WA (2)	Highly articulated passive movements	Stoutamire (1981)
Leporella (N)	1 (1)	WA, AN (1)	Similar to *Caladenia*, but lip hinged	Woolcock (1980)
Pterostylis (N)	51 (7)	FG (7)	Trap flower, irritable labellum with dummy appendages	Beardsell and Bernhardt (1982)
Spiculea (N)	1 (1)	WA (1)	Labellum hinged	Stoutamire (1986)

[a]See Table I for subfamily and pollinator abbreviations.

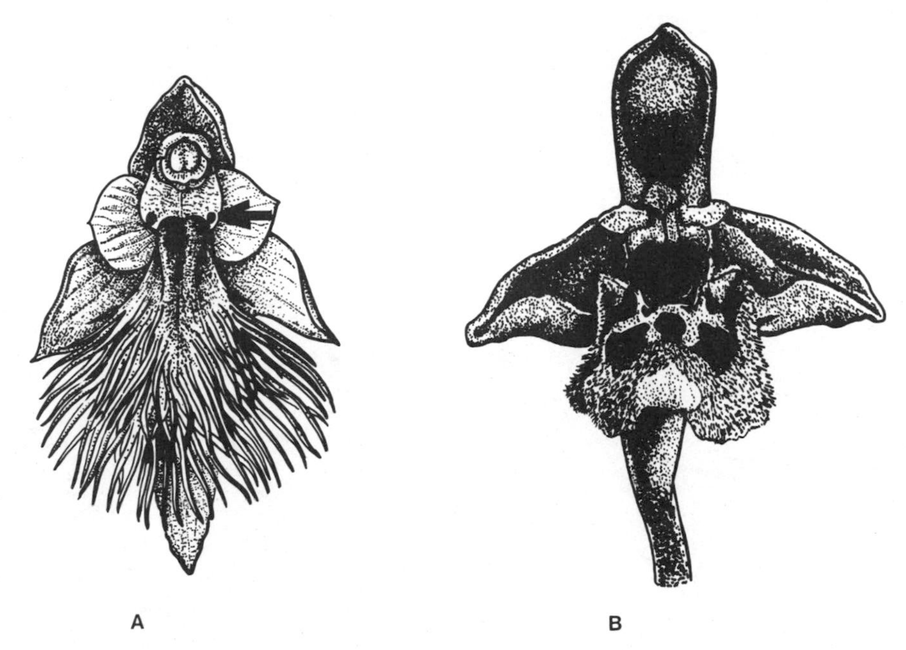

FIG. 6. Convergence of flowers of Australian *Calochilus* species (Neottioideae) and Mediterranean *Ophrys* species (Orchidoideae) toward mimesis of winged, female Hymenoptera. (A) *Calochilus campestris* (×7); (B) *Ophrys holosericea* (×4). The labellum petal of both species is vestate and bears a shiny speculum at its base (see text), but lacks a motile hinge. The vestiture of *C. campestris* consists of many elongated, vasculated appendages with each appendage bearing glandular papillae. The dense trichomes of *Ophrys* spp. arise directly from the epidermis, giving the surface its furry texture. The dummy eyes of *Calochilus* spp. are located on opposite bases of the column (staminodal collar; arrow). Proportionately fewer *Ophrys* spp. bear sham eyes (e.g., *O. insectifera*) and these structures, in contrast, are attached to the base of the labellum.

hairy abdomens. This is copied by the labella of *Ophrys* species in the Mediterranean and *Calochilus* species in Australia (Fig. 6) (Kullenberg, 1961; Agren *et al.*, 1984; Jones, 1981; Barth, 1985).

None of the pseudocopulatory species is known to offer an edible reward. Olfactory stimulus, though, appears to be crucial to *Ophrys* species (Kullenberg, 1961) as well as in southern Australian species (Coleman, 1930; Stoutamire, 1974, 1976; Wallace, 1978). In the true pseudocopulatory species, the floral fragrance is usually not discernible to the human nose. A thorough study of the chemical composition of the odor of *Ophrys* species in comparison with their insect pollinators has been completed by Kullenberg and his co-workers (Kullenberg and Bergstroem, 1975, 1976; Kullenberg *et al.*, 1984*a,b;* Borg-Karlson and Tengoe, 1986; Argen and Borg-Karlson, 1984). In some cases a chemical con-

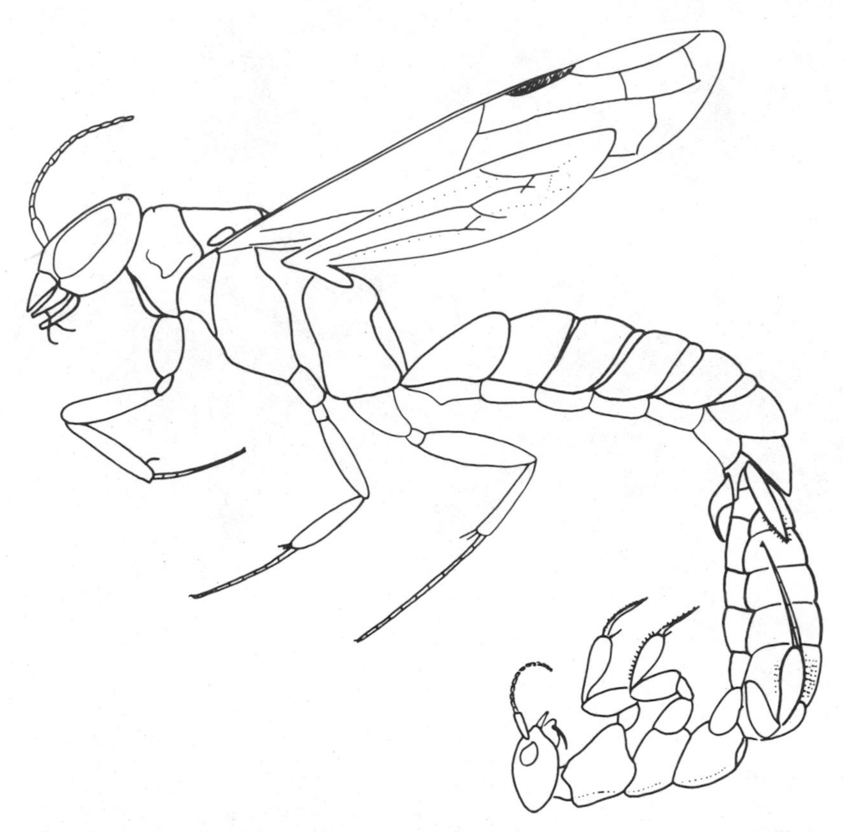

FIG. 7. Nuptial flight of *Dimorphothynnus haemorrhoidalis* (×5) with the winged male carrying the wingless female.

vergence was found between *Ophrys* species and their insect males (Borg-Karlson and Tengoe, 1986; Borg-Karlson, 1985), confirming the Kullenberg (1961) hypothesis of chemical mimesis.

It appears that some pseudocopulatory species (e.g., *Ophrys* and *Cryptostylis* species) flower while the male insects emerge, but these orchids are in flower before the female insects emerge (Ames, 1937; Coleman, 1927; Erickson, 1951; Pouyanne, 1917; Vogel, 1975; Wallace, 1978). This suggests that the flowers exploit the naiveté and/or sexual maturity of the male insect, which appears functional prior to female emergence.

The taxa of bees, wasps, and sawflies which are regularly deceived by pseudocopulatory orchids are not made up of territorial insects and these often seek mates in flowers or on plant stems (e.g., Tiphiidae). The insect females tend to have a diffuse nesting pattern and this has been used to explain the highly disjunctive pattern of distribution in the pseudocopulatory orchid species (C.

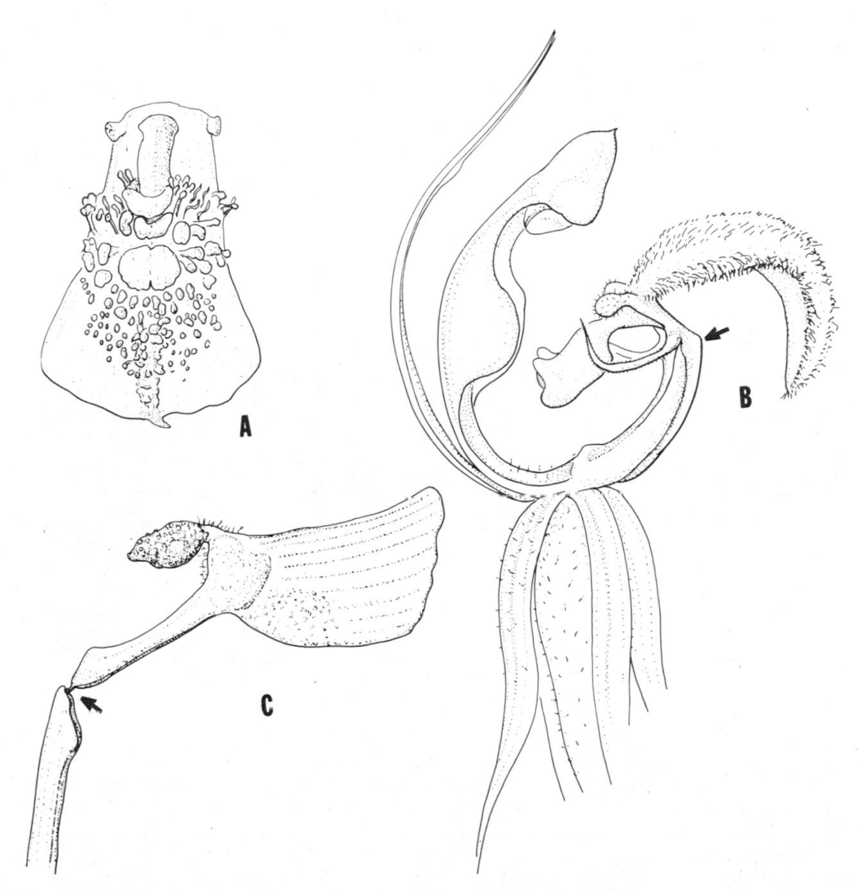

FIG. 8. Labella of some Australian orchids modified for mimesis of female tiphiid wasps. (A) *Chiloglottis reflexa* (×19); (B) *Caladenia barbarossa* (×12); (C) *Drakea glyptodon* (×7). In each genus there is a sculpted callous with one or more appendages forming the body of the dummy female. The labella of *C. barbarossa* and *Drakea* spp. are hinged (indicated by arrows).

O'Toole and A. Dafni, cited in Dafni, 1984). Raceme length in flowers of *Ophrys* species may be predetermined, as the insects that pollinate *Ophrys* species tend to copulate on vegetation up to 50 cm above ground level. Paulus and Gack (1981) have shown that a change in the height of the flower of an *Ophrys* species dramatically influences its chances for successful pollination via pseudocopulation.

Ophrys species are pollinated by eight genera of bees and to a lesser extent by solitary wasps (two genera in Scoliidae and Sphecidae, respectively) and one scarab beetle (Borg-Karlson, 1985). In Australia, pseudocopulatory syndromes

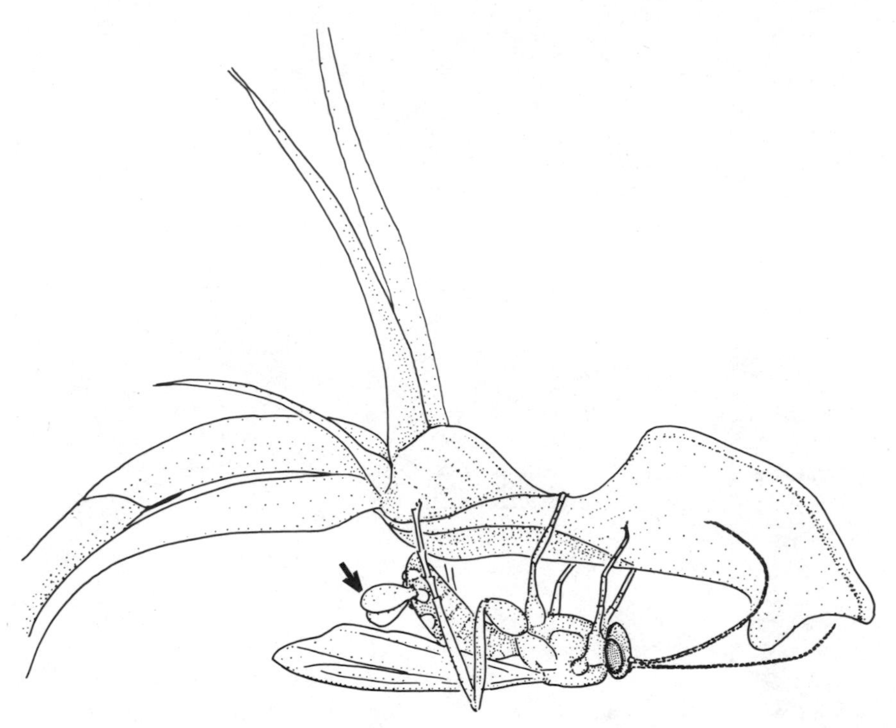

FIG. 9. Pseudocopulation in *Cryptostylis ovata* by *Lissopimpla semipunctata* (×3). The copulating wasp tends to place the posterior of its abdomen within the labellum depression under the short column (not visible) and receives a dorsal deposition of the soft pollinia (arrow) when it withdraws its genitalia. In some cases of alignment the wasp places its head under the column and receives dorsal depositions of pollinia on the head or thorax.

span two insect orders (Diptera and Hymenoptera) and two orchid subfamilies (Spiranthoideae and Neottioideae). Exploitation of male flies is restricted to *Pterostylis* (Neottioideae), which attracts male mycetophilids (Table VI). This is the only orchid genus in southern Australia with irritable labella in which motility is determined by internal biochemistry and not exclusively by the weight or physical strength of the contacting insect.

The exploitation of male Hymenoptera in southern Australia includes wasps, sawflies, and one genus of ponerine (wasplike) ants (Table VI) (Peakall *et al.*, 1987). Male wasp pollination dominates in the pseudocopulatory taxa, with the majority of orchid species pollinated, at least part of the time, by male Tiphiidae. Labellum motility in the pseudocopulatory wasp flowers of southern Australia appears to relate in part to the dependence of the orchid species on the courtship of male tiphiids. The wingless females are carried off by the males

(Fig. 7). Copulation occurs in the air and the male often carries the female to nectariferous sources (Stoutamire, 1974, 1976; Bernhardt, 1987). Consequently, a portion of the claw of the orchid's labellum is often modified to form a thin, flexible, and elastic hinge which connects the lamina to the claw apex (e.g., *Caladenia* Sec. Calonema) or to the base of the column (e.g., *Spiculea*) or to an elongated column foot (e.g., *Drakea, Arthrochilus*). When the male tiphiid attempts to fly away after grasping the dummy female the insect is flipped into the receptive tip of the column, where it may be temporarily held between two broad column wings while the viscidium is fixed to the wasp's exoskeleton (Fig. 8). As male scoliid, ichneumonid, and sphecid wasps do not engage in the same aerial copulations with their winged mates, the labella of *Calochilus, Cryptostylis,* and *Ophrys* species remain stationary (Figs. 6 and 9). The labella of *Chiloglottis, Lyperanthus,* and some pseudocopulatory *Caladenia* (Sec. Calonema) species appear intermediary between motile and nonmotile systems, but pseudocopulation in these genera may combine wasps from different families or several general of tiphiids with varying modes of courtship (Stoutamire, 1975, 1983).

Autogamy

Self-pollination which leads to viable seedset is common in many of the orchids of the Mediterranean and southern Australia. Darwin (1877) lists nine self-pollinating species in Britain, but distinguishes between species which regularly self-pollinate when insects fail to visit the flower, e.g., *Neottia nidus-avis,* and species which set seed exclusively by self-pollination and were never observed to receive insect visitation, e.g., *Ophrys apifera.* Today we would say that *N. nidus-avis* is facultatively autogamous while British populations of *O. apifera* are obligately autogamous.

Examples of self-pollination in the orchids of the Mediterranean have undergone a wide expansion due to reports by Reinhard (1977), who describes an additional 43 cases of self-pollination. This includes orchids bearing modified flowers which never open (obligate cleistogams) grading into flowers which self-pollinate in the bud or as the open flower ages. Furthermore, self-pollination in self-compatible, outcrossing species may be induced by external agents (insects, wind/rain). These agents effect autogamy and/or geitonogamy (Reinhard, 1977).

In the Mediterranean region, however, spontaneous obligate autogamy will most likely follow one or two patterns of environmental pressure and population distribution. Mechanical autogamy is expected in species which inhabit harsh environments featuring brief flowering seasons and a paucity of pollinators (e.g., *Chamorchis alpina, Leucorchis albida, Coeloglossum viride,* and *Platanthera hyperborea*). As the species names suggest, these taxa are distributed through alpine habitats and/or circumpolar latitudes (Hagerup, 1952). Mechanical autogamy is predicted for "weedy orchids" which have wide, disjunctive distributions but show colonizer habits, e.g., *Neotinea maculata* and *Ophrys apifera* (S. Vogel, cited in Dafni, 1987). Regional populations are often based on the successful establishment of one self-pollinating individual (Baker, 1974).

The Mediterranean species of the genus *Epipactis* show a gradation of the various modes of self-pollination, and this is reflected in the distribution of taxa and the ability of taxa to become established. Breeding systems in 21 out of a possible 22 *Epipactis* taxa have been studied. Only five may be considered allogamous. Eight taxa balance allogamy with autogamy, while eight appear dominated by autogamous and/or cleistogamous flowers (Richards, 1982; Burns-Balogh *et al.*, 1987).

Epipactis taxa showing high levels of allogamy reflect a panmictic trend and continuous populations. Allogamous *Epipactus* species do not have disjunctive populations broken down into localized races by taxonomists.

Epipactis species dominated by autogamous seedset show a clearly disjunctive distribution, e.g., *E. muelleri* (Richards, 1982). If the distribution of the species is clearly nondisjunctive, though, localized populations appear to break down into isolated, self-perpetuating, local races, e.g., *E. dunesis, E. confusa,* and *E. youngyana*. Autogamous flowers of *Epipactis* show a suite of characters, including powdery pollen, degeneration of the rostellum, and a paucity of discernible odor and nectar secretions (Robatsch, 1983). It is assumed that 60% of all *Epipactis* species occupy habitats deficient in potential pollinators [see also Proctor and Yeo (1973), who place *Cephalanthera* within the same trend toward autogamy].

As all *Epipactis* species reproduce vegetatively, ramet increase or autogamous mutants will reproductively isolate distinct genotypes in the same population (Richards, 1982). In *E. microphylla,* for example, populations contain two differing reproductive forms, obligately autogamous plants with small, scentless flowers versus allogamous plants with large, fragrant flowers (Wiefelspuetz, 1964). Intermediate forms between obligately autogamous plants and allogamous forms are recorded for *E. phyllanthes* (Robatsch, 1983), indicating some continued gene flow. Some orchid species include variable populations of plants reproducing by sexual reproduction with apomicts, e.g., *Listera ovata, Cephalanthera damsonium,* and *C. longiflora* (Prochazka and Velisek, 1983).

The publications of Fitzgerald (1875–1895) and Cheeseman (1881) are among the earliest records describing self-pollination in the terrestrial Orchidaceae of southern Australia and New Zealand. It would not be an exaggeration to state that mechanical autogamy is suspected, or has been observed, in at least one species of all major Australian terrestrial genera, with the exception of self-incompatible *Cryptostylis* species (Cady and Rotherham, 1970; Beardsell and Bernhardt, 1982; Bernhardt and Burns-Balogh, 1983).

Modes of self-pollination (*sensu* Reinhard, 1977) in Australian orchids tend to surpass modes proposed originally by Fitzgerald and Cheeseman. As the majority of taxa studied prove to be self-compatible, the prospect for self-pollination changes according to floral morphology, inflorescence structure, flower size, and pollinator patronization. Insect-mediated autogamy has been observed and photographed in *Calochilus* (Jones and Gray, 1974; Jones, 1981) and insect-mediated geitonogamy may be common in *Prasophyllum* (Jones, 1972; Bernhardt and Burns-Balogh, 1986*b*).

It is undoubtedly true that the majority of Australian orchids will self-pollinated mechanically in the absence of appropriate pollinators. However, the vast majority of south Australian terrestrial orchids are vernal flowering plants and the fail-safe mechanism of delayed autogamy has been well described in many other plant families with vernal herbs (Schemske *et al.*, 1978). Furthermore, mechanical autogamy in the absence of suitable pollinators is a well-known phenomenon in both tropical and temperate Orchidaceae. In fact, many monandrous orchids with friable pollinia would appear to be preadapted for mechanical autogamy as the anther is above or parallel to the stigma and the majority of genera appear to show a trend toward self-compatibility (Burns-Balogh and Bernhardt, 1985).

SYNECOLOGICAL AND EVOLUTIONARY CONSEQUENCES

Ecological Constraints, Pollinator Behavior, and Population Structure

Food Mimics

In both regions reproductive success of food mimic species is highly dependent on the balance between population density of the mimic taxon and the ethology of the primary pollinators. Further constraints occur according to which syndrome of food mimesis is expressed by the orchid taxon.

In Batesian floral mimicry the interrelations between the pollinators and the flowers and between the model and mimic pose limitations on the population size of the mimic in comparison to that of the model. A high proportion of the mimics could enhance the recognition abilities of the insect (Ducke, 1901; Macior, 1970; Yeo, 1972; Beardsell *et al.*, 1986), causing the potential pollinator to learn rapidly and avoid the mimic. This probably explains the steep reduction of capsule production in mimectic orchids when they flower in the absence of models (Dafni and Ivri, 1981*a*,*b*).

Dafni (1983) noted that bees pollinating *Orchis caspia* (*Anthophora* sp.,

Eucera clypeata, E. nigrifacies, and *Melechta mediterranea*) were misled several times by the same orchid species, and an individual bee could bear as many as 22 pollinia. It was concluded that such bees had a poor ability to discriminate between model and orchid.

Nonmimic deception, as previously discussed (p. 209), is based on inexperienced insects. Nilsson (1980) showed that *Dactylorhiza sambucina* exploits naive *Bombus* sp. queens which emerge after hibernation. The short period in which the plant exploits the pollinator occurs before the bees have established foraging routes. The bees' naiveté was stressed in similar studies on other taxa having the same pollination system (Nillson, 1979; Stoutamire, 1971; Ackerman, 1981; Lock and Profita, 1975; Dafni and Woodell, 1986). Therefore, seedset in the orchid mimic should be highest toward the beginning of the flowering season (Nilsson, 1981). However, there is no limitation on the population size of a nonmimic deception taxon, since it is not dependent on the overwhelming density of any particular model.

Ecological constraints and population structure for guild mimics should fall in between Batesian and nonmimicry deception, depending on how closely the orchid mimics resemble two or more models (Beardsell *et al.*, 1986). *Thelymitra nuda* exploits an aspect of pollen foraging behavior that is probably unique to those orchids that mimic nectarless flowers that release copious pollen via thoracic vibration. Buchmann (1983) has referred to this as the "carryover phenomenon." Bees that apply thoracic vibration to a series of nectarless, poricidal flowers will continue to carry over this foraging behavior to the first few flowers they visit that do not bear a true buzz-pollination form of presentation. This has been observed in the female of *Lasioglossum* species, which attempt to forage on the ornamented hood of *T. nuda* after the insect has removed pollen from the poricidal flowers of coblooming Liliaceae *sensu lato* (Bernhardt and Burns-Balogh, 1986*a*).

Pseudocopulation

There is surprisingly less consensus regarding the ecological constraints influencing the success of pseudocopulatory syndromes. The majority of recent challenges derive primarily from field experimentation in the Mediterranean.

For example, pseudocopulation is commonly thought to be based on the action of reproductive stimuli. The chemical trigger acts directly on the pollinator's sensory system and this releases a pattern of innate behavior that is not susceptible to either individual experience or true learning (Vogel, 1975; Yeo, 1972; Dafni, 1984; B. Kullenberg, cited in Burns-Balogh, 1985; Stowe, 1988). Barth (1985) has even argued that successful pollination of *Ophrys insectifera* is due primarily to the fragrance lure, since the labellum fails to duplicate some essential visual cues, such as the distinctive cross-stripes found on the abdomens of the female *Gorytes* sp. wasps. Paulus *et al.* (1983) challenged this attitude and supplied evidence which showed that the frequency of male insect visits declined as a function of the duration of exposure to the stimuli. Flowers were replaced frequently to maintain fragrance production. They concluded that males had learned to avoid flowers. Two points must be made here. First, the male insects were not marked individually, so there is no way to assess the individual's ability to distinguish the mimic. Second, the males may regularly alter their activity levels on the flowers depending considerably on such factors as the time of day and/or due to slight changes in the weather, especially direct irradiation and temperature (Goelz and Reinhard, 1977; Kullenberg *et al.*, 1984*a*).

Pseudocopulation is classically defined as a form of pollination by deceit (Wiens, 1978, Vogel, 1975; Dafni, 1984). Recently, though, even this prevailing interpretation has been challenged. B. Kullenberg (in Burns-Balogh, 1985) argues that the pollination of *Ophrys* species should be regarded as an environmental resource that actually benefits the pollinator. The flower odor may keep the patrolling males within the habitat of the flowers of the female insect. Second, the copulatory instinct of the male may also be reinforced by the fragrance of the *Ophrys* species and this might cause a more active search for the receptive females.

The first argument implies that a dense population of *Ophrys* species is more beneficial to both the flowers and the insects. It would be expected, therefore, that relative forces favor dense populations of *Ophrys* species rather than the wide dispersal of a few isolated plants. In the Middle East, though, most of the *Ophrys* species appear in small, scattered populations. Wolf (1950) reports that the lowest rates of pollination in *Ophrys* species occur in large populations located in shady areas. In regard to benefits of floral scent to the insect, Paulus and Gack (1983) regard *Ophrys* species as competitors for the real female and note that too many visits to the flowers may reduce the reproductive success of the insect. They note that there is a competition between plants of *Ophrys* species for males, which implies a better chance for seedset in a small population. Dafni (1984) argues that the generally low density of *Ophrys* species populations actually mimics the odor of female insects, which have typically diffuse nesting patterns. These conflicting hypotheses of the Kullenberg versus Paulus schools opens the door for further fieldwork and laboratory authentification.

Paulus and Gack (1980), working on *Campsocilia ciliata,* the wasp pollinator of *Ophrys speculum,* found that males fly at a height of 8–10 cm above the ground (see also Goelz and Reinhard, 1977). Flowers placed on the soil surface were visited more frequently than were those at their normal height, while those which were raised above 15 cm were visited less frequently. They concluded that there may be selection for an optimal plant height. However, Jacobsen and Rassmussen (1976) found that pseudocopulation of *O. speculum* occurred even when the flowers were 1.5 cm above the soil surface. Furthermore, flowers placed on the soil had the same attraction ability as those left at their normal heights.

Pseudocopulatory systems dictate a comparatively narrow range of distribution for the plant in regard to the distribution of its pollinator (Proctor and Yeo, 1973) and a flowering period which must not deviate from the emergence of male insects (Ames, 1937). Abiotic factors may further restrict the ability of the pseudocopulatory taxon to follow the range of its pollinators. Ames (1937) was one of the first to note that the distribution of *Cryptostylis* species in Australia was limited to only a fraction of the total distribution of *Lissopimpla semipunctata,* the sole pollinator of the genus. Continental aridity probably has a far more severe impact on terrestrial orchids than on some insects. Australian *Cryptostylis* is believed to have colonized New Zealand within the last century, according to Austral phytogeographers. However, the colonization of the orchid genus followed the arrival and establishment of its pollinator (Johns and Molloy, 1985).

Kullenberg *et al.* (1984*b*) has shown that several species of *Ophrys* (Sec. Bombiliflorae and Fuciflorae) were able to extend their range southward, but they move into areas in which their pollinators exist. For an *Ophrys* species to move outside the range of its pollinator implies either changing pollinators or a trend toward obligate autogamy. The latter has been true in the northward expansion of *Ophrys apifera* in the absence of the bee, *Eucera tuberculata* (S. Vogel and C. O'Toole, cited in Dafni, 1987; Kullenberg *et al.*, 1984*a*).

The comparatively long flowering season (3–8 weeks) of an *Ophrys* species has been interpreted as a phenological adaptation ensuring some degree of overlap with pollinator emergence (Proctor and Yeo, 1973). The presence of a few long-lived, simultaneously opening flowers on one inflorescence appears to establish a long flowering period in *Ophrys* as well as the southern Australian *Calochilus* and *Cryptostylis* species (Jones and Gray, 1974; Wallace, 1978).

Breeding Systems, Hybridization, and Speciation

Isolation Barriers

Nearly a century of horticultural experiments has shown that taxa within the Orchidaceae may be crossed by hand to produce a wide range of interspecific and intergeneric hybrids of variable fecundity. The taxa of southern Australia and the Mediterranean regions do not appear to be exceptions to this phenomenon (Dressler, 1981; Bernhardt and Burns-Balogh, 1987). The taxa of both regions employ a range of isolation barriers which lowers the potential for interspecific recombination under natural circumstances. All barriers should be regarded as partial when addressed to each region. The range of isolation barriers in the

Mediterranean region may be especially weak due to a millenium of human history in the area, which has broken down habitats and led to the creation of disturbed sites promoting hybridization (Kullenberg and Bergstroem, 1976; Dafni and Baumann, 1982). In general, ecological and ethological barriers are of greatest importance to the orchid genera of the Mediterranean. Morphological and genetic barriers play more important roles in some of the genera of southern Australia. Of course, the relative importance of each type of isolation barrier differs among genera within the same region.

Ecological Barriers. Different habitat requirements, geographic distributions, and reproductive phenologies may serve jointly or separately as efficient barriers in the Mediterranean region. Conversely, it is the breakdown of habitat identity that juxtaposes species formerly separated by divergent environmental parameters (Dafni and Baumann, 1982). Within the genus *Ophrys,* separate flowering periods appear to isolate some of the more closely allied species (Kullenberg and Bergstroem, 1973; Paulus and Gack, 1980, 1981; Voeth, 1984). The importance of geographic isolation in this genus was discussed by Sundermann (1977) (also see Borg-Karlson, 1985).

Allopatry plays a more limited role in southern Australia. First, the subalpine zone and arid central regions of the continent are notoriously depauperate in orchid taxa (Weber, 1981). The few species distributed through these regions (e.g., subalpine *Thelymitra venosa* and *T. sargentii* are found within the arid zone) will remain isolated from their many coblooming congeners until the range of these hardier species extends toward true temperate zones (Burns-Balogh and Bernhardt, 1988). Second, a high degree of endemism is found in southwestern Australia due to its soil mosaic (Hopper and Maslin, 1978). This leads to a highly disjunctive distribution of congeners, with different edaphic requirements. Consequently, this comparatively small area of southern Australia is regarded as a center of diversity for terrestrial orchids (Nicholls, 1969).

Within the otherwise vernal flowering genera *Pterostylis* and *Eriochilus,* a few species flower only in autumn through early winter (e.g., *P. parviflora, E. cucculatus*). These species will be rigidly isolated from the vast majority of congeners. Even in large, primarily vernal-flowering genera like *Prasophyllum, Thelymitra,* and *Caladenia* some temporal isolation is still possible, as those species flowering in early spring or late winter will not overlap with those species flowering from late spring into early summer. Since cyclical fire is the major flowering stimulus for some south Australian species, we suggest that sympatric species (e.g., *Lyperanthus* species) may remain isolated according to whether their floral phenology is dependent on summer burnoffs for maximum flowering.

Morphological Barriers. Flower shape, size, and color dictate the size and variety of potential pollinators. Nilsson (1978) has shown that interspecific isolation between *Platanthera chlorantha* and *P. bifoliata,* which hybridize infrequently, is due, in part, to differing column structures. Pollinia of *P. bifoliata* are attached to the eyes of noctuid moths. Differing spur lengths between the two *Platanthera* species may also play a major role in limiting moth foraging, and this may limit hybridization, as has also been suggested for *Orchis* (Summerhayes, 1968; Dafni, 1987).

Floral size may lower the frequency of interspecific hybridization between *Ophrys* species to a limited extent (Borg-Karlson, 1985). However, specific labellum vestiture and pilosity appear to have a most important role in encouraging assortative pollination (Kullenberg, 1961; Agren *et al.*, 1984). The gross morphology within most genera of orchids in southern Australia is usually so self-consistent that there is little evidence that interspecific isolation may be based on variations in floral anatomy, with two exceptions. The sections Calonema and Eucaladenia of the genus *Caladenia* each present labella of such divergent pigmentation, ornamentation, and motility that two different sets of pollinators tend to visit the two sections (Stoutamire, 1983; Bates, 1985*b*). In *Thelymitra* the trend toward actinomorphy coupled with interspecific variation in the morphology and ornamentation of the column mitra will orient the bee pollinator's body in relation to contact with the viscidium. Consequently, the pollinia of some *Thelymitra* species may be deposited consistently on the underside of the head, thorax, or abdomen (Dafni and Calder, 1987), while in other species pollinia are deposited dorsally on the abdomen (Burns-Balogh and Bernhardt, 1988). Comparatively few bee genera pollinate *Thelymitra* species. Therefore, since the floral phenology of many *Thelymitra* species overlap broadly, it is conceivable that sympatric species share the same bee pollinator (Burns-Balogh and Bernhardt, 1988). Opportunities for interspecific hybridization may be lowered, though, if the bee receives pollinia of one species ventrally on her head but receives the pollinia of the second species dorsally on her abdomen, e.g., *T. antennifera* (Dafni and Calder, 1987) versus *T. nuda* (Bernhardt and Burns-Balogh, 1986*a*). In this respect, morphological isolation in *Thelymitra* species appears to converge with neotropical epidendroid genera which share the same euglossine bee but avoid intergeneric crosses by adhering the viscidia to different parts of the bee's body (Dressler, 1981).

Conversely, the floral morphology of different Australian genera is so distinctive that floral characters have been retained as the dominant set of taxonomic characters at the generic level. For example, recent analyses of *Prasophyllum sensu lato* suggested not only that separate sections be divided into different genera, namely *Prasophyllum sensu stricto* and *Genoplesium sensu stricto,* but into spiranthoid and neottioid subfamilies, respectively (Burns-Balogh and Funk, 1986). As we have seen, coblooming genera actively share or compete for certain pollinators, especially short-tongued bees, syrphid flies, and tiphiid wasps. Intergeneric variation in floral morphology and pollinia deposition undoubtedly restricts intergeneric hybridization in the Australian orchids. In some *Caladenia* species (Sec. Cyanicula *sensu* S. Hopper, personal communication) floral presentation, column morphology, and labellum position overlap with the satellite genera *Glossodia* and *Elythranthera,* leading to disjunctive, intergeneric hybrids (p. 232).

Ethological Barriers. The behavior of the dominant pollinators plays an important role in preventing hybridization in the Mediterranean region. In *Ophrys* species this behavior is dictated by the blending of floral odor. The subject was studied admirably by Kullenberg and his co-workers (Kullenberg, 1961; Kullenberg and Bergstroem, 1975, 1976; Tengoe, 1979; Borg-Karlson, 1979, as cited in Borg-Karlson, 1985; Agren and Borg-Karlson, 1984; Kullenberg *et al.,* 1984*b;* Warncke and Kullenberg, 1984; Borg-Karlson and Tengoe, 1986). In most *Ophrys* species the pollinia are carried on the insect's head. However, in the *O. fusca*/*O. lutea* complex the pollinia are deposited dorsally on the abdomen (Kullenberg and Bergstroem, 1973; Sundermann, 1977; Paulus and Gack, 1983). This divergence in deposition has been attributed to the way in which the insect orients itself to the coloration patterns of the labellum.

Ethological isolation may occur at the intraspecific level in *Ophrys.* Populations of *O. fusca* in southern France and Spain are dominated by either large- or small-flowered forms. These two forms are pollinated by two different species of pollinators even though the orchids are still regarded as the same species (Paulus and Gack, 1980, 1983). Similar results were found for the two forms of *O. oestrifera* ssp. *oestrifera* in Crete (Voeth, 1984).

Hybridization usually occurs between orchid species that are closely allied both morphologically and chemically (Borg-Karlson, 1985). Borg-Karlson (1985) notes that hybridization in pseudocopulatory orchids may be the result of a highly excited male insect which has a low behavioral threshold for sexually attractive odors. The odor from a more attractive species may cover the less attractive scent of a second species. If this occurs, then only the visual and tactile stimuli offered by the labellum of the second species will be decisive. However, Paulus and Gack (1980) suggest that there are more opportunities to prevent hybridization between *Ophrys* species, as the pollinators of two sympatric species are often not related to each other and two different sets of pollinators respond to two quite different blends of floral odors (Borg-Karlson, 1985).

The full extent of ethological isolation within the genera of most Australian orchids is still not understood due to the low priority given to the taxonomic study of endemic wasps. Some orchid taxa appear to be well isolated by ethological mechanisms, while the pollination systems of their congeners may overlap broadly. This dichotomy is particularly obvious in flowers pollinated by pseudocopulatory syndromes. Such genera as *Caladenia, Chiloglottis, Drakea,* and *Spiculea* are pollinated by different wasp genera which all belong to the Tiphiidae (Stoutamire, 1981, 1983, 1986). This may limit crosses between genera, but some *Caladenia* species share one or more of the same wasp genera and hybridization occurs when the distributions of *Caladenia* species overlap (Stoutamire, 1983). Ethological isolation appears unimportant within the genus *Cryptostylis*. All *Cryptostylis* species studied thus far are pollinated exclusively by *Lissopimpla excelsa* (Leonard, 1970). Interspecific isolation in this genus has a different and more efficient basis (see below).

Artificial crosses suggest that the genera of southern Australia show a capacity for successful recombination between highly divergent species. For example, viable seed set may be achieved by crossing *Thelymitra nuda* with *Calochilus robertsonii* (Bates, 1985*a*). These two species are highly sympatric and coblooming, but an F_1 generation has never been found between them. A natural cross is unlikely, as respective floral presentations attract completely different pollinators (Tables V and VI) and pollinia are deposited on opposite ends of the insects' bodies (Fordham, 1946; Bernhardt and Burns-Balogh, 1986*a*).

Genetic Barriers. Due to the high level of infra- and intergeneric hybrids in natural populations of the Orchidinae of the Mediterranean region (Potucek, 1968; Sundermann, 1980; Dressler, 1981; Baumann and Kuenkele, 1982; Dafni, 1987), genetic barriers are probably very weak. Of course, certain genera are more likely to hybridize with each other (Dressler, 1981; Dafni, 1987) (see p. 230).

Genetic barriers in the terrestrial orchids of southern Australia have been shown only in the genus *Cryptostylis* (Leonard, 1970). *Cryptostylis* species remain highly self-incompatible and this correlates with their ability to recognize and reject the viable pollen of congengers in contrast to sympatric genera with high levels of self-compatibility (Levin, 1978; Bernhardt and Calder, 1981). Conversely, southern Australian taxa that are successful inbreeders are involved in a high proportion of known interspecific hybridizations (Bernhardt and Burns-Balogh, 1988) (see below).

Hybridization

Excluding the genetic barrier described above, virtually all the isolation barriers discussed previously are only partially successful in both regions. Of course, hybridization records in the Mediterranean are far more extensive than in southern Australia. Furthermore, intergeneric hybrids are far commoner in the Mediterranean. Hybridization in the Mediterranean region occurs most commonly within the larger genera, with *Ophrys* species more likely to form infrageneric hybrids and species of genera like *Orchis* more likely to form intergeneric hybrids.

Infrageneric Hybridization. *Orchis, Dactylorhiza, Ophrys,* and *Serapias* species are notorious for forming infrageneric hybrids within the Mediterranean region. At least 50 hybrids have been recorded within the genus *Ophrys* (Dafni, 1987). In *Ophrys,* 100 hybrids were recorded for 50 species (Baumann and

Kuenkele, 1982). Infrageneric hybrids are comparatively rare in the nectariferous genera, but there are comparatively few large nectariferous genera in the Mediterranean region.

In the nectariferous Neottieae (Burns-Balogh *et al.*, 1987), for example, *Epipactis* is known to set few infrageneric hybrids. This, however, could be related to a trend toward obligate autogamy within *Epipactis* (Reinhard, 1977; Robatsch, 1983). Nectarless *Cephalanthera* (Neottioideae), on the other hand, shows at least seven hybrids, although most combination involve *C. longifolia,* a widespread species, as one of the parents (Burns-Balogh *et al.*, 1987).

In southern Australia, recurrent hybrids and/or hybrid swarms have been identified in such large genera as *Caladenia* (Clements, 1982), *Diuris* (Beardsell, 1975), *Thelymitra* (Burns-Balogh and Bernhardt, 1988), *Pterostylis* (Clements, 1982), and *Microtis* (Bates, 1981). There are comparatively few authenticated reports of intrageneric hybridization in the large, nectariferous genus *Prasophyllum sensu stricto* (Bates and Weber, 1980). This not surprising, but it probably reflects the status of research on the genus and not the natural dynamics of the taxa. We must remember that the recording of hybrids reflects a comparatively recent goal in Australian orchidology. Due to high levels of intraspecific variation, taxonomists tend to confine themselves to the morphological parameters that can be used to delimit species (Nicholls, 1969; Burns-Balogh and Bernhardt, 1988). Until recently, recurrent hybrids were often described as rare species with disjunct distributions [compare Willis (1970) with Clements (1982) and/or Nicholls (1969) versus Burns-Balogh and Bernhardt (1988)]. The comparative absence of known hybrids in *Prasophyllum* merely reflects a greater interest in mimectic genera over nectariferous genera in Australia.

Recurrent hybridization and backcrossing in Australian orchids is commonly attributed to sympatry and generalist foraging patterns of principal pollinators. Bates (1985*a*), for example, recently reported seeing a single bee visit three *Thelymitra* species during a single foraging bout. Bates has also offered the novel idea that interspecific hybridization may occur in *Thelymitra* species when syrphid flies drink stigmatic fluid from previously self-pollinated flowers, and the flies pass viable fragments of friable pollinia to the stigmas of congeners. This would certainly explain the high level of infrageneric hybridization between *Thelymitra* species with subcleistogamous biotypes, e.g., *T. carnea* and *T. pauciflora.* The flowers of these biotypes open only on hot spring days after the pollinia have already fallen onto their receptive stigmas (Beardsell and Bernhardt, 1982). Syrphids have been observed sucking the convex stigmatic droplets secreted by the stigmas of *Thelymitra* species (Bernhardt and Burns-Balogh, 1986*a;* see photos in Cady and Rotherham, 1970).

In *Caladenia,* infrageneric hybrids occur most often in sections sharing the same pollination syndrome. Consequently, intersectional hybrids are more uncommon (Heberle, 1982). The hybrids between *C. patersonii* (Sec. Calonema) and bee-pollinated species of Sec. Eucaladenia occur, as *C. patersonii* remains a nonmodel mimic deception system pollinated by a wide range of foragers, including bees.

Intergeneric Hybridization. Intergeneric hybrids are common in Mediterranean populations of the subtribe Orchidinae. Consequently, some authorities have questioned the alliance and veracity of generic parameters within the subtribe (Summerhayes, 1968). Dressler (1981) establishes two alliances within the Orchidinae: the *Orchis* alliance (also includes *Aceras, Anacamptis,* and *Himantoglossum*) and the *Serapias–Dactylorhiza* alliance (also includes *Coeloglossum, Gymandenia, Nigritella, Platanthera,* and *Pseudorchis*). Hybrids between genera of the same alliance occur with greater frequency compared to hybrids between genera of different alliances. Intergeneric hybrids within the same alliance show a higher F_1 fertility.

Analyses of known intergeneric combinations in the Orchidinae (Potucek, 1968; Willing and Willing, 1977; Sundermann, 1980; Baumann and Kenkele, 1982) show that three genera are involved in the majority of intergeneric combinations: *Gymnadenia, Orchis,* and *Dactylorhiza.* In contrast, the genera *Ophrys, Barlia, Himantoglossum, Chamorchis,* and *Traunsteinera* form infrequent intergeneric crosses.

To be more precise, certain species within a genus are more likely to form intergeneric hybrids than others. For example, *Gymnadenia* consists of about ten nectariferous species, but only two are Mediterranean in distribution (Nilsson, 1979). The species are pollinated primarily by sphingid moths or diurnal Lepidoptera (van der Pijl and Dodson, 1966). *Gymnadenia conopsea* is more likely to form intergeneric hybrids with nine other genera, namely *Dactylorhiza, Coeloglossum, Anacamptis, Platanthera, Nigritella, Traunsteinera, Pseudorchis, Cephalanthera,* and *Epipactis.* However, *G. conopsea* is the most widely distributed species in its genus throughout Eurasia southward to the Mediterranean and east through Iran (Baumann and Kuenkele, 1982). The potential for intergeneric hybrids between *G. conopsea* and other taxa remains high because this species shows considerable tolerance for a wide variety of habitats. Although it is primarily a species of comparatively dry habitats, it invades wet, boggy places in Britain (Summerhayes, 1968) and is widespread in disturbed sites such as pastures and chalk pits (Nilsson, 1979).

Orchis hybridizes with nectariferous genera, e.g., *Gymnadenia* and *Anacamptis,* and with nectarless genera, *Serapias* and *Aceras.* In general, such hybrids are infrequent.

Dactylorhiza shows a hybridization pattern different from that of *Orchis.* It hybridizes with *Gymnadenia, Coeloglossum, Nigritella,* and *Pseudorchis.* Hybridization between *Orchis* and *Dactylorhiza* remains controversial. Vermeulen (1947, 1972) concluded that such hybrids were extremely rare, while according to Baumann and Kuenkele (1982) the two genera do not hybridize at all. Potucek (1968) recorded 28–30 examples of *Orchis* × *Dactylorhiza!*

The genus *Serapias* hybridizes only with *Orchis,* and about 15 hybrids are known (Baumann and Kuenkele, 1982). Many solitary bees, which are the main pollinators of *Orchis,* are often found sleeping in *Serapias* sp. flowers (Dafni *et al.,* 1981). The main isolating mechanism between the two genera appears to be the relative position of the stigmatic surfaces, horizontal in *Serapias* and subvertical in *Orchis.* In Israel, both genera sometimes appear in dense sympatric populations, but hybrids remain extremely rare.

Ophrys is regarded as closely allied to *Orchis* and *Serapias* (Kullenberg, 1961). This view is sustained by Greilhuber and Ehrendorfer (1975), who found a chromosomal affinity ($x = 18$) among *Orchis* (Sec. Murianthae), *Serapias Anacamptis, Himantoglossum,* and *Ophrys.* They postulated that *Serapias* is an intermediate stage between *Orchis,* the ancient stock, and *Ophrys.* Although pollinators of *Ophrys* also visit species of *Orchis* and *Serapias* (A. Dafni, unpublished), it is questionable whether hybrids are produced.

Nelson (1962) mentioned a hybrid between *Ophrys sphegodes* ssp. *mammosa* and *Serapias lingua* with a question mark, but this was omitted in a later publication (Nelson, 1968). Moore (1980) has stated that "all species (of *Serapias*) appear to hybridize readily with *Ophrys.*" This statement requires further substantiation.

Ophrys is an isolated genus (Vermeulen, 1947) even though it is regarded as "very young" from an evolutionary point of view (Kullenberg and Bergstroem, 1973; Sundermann, 1977). It seems that the main reason for this isolation lies in the specific volatile compounds produced by this genus (Kullenberg and Bergstroem, 1973; Bergstroem, 1978; Borg-Karlson, 1985). The olfactory cues are specific enough to prevent intergeneric hybridizations, but are chemically close at the specific level. This leads to the production of numerous infrageneric hybrids. The olfactory factor is central to the isolation of many tropical orchids (Dodson *et al.,* 1969; Williams and Dodson, 1972; Hills *et al.,* 1972). Thus, *Ophrys* is not an exceptional genus within the Orchidaceae. In principle, olfactory deception in *Ophrys* ensures superior isolation at the generic level, while olfactory deception on *Orchis galilaea* ensures interspecific isolation. In both taxa, olfactory deception is a more important mechanism than visual deception (Dafni, 1987).

Intergeneric hybrids of Mediterranean Neottieae are extremely rare. There is one old record of *Cephalanthera alba* (= *damasonium*) × *Epipactis rubiginosa* (Wettstein, 1889). Possible intertribal hybridization between *Cephalanthera grandiflora* (= *damasonium*) and *Gymnadenia conopsea* (Orchidinae, *sensu* Burns-Balogh and Funk, 1986) reflects an interesting rare case, but it seems far too early to draw any new conclusions regarding generic affinities (but see Dressler, 1981). Dressler (1981) also reports *Epipactis palustris* × *Gymnadenia conopsea.*

The relative absence of natural levels of intergeneric hybridization in southern Australia provides an interesting contrast to the Mediterranean melange.

Thus far, intergeneric recombinations are restricted to small, blue-flowered species in the genus *Caladenia* (*Cyanicula* species; S. Hopper, personal communication) with their satellite genera *Glossodia* (Willis, 1970) and *Elythranthera* (see cover of Dixon and Buirchell, 1986): *Caladenia deformis* × *Glossodia major* (*C.* × *tutelar*) and *C. sericea* × *Elythranthera brunonis*. P. Bernhardt (unpublished) examined living plants of *C.* × *tutelar* during the spring of 1983 in Angelsea Victoria. No flower on these putative F_1's bore a functional column. Records of intergeneric hybrids will jump in a few years, though, as S. Hopper (personal communication) plans to subdivide *Caladenia sensu lato* into at least five smaller genera.

Speciation

A process of rapid speciation has been proposed for the epiphytic Orchidaceae of the Neotropics by Gentry and Dodson (1987). This model for adaptive radiation and diversification of epiphytic orchids is based on such combined characters as high seedset, successful dispersal patterns, exploitation/colonization of microhabitats or refugia, and genetic transilience associated with the founder effect. Gentry and Dodson (1987) note that this model applies in other floristic regions and we suggest that there is probably more compelling direct evidence of the process in the Orchidaceae of the Mediterranean and southern Australia. Sympatric, allopatric, and introgressive patterns of speciation are suggested for the orchid taxa of both regions, although orchidologists remain cautious as to the influence of introgressive hybridization in southern Australia.

In the Mediterranean region sympatric speciation has been suggested in the genus *Ophrys* (Kullenberg and Bergstroem, 1973). Divergence may be promoted due to natural levels of variation in the production of specific blends of scent (Borg-Karlson *et al.*, 1985; Kullenberg and Bergstroem, 1975). Flower size may also confer differential pollination within a population (Paulus and Gack, 1980; Voeth, 1984). Hybridization has also been regarded as an important factor in *Ophrys* speciation (Stebbins and Ferlan, 1956; Del Prete, 1984), especially if it is accompanied by geographic isolation (Sundermann, 1977).

Allopatric speciation was suggested for *Orchis, Dactylorhiza* (Dafni, 1987), and *Epipactis* (Burns-Balogh *et al.*, 1987), based on the distribution of endemics along the fringes of distribution of the genus. Dafni (1987) attempted to attribute the different speciation patterns of *Orchis* and *Ophrys* to the relative importance of odor variation in comparison to floral morphology and its visual cues (Kullenberg and Bergstroem, 1976) in interspecific isolation. In general, *Orchis* and *Dactylorhiza* are characterized more by wide variation in morphology (Soó, 1980) compared to complexity of floral fragrance [excluding *O. galilaea* (Bino *et al.*, 1982)].

Introgressive hybridization was found between *Orchis caspia* and *O. isra-*

elitica. Interspecific isolation broke down as a direct result of habitat disruptions due to human activity (Dafni and Baumann, 1982).

Sympatric speciation should be considered in the evolution of the pseudocopulatory *Caladenia* species and the possible derivation of satellite genera such as *Arthrochilus, Drakea,* or *Spiculea*. As the pollination systems of these plants have grown increasingly specialized due to the restricted attraction and exploitation of specific male insects, gene flow between allied populations may decline due to canalization of the ethological barrier (Stoutamire, 1981, 1983). A similar example of sympatric speciation may reflect adaptive radiation and overlapping distributions in *Prasophyllum sensu lato* in which wasps or beetles or bees and syrphid flies dominate as pollinators (Beardsell and Bernhardt, 1982; Bates, 1984*b;* Bernhardt and Burns-Balogh, 1986*a*).

The overwhelming trend toward mechanical autogamy in the majority of Australian terrestrial orchid genera (Beardsell and Bernhardt, 1982) may also be considered as a partial mechanism promoting sympatric speciation given the high degree of intraspecific variation in genera such as *Diuris* and *Thelymitra*. There appears to be no genetic barrier to interspecific hybridization in *Thelymitra* (Burns-Balogh and Bernhardt, 1988), but many species are dominated by populations of subcleistogamous or tardily opening individuals (p. 221), which obviously slow the process of gene flow within populations [e.g., *T. venosa* (Jones, 1971)]. Similar trends may occur in the genera *Calochilus, Prasophyllum, Genoplesium,* and *Pterostylis* (Beardsell and Bernhardt, 1982). Trends toward apomixis have been recorded within populations of *Prasophyllum* (Jones, 1977; Bates, 1984*b*) and in *Paracaleana* (Beardsell and Bernhardt, 1982).

Allopatric speciation has been suggested in *Thelymitra* by Bernhardt and Burns-Balogh (1987), but it may occur extensively in other genera of Australian neottioids. Putative evidence would combine the high degree of endemism and diversity of orchid taxa in southwestern Australia (Leigh *et al.*, 1984) with the repeated success of long-distance seed dispersal in such genera as *Calochilus, Corybas, Cryptostylis, Chiloglottis, Pterostylis,* and *Thelymitra* (Burns-Balogh and Bernhardt, 1985; Johns and Molloy, 1985). Due to the high degree of intraspecific variation, especially in food mimics, speciation may occur if successful long-distance dispersal is followed by genetic drift as maintained by self-pollination (Burns-Balogh and Bernhardt, 1988). Geographic isolation and establishment would naturally account for the distinct races or biotypes recorded for such widely distributed species as *Thelymitra fusco-lutea, T. nuda,* and *T. pauciflora* (Nicholls, 1969). Vicariance may be invoked in *Cryptostylis* species, which all share the same pollinator species, but remain genetically isolated. In general, the distribution of terrestrial orchids in southern Australia would appear to have grown increasingly disjunct due to climatic fluctuations and the subsequent invasion of the dominant sclerophyllous vegetation (Leigh *et al.*, 1984; James and Hopper, 1981; Barlow, 1981).

Introgressive hybridization remains a controversial mechanism for speciation in the Australian orchids. Recognition of hybrid swarms automatically established the validity of some degree of backcrossing (Beardsell, 1975) and some species have been redefined as F_2, F_3 populations [e.g., *Thelymitra truncata* (Burns-Balogh and Bernhardt, 1988)]. Intergradation based on introgression has been demonstrated morphometrically between *Caladenia alba* and *C. carnea*, suggesting gene migration into a population of *C. alba* (Morrison and Weston, 1985). However, no authorities are willing to state that any extant species in southern Australia is the direct consequence of introgression. Burns-Balogh and Bernhardt (1988) do suggest, however, that introgressive speciation will ultimately account for the intense morphological intergradation in the mitras (staminode–fertile stamen filament complex) of *Thelymitra* species.

Furthermore, microspeciation (*sensu* Grant, 1971) may occur in eastern populations of the recurrent hybrid *Thelymitra* × *macmillanii* (*T. antennifera* × *T. luteocilium*). The hybrid is believed to undergo Mendelian segregation when selfed and there are reports of backcrosses to one of the parent species (Bates, 1984*a*, 1985*a*). However, anecdotal material provided by Bates (1984*a*) and Beardsell and Bernhardt (1982) suggests that F_1 populations may increase their size and range by vigorous clonal reproduction via natural rates of vegetative propagation of root–stem tuberoids. Furthermore, *T.* × *macmillanii* is reputed to replace either of its parent species over time and may be found outside the distribution of one or both of the parent species. Despite field experimentation by Bates (1984*a*), the genetic evidence remains highly controversial.

If Australian orchidologists seem to be skeptical of the role of hybridization in speciation, it may be seen within an important historical context. Fitzgerald (1875–1895), regarded as the father of Australian orchidology, made some rather wild claims regarding the origin of Australian genera, breaking with the Darwinian concepts of reproductive isolation and adaptive floral morphology via modification. He discounted the adaptive significance of floral structures in encouraging cross-pollination for the majority of Australian taxa. Orchid species, in Fitzgerald's scenarios, appeared spontaneously following chance hybridization between extremely different taxa. The new species were orchidaceous satyrs and centaurs self-maintained by continuous self-pollination. Although this provides a unique and early version of punctuated equilibria in angiosperms, it must be discounted entirely in the light of fieldwork on cross-pollination in some of Fitzgerald's favorite examples, namely Mendelian segregation and the general absence of fertile intergeneric hybrids in Australia.

The Evolution from Reward to Deception

The phylogenetic trend from ancestral flowers offering rewards to derivative taxa which mimic rewards (edible/reproductive) appears to be polyphyletic in origin (Vogel, 1978). Evolutionary biologists have agreed generally that this trend shows its widest expression within the Orchidaceae. However, since the trend is polyphytic through the angiosperms, there is no reason to presume it must follow the same pathway through the Orchidaceae. We suggest that the genealogies of pollination syndromes may vary among the taxa of southern Australia and Mediterranean regions especially in regard to the origin of pseudocopulatory syndromes.

The transition from a general imitation of nectariferous plants in Mediterranean taxa, e.g., *Orchis caspia,* to species-specific flora mimicry is accompanied by decreased floral variability and a subsequent trend from oligophily towards monophily (Dafni, 1987). True nectariferous species (e.g., *Orchis coriophora*) are polyphilic. The loss of nectar production in *O. israelitica* and its resemblance to a specific model may be selectively advantageous, as it reduces the rate of wasted pollinia. The pollinators of *O. coriophora* and *O. caspia,* by comparison, lack the same degree of pollinator specificity, so one assumes that they lose pollinia to inconstant foragers. This hypothesis for the acquisition of greater specificity of the primary pollinator coupled with a higher pollination efficiency by species-specific floral mimicry requires further verification by field experimentation.

It is suggested that the chances for successful cross-pollination probably increase when optical cues are discernible to the pollinator from great distances. In closed, darkened habitats (woods and forests), terrestrial herbs, especially those with mimectic flowers, have supposedly fewer chances of being located by the pollinator. On the other hand, one may also assume that distribution of forest herbs is also controlled by low light intensity, which lowers photosynthate production. This may be one of the selective pressures determining the restriction of most species of *Orchis, Dactylorhiza, Serapias,* and *Ophrys* to more open habitats. Some of these species, especially those confined to damp, open sites, produce large populations in relatively confined areas, which are more attractive to naive pollinators than are widely scattered individuals. This generalization is supported, to some extent, by the fact that many species of *Orchis* and *Dactylorhiza* emit only a faint fragrance (Nilsson, 1981), suggesting that the olfactory cue may be a lesser importance compared to the massed visual cue.

Although our knowledge of the pollination ecology of *Orchis sensu lato* is still fragmentary, the available evidence, supplemented by data from closely related genera, such as *Ophrys* and *Serapias,* enables us to draw a general overview. Two main evolutionary lines are detected in this generic complex. One is a transition from nectariferous flowers to sexual deceit in which the crucial selective pressures are exerted through the production of specific volatile compounds. The second is a transition from rewarding flowers to food mimicry and shelter imitation in which visual cues remain the primary stimuli (Dafni, 1987).

In the tribe Neottieae, the Cephalantherinae is the only subtribe which does not offer nectar as a reward, and evidence suggests that both Cephalanthera (Mediterranean) and *Rhizanthella* (southern Australian) are pollinated by deceit (Burns-Balogh *et al.,* 1987). Therefore, the evidence suggests that within the subfamily Neottioideae, the more primitive taxa offer nectar as a reward, while the more advanced taxa show a trend toward deception. The loss of the rostellum (no visicidium present) and the tubular flowers hints that the Cephalantherinae are more advanced than the nectariferous Limodorinae or Neottinae in their respective pollination syndromes (Burns-Balogh *et al.,* 1987).

Floral mimesis is such a dominant trend in the Australian terrestrial orchids that the interrelationships between taxa with different mimectic syndromes seems clearer than interrelationships between nectariferous and mimectic taxa. No extant taxa in the Australian Neottioideae combine morphological modifications for mimectic syndromes with functional floral nectaries, unless one accepts the more classical interpretation of *Corybas* by Dressler (1981) and retains the genus within Diurideae *sensu stricto.* Of course, it has been suggested that a number of Australian neottioid genera (*sensu* Burns-Balogh and Funk, 1986) offer stigmatic secretions as a nutritious nectar substitute (p. 200). However, even if this is true, the use of stigmatic secretions as an edible reward reflects a derived condition in the floral evolution of orchids that would follow the suppression of floral nectaries.

It is plausible to suggest that food mimics ultimately derive from extinct ancestors bearing fundamental nectaries in Australia, as has been suggested above for the orchid flora of the Mediterranean and for the epiphytic epidendroids of the neotropics (van der Pijl and Dodson, 1966).

The nectariferous reward system occurs in the spiranthoid genera *Prasophyllum sensu stricto* and *Spiranthes sensu stricto.* This ancestral syndrome has been eliminated in *Diuris* with a subsequent trend toward mimesis of the vernal legume guild. Functional nectaries have been described often in the Spiranthoideae of the northern hemisphere and neotropics (Luer, 1972, 1975).

Much the same argument may be made for the terrestrial epidendroids in southern Australia. *Gastrodia* and *Dipodium* species are nonmimic deceptors, and the center for both genera is in tropical Asia, where there are many nectariferous epidendroids. Burns-Balogh and Funk (1986) have tentatively placed the Acianthinae (*Acianthus, Corybas, Stigmatodactylus,* and *Towsonia*) within the Epidendroideae. The derivation of the oviposition mimic in *Corybas* is quite clear, as it still appears to bear semifunctional nectar glands. We may presume that *Corybas* is derived from an ancestor similar to modern *Acianthus,* which is nectariferous but lacks sporocarp mimesis and is pollinated by *both* male and female mycetophilids (Table II). Specialized mimesis in *Corybas* permits three orchid genera (*Acianthus, Corybas,* and the neottioid *Pterostylis*) to share the single pollinator resource in late winter to early spring as each genus exploits the same insect family in an entirely different way. This trend parallels the exploitation of euglossine bees in neotropical epiphytes (Williams, 1982).

Evolutionary trends in the Australian Neottioideae are far less obvious, as the group lacks a smooth intergradation from reward to reward/mimic systems present in the Acianthinae, nor do we have nectariferous systems outnumbering food mimic taxa as in Australian Spiranthoideae. *Microtis,* and possibly *Genoplesium,* remain the only genera of Australian neottioids that offer nectar (Table II). Column morphology in *Microtis* (about ten species) is quite anomalous for the Australian neottioids in general (Burns-Balogh and Funk, 1986), and the pollination of the flowers by worker ants (Table II) is not repeated in any other genus of neottioids in southern Australia.

However, one should remember that nectariferous neottioids are common in the northern hemisphere (e.g., *Listera, Epipactis, Limodorum*) and such genera are regarded as among the most primitive of monandrous orchids (Nilsson, 1978, 1981; Burns-Balogh *et al.*, 1987), although they are not in the same tribes as the neottioids of southern Australia. Temperate South America provides a more convincing link between nectariferous and food mimic neottioids. The largest genera, *Chlorea* and *Gavilea,* contain some nectariferous species (Gumprecht, 1980). *Chlorea* and *Gavilea* are typically placed within the same tribe, Geoblasteae, as the Australian subtribe Caladeniinae containing food mimics *Caladenia sensu lato, Elythranthera, Eirochilus,* and *Glossodia* (Table V). Dressler (1981) has hypothesized that the tribe Geoblasteae *sensu stricto* reached Australia from South America via the Antarctic Corridor as late as the mid-Tertiary (Raven and Axelrod, 1974). Furthermore, Dressler (1981) has interpreted the Chloraeinae as richer in ancestral characters than the Caladeniinae. The food mimesis of Australian genera would have derived from South American stock that was nectariferous or, more likely, combined food mimic species (e.g., *Codonorchis* and *Megastylis*) and true nectariferous species.

Chlorea and *Gavilea* species bear stalked or paired nectaries that are usually located on the column or between the labellum claw and style base (Correa, 1956, 1969). *Caladenia* species often bear paired appendages on the style base (Woolcock and Woolcock, 1984), but they are rounded, ornamented, and nonsecretory. Further studies are therefore required to determine if the anomalous genus *Microtis* (Burns-Balogh and Funk, 1986) represents a regression toward the ancestral nectariferous syndrome or if it is simply more closely allied to *Gavilea* (P. Burns-Balogh, personal communication) than previously anticipated.

In general, the majority of food mimics in southern Australia reflect a trend toward the exploitation of polylectic and/or polytrophic foragers. *Caladenia* species exploit a wide range of naive, short-tongued Hymenoptera and Diptera. *Thelymitra* species appear to be slightly more specialized, with the majority of species mimicking the guild of nectarless pollen flowers (*sensu* Buchmann, 1983), which often require thoracic vibration to harvest pollen. Consequently, these *Thelymitra* sp. flowers are pollinated by just a few genera of polylectic bees. The mimicry of papilionoid legumes by *Diuris* species leads to an interesting dichotomy in the spectrum of pollinia vectors. *Diuris* flowers are pollinated by naive, generalist wasps and halictid bees, but these orchids are also pollinated by specialized female colletids, which feed their young primarily on legume pollen (Beardsell *et al.*, 1986; Armstrong, 1979).

In contrast, the derivation of pseudocopulatory systems exploiting winged Hymenoptera seems strikingly clear in Australian taxa. The exploitation of at least three families of wasps, one family of sawflies, and possibly one family of bees (Halictidae) has been derived repeatedly from food mimic ancestors. The shift from floral mimesis to sexual mimesis in *Caladenia* (Sec. Calonema) and its satellites (*Caleana, Drakea, Spiculea,* etc.) reflects nothing less than a canalization of a floral syndrome emphasizing lower diversity of potential pollinators. *Caladenia patersonii,* as discussed previously (p. 213), represents an intermediate case in which floral attractants and morphological modifications are still sufficiently generalized to exploit both male tiphiids and insects in other families and/or orders. Changes in pigmentation, fragrance chemistry, and the labellum claw [via the evolution of a hinge (Woolcock and Woolcock, 1984)] permit the shift toward vector specialization within the genus (Fig. 10).

This trend from a food mimic which was pollinated by a wide range of insects to a pseudocopulatory system that narrows the range of pollinators to male flower wasps is seen again in the Thelymitreae of southern Australia. *Calochilus* may be derived ultimately from a *Thelymitra*-type ancestor. This morphological trend toward the narrow exploitation of male scoliid wasps has involved the reduction of the column hood coupled with the development of dummy eyes on the expanded basal lobes of the hood (Fig. 4 versus Fig. 6). The *Calochilus* labellum is interpreted as a reversion to the ornamented lamina condition with a surfeit of long, darkly pigmented, vasculated calli obviously derived from ancestors in common with the Caladeniinae (Fig. 4 versus Fig. 10). Such minor modifications are easily appreciated by viewing the recurrent peloric forms of *Calochilus robertsonii* [*C. imberbis* (see Nicholls, 1969)], as the flowers are almost indistinguishable from less-advanced *Thelymitra* species [e.g., *T. antennifera, T. carnea, T. venosa* (Burns-Balogh and Bernhardt, 1988)]. To shamelessly anthropomorphize this trend, *Calochilus* is merely a dull-colored *Thelymitra* that has lost its gaudy bonnet but now wears goggles and has grown a beard! This shows a reversion toward the ancestral and dominant condition of floral zygomorphy in monandrous orchids in general.

The pseudocopulatory syndrome expressed by neottioid *Pterostylis* species remains obscure. One presumes that flowers with irritable labella must be derived from ancestors with nonmotile but hinged labella. The column structure, sepal morphology, and labellum callus ornamentation of *Pterostylis* species

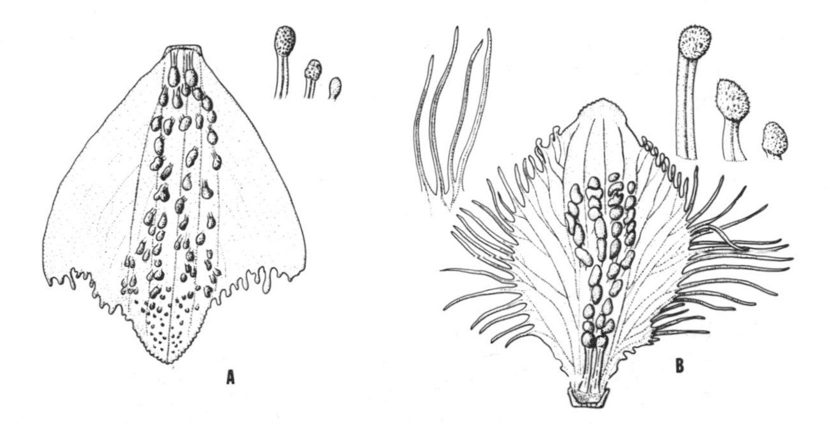

FIG. 10. Food versus sexual mimesis in *Caledenia* species. (A) *Caladenia lyalii* labellum (×23) and vasculated calli (×30). (B) *Caladenia discoidea* labellum (×14), marginal vasculated teeth, and central calli (×30). The labellum of the food mimic *C. lyalii* tends to be trilobate, but has reduced teeth and white-yellow calli. The calli have globose, pappillate heads often contrasting in color with the labellum lamina, forming "pseudo stamens" which may attract pollen-eating syrphid flies, wasps, or pophylectic bees. The trilobate condition is suppressed in the pseudocopulatory labellum of *C. discoidea.* The teeth are elongated and dark in color. The dark maroon heads of the calli bear additional trichomes and the heads tend to overlap, forming a compound, grossular structure imitating the body of the female wasp. Such labella are pollinated primarily by selected genera of male Tiphiidae. In some pseudocopulatory *Caladenia* species the base of the column bears two faceted dummy eyes, which are never present in food-mimic *Caladenia* species.

show an obvious parallel to genera in the Caladeniinae (Woolcock and Woolcock, 1984). Even orchidologists who criticize each other's phylogeny tend to agree that *Caladeniinae* and *Pterostylidinae* share a common origin (Dressler, 1981; Burns-Balogh and Funk, 1986) and may be sister taxa (P. Burns-Balogh, personal communication), although *Pterostylis* is considered as more derived in floral morphology. How does one derive a female-fungus-gnat mimic (e.g., *Pterostylis*) from an ancestor that was a mimic of female wasps? We suggest that Caladeniinae that flower early in the spring and bear small, hinged labella with dummy females have, or had, a latent selection potential (Hartl and Dykhuizen, 1985) for pollination by male fungus gnats. *Arthrochilus* and *Spiculea* species, for example, bear labella less than 8 mm in length and their pseudocopulatory syndromes attract the physically smaller genera of tiphiids (Stoutamire, 1986). Due to morphological preadaptations, ancestral Caladeniinae, pollinated by tiny male wasps, may have had a greater potential for minor morphological and biochemical modifications in order to exploit the male mycetophilid resource.

Distinctions between the floral topology of *Pterostylis* species and genera of Caladeniinae pollinated by small tiphiids appears to rest in the comparative architecture and fusion of their perianths in contrast to respective similarities in the column–labellum complex (P. Bernhardt and P. Burns-Balogh, unpublished observations).

Nectar is costly to produce (Southwick, 1984). The trend toward mimicry in the terrestrial orchids of southern Australia and the Mediterranean region would obviously conserve water and carbohydrate resources. This is selectively advantageous in perennial plants that must withstand seasonal aridity (Gentry, 1974; Brown and Kodric-Brown, 1979; Heinrich, 1979; Boyden, 1980; Nilsson, 1983*b;* Ackerman, 1986). Field experimentation must be employed to demonstrate the adaptive significance of the nectarless condition in deceptive orchids.

It has also been suggested that pollination by deceit is selectively advantageous when a species habitually shows low population densities throughout its range (Macior, 1971; Ackerman, 1986). A population consisting, say, of a dozen nectariferous orchids may not be able to compete for a limited pollinator guild when the floral phenology of the orchids overlaps with many shrubs, trees, and herbs that offer pollen and nectar or pollen or nectar. As mentioned above, orchids in general tend to exploit only a fraction of the potential guild of pollinators within a habitat. Consider the many orchids pollinated largely by halictid bees. In southern Australia orchids must share the halictids with coblooming, massively floriferous *Acacia* species, Myrtaceae, and rhizomatous composites (Bernhardt and Walker, 1984, 1985; Bernhardt, 1987).

Using low density to explain the trend toward deceit in orchid flowers of both Mediterranean and southern Australia regions undoubtedly has merit, but there is an important counterargument. Population size in the orchids of these regions does not always parallel the same low densities associated with midmontane epiphytes of the tropics. Unlike the epiphytic epidendroids, many of the terrestrial taxa previously discussed have large regional populations that have increased exponentially due to a combination of vegetative reproduction (Beardsell *et al.*, 1986; Dafni and Calder, 1987) and the fail-safe mechanism of self-pollination (p. 219). Therefore, trends toward deceit may be less important in the long-term survival of small fragmented populations [e.g., *Thelymitra venosa* (Jones, 1971; Bates, 1984*a*)], but very important in the survival of a species requiring certain levels of gene migration and heterozygosity [e.g., *Orchis mascula* (Nilsson, 1983*a;* Dafni, 1987)] required to alter the gene pool as environmental pressures fluctuate or new sites are colonized. Dressler (1981) has previously commented on deceit stimulating higher levels of xenogamy. In this respect, the trend toward deceit in orchids may actually converge with food and oviposition site mimesis recorded in many relictual magnolioids and in the genus *Cyclanthus* (Bernhardt and Thien, 1987).

DISCUSSION AND CONCLUSIONS

Mediterranean Region versus Southern Australia—Divergence and Convergence

Despite considerable geographic and phylogenetic isolation, it is obvious that parallelisms can be noted between the two orchid floras. Obviously, a climatic trend toward xerism has been a powerful selective pressure directly upon orchid diversity, vegetative morphology, and phenology. Furthermore, the mosaic of soils in the Mediterranean basin and southwestern Australia has contributed to wide disjuncts. Consequently, the orchids of both floristic regions show a higher density in open shrublands and woodlands, a floral phenology that is partially dependent on cyclical fires, and a flowering season that concentrates diversity from early spring through early summer. Of the two floristic regions, the orchid flora of southern Australia appears more variable, with a flowering season that properly begins in late winter and continues, at a most restricted level, through early autumn.

We further suggest that it is this combination of abiotic factors that is responsible, in part, for the shared trend toward monotipism and small satellite genera in the orchid floras of both regions. We define a small genus as one in which there are only four or less species, e.g., *Eriochilus, Glossodia, Limdorum, Nigritella,* and *Paracalaeana*. It remains to be seen whether mass extinction or punctuated equilibria models better account for the shared macroevolutionary trends.

In pollination systems, although the orchid floras must obviously exploit different insect species, the actual syndromes expressed in both regions show extensive convergences. For example, we do not see any trend toward bird pollination in either region. This syndrome is uncommon in the Orchidaceae and, when present, is concentrated in the tropics (van der Pijl and Dodson, 1966). However, Lepidoptera appear to play a minor role as pollinators in both regions. Furthermore, the orchid floras of both regions do contain a few taxa pollinated, at least in part, by beetles.

It must be emphasized, though, that while pollination systems overlap, the relative proportions of each syndrome diverge sharply between the two regions. An obvious example can be seen when we contrast the number of taxa offering nectar rewards. Nectariferous flowers, in general, are considered uncommon in the Orchidaceae, but one commonly finds nectariferous genera within some primitive subfamilies, e.g., Spiranthoideae, the tribe Orchideae and subtribes, Listerinae, and Limodorinae (*sensu* Burns-Balogh and Funk, 1986). The above taxa are more common in Europe and the Middle East than in southern Australia.

Proportionately, the diversity of nectariferous orchid flowers in the Mediterranean region is almost double that of southern Australia.

It is also suspected that genera of orchids in southern Australia bearing floral nectaries have a wider range of potential pollinators than the majority of nectariferous species throughout the Mediterranean. Although the nectariferous orchids from the latter regions receive an extremely wide range of foragers, comparatively few of the insect visitors are consistent and accurate dispersers of pollinia. The comparative trend toward generalist entomorphily hinge, in part, on floral morphology. The nectariferous species of southern Australia lack the pouches and spurs that conceal nectar in a number of Mediterranean genera, so the edible reward produced by *Prasophyllum* and *Microtis* species is accessible to a wider variety of foragers, and pollination is not restricted to insects with long tongues. We also suggest that the natural diversity of nectar-feeding insects is higher in southern Australia. Even the Mediterranean species offering nectar in shallow cups may still have a smaller spectrum of potential pollinators compared to *Prasophyllum* and *Microtis* species (p. 202).

The mimesis of edible rewards, pollen and/or nectar, occurs extensively in both regions, but here again, the proportionate number of species showing each subdivision of floral mimesis may diverge due to independent modes of adaptive radiation. In fact, there is an obvious dichotomy. Guild mimesis is common in Australia, but rare in the Mediterranean. Classic Batesian mimicry is well defined in the Mediterranean, but may not even exist in southern Australia. The comparative importance of nonmimic deception requires further research. If *Caladenia* and its satellite genera do prove to be nonmimic deceptors, this mode of mimesis will be of equal importance in southern Australia as it is interpreted in the Mediterranean based on the diversity of the genus *Orchis*.

While both regions have orchid taxa showing brood-site deception of dipterans, two different fly families are exploited. Australian *Corybas* employs fungal mimesis, while the labellum of *Epipactis* mimics the bodies of aphids. The only shelter flower belongs to *Serapias*, and it is confined to the Mediterranean.

Pseudocopulatory systems and sexual mimesis, in general, are well expressed within both orchid floras, but once again, proportions vary. Adaptive radiation has been far greater in southern Australia, where ten genera in two orchid subfamilies express some aspect of the syndrome. In the Mediterranean, though, pseudocopulatory flowers are limited to the large genus *Ophrys*, which exploits at least three insect superfamilies. Compare this to the seven superfamilies of insects, representing two orders (Diptera and Hymenoptera), exploited in southern Australia. Of course, the vast majority of southern Australian orchids with pseudocopulatory syndromes reflect intensive coevolution with the single family, Tiphiidae.

Ironically, the two orchid floras show the greatest degree of convergence in their shared trends toward self-pollination, whether mechanical or insect-mediated. Self-pollinating races are so common to some of the genera of both regions that they have commonly been granted specific/subspecific status and probably account for successful establishment following long-distance dispersal. Both regions encourage the evolution of autogamous mechanisms at their margins due to the harshness of abiotic factors and/or depauperate populations of pollinators over time.

We should qualify the degree of convergence of interspecific isolation mechanisms by noting that pre- and postzygotic barriers appear more efficacious in the Australian flora. Specifically, intergeneric F_1's are more commonly found in the Mediterranean region, and such hybrids show varying levels of fertility. Even if we insist that the European genera are too finely divided (Dressler, 1981), we would still be left with the fact that species with highly dissimilar pollination systems show a high capacity for recombination. *Orchis* and *Serapias* offer quite different attractants and mimectic syndromes, but they are pollinated by the same bees and risk hybridization when sympatric. Furthermore, none of the Mediterranean orchids shows the stigmatic recognition/rejection system for the pollen of other species described for the genus *Cryptostylis* in southern Australia.

This is not to suggest that interspecific hybridization in some southern Australian genera is infrequent or negligible as a source of gene migration. As both regions are subject to habitat disruption and the subsequent hybridization of the habitat, in all probability, introgressive speciation plays a similar and important role in selected genera of both orchid floras. Frankly, in view of the similarities between the two regions of prevailing climate, edaphic mosaics, competing shrubby vegetations, and habitat disruptions, both orchid floras may show rapid, California-type speciation as reviewed by Stebbins (1974).

Conclusions

The orchid floras of southern Australia and the Mediterranean region belong to different monandrous tribes and are dependent on different insect species for pollination. However, the pollination systems of both regions show broad convergence. The pollinating insects of both continents undergo divergent life cycles with particular regard to courtship, mating, and provisions for eggs and offspring. Therefore, while the same major pollination syndromes are found on both continents, the subdivisions of each syndrome are often highly divergent when shared syndromes are contrasted.

Pollination systems within the Orchidaceae emphasize mimicry over edible rewards. This pattern is typical of both regions and features the mimesis of nectariferous and/or pollen-rich flowers, oviposition sites, shelter, and prospective mates. Pseudocopulatory syndromes show a much wider adaptive radiation in southern Australia than in the Mediterranean region. Shelter flowers have not been described in Australia.

The two orchid floras occupy habitats subjected to fragmentation and disruption by human and natural factors. Species survival appears to be maintained largely through a high degree of intraspecific variation. Speciation is probably rapid due to fluctuating environmental pressures which encourage allopatric, sympatric, and introgressive speciation. These modes of speciation in both orchid floras may be aided by convergent trends toward self-pollination.

REFERENCES

Ackerman, J. D., 1981, Pollination biology of *Calypso bulbosa* var. *occidentalis* (Orchidaceae): A food-deception system, *Madrono* **28:**101–110.

Ackerman, J. D., 1986, Mechanisms and evolution of food-deceptive pollination systems in orchids, *Lindelyana* **1:**108–113.

Agren, L., and Borg-Karlson, A. K., 1984, Responses of *Agrogorytes* (Hymenoptera: Sphecidae) males to odour signals from *Ophrys insectifera* (Orchidaceae). Preliminary EAG and chemical investigation, *Nova Acta Reg. Soc. Sci. Ups. Ser. V* **3:**111–117.

Agren, L., Kullenberg, B., and Sensenbaugh, T., 1984, Congruences in pilosity between three species of *Ophrys* (Orchidaceae) and their hymenoptrean pollinators, *Nova Acta Reg. Soc. Sci. Ups. Ser. V* **3:**15–25.

Allan, N., 1977, *Darwin and His Flowers,* Faber & Faber, London.

Ames, O., 1937, Pollination of orchids through pseudocopulation, *Bot. Mus. Leafl. Harv. Univ.* **5:**1–24.

Armstrong, J. D., 1979, Biotic pollination mechanisms in the Australian flora—A review, *N. Z. J. Bot.* **17:**467–508.

Aschmann, H., 1973, Distribution and peculiarity of Mediterranean ecosystems, in: *Mediterranean Type Ecosystems* (F. di Castri and H. Mooney, eds.), pp. 11–19, Springer, Berlin.

Baker, H. G., 1974, The evolution of weeds, *Annu. Rev. Ecol. Syst.* **5:**1–24.

Barlow, B., 1981, The Australian flora: Its origin and evolution, in: *Flora of Australia,* Vol. 1, introduction, pp. 25–76, Griffin Press, Netley, South Australia.

Barth, F. G., 1985, *Insects and Flowers: The Biology of a Partnership,* Princeton University Press, Princeton, New Jersey.

Bates, R., 1981, Observations of pollen vectors on a putative hybrid swarm of *Microtis* R. B. R., *Orchadian* **7:**14.

Bates, R., 1984*a,* Australia's colourful sun-orchids. *Thelymitra, Aust. Orchid Rev.* **49:**109–112.

Bates, R., 1984*b,* Pollination of *Prasophyllum elatum* R. Br. (with notes on associated biology), *Orchadian* **10:**14–17.

Bates, R., 1985*a,* Colorful *Thelymitra*-hybrids, *Orchadian* **8:**119–121.

Bates, R., 1985*b,* Colorful *Caladenias, Aust. Orchid Rev.* **50:**6–11.

Bates, R., 1986, Delightful *Diuris:* The Donkey Orchid, *Aust. Orchid Rev.* **51:**12–18.

Bates, R., and Weber, J. Z., 1980, A putative hybrid between *Prasophyllum archeri* and *P. despectans, Orchadian* **6:**188–189.

Baumann, H., and Kuenkele, S., 1982, *Die wildwachsenden Orchideen Europas,* Kosmos Verlag, Stuttgart.

Beardsell, D., 1975, Remnants of a *Diuris* series of Warrandyte, *Vic. Nat.* **92:**244–246.

Beardsell, D., and Bernhardt, P., 1982, Pollination biology of Australian terrestrial orchids, in: *Pollination '82* (E. G. Williams, R. B. Knox, J. H. Gilbert, and P. Bernhardt, eds.), pp. 166–183, University of Melbourne Press, Parkville, Australia.

Beardsell, D. V., Clements, M. A., Hutchinson, J. F., and Williams, E. G., 1986, Pollination of *Diuris maculata* R. Br. (Orchidaceae) by floral mimicry of the native legumes, *Daviesia* spp. and *Pultenaea scabra* R. Br., *Aust. J. Bot.* **34:**165–173.

Bergstroem, G., 1978, Role of volatile chemicals in *Ophrys*–pollinator interactions, in: *Biochemical Aspects of Plant and Animal Co-evolution* (J. B. Harborne, ed.), pp. 207–232, Academic Press, London.

Bernhardt, P., 1986, Orchidelirium, *Garden* **10:**6–11.

Bernhardt, P., 1987, A comparison of the diversity density and foraging behavior of bees and wasps on Australian *Acacia* spp. *Ann. Mo. Bot. Gard.* **74:**42–50.

Bernhardt, P., and Burns-Balogh, P., 1983, Pollination and pollinarium of *Dipodium punctatum* (Sm.) R. Br., *Vict. Nat.* **100:**197–199.

Bernhardt, P., and Burns-Balogh, P., 1986*a*, Floral mimesis in *Thelymitra nuda* (Orchidaceae), *Plant Syst. Evol.* **151:**187–202.

Bernhardt, P., and Burns-Balogh, P., 1986*b*, Observations on the floral biology of *Prasophyllum odoratum* (Orchidaceae, Spiranthoideae), *Plant Syst. Evol.* **153:**65–76.

Bernhardt, P., and Calder, D. M., 1981, Hybridization between *Amyema pendulum* and *Amyema fuandang* (Loranthaceae), *Bull. Torr. Bot. Club* **108:**213–230.

Bernhardt, P., and Thien, L. B., 1987, Self isolation and insect pollination in the primitive angiosperms: New evaluations of older hypotheses, *Plant Syst. Evol.*, **156:**159–176.

Bernhardt, P., and Walker, K., 1984, Bee foraging on three sympatric species of Australian *Acacia*, *Int. J. Entomol.* **32:**322–330.

Bernhardt, P., and Walker, K., 1985, Insect foraging on *Acacia retinodes* var. *retinodes* in Victoria, Australia, *Int. J. Entomol.* **27:**97–101.

Bino, R. J., Dafni, A., and Meeuse, A. D. J., 1982, The pollination ecology of *Orchis galilaea* (Bornm. et Schulze) Schltr. (Orchidaceae), *New Phytol.* **90:**315–319.

Boyden, T. C., 1980, Floral mimicry by *Epindendrum ibaguense* (Orchidaceae) in Panama, *Evolution* **34:**135–136.

Borg-Karlson, A. K., 1979, Kemisk beteendestimulation hos *Eucera longicornis* hanar, *Entomol. Tidskr.* **100:**125–128.

Borg-Karlson, A. K., 1985, Chemical and behavioral studies of pollination in the genus *Ophrys* L. (Orchidaceae), Ph.D. dissertation, Department of Organic Chemistry, Royal Institute of Technology, Stockholm, Sweden.

Borg-Karlson, A. K., and Tengoe, J., 1986, Odour mimetism? Key substances in the *Ophrys lutea–Andrena* pollination relationship (Orchidaceae–Andrenidae), *J. Chem. Ecol.* **12:**1927–1941.

Borg-Karlson, A. K., Bergstroem, G., and Kullenberg, B., 1985, Chemical basis for the relationship between *Ophrys* orchids and their pollinators II: Volatile compounds of *Ophrys insectifera* and *O. speculum* as insect mimetic attractants/excitants, Chapter III in A. K. Borg-Karlson, Chemical and behavioral studies of pollination in the genus *Ophrys* L. (Orchidaceae), Ph.D. dissertation, Department of Organic Chemistry, Royal Institute of Technology, Stockholm, Sweden (1985).

Brantjes, N. B. M., 1981, Ant, bee and fly pollination in *Epipactis palustris* (L.) Crantz, *Acta Bot. Neerl.* **30:**53–68.

Brown, J. H., and Kodric-Brown, A., 1979, Convergence, competition, and mimicry in a temperate community of hummingbird-pollinated flowers, *Ecology* **60:**1032–1035.

Buchman, S. L., 1983, Buzz pollination in angiosperms, in: *Handbook of Experimental Pollination Biology* (C. E. Jones and R. J. Little, eds.), pp. 73–113, Van Nostrand Reinhold, New York.

Burns-Balogh, P., 1985, Evolution of the monandrous Orchidaceae VI. Evolution and pollination mechanisms in the subfamily orchidoideae, *Can: Orchid. J.* **3:**29–56.

Burns-Balogh, P., and Bernhardt, P., 1985, Evolutionary trends in the androecium of the Orchidaceae, *Plant Syst. Evol.* **149:**119–134.

Burns-Balogh, P., and Funk, V. A., 1986, *A Phylogeny Analysis of the Orchidaceae,* Smithsonian Contributions to Botany, No. 61.

Burns-Balogh, P., and Bernhardt, P., 1988, Floral evolution and phylogeny in the Tribe Thelymitreae, *Plant Syst. Evol.* **151:**187–202.

Burns-Balogh, P., Szlachetko, D.L., and Dafni, A., 1987, Evolution, pollination, and systematics of the tribe Neottieae (Orchidaceae), *Plant Syst. Evol.* **156:**91–115.

Cady, L., 1965, Notes on the pollination of *Caleana major* R. Br., *Orchadian* **2:**34–35.

Cady, L., and Rotherham, E. R., 1970, *Australian Native Orchids in Colour,* Tuttle, Rutland, Vermont.

Calder, D. M., Adams, P. B., and Slater, A. T., 1982, The floral biology and breeding system of *Dendrobium speciosum* Sm., in: *Pollination '82* (E. G. Williams, R. B. Knox, J. H. Gilbert, and P. Bernhardt, eds.), pp. 84–92, University of Melbourne Press, Parkville, Australia.

Carlquist, S., 1979, *Stylidium* in Arnhem Land, Australia New species, mode of speciation on the Sandstone Plateau and comments on floral mimicry, *Aliso* **9:**411–461.

Cheeseman, T. F., 1981, On the fertilization of *Thelymitra, Trans. N. Z. Inst.* **13:**291–296.

Clements, M. A., 1982, *Preliminary Checklist of Australian Orchidaceae,* Department of Capital Territory, Canberra, Australia.

Coleman, E., 1927, Pollination of an Australian orchid *Cryptostyllis leptochila* F. V. Muell., *Vict. Nat.* **44:**333–340.

Coleman, E., 1928, Pollination of an Australian orchid by male ichneumonid *Lissopimpla semipunctata* Kirby, *Trans. R. Entomol. Soc. Lond.* **76:**533–539.

Coleman, E., 1929, Pollination of *Cryptostylis subulata* (labill.) Reichh., *Vict. Nat.* **46:**62–66.

Coleman, E., 1930, Pollination of *Cryptostylis erecta* R. Br., *Vict. Nat.* **46:**62–66.

Coleman, E., 1932, Pollination of *Diuris pedunculata* R. Br., *Vict. Nat.* **49:**179–186.

Coleman, E., 1933*a,* Pollination of orchids: genus *Prasophyllum, Vict. Nat.* **49:**214–221.

Coleman, E., 1933*b,* Pollination of *Diuris sulphurea* R. Br., *Vict. Nat.* **50:**3–8.

Correa, M. N., 1956, Las especies argentinas del genero *Gavilea, Bol. Soc. Arg. Bot.* **6:**73–86.

Correa, M. N., 1969, *Chloraea,* genero sudamericano de Orchidaceae, *Darwiniana* **15:**374–500.

Crowson, R. A., 1981, *The Biology of Coleoptera,* Academic Press, New York.

Dafni, A., 1983, Pollination of *Orchis caspia*—A nectarless plant species which deceives the pollinators of nectariferous species from other plant families, *J. Ecol.* **71:**467–474.

Dafni, A., 1984, Mimicry and deception in pollination, *Annu. Rev. Ecol. Syst.* **15:**259–278.

Dafni, A., 1986, Floral mimicry—Mutualism and unidirectional exploitation of insects by plants, in: *Plant Surface and Insects* (T. R. F. Southwood and B. E. Juniper, eds.), pp. 86–94, Arnold, London.

Dafni, A., 1987, Pollination in *Orchis* and related genera: Evolution from reward to deception, in: *Orchid Biology: Reviews and Perspectives 4* (J. Arditti, ed.), pp. 79–104, Cornell University Press, Ithaca.

Dafni, A., and Baumann, H., 1982, Biometrical analysis in populations of *Orchis israelitica* Baumann and Dafni, *O. caspia* Trautv, and their hybrids, *Plant Syst. Evol.* **140:**87–94.

Dafni, A., and Calder, D. M., 1987, Pollination by deceit—Floral mimesis in *Thelymitra antennifera* Hook f., (Orchidaceae), *Plant Syst. Evol.* **158:**11–22.

Dafni, A., and Ivri, Y., 1979, Pollination ecology of, and hybridization between, *Orchis coriophora* L. and *O. collina* Sol. ex Russ. (Orchidaceae) in Israel, *New Phytol.* **83:**181–187.

Dafni, A., and Ivri, Y., 1981*a,* The flower biology of *Cephalanthera longifolia* (Orchidaceae)—Pollen imitation and facultative floral mimicry, *Plant Syst. Evol.* **137:**229–240.

Dafni, A., and Ivri, Y., 1981*b,* Floral mimicry between *Orchis israelitica* Baumann and Dafni (Orchidaceae) and *Bellevalia flexuosa* Boiss. (Liliaceae), *Oecologia* **49:**229–232.

Dafni, A., and Woodell, S. R. J., 1986, Stigmatic exudate and the pollination of *Dactylorhiza fuchsii* (Druce) Soo, *Flora* **178:**343–350.

Dafni, A., Ivri, Y., and Brantjes, N. B. M., 1981, Pollination of *Serapias vomeracea* Briq. (Orchidaceae) by imitation of holes for sleeping solitary male bees (Hymenoptera), *Acta Bot. Neerl.* **30:**69–73.

Darwin, C., 1877, *The Various Contrivances by Which Orchids Are Fertilized,* 2nd ed., John Murray, London.

Del Prete, C., 1984, The genus "*Ophrys*" L. (Orchidaceae): A new taxonomic approach, *Webbia* **38:**209–220.

Dixon, K., 1986, Tropical terrestrial orchids: The orchid flora of the Kimberley Region, in: *Orchids of Western Australia: Cultivation and Natural History* (K. W. Dixon and B. Buirchell, eds.), pp. 85–95. Western Australian Native Orchid Study and Conservation Group, Western Australia.

Dixon, K. W., and Buirchell, B., eds., 1986, *Orchids of Western Australia: Cultivation and Natural History,* Western Australian Native Orchid Study and Conservation Group, Western Australia.

Dockrill, A. W., 1956, The cross-pollination of *Chiloglottis formicifera* Fitzgs., *Aust. Orchid Rev.* **21:**26–27.

Dockrill, A. W., 1969, *Australian Indigenous Orchids,* Volume 1. *The Epiphytes, The Tropical Terrestrial Species,* Halstead Press, Sydney, Australia.

Dodson, C. H., 1962, The importance of pollination in the evolution of the orchids of tropical America, *Am. Orchid Soc. Bull.* **31:**525–534, 641–649, 731–735.

Dodson, C. H., Dressler, R. L., Hills, H. G., Adams, R. M., and Williams, N. H., 1969, Biologically active components of orchid fragrances, *Science* **164:**1243–1249.

Dressler, R. L., 1968, Pollination by euglossine bees, *Evolution* **22:**202–210.

Dressler, R. L., 1981, *The Orchids—Natural History and Classification,* Harvard University Press, Cambridge, Massachusetts.

Ducke, A., 1901, Beobachtungen ueber Bluetenbesuch, Erscheinumgseit etc. der bei Paravorkommenden Bienen, *Z. Syst. Hymenopt. Dipt.* **1:**25–32.

Erickson, R., 1951, *Orchids of the West,* Paterson Brokensha, Perth, Australia.

Faegri, K., and van der Pijl, L., 1979, *The Principles of Pollination Ecology,* 3rd ed., Pergamon Press, Oxford.

Fitzgerald, R. D., 1875–1895, *Australian Orchids,* Vols. 1 and 2, Government Printer, Sydney, Australia.

Fordham, F., 1946, Pollination of *Calochilus campestris,* Introductory note by H. M. R. Rupp, *Vict. Nat.* **62:**199–201.

Gentry, A. H., 1974, Flowering phenology and diversity in tropical Bignoniaceae, *Biotropica* **6:**64–68.

Gentry, A. H., and Dodson, C. H., 1987, Diversity and biogeography of neotropical vascular epiphytes, *Ann. Missouri Bot. Gard.* **74:**205–233.

Gill, A. M., 1981, Coping with fire, in: *The Biology of Australian Plants* (J. S. Pate and A. J. McComb, eds.), pp. 65–85, University of Western Australia Press, Nedlands, Western Australia.

Godfrey, M. J., 1925, The fertilization of *Ophrys speculum, O. lutea* and *O. fusca, J. Bot.* **65:**350–351.

Godfrey, M. J., 1931, The pollination of *Coeloglossum, Nigritella, Serapias,* etc. *J. Bot.* **59:**129–130.

Goelz, P., and Reinhard, H. R., 1977, Weitere Beabachtungen ueber die Bestaeunbung von *Ophrys speculum* Link., *Orchidee* **28:**147–148.

Grant, V., 1971, *Plant Speciation,* Columbia University Press, New York.

Greilhuber J., and Ehrendorfer, F., 1975, Chromosome number and evolution in *Ophrys* (Orchidaceae), *Plant Syst. Evol.* **124:**125–138.

Gumprecht, R., 1977, Seltsame Bestaeubungsvorgaenge bei Orchideen, *Orchidee* **1977**(Beilage zu heft 3):1–32.

Gumprecht, R., 1980, Blossom-structure and pollination mechanism in endemic orchids of South America, *Medio Ambiente* **4:**99–102.

Hagerup, O., 1952, The morphology and biology of some primitive orchid flowers, *Phytomorphology* **2:**134–138.

Hammersted, O., 1980, Metallfyn som pollinatorer av gronvit hattviel. *Entomol. Tidskr.* **101:**115–118.

Hartl, D. L., and Dykhuizen, D. E., 1985, The neutral theory and the molecular basis of preadaptation, in: *Population Genetics and Molecular Evolution* (T. Ohta and K. Aoki, eds.), pp. 107–124, Japanese Science Society Press, Tokyo/Springer-Verlag, Berlin.

Hawkeswood, T. J., 1981, Insect pollination of *Angophora woodsiana* F. M. Bail. South-East Queensland, *Vic. Nat.* **98:**120–129.

Hawkeswood, T. J., 1987, Pollination of *Leptospermum flavescens* Sm. (Myrtaceae) by beetles (Coleoptera) in the Blue Mountains, New South Wales, Australia, *J. Ital. Entomol.* **3:**261–269.

Heberle, R. L., 1982, *Caladenia* in Western Australia and natural hybridization, *Orchadian* **7:**78–83.

Heinrich, B., 1977, Pollination energetics: An ecosystem approach, in: *The Role of Arthropods in Forest Ecosystems* (W. J. Mattson, ed.), pp. 41–46, Springer-Verlag, New York.

Heinrich, B., 1979, *Bumblebee Economics,* Harvard University Press, Cambridge, Massachusetts.

Hills, H. G., Williams, N. H., and Dodson, C. H., 1972, Floral fragrances and isolating mechanisms in the genus *Catasetum* (Orchidaceae), *Biotropica* **4:**61–76.

Hopper, S., and Maslin, B., 1978, Phytogeography of *Acacia* in Western Australia, *Aust. J. Bot.* **26:** 63–78.

Ivri, Y., and Dafni, A., 1977, The pollination ecology of *Epipactis consimilis* (Orchidaceae) in Israel, *New Phytol.* **79:**173–178.

Jacobsen, N. V., and Rasmussen, F. N., 1976, Ueber die Bestaeunbung von *Ophrys speculum* Link auf Mallorca, *Orchidee* **27:**64–66.

James, S. H., and Hopper, S. D., 1981, Speciation in the Australian flora, in: *The Biology of Australian Plants* (J. S. Pate and A. J. McComb, eds.), pp. 361–381, University of Western Australia Press, Nedlands, Western Australia.

Johns, J., and Molloy, B., 1985, *Native Orchids of New Zealand,* Reed, Wellington, New Zealand.

Jones, D. L., 1970, The pollination of *Corybas diemenicus* H. M. R. Rupp, W. H. Nichols, *Vict. Nat.* **87:**372–374.

Jones, D. L., 1971, A study of the self pollination of *Thelymitra venosa* R. Br. and some notes on its implications, *Vict. Nat.* **88:**217–228.

Jones, D. L., 1972, The pollination of *Prasophyllom alpinum* P. B., *Vict. Nat.* **89:**260–263.

Jones, D. L., 1974, The pollination of *Acianthus candatus* R. Br., *Vict. Nat.* **91:**272–274.

Jones, D. L., 1975, The pollination of *Microtis parviflora* R. Br., *Ann. Bot.* **39:**585–589.

Jones, D. L., 1977, Miscellaneous notes on Australian Orchidaceae II. Reduction of six teratological forms to synonymy, *Orchadian* **5:**126–128.

Jones, D. L., 1981, The pollination of selected Australian orchids, *Proc. Orchid Symp. 13th Int. Bot. Cong.* **1981:**40–43.

Jones, D. L., and Gray, B., 1974, The pollination of *Calochilus holtzei* F. Muell., *Am. Orchid Soc. Bull.* **43:**604–606.

Kullenberg, B., 1961, Studies in *Ophrys* pollination, *Zool. Bidrag.* (Uppsala) **34:**1–340.

Kullenberg, B., and Bergstroem, G., 1973, The pollination of *Ophrys* orchids, in: *Chemistry in*

Botanical Research (G. Bendz and G. Santesson, eds.), pp. 253–258, Nobel Foundation, Stockholm, Sweden.

Kullenberg, B., and Bergstroem, G., 1975, Chemical communication between living organisms, *Endeavor* **34:**59–66.

Kullenberg, B., and Bergstroem, G., 1976, The pollination of *Ophrys* orchids, *Bot. Not.* **129:**11–19.

Kullenberg, B., Borg-Karlson, A. K., and Kullenberg, A. Z., 1984*a,* Field studies on the behaviour of the *Eucera nigrilabris* male in the odour flow from flower labellum extract of *Ophrys tenthredinifera, Nova Acta Reg. Soc. Ups. Ser. V* **3:**79–110.

Kullenberg, B., Buel, H., and Tkalcu, B., 1984*b,* Uebersicht von Beobachtungen ueber Besuche von *Eucera* und *Tetralonia* = Maennchen auf *Ophrys* Blueten (Orchidaceae), *Nova Acta Reg. Soc. Ups. Ser. V* **3:**27–40.

Lavarack, P. S., 1981, Origins and affinities of the orchid flora of Cape York peninsula, *Proc. Orchid Symp. 13th Int. Bot. Cong.* **1981:**17–26.

Leigh, J., Boden, R., and Briggs, J., 1984, *Extinct and Endangered Plants of Australia,* Macmillan, Melbourne.

Leonard, D. R., 1970, A recent observation of the pollination of *Cryptostylis erecta* by a wasp, *Orchadian* **3:**111.

Levin, D. A., 1978, The origin of isolating mechanisms in flowering plants, *Evol. Biol.* **11:**185–317.

Little, R. J., 1980, Floral mimicry between two desert annuals, *Mohavea conferiflora* (Scrophulariaceae) and *Mentzelia involucrata* (Loasaceae), Ph.D. dissertation, Claremont Graduate School, Claremont, California.

Little, R. J., 1983, A review of floral food deception mimicries with comments on floral mutualism, in: *Handbook of Experimental Pollination Biology* (C. E. Jones and R. J. Little, eds.), pp. 294–309, S. and E. Scientific and Academic Editions, New York.

Lock, J. M., and Profita, J. C., 1975, Pollination of *Eulophia cristata* (s.w.) Steud. (Orchidaceae) in southern Ghana, *Acta Bot. Neerl.* **24:**135–138.

Luer, C. A., 1972, *The Native Orchids of Florida,* New York Botanical Garden, New York.

Luer, C. A., 1975, *The Native Orchids of the United States and Canada, Excluding Florida,* New York Botanical Garden, New York.

Macior, L. W., 1970, Pollination ecology of *Dodecatheon amethystinum* (Primulaceae), *Bull. Torrey Bot. Club* **97:**150–153.

Macior, L. W., 1971, Co-evolution of plants and animals—Systematic insight from plant–insect interaction, *Taxon* **20:**17–28.

Michener, C. D., 1974, *The Social Behavior of the Bees,* Harvard University Press, Cambridge, Massachusetts.

Michener, C. D., 1979, Biogeography of the bees, *Ann. Mo. Bot. Gard.* **66:**277–347.

Moore, D. M., 1980, *Serapias,* in: *Flora Europaea,* Vol. *5* (T. G. Tutin, V. H. Heywood, N. A. Burges, D. M. Moore, D. H. Valentine, S. M. Walters, and D. A. Webb, eds.), pp. 343–344, Cambridge University Press, Cambridge, Massachusetts.

Morrison, D., and Weston, P., 1985, Analysis of morphological variation in a field sample of *Caladenia catenata* (Smith) Druce (Orchidaceae), *Aust. J. Bot.* **33:**185–195.

Nelson, E., 1962, *Gestaltwandel und Artbildung eroertet am Beispiel der Orchidaceen Europas und der Mittelmeerlaender, insbesondere der Gattung Ophrys,* E. Nelson, Chernez-Montreaux.

Nelson, E., 1968, *Nonographie und Ikonographie der Orchidaceen-Gattungen Serapias, Aceras, Loroglossum, Barlia,* E. Nelson, Chernex-Montreaux.

Nicholls, W. H., 1969, *Orchids of Australia,* Thomas Nelson, Melbourne, Australia.

Nilsson, L. A., 1978, Pollination ecology and adaptation in *Platanthera chlorantha* (Orchidaceae), *Bot. Not.* **131:**35–51.

Nilsson, L. A., 1979, The pollination ecology of *Herminiam monorchis* (Orchidaceae), *Bot. Not.* **132:**537–549.

Nilsson, L. A., 1980, The pollination ecology of *Dactylorhiza sambucina* (Orchidaceae), *Bot. Not.* **133:**367–385.

Nilsson, L. A., 1981, Pollination ecology and evolutionary processes in six species of orchids, *Abstr. Upps. Diss. Fac. Sci.* (Uppsala) **1981:**593.

Nilsson, L. A., 1983*a,* Anthecology of *Orchis mascula* (Orchidaceae), *Nord. J. Bot.* **3:**157–179.

Nilsson, L. A., 1983*b, Mimesis of bellflower (Campanula)* by the red helleborine orchid *Cephalanthera rubra, Nature* **305:**799–800.

Nilsson, L. A., 1984, Anthecology of *Orchis morio* (Orchidaceae) at its outpost in the north, *Nova Acta Reg. Soc. Sci. Ups. Ser. V* **3:**167–175.

Pasteur, G., 1982, A classificatory review of mimicry systems, *Annu. Rev. Ecol. Syst.* **13:**169–199.

Pate, J. S., and Dixon, K. W., 1981, Plants with fleshy underground storage organs—A western Australian survey, in: *Biology of Australian Plants* (J. S. Pate and A.J. McComb, eds.), pp. 181–215, University of Western Australia Press, Perth, Australia.

Patterson, G., and Nilsson, L. A., 1983, Pollinationsekologin hos Adam och Eva pa stora Karlso, *Svensk Bot. Tidsk.* **77:**123–132.

Paulus, H. F., and Gack, C., 1980, Beobachtungen und Untersuchungen zur Bestauebungsbiologie suedspanischer *Ophrys*-Arten, in: *Probleme der Evolution ei europaeischen and mediterranen Orchideen* (K. Senghas and H. Sundermann, eds.), pp. 55–68, *Jahresber. Naturwiss. Ver. Wuppertal,* No. 33.

Paulus, H. F., and Gack, C., 1981, Neue Beobachtungen zur Bestaeubung von *Ophrys* (Orchidaceae) in Sued-Spanian mit besonderer Beruecksichtigung des Formenkreises *Ophrys fusca* agg., *Plant Syst. Evol.* **137:**73–79.

Paulus, H. F., and Gack, C., 1983, Beobachtungen und Experimente zum Pseudokopulationsverhalten an *Ophrys*—Das Lernverhalten von *Eucera barbiventris* an *Ophrys scolopax* in Sued-Spanien, *Orchidee* **33:**73–79.

Peakall, R., Beattie, A. J., and James, S. H., 1987, Pseudocopulation of an orchid by male ants: a test of two hypotheses accounting for the rarity of ant pollination, *Oecologia* **73:**52204.

Peisl, P., and Forster, J., 1975, Zur Bestauebunsbiologie des Knabenkrautes *Orchis coriophora* L. ssp. *fragrans, Orchidee* **26:**172–173.

Potucek, O., 1968, Intergenerische Hybriden der Gattung *Dactylorhize,* in: *Probleme der Orchideengattung Dactylorhiza* (K. Senghas and H. Sundermann, eds.), pp. 102–106, *Jahresber. Naturwiss. Ver. Wuppertal,* Nos. 21–22.

Pouyanne, A., 1917, La fecondation des *Ophrys* par les insectes, *Bull. Soc. Hist. Nat. Afr. Nord* **8:** 6–7.

Prochazka, F., and Velisek, V., 1983, *Orchideje Nasi Prirody,* Prague.

Proctor, M., and Yeo, P. F., 1973, *The Pollination of Flowers,* Collins, London.

Raven, P. H., and Axelrod, D. I., 1974, Angiosperm biogeography and past continental movements, *Ann. Mo. Bot. Gard.* **61:**539–673.

Reinhard, H. R., 1977, Autogamie bei europaeischen Orchideen, *Orchidee* **28:**178–182.

Richards, A. J., 1982, The influence of minor structural changes in the flower on the breeding systems and speciation in *Epipactis* (Zinn.) Orchidaceae, in: *Pollination and Evolution* (J. A. Armstrong, J. M. Powell, and A. J. Richards, eds.), pp. 47–53, Royal Botanic Gardens, Sydney, Australia.

Robatsch, K., 1983, Bluetenbiologie und Autogamie der Gattung *Epipactis, Jahresber. Naturwiss. Ver. Wuppertal* **36:**25–32.

Rogers, R. S., 1931, Pollination of *Caladenia deformis* R. Br., *Trans. R. Soc. S. Aust.* **44:**143–146.

Rotherham, E. R., 1968, Pollination of *Spiculea huntiana* (elbow orchid), *Vict. Nat.* **85:**7–8.

Schemske, D. W., 1981, Floral convergence and pollinator sharing in two bee pollinated tropical herbs, *Ecology* **62:**946–954.

Schemske, D. W., Willson, M. F., Melampy, M. N., Miller, L. J., Verner, L., Schemske, R. M., and Best, L. B., 1978, Flowering ecology of some spring woodland herbs, *Ecology* **59:**351–366.

Simpson, B. B., and Neff, J. L., 1981, Floral rewards: Alternatives to pollen and nectar, *Ann. Mo. Bot. Gard.* **68:**301–322.

Soó, R. de, 1980, *Dactylorhiza, Orchis,* in: *Flora Europaea,* Vol. 5 (T. G. Tutin, V. H. Heywood, N. A. Burges, D. M. Moore, D. H. Valentine, S. M. Walters, and D. A. Webb, eds.), pp. 333–337, 337–342, Cambridge University Press, Cambridge, Massachusetts.

Southwick, E. E., 1984, Photosynthate allocation to floral nectar: A neglected energy investment, *Ecology* **65:**1775–1779.

Stebbins, G. L., 1974, *Flowering Plants: Evolution above the Species Level,* Harvard University Press, Cambridge, Massachusetts.

Stebbins, G. L., and Ferlan, L., 1956, Population variability, hybridization, and introgression in some species of *Ophrys, Evolution* **10:**32–46.

Stoutamire, W. P., 1971, Pollination in temperate American orchids, in: *Proceedings 6th World Orchid Conference* (M. J. G. Corrigan, ed.), pp. 233–243, Halstead, Sydney, Australia.

Stoutamire, W. P., 1974, Australian terrestrial orchids, thynnid wasps and pseudocopulation, *Am. Orchid Soc. Bull.* **43:**13–18.

Stoutamire, W. P., 1975, Pseudocopulation in Australian terrestrial orchids, *Am. Orchid Soc. Bull.* **44:**226–233.

Stoutamire, W. P., 1976, Pollination strategies in orchids of southern Australia, in: *1st Symposium on the Scientific Aspects of Orchids* (H. H. Szmant and J. Wemple, eds.), pp. 27–33, Southfield, Michigan.

Stoutamire, W. P., 1981, Pollination studies in Australia terrestrial orchids, *Natl. Geog. Soc. Res. Rep.* **13:**591–598.

Stoutamire, W. P., 1983, Wasp pollinated species of *Caladenia* (Orchidaceae) in south-western Australia, *Aust. J. Bot.* **31:**383–394.

Stoutamire, W. P., 1986, *Spiculea cileata* the Australian elbow orchid, *Can. Orchid J.* **3:**19–23.

Stowe, M. K., 1988, Chemical mimicry, in: *The Chemical Mediation Of Coevolution* (K, Spencer, ed.), Pergamon Press, New York (in press).

Summerhayes, V. S., 1968, *Wild Orchids of Britain,* 2nd ed., Collins, London.

Sundermann, H., 1977, The genus *Ophrys*—An example of the importance of isolation for speciation, *Am. Orchid Soc. Bull.* **46:**825–831.

Sundermann, H., 1980, *Europaeische und mediterrane Orchideen, Eine Bestimmungsflora,* 3rd ed., Kurt Schmersov, Hannover.

Tengoe, J., 1979, Odour released behavior in *Andrena* male bees (Apoidea, Hymenoptera), *Zoon* **7:** 15–48.

Teschner, W., 1975, The orchids of Europe—Some facts and problems, *Am. Orchid Soc. Bull.* **44:** 288–291.

Van der Pijl, L., and Dodson, C. H., 1966, *Orchid Flowers: Their Pollination and Evolution,* University of Miami Press, Coral Gables, Florida.

Vermeulen, P., 1947, Studies on Dactylorchids, Ph.D. dissertation, Schotanus and Jens, Utrecht.

Vermeulen, P., 1972, Uebersicht zur Systematik und Taxonomie der Gattung *Orchis s. str.,* in: *Probleme der Orchideengattung Orchis* (K. Senghas and H. Sundermann, eds.), pp. 22–36, *Jahresber. Naturwiss. Ver. Wuppertal,* No. 25.

Voeth, W., 1975, *Trielis villosa* var *rubra.* Bestaueber auf Kreta, *Jahresber. Naturwiss, Ver. Wuppertal* **29:**131–139.

Voeth, W., 1980, Koennen *Serapias* blueten Nesttaeublumen sein?, *Orchidee* **30:**159–162.
Voeth, W., 1982, Die "ausgenborgten" Bestaueber von *Orchis pallens L.*, *Orchidee* **33:**196–203.
Voeth, W., 1983, Bluetenbockkaefer (Cerambycidae) als Bestaueber von *Dactylorhiza maculata* (L.) Soo subsp. *meyeri* (Rchb.f.) Tournay, *Mitt. Bl. Arb. Kr. Heim. Orch. Baden-Wuertt.* **15:** 305–330.
Voeth, W., 1984, Bestauebungsbiologische Beobachtungen an griechischen *Ophrys*—Arten, *Mitt. Bl. Arb. Kr. Heim. Orch. Baden-Wuertt.* **16:**1–20.
Vogel, S., 1975, Mutualismus und Parasitismus in der Nutzung von Pollentraegern, *Verh. Dtsch. Zool. Ges.* **1975:**105–110.
Vogel, S., 1978, Pilzmueckenblumen als Pilzmimeten, *Flora* **167:**329–398.
Wallace, B. J., 1978, On *Cryptostylis* pollination and pseudocopulation, *Orchadian* **5:**168–169.
Wallace, B. J., 1980, Cantharophily and the pollination of *Peristeranthus hillii*, *Orchadian* **6:**214–215.
Warncke, K., and Kullenberg, B., 1984, Uebersicht von Beobachtungen ueber Besuche von *Andrena* und *Colletes cornicularius* Muennchen auf *Ophrys* Blueten (Orchidaceae), *Nova Acta Reg. Soc. Sci. Ups. Ser. V* **3:**41–55.
Weber, J. Z., 1981, Orchidacae, in: *Flora of Central Australia* (J. Jessop, ed.), pp. 515–516, Reed, Sydney, Australia.
Wettstein, R. V., 1889, Untersuchungen bei *Nigritella angustifolia* Rich., *Ber. Dtsch. Bot. Ges.* **7:** 306–317.
Wiefelspuetz, W., 1964, Ueber die Selbstfruchtung bei *Ophrys apifera*, *Jahresber. Naturwiss. Ver. Wuppertal* **19:**56–62.
Wiens, E., 1978, Mimicry in plants, *Evol. Biol.* **11:**365–403.
Williams, N. H., 1982, The biology of orchids and euglossine bees, in: *Orchid Biology: Reviews and Perspectives,* Vol. 2 (J. Arditti, ed.), pp. 119–172, Cornell University Press, Ithaca, New York.
Williams, N. H., and Dodson, C. H., 1972, Selective attraction of male euglossine bees to orchid fragrances and its importance in long-distance pollen flow, *Evolution* **26:**84–95.
Willing, B., and Willing, E., 1977, *Bibiographie ueber die Orchideen Europas und des Mittelmerlaender, 1744–1976,* No. 11, Willdenowia.
Willis, J. H., 1970, A Handbook to Plants in Victoria, Vol. 1, 2nd ed., Melbourne University Press, Melbourne.
Wolf, T., 1950, Pollination and fertilization of fly-orchid *Ophrys insectifera* L. in Allindellille fredskov, Denmark, *Oikos* **2:**20–59.
Woolcock, C., 1980, Pollination of *Leporella fimbriata*, *Orchadian* **6:**157.
Woolcock, C., and Woolcock, D., 1984, *Australian Terrestrial Orchids,* Nelson, Melbourne, Australia.
Yeo, P. F., 1972, Floral allurements for pollinating insects, in: *Insect–Plant Relationships* (H. F. Van Emden, ed.), pp. 51–57, Blackwell, Oxford.

7

The Biogeography of Yeasts Associated with Decaying Cactus Tissue in North America, the Caribbean, and Northern Venezuela

WILLIAM T. STARMER, MARC-ANDRE LACHANCE, HERMAN J. PHAFF, and WILLIAM B. HEED

INTRODUCTION

Yeasts are associated with a wide range of insects and plants (Phaff and Starmer, 1987). In many cases the yeasts associated with insects are also found growing in or on the plants that the insects use for feeding and breeding. In these situations the three-way association of insect, yeast, and plant forms an interactive system that is useful for the study of ecology and evolution. We have been studying one such system that includes species of *Drosophila,* cactus, and yeast (Barker and Starmer, 1982). This system is characterized by the interdependence of yeast and *Drosophila* on the decaying tissues of cactus. It is a saprophytic system because the cactus tissue (fruit, stem, or pad) is dead and decaying under the action of soft-rot bacterial enzymes. The yeasts apparently serve two general purposes for the *Drosophila:* (1) they grow vigorously in the decaying plant tissue and provide essential nutrients such as proteins, sterols, and vitamins for the larval and adult

---

WILLIAM T. STARMER • Department of Biology, Syracuse University, Syracuse, New York 13244. MARC-ANDRE LACHANCE • Department of Plant Sciences, University of Western Ontario, London, Ontario N6A 5B7, Canada. HERMAN J. PHAFF • Department of Food Science and Technology, University of California at Davis, Davis, California 95616. WILLIAM B. HEED • Department of Ecology and Evolutionary Biology, University of Arizona, Tucson, Arizona 85721.

stage of the drosophilids (Sang, 1978), and (2) they produce volatile cues to adults in transit from one habitat to another, and selective stimuli for the larvae foraging on the yeast-rich necrosis (Fogleman *et al.*, 1981, 1982; Fogleman, 1982). The drosophilids are important to the yeast because they provide a means of dispersal to new habitats (i.e., the adults vector them), and the larval stages spread yeast cells within a new habitat once they have successfully colonized.

In this chapter we wish to describe the distribution of cactus-associated yeasts according to their host plants and their geography. Previous taxonomic work on cactus-associated yeasts has revealed a spectrum of host plant specificity. Some species, such as *Pichia heedii* and *Pichia antillensis,* are specific to only one or two cactus species (Phaff *et al.*, 1978; Starmer *et al.*, 1984). Other species appear to be restricted to larger taxonomic categories [e.g., *Candida deserticola* is found predominantly in species of the genus *Stenocereus* (Phaff *et al.*, 1985)], while several species, such as *Pichia cactophila* and *Candida sonorensis,* are found in a broad range of cactus genera (*Opuntia, Stenocereus, Pachycereus, Cephalocereus, Ferocactus,* and others). Yeast species also appear to be separated according to tissue type. Decaying fruits and pads (cladodes) of the same prickly pear plant have distinctly different yeast communities (Starmer *et al.*, 1988) and yeasts found in fruits appear to be subjected to additional influences from the extrinsic environment (Starmer *et al.*, 1987*a*).

The taxonomic work on yeasts also has revealed distinct as well as diffuse geographic distributions. Common cactus yeasts such as *Pichia cactophila* and *Candida sonorensis* are found in almost all regions and localities where yeasts have been collected from cactus tissues. Other species have more limited distributions. *Pichia thermotolerans* has been found only in the North American Sonoran Desert, whereas *Pichia pseudocactophila* is found in the Sonoran Desert as well as southern Mexico (Holzschu *et al.*, 1983). To some extent the geographic distribution and host plant specificity can be confounded. This is true for species such as *Pichia antillensis,* which has been primarily collected from rotting stems of *Cephalocereus royenii,* which is restricted to islands in the Caribbean Sea (Starmer *et al.*, 1984). One of the goals of this chapter is to bring together the separately published records on the distribution of related yeasts species and merge them with more recent unpublished findings for the analysis of biogeographic patterns and host plant utilization. This analysis should give us some insights into the evolutionary divergence of cactus-inhabiting yeasts and also an overview of the likely forces involved in the speciation of yeasts.

YEAST COLLECTIONS

Most of our yeast collections originated from semiarid regions of North America, northern South America, and on islands in the Caribbean Sea. In addition, efforts on introduced cactus yeast communities in Australia (Barker *et al.*, 1983, 1984; Starmer *et al.*, 1987*a*), the Mediterranean (F. Peris, unpublished data), and the Hawaiian Islands (W. T. Starmer, H. J. Phaff, and M.-A. Lachance, unpublished data) have been conducted. The yeasts found in their native habitats are the primary interest of this chapter, and the introduced communities (their distributions and specificities) will not be discussed in great length.

A total of 1885 samples of rotting cactus tissue have been analyzed for the presences of yeast species. These samples included stem or pad rots (1704 samples) of 50 different species of cactus, and fruit rots (181 samples) of 10 different species of cactus. The stem or pad rots have been merged into three distinct and one indistinct cactus type. These are Stenocereinae species [484 samples from species of the genera *Escontria, Myrtillocactus,* and *Stenocereus,* all members of the subtribe Stenocereinae *sensu* Gibson and Horak (1978)], Pachycereinae species [250 samples of species of the genera *Backebergia, Carnegiea,* Mexican *Cephalocereus* (including *C. royenii*), *Lophocereus, Neobuxbaumia,* and *Pachycereus,* all members of the subtribe Pachycereinae *sensu* Gibson and Horak (1978)], Opuntieae species (867 samples of the subgenus Platyopuntia and *Nopalea,* both members of the tribe Opuntieae), and lastly other species [103 samples of the genera *Acanthocereus, Cephalocereus* (only *C. lanuginosus*), *Cereus, Ferocactus, Melocactus,* and *Neoabbottia*].

The collections were conducted in 121 different geographic localities, including regions of Arizona, California, Texas, and Florida (United States); Baja California Norte, Baja California Sur, Chiapas, Guerrero, Jalisco, Michoacan, Oaxaca, Sinaloa, and Sonora (Mexico); and Anzoategui, Falcon, Lara, Miranda, Sucre, Truillo, and Zulia (Venezuela). Islands that have been sampled in the Caribbean Sea and the Bahamas are the Dominican Republic, Haiti, Jamaica, Monserrat; Little Conception and Great Inagua (Bahamas), Tortola and Virgin Gorda (British Virgin Islands); Cayman Brac, Little Cayman and Grand Cayman (Cayman Islands); Curaçao and Sanare (Netherlands); Navassa (United States); and Los Roques (Venezuela). Localities in these areas were merged according to geographic proximity to form the following five geographic regions: (1) Caribbean (including Florida but not Curaçao, Sanare, or Los Roques), (2) Sonoran Desert (Baja California, Sinaloa, and Sonora), (3) southern Mexico (Chiapas, Guerrero, Jalisco, Michoacan, and Oaxaca), (4) southwestern United States (Arizona, California, Texas, and the extreme northern region of Baja California, Mexico), and (5) Venezuela, including Curaçao, Sanare, and Los Roques. It should be noted that these regional definitions are mainly geographic and that some localities in these regions overlap with climatic realms (i.e., some localities in the southwestern United States are found in the northern parts of the Sonoran Desert). In addition, some localities were included in regions because their proximity to areas where sampling was more extensive (i.e., localities in Florida and the Bahamas were included in the Caribbean region).

YEAST DISTRIBUTIONS

Table I shows the number of localities sampled for each of the above regions along with the number of samples for each of the five habitat types taken from those regions. Table II shows the taxonomic designation of the yeast species along with their distribution as a function of host plant type. The companion Table III contains a list of the same yeast species according to their region of isolation. In both Tables II and III the number of times is given that a particular yeast species was recovered in the sample of plants of that category. The survey includes 92 species of yeast or yeastlike organisms and 3701 strains. Identification of these strains and their taxonomic affinities are discussed by Lachance *et al.* (1988). Table IV contains a list of the five host plant categories that were sampled and the number of each species sampled in each region.

Two simple statistics were used for describing the relative occurrence of yeasts in either habitats or regions. One is the quotient of the number of isolates of a particular yeast and the total number of isolates of all species in a particular habitat or region. This provides a relative measure of the contribution of that yeast to the yeast community found in that habitat or region. Another similar proportion is calculated by dividing the number of isolates by the number of samples taken from that habitat or region. This also gives an insight into the relative importance and distribution of yeast species, but because yeasts occur together, it does not always give an indication of the relative importance to community membership. In the descriptions that follow we shall use the former measure as a general descriptive statistic and we shall occasionally refer to the proportion in the sample as well. In calculating the proportions of yeasts in the communities, we have not included *Geotrichum* isolates in the total because these forms were not always counted or isolated in some surveys. Host and geographic diversity were calculated by the formula $(\text{diversity})_j = 1 - \Sigma_i p_{ij}^2$, where $p_{ij} = x_{ij}/\Sigma_j x_{ij}$ and x_{ij} is the number of isolates of yeast j from host or region i and $\Sigma_j x_{ij}$ is the total number of yeast isolates for that host or region. This measure of diversity is similar to the estimate of the effective number of species ($e_j = 1/\Sigma_i {p_{ij}}^2$) that has been used in previous work on yeast ecology (Lachance *et al.*, 1982; Starmer, 1982*a;* Fogleman and Starmer, 1985).

There are five distinct complexes of cactophilic yeast in the genus *Pichia*. These complexes are interrelated insofar as they belong to the same genus. Membership of the five complexes is given in Tables V and VI along with their percent contribution and diversity for host (Table V) and geographic yeast communities (Table VI). Two other taxa listed in Tables I and III probably represent distinct complexes of interrelated species or varieties. These are *Cryptococcus cereanus* and *Pichia mexicana*. Both taxa show either physiological or nuclear DNA composition differences among isolates obtained from different regions or substrates. We have not yet discerned the salient properties of the members in each of these two complexes.

TABLE I. The Number of Localities Sampled for Each Region and the Number of Plants Sampled for Each Habitat Type from That Region

	Number of plants sampled for given host type						
Region	Number of localities	Stenocereinae	Pachycereinae	Opuntieae	Cactus fruit	Other cactus	Total
Southwestern United States	18	14	41	402	0	13	470
Sonoran Desert	46	181	102	46	19	0	348
Southern Mexico	19	117	11	64	34	0	226
Caribbean	28	121	96	258	126	22	623
Venezuela	10	51	0	97	2	68	218
Total	121	484	250	867	181	103	1885

TABLE II. The Distribution of Yeast Species According to Habitat Type[a]

Yeast species	ST $N = 484$	PA $N = 250$	OP $N = 867$	FR $N = 181$	OT $N = 103$	Total $N = 1885$
Candida boidinii Ramirez	1	1	18			20
"*C. caseinolytica*"	5		14			19
C. catenulata Diddens et Lodder			1			1
C. deserticola Phaff, Starmer, Tredick et Miranda	72	2	2	1	1	78
C. famata (Harrison) Meyer et Yarrow			3	1		4
C. guilliermondii (Castellani) Langeron et Guerra	3		9	7		19
"*C. guilliermondii*-like"				3		3
C. inconspicua (Lodder et Kreger-van Rij) Meyer et Yarrow			1			1
C. ingens van der Walt et van Kerken	76	32	9		10	127
C. krusei (Castellani) Berkhout			2	15		17
C. maltosa Komagata, Nakase et Katsuya				3		3
C. mucilagina Phaff, Starmer, Miranda et Miller	16		92		1	109
C. parapsilosis (Ashford) Langeron et Talice				1		1
C. sake (Saito et Ota) van Uden et Buckley				1		1
C. sonorensis (Miller, Phaff, Miranda, Heed et Starmer) Meyer et Yarrow	141	41	264	7	22	475
C. stellata (Kroemer et Krumbholz) Meyer et Yarrow				3		3
C. tropicalis (Castellani) Berkhout			1			1
C. valida (Leberle) van Uden et Buckley	13		4	9	13	39
C. zeylanoides (Castellani) Langeron et Guerra		1		1		2
Candida sp.	1	1	1	16		19
Clavispora lusitaniae Rodrigues de Miranda		6		4		10
Cl. opuntiae Phaff, Miranda, Starmer, Tredick et Barker	14	3	88	10	4	119
Cryptococcus albidus (Saito) Skinner	13	4	16		1	34
Cr. cereanus Phaff, Miller, Miranda, Heed et Starmer	102	41	325		17	485
Cr. curvatus				1		1
Cr. flavus (Saito) Phaff et Fell			1			1
Cr. laurentii (Kufferath) Skinner	3	4	12	2		21
Cr. luteolus (Saito) Skinner			3			3

Cr. magnus (Lodder et Kreger-van Rij) Baptist et Kurtzman	1					1
Cr. skinneri Phaff et do Carmo-Sousa	1	1				2
Cryptococcus sp.			9		6	15
Debaryomyces hansenii (Zopf) Lodder et Kreger-van Rij		1				1
D. melissophilus (van der Walt et van der Klift) Kurtzman et Kreger-van Rij				1		1
Geotrichum (not always picked) sp.	3		148	3	1	155
Hansenula polymorpha de Morais et Maia (syn. *Pichia angusta*)			66			66
Hanseniaspora guilliermondii Pijper				4		4
Hanseniaspora uvarum (Niehaus) Shehata, Mrak et Phaff			5			5
Hanseniaspora sp.				1		1
Issatchenkia terricola (van der Walt) Kurtzman, Smiley et Johnson				23		23
Kluyveromyces marxianus (Hansen) van der Walt	2	2	6	1		11
Kloeckera apiculata (Reess emend. Klocker) Janke		1	2	28		31
K. lapis Lavie ex Smith, Simione et Meyer	3		3	50		56
Pichia amethionina var *amethionina* Starmer, Phaff, Miranda et Miller	30	2	25		2	59
"*P. amethionina* var *fermentans*"	2	18	38	9	5	72
P. amethionina var *pachycereana* Starmer, Phaff, Miranda et Miller	1	17	13		1	32
P. antillensis Starmer, Phaff, Miranda et Miller		19			1	20
P. barkeri Phaff, Starmer, Tredick et Aberdeen	3		12	19		34
P. cactophila Starmer, Phaff, Miranda et Miller	304	112	339	13	65	833
P. deserticola Phaff, Starmer, Tredick et Miranda		3	29			32
P. farinosa (Lindner) Hansen		1				1
P. fermentans Lodder				1		1
P. heedii Phaff, Starmer, Miranda et Miller	2	64				66
P. kluyveri var. *kluyveri* Bedford ex Kudriavzev	2	2	18	10		32
P. kluyveri var *cephalocereana* Phaff, Starmer et Tredick-Kline		5				5

(*continued*)

TABLE II. *(Continued)*

Yeast species	ST $N = 484$	PA $N = 250$	OP $N = 867$	FR $N = 181$	OT $N = 103$	Total $N = 1885$
P. kluyveri var *eremophila* Phaff, Starmer et Tredick-Kline	32	2	52			86
P. membranaefaciens Hansens			5	9		14
P. mexicana Miranda, Holzschu, Phaff et Starmer	24	3	45	1	8	81
P. norvegensis Leask et Yarrow		2	31	11		44
P. onychis Yarrow	1	1				2
P. opuntiae Starmer, Phaff, Miranda, Miller et Barker			1			1
P. pseudocactophila Holzschu, Phaff, Tredick et Hedgecock	4	18	1			23
P. strassburgensis (Ramirez et Boidin) Phaff	1					1
P. thermotolerans Holzschu, Phaff, Tredick et Hedgecock		22	1			23
Atypical *Pichia* sp.	3	4	6	1	1	15
Prototheca sp.	34	13	136	1	5	189
Rhodotorula glutinis (Fresenius) Harrison					1	1
R. graminis DiMenna	2		7	1	1	11
R. marina Phaff, Mrak et Williams	1					1
R. minuta (Saito) Harrison	1	1	7		1	10
R. pallida Lodder			1			1
R. rubra (Demme) Lodder	3		4		1	8
Rhodotorula sp.			1			1
Saccharomyces cerevisae Meyen ex Hansen				1		1
Schwanniomycers sp.					1	1
Trichosporon cutaneum (deBeurm., Gougerot et Vaucher) Ota			1		2	3
Trichosporon sp.	1					1
Torulaspora delbrueckii (Lindner) Lindner				1		1
Unknown	1		4	1		6
Zygosaccharomyces fermentati Naganishi				1		1
Total number of isolates	922	450	1882	276	171	3701

[a]The number of isolates from the number N of plants sampled is given. Names in quotations are undescribed species or varieties. Habitat type: ST, Stenocereinae species; PA, Pachycereinae species; OP, Opuntieae species; FR, fruit of cactus; OT, other cactus species.

TABLE III. The Distribution of Yeast Species According to Geographic Region[a]

	US N = 470	SD N = 348	SM N = 226	CA N = 623	VZ N = 218	Total N = 1885
Candida boidinii	14	2		4		20
"C. caseinolytica"	18		1			19
C. catenulata				1		1
C. deserticola	20	14	44			78
C. famata	1		1	1	1	4
C. guilliermondii	4	2	1	8	4	19
"*C. guilliermondii*-like"				3		3
C. inconspicua	1					1
C. ingens	20	45	21	19	22	127
C. krusei	1		2	14		17
C. maltosa				3		3
C. mucilagina	80	11	1	13	4	109
C. parapsilosis				1		1
C. sake			1			1
C. sonorensis	125	100	30	160	60	475
C. stellata			1	2		3
C. tropicalis	1					1
C. valida				8	31	39
C. zeylanoides				2		2
Candida sp.			10	8	1	19
Clavispora lusitaniae		2	4	4		10
Cl. opuntiae	27	8	18	45	21	119
Cryptococcus albidus	14	8	5	2	5	34
Cr. cereanus	228	34	49	128	46	485
Cr. curvatus				1		1
Cr. flavus	1					1
Cr. laurentii	8	1	3	7	2	21
Cr. luteolus				3		3
Cr. magnus			1			1
Cr. skinneri	1		1			2
Cryptococcus sp.					15	15
Debaryomyces hansenii				1		1
D. melissophilus			1			1
Geotrichum	134		2	19		155
Hansenula polymorpha	64	1		1		66
Hanseniaspora guilliermondii				4		4
Hanseniaspora uvarum	5					5
Hanseniaspora sp.				1		1
Issatchenkia terricola		13		10		23
Kluyveromyces marxianus	8	2		1		11
Kloeckera apiculata		19		12		31
Kl. apis			17	39		56
Pichia amethionina var. *amethionina*	30	29				59

(*continued*)

TABLE III. (*Continued*)

	US $N = 470$	SD $N = 348$	SM $N = 226$	CA $N = 623$	VZ $N = 218$	Total $N = 1885$
"*P. amethionina* var *fermentans*"	6		2	59	5	72
P. amethionina var *pachycereana*	20	12				32
P. antillensis				20		20
P. barkeri			4	30		34
P. cactophila	178	156	77	292	130	833
P. deserticola	32					32
P. farinosa	1					1
P. fermentans					1	1
P. heedii	20	46				66
P. kluyveri var. *kluyveri*	1	5	9	15	2	32
P. kluyveri var. *cephalocereana*				5		5
P. kluyveri var. *eremophila*	55	21	10			86
P. membranaefaciens	1		5	7	1	14
P. mexicana	14	19	2	14	32	81
P. norvegensis	2	6		36		44
P. onychis		1		1		2
P. opuntiae		1				1
P. pseudocactophila	6	12	5			23
P. strassburgensis		1				1
P. thermotolerans	14	9				23
Atypical *Pichia* sp.	3		7	3	2	15
Prototheca	73	5	6	70	35	189
Rhodotorula glutinis					1	1
R. graminis	3		1	6	1	11
R. marina				1		1
R. minuta	7	2			1	10
R. pallida	1					1
R. rubra			1	1	6	8
Rhodotorula sp.				1		1
Saccharomyces cerevisae				1		1
Schwanniomyces sp.				1		1
Trichosporon cutaneum					3	3
Trichosporon sp.		1				1
Torulaspora delbrueckii				1		1
Unknown			1	5		6
Zygosaccharomyces fermentati				1		1
Total	1222	594	314	1139	432	3701

[a]The number of isolates from the number N of samples is given. Region: US, southwestern U. S.; SD, Sonoran Desert; SM, Southern Mexico; CA, islands in the Caribbean Sea; VZ, Venezuela. See section on yeast collections for regional definitions.

TABLE IV. Taxonomic Designation of the Host Cacti[a]

	US	SD	SM	CA	VZ	Total
Species belonging to the subtribe Stenocereinae						
Escontria chiotilla (Weber) Rose			16			16
Myrtillocactus chocal (Orcutt) Britton et Rose	3	12				15
M. geometrizans (Martius) Console			1			1
M. schenckii (Purpos) Britton et Rose			8			8
Stenocereus alamosensis (Coulter) Gibson et Horak		9				9
S. dumortieri (Scheidweiler) Buxbaum			5			5
S. griseus (Hawthorn) Buxbaum					51	51
S. gummosus (Englemann) Gibson et Horak		115				115
S. hystrix (Hawthorn) Buxbaum				121		121
S. pruinosus (Otto) Buxbaum			56			56
S. A. Berg sp.			6			6
S. stellatus (Pfeiffer) Riccobono			6			6
S. thurberi (Engelmann) Buxbaum	11	45				56
"*S. thurberi*-like"			3			3
S. treleasei (Britton et Rose) Backeberg			16			16
Total	14	181	117	121	51	484
Species belonging to the subtribe Pachycereinae						
Backebergia militaris (Audot) Bravo ex Sanchez Mejorada			1			1
Carnegiea gigantea (Engelmann) Britton et Rose	40					40
Cephalocereus chrysacanthus (Weber) Britton et Rose			2			2
C. hoppenstedtii (Weber) K. Schum.			2			2
C. royenii (Linnaeus) Britton et Rose				96		96
Lophocereus schottii (Engelmann) Britton et Rose	1	70				71
Neobuxbaumia tetetzo (Weber) Backeberg			1			1
Pachycereus hollianus (Weber) Buxbaum			1			1
P. marginatus (DeCandolle) Britton et Rose			3			3
P. pecten-aboriginum (Engelmann) Britton et Rose		2	1			3
P. pringlei (S. Watson) Britton et Rose		30				30
Total	41	102	11	96	0	250
Species belonging to the tribe Opuntieae						
Nopalea Salm-Dyck sp.			40			40
Opuntia basilaris Englemann et Bigelow	2					2
O. boldinghii Britton et Rose					1	1
O. elatior Miller					9	9
O. ficus-indica (Linnaeus) Miller	52	9		15	4	80
O. humifusa Rafinesque				4		4
O. lindheimeri Englemann	79					79
O. moniliformis (Linnaeus) Haworth				34		34
O. oricola Philbrick		10				10
O. phaeacantha Engelmann	269					269
O. pilifera Weber			10			10

(continued)

TABLE IV. (*Continued*)

	US	SD	SM	CA	VZ	Total
O. sp.		25	5	1		31
O. stricta Haworth				204		204
O. wentiana Britton et Rose					83	83
O. wilcoxii Britton et Rose		2	9			11
Total	402	46	64	258	97	867
Other species						
Acanthocereus (Berger) Britton et Rose sp.					3	3
Cephalocereus lanuginosus (Linnaeus) Britton et Rose					18	18
Cephalocereus russelianus (Otto) Rose					3	3
Cereus repandus (Linnaeus) Britton et Rose					36	36
Ferocactus acanthodes Lemaire	11					11
F. wislizeni (Engelmann) Britton et Rose	2					2
Melocactus sp.					8	8
Melocactus intortus Urban				10		10
Neoabbottia paniculata (Lamarck) Britton et Rose				12		12
Total	13	0	0	22	68	103
Fruit of						
Cephalocereus lanuginosis					1	1
Ce. royenii				1		1
Nopalea sp.			22			22
Opuntia ficus-indica				1		1
O. moniliformis				1		1
O. stricta				90		90
Stenocereus griseus					1	1
S. gumosus		19				19
S. hystrix				33		33
S. pruinosus			12			12
Total	0	19	34	126	2	181
Overall total	470	348	226	625	218	1885

[a]Numbers of plants sampled in each region and the total number of samples for each cactus host type are given. Geographic region: US, southwestern United States; SD, Sonoran Desert; SM, southern Mexico; CA, islands in the Caribbean Sea; VZ, Venezuela. (See section on yeast collections for regional definitions).

TABLE V. Distribution of Complexes of Cactophilic Yeasts of the Genus *Pichia* and Other Common or Frequently Isolated Cactus Yeasts According to Host Plant Category

	Host[a]					Host diversity[b]
	ST	PA	OP	FR	OT	
Cactophila complex						
Pichia cactophila	33.1	24.9	19.6	4.8	37.8	0.75
P. pseudocactophila	0.4	4.0	0.1			0.19
P. norvegensis		0.4	1.8	4.1		0.49
Amethionina complex						
P. amethionina var. *amethionina*	3.3	0.4	1.4		1.2	0.64
P. amethionina var. *pachycereana*	0.1	3.8	0.7		0.6	0.43
"*P. amethionina* var. *fermentans*"	0.2	4.0	2.2	3.3	2.9	0.75
"*P. amethionina* var *australensis*"		Found in Australia				
Opuntiae complex						
P. antillensis		4.2			0.6	0.21
P. thermotolerans		4.9	0.1			0.02
P. opuntiae		Found in Australia				
"*P. opuntiae* var. hem"		Found in Australia				
Kluyveri complex						
P. kluyveri var. *kluyveri*	0.2	0.4	1.0	3.3		0.50
P. kluyveri var. *eremophila*	3.5	0.4	3.0			0.55
P. kluyveri var. *cephalocereana*		1.1				0.00
Deserticola complex						
P. deserticola		0.7	1.7			0.41
Candida deserticola	7.8	0.4	0.1	0.4	0.6	0.28
Common species						
Cryptococcus cereanus	11.1	9.1	18.7		9.9	0.73
Candida sonorensis	15.3	9.1	15.2	2.6	12.8	0.76
Frequent species						
Candida ingens	8.3	7.1	0.5		5.8	0.67
C. mucilagina	1.7		5.3		0.6	0.46
Clavispora opuntiae	1.5	0.7	5.1	3.7	2.3	0.73
Hansenula polymorpha			3.8			0.00
Pichia heedii	0.2	14.2				0.03
P. mexicana	2.6	0.7	2.6	0.4	4.7	0.70
Prototheca sp.	3.7	2.9	7.8	0.4	2.9	0.71
Total percentage of community	93.0	93.4	90.7	23.0	82.7	

[a]*Geothricum* is not included. Hosts: ST, Stenocereinae species; PA, Pachycereinae species; OP, Opuntieae species; FR, cactus fruit; OT, species of other cacti.
[b]Host diversity was calculated as $1 - \Sigma$ (percent/row total)^2.

TABLE VI. Distribution of Complexes of Cactophilic Yeasts of the Genus *Pichia* and Other Common or Frequently Isolated Cactus Yeasts According to Geographic Region

	Region[a]					Geographic diversity[b]
	US	SD	SM	CA	VZ	
Cactophila complex						
Pichia cactophila	16.4	26.2	24.7	26.1	30.1	0.79
P. pseudocactophila	0.6	2.0	1.6			0.61
P. norvegensis	0.2	1.0		3.2		0.41
Amethionina complex						
P. amethionina var. *amethionina*	2.8	4.9				0.46
P. amethionina var. *pachycereana*	1.8	2.0				0.50
"*P. amethionina* var. *fermentans*"	0.6		0.6	5.3	1.2	0.49
"*P. amethionina* var *australensis*"		Found in Australia				
Opuntiae complex						
P. antillensis				1.8		0.00
P. thermotolerans	1.3	1.5				0.50
P. opuntiae		Found in Australia				
"*P. opuntiae* var. hem"		Found in Australia				
Kluyveri complex						
P. kluyveri var. *kluyveri*	0.1	0.8	2.9	1.3	0.5	0.65
P. kluyveri var. *eremophila*	5.1	3.5	3.2			0.65
P. kluyveri var. *cephalocereana*				0.4		0.00
Deserticola complex						
P. deserticola	2.9					0.00
Candida deserticola		3.4	4.5	3.9		0.66
Common species						
Cryptococcus cereanus	21.0	5.7	15.7	11.4	10.6	0.77
Candida sonorensis	11.5	16.8	9.6	14.3	13.9	0.79
Frequent species						
Candida ingens	1.8	7.6	6.7	1.7	5.1	0.74
C. mucilagina	7.4	1.8	0.3	1.2	0.9	0.55
Clavispora opuntiae	2.5	1.3	5.8	4.0	4.9	0.76
Hansenula polymorpha	5.9	0.2		0.1		0.08
Pichia heedii	1.8	7.7				0.31
P. mexicana	1.3	3.2	0.6	1.3	7.4	0.64
Prototheca sp.	6.7	0.8	1.9	6.3	8.1	0.73
Total percentage of community	91.7	90.4	78.1	82.3	82.7	

[a] *Geothricum* is not included. Regions: US, southwestern United States; SD, Sonoran Desert; SM, southern Mexico; CA, islands in the Caribbean; VZ, Venezuela. (See yeast collections for regional definitions).

[b] Host diversity was calculated as $1 - \Sigma\ (\text{percent/row total})^2$.

Cactophila Complex

The cactophila complex is composed of three species, *Pichia cactophila, P. pseudocactophila,* and *P. norvegensis. Pichia cactophila* (geographic distribution given in Fig. 1) is the most commonly isolated cactus yeast (833 isolates from 1885 plants sampled), has a broad host plant distribution (isolated at least once from 40 hosts of a possible 50 host species sampled), and has been found in all regions of the world where cacti have been sampled, including regions where cacti have been introduced (Australia, Hawaii, Spain, and South Africa). This species is most abundant in columnar cacti (25–33% of the total number of isolates from the Stenocereinae or Pachycereinae categories). It is commonly found in Opuntieae necroses (20% of the Opuntieae isolates), but is infrequent in the cactus fruit yeast community (<5% of the fruit isolates). *Pichia pseudocactophila* is not frequently encountered (<1% of the total number of yeasts isolated) and is almost entirely restricted to *Pachycereus* and related species in Mexico and the United States, where it makes up 4% of the Pachycereinae yeast community. *Pichia norvegensis* is found primarily in Opuntieae cactus (4% of the Opuntieae yeast isolates) and appears to occupy two regions, the Cape region of the Baja California peninsula and islands in the Caribbean. This species has also been recovered from *Opuntia* species in Hawaii. Both *P. norvegensis* and *P.*

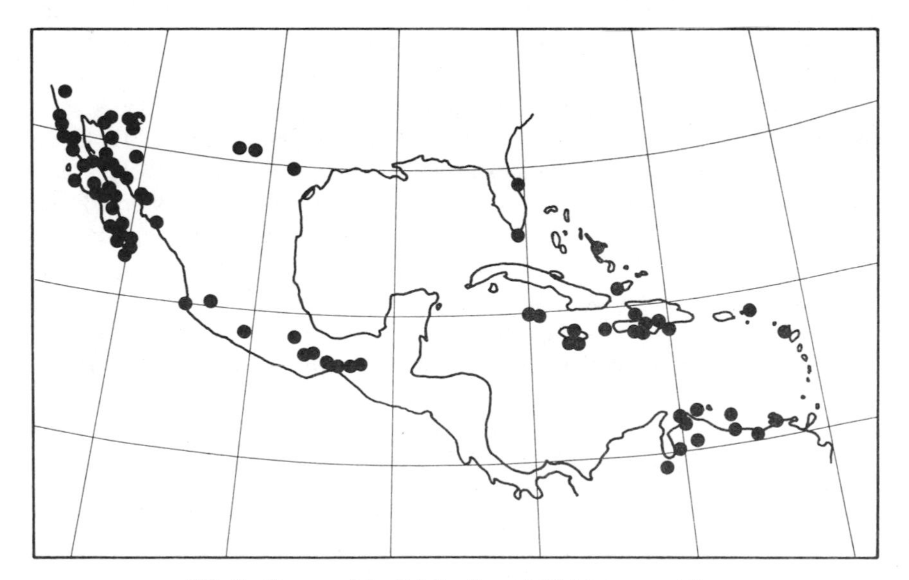

FIG. 1. Geographic distribution of *Pichia cactophila.*

pseudocactophila (Fig. 2) are found within in the distribution of the more widespread *P. cactophila. Pichia norvegensis* and *P. pseudocactophila* have slightly overlapping distributions and each occupies one region that the other does not. This distributional pattern argues for an evolutionary pattern with *P. cactophila* as the extant representative of a possible ancestral form, and the other two species representing derivative forms. *Pichia pseudocactophila* apparently has specialized on Mexican Pachycereinae species, while *P. norvegensis* occupies mainly Opuntieae tissue and cactus fruit. *Pichia norvegensis* is also known as a yeast that, prior to the cactus isolates, had been isolated from clinical samples (sputum, abscesses, vaginal smears, etc.). It is most likely that the relatively high temperature limits of growth for this species has allowed it to gain access to the warm-blooded animal habitat.

Amethionina Complex

The amethionina complex is composed of four varieties (three are listed in Table II; the fourth, "*P. amethionina* var. *australensis*" has been isolated from *Opuntia* spp. in eastern Australia). The geographic distribution for the three North American and Caribbean varieties is displayed in Fig. 3. The complex is characterized by the fact that all members are naturally occurring auxotrophs that have an absolute requirement for L-methionine or L-cysteine that can be substituted by sodium thiosulfate. The undescribed variety "*P. amethionina* var. *fermentans*" is geographically the most widespread member (Table VI), but it never occurs in more than 2–4% of any yeast community of the five habitat types (Table V). It has been found in all regions except in the Sonoran Desert. It is most often collected on certain Caribbean islands, where it was found in 9.5% of the samples and accounts for 5.3% of the regional yeast community (Table VI). *Pichia amethionina* var. *amethionina* and *P. amethionina* var. *pachycereana* have completely overlapping distributions. Both are found in the Sonoran Desert and the southwestern United States, but *P. a.* var. *amethionina* appears to be more common in the Sonoran Desert, especially the Baja California peninsula (Table VI). Table V shows that these two varieties reside primarily in either Pachycereinae hosts (*P. amethionina* var. *pachycereana*) or Stenocereinae hosts (*P. amethionina* var. *amethionina*). Taxonomic identification of the two varieties is probabilistic and can sometimes lead to misclassification (Holzschu and Phaff, 1982; Lachance *et al.*, 1988); thus, the extent of geographic overlap and host plant specificity for these varieties is only certain for the Baja California region in the Sonoran Desert. The undescribed variety "*P. amethionina* var. *australensis*" is found in *Opuntia* sp. tissue in eastern Australia. It is closely related to *P. amethionina* var. *amethionina* (ca. 87% DNA complementarity), but the origin of its introduction into Australia is unknown.

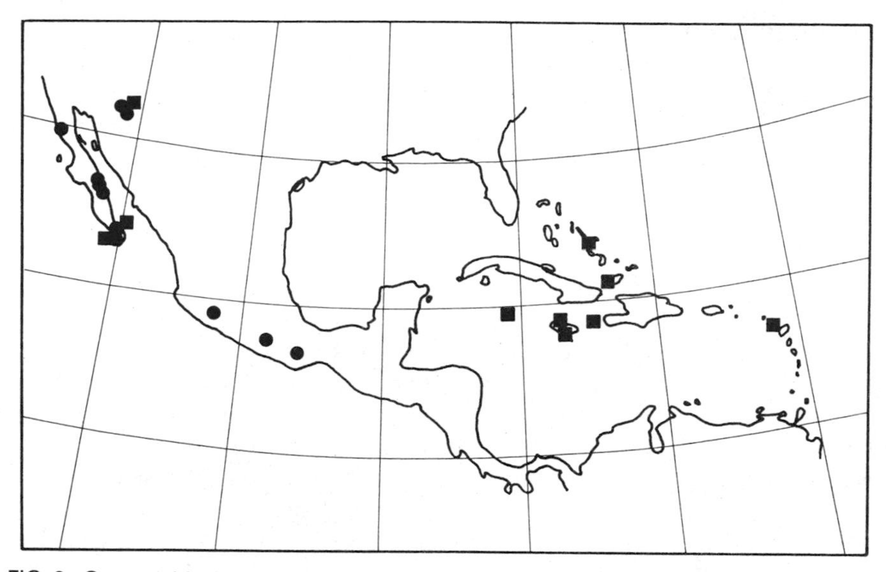

FIG. 2. Geographic distribution of *Pichia pseudocactophila* (circles) and *P. norvegensis* (squares).

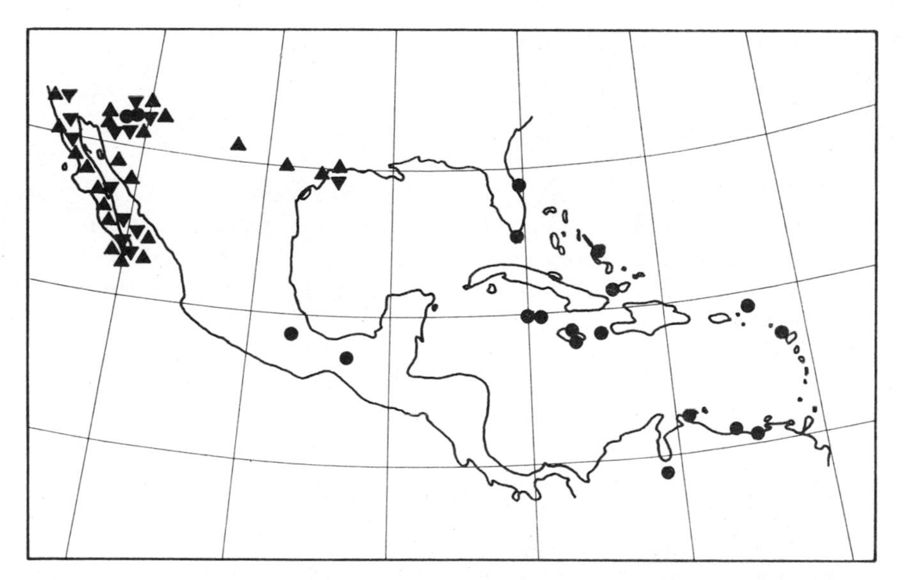

FIG. 3. Geographic distribution of the amethionina complex, *P. amethionina* var. *amethionina* (triangles pointing up), *P. amethionina* var. *pachycereana* (triangles pointing down), and "*P. amethionina* var. *fermentans*" (circles).

Opuntiae Complex

The opuntiae complex is composed of four taxa (*Pichia opuntiae*, ''*P. opuntiae* var. hem,'' *P. antillensis*, and *P. thermotolerans*). All of these are host and region specific. We know little of the origins of *P. opuntiae* and ''*P. opuntiae* var. hem.'' Both have been isolated from *Opuntia* sp. necroses in eastern Australia, but ''*P. opuntiae* var. hem'' is restricted to one locality in the vicinity of Brisbane, Queensland. *Pichia antillensis* has been isolated almost exclusively from decaying stems of *Cephalocereus royenii* wherever the host occurs on islands in the Lesser Antilles. In a parallel manner, *Pichia thermotolerans* is found in the Sonoran Desert and the southwest of the United States wherever Pachycereinae species (*Carnegiea gigantea* and *Pachycereus pringlei*) occur. Both *P. antillensis* and *P. thermotolerans* occur at low frequency, accounting for 4–5% of the Pachycereinae yeast community. Their disjunct geographic distribution is shown in Fig. 4. DNA–DNA reassociation studies have shown that *P. antillensis* is more closely related to *P. opuntiae* (about 55% DNA complementarity) than to *P. thermotolerans* [about 25% DNA complementarity (Starmer *et al.*, 1984)], and that *P. thermotolerans* is more closely related to ''*P. opuntiae* var. hem'' [about 52% DNA complementarity (H. J. Phaff, unpublished data)] than to *P. opuntiae* or *P. antillensis* [about 26–34% DNA complementarity (H. J. Phaff, unpublished data)]. It is suspected that *P. opuntiae* and perhaps other species of this complex may be found in South America (Argentina) when surveys are extended into that region.

Kluyveri Complex

The *Pichia kluyveri* complex (geographic distribution in Fig. 5) is composed of four taxa (Phaff *et al.*, 1987). Three are heterothallic and have limited interfertility, while the fourth can be asexual. *Pichia kluyveri* var. *kluyveri* is found also outside of the cactus habitat, as it occurs in citrus fruits (Vacek *et al.*, 1979) and tomatoes (de Camargo and Phaff, 1957) as well as other habitats (Lodder, 1970). In the cactus habitat it occurs as 3.3% of the fruit yeast community and is found in the other habitat types at 1% or less of the total number of isolates of a category. *Pichia kluyveri* var. *kluyveri* has been found in all regions considered here and has been recovered also from eastern Australia growing in cladode and fruit rots of *Opuntia stricta* (Starmer *et al.*, 1987*a*). *Pichia kluyveri* var. *eremophila* is found primarily in Stenocereinae yeast communities (3.5% of the isolates) and Opuntieae yeast communities (3.0% of the isolates), is rarely isolated from Pachycereinae yeast communities (0.4% of the isolates), and has not been isolated from fruit or ''other yeast'' communities. It is found throughout Mexico and the southwestern United States. There is some indication that the

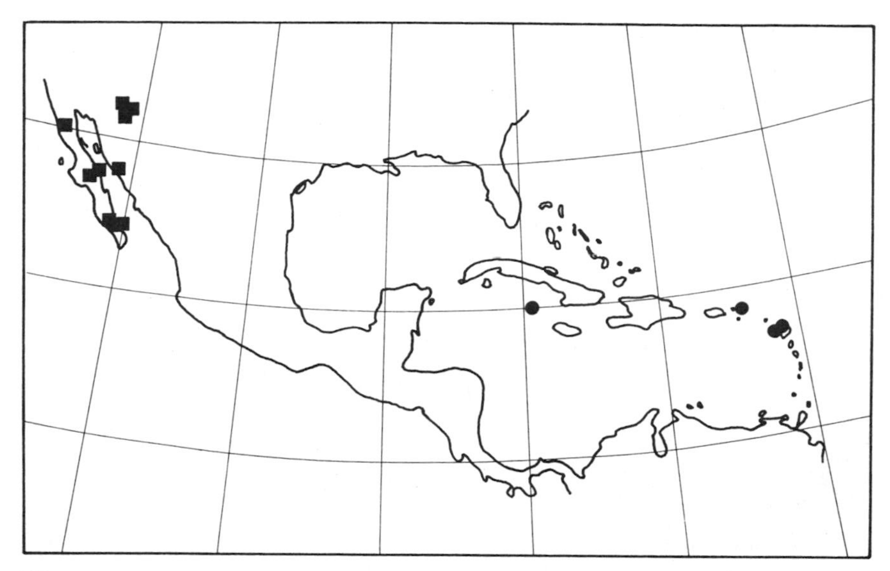

FIG. 4. Geographic distribution of *Pichia antillensis* (circles) and *P. thermotolerans* (squares).

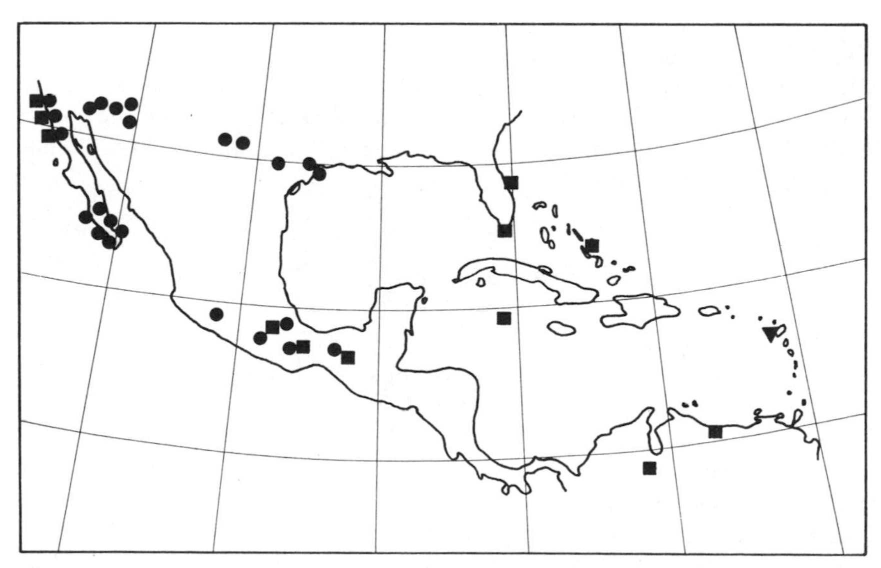

FIG. 5. Geographic distribution of the kluyveri complex, *P. kluyveri* var. *kluyveri* (squares), *P. kluyveri* var. *eremophila* (circles), and *P. kluyveri* var. *cephalocereana* (triangle).

anamorph (*Candida eremophila*) is found more frequently in Stenocereinae habitats, while the teleomorph is more readily recovered from Opuntieae pad necroses. *Pichia kluyveri* var. *eremophila* and *Candida eremophila* have been merged together for analysis. This was done because the anamorph (*C. eremophila*) occurs in nature either as a haploid mating type or as an asexual form (does not sporulate or mate) and the distinction between these two types of asexuality was not always possible to determine. *Pichia kluyveri* var. *cephalocereana* was found in rotting arms of the columnar cactus *Cephalocereus royenii* on the island of Montserrat in the Caribbean. Recently, additional strains were isolated from the same host on the island of St. Martin, also in the West Indies. The mating types of *P. kluyveri* are interfertile and their mutual DNA relatedness values ranges from 66 to 72% relative binding (Phaff *et al.*, 1987).

Deserticola Complex

Pichia deserticola occurs primarily in Opuntieae pad rots (1.7% of this yeast community) in the southwestern United States, while the anamorph, *Candida deserticola,* is commonly found in Stenocereinae stem rots (7.8% of this yeast community) in the Caribbean, Mexico, and the Sonoran Desert (Phaff *et al.*, 1985). These two forms are disjunct in their distribution (Fig. 6) and are similar phenotypically to the commonly encountered food spoilage yeast *Candida valida.* Some of the *C. valida* isolates listed in Tables II and III from Venezuela may actually be *C. deserticola,* but the present information on the Venezuelan isolates does not allow us to place them clearly in that taxon. The guanine + cytosine (G+C) contents of the nuclear DNAs, however, are greatly different: 27.5% for *P. deserticola* and *C. deserticola,* while that for *C. valida* ranges from 42 to 44 mole %.

Common Cactus-Associated Yeasts

Pichia cactophila, Cryptococcus cereanus, and *Candida sonorensis* are commonly found in the cactus yeast community. Together they account for almost 50% of all of the yeasts isolated from the cactus habitat. *Candida sonorensis* (Fig. 7) accounts for 9–15% of the yeast communities of cactus habitats other than fruit, where it represents 2.6% of the community. It is found in all regions sampled and is common to regions where cacti have been introduced (Australia, Hawaii, Spain, and South Africa). *Cryptococcus cereanus* (teleomorph, *Sporopachydermia cereana*) is composed of two or more species based on G+C content and DNA–DNA reassociation experiments (H. J. Phaff, unpublished data) and forms a complex (Lachance *et al.*, 1988) that is geograph-

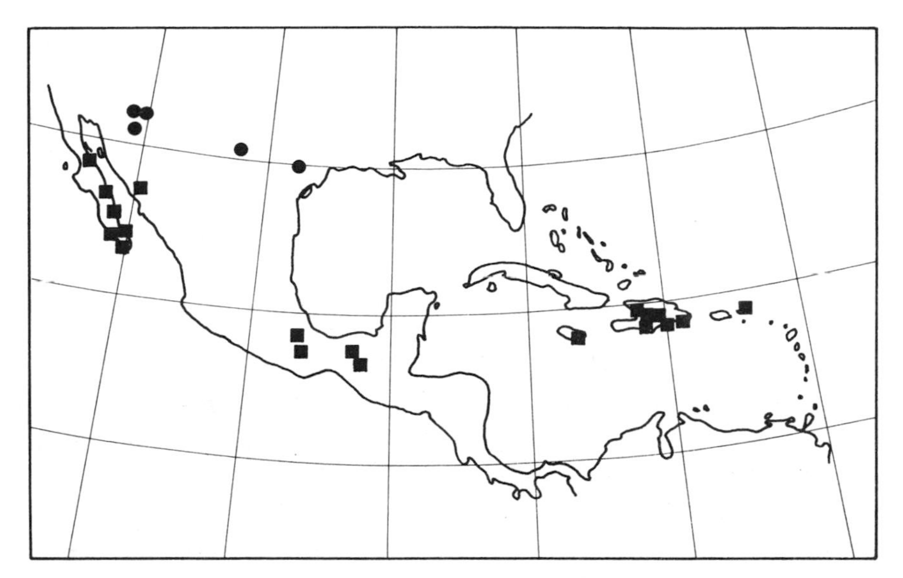

FIG. 6. Geographic distribution of the deserticola complex, *Pichia deserticola* (circles) and *Candida deserticola* (squares).

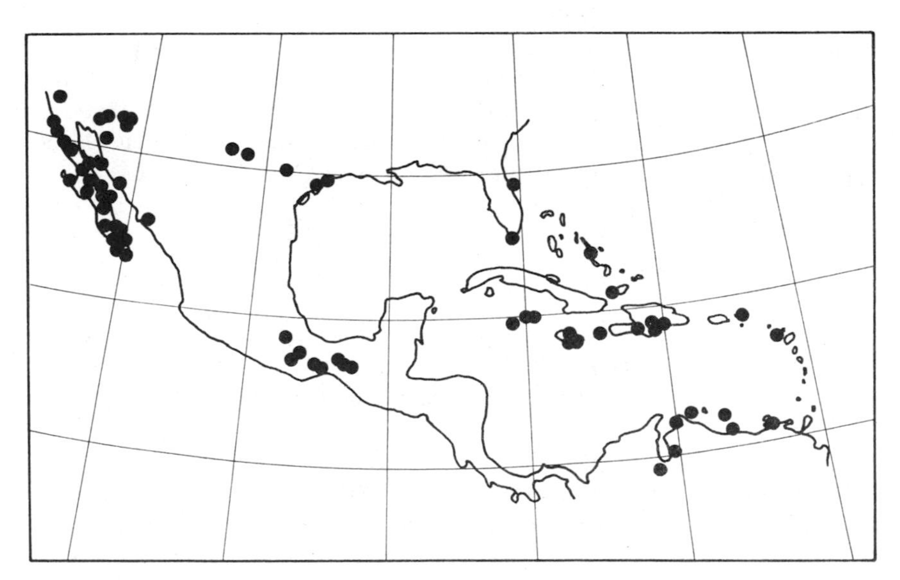

FIG. 7. Geographic distribution of *Candida sonorensis*.

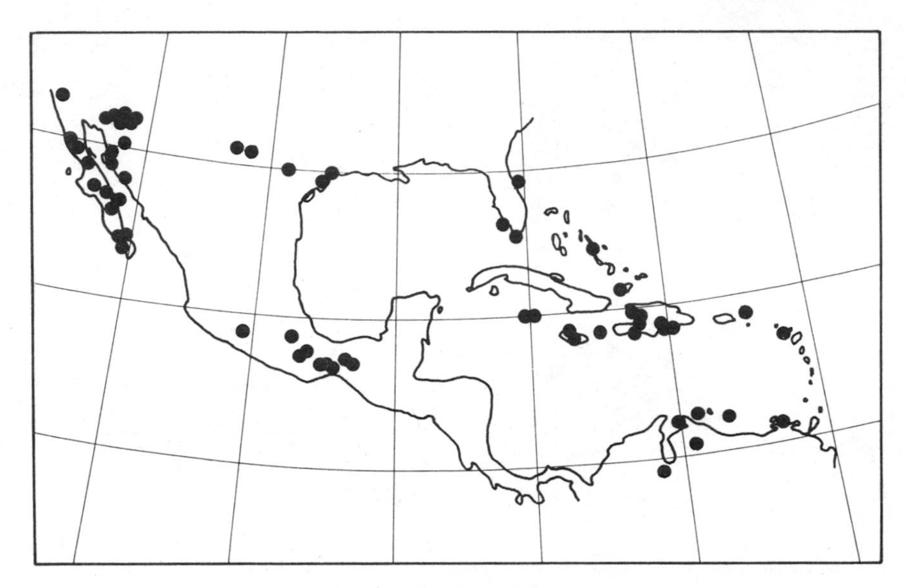

FIG. 8. Geographic distribution of *Cryptococcus cereanus.*

ically widespread (Fig. 8). Most commonly found in Opuntieae yeast communities (18.7% of the isolates of this community), *Cryptococcus cereanus* is also frequently found as a member of the "other cactus" habitats (9–11%), but it has never been isolated from the cactus fruit habitat (Table 5). *Cryptococcus cereanus* has been recovered from decaying cacti in all regions of the world where cacti occur. *Pichia cactophila, Cryptococcus cereanus,* and *Candida sonorensis* are characteristic of the cactus yeast community and form the nucleus of most community types other than the fruit yeast community.

Frequently Isolated Cactus Yeasts or Yeastlike Organisms

Seven species of yeast or yeastlike organisms that are not members of the complexes discussed above individually account for 1.8–3.4% of the overall yeast isolates from cactus rots. These species are *Candida ingens* (Fig. 9), *C. mucilagina* (*Myxozyma mucilagina*) (Fig. 10), *Clavispora opuntiae* (Fig. 11), *Hansenula polymorpha* (*Pichia angusta*) (Fig. 12), *Pichia heedii* (Fig. 13), *Pichia mexicana* (Fig. 14), and *Prototheca* sp. (Fig. 15). Three of these species (*C. mucilagina, H. polymorph,* and *P. heedii*) show low host diversity (Table V) and thus in general account for more community composition in the individual habitat types (Table VII, Stenocereinae; Table VIII, Pachycereinae; Table IX,

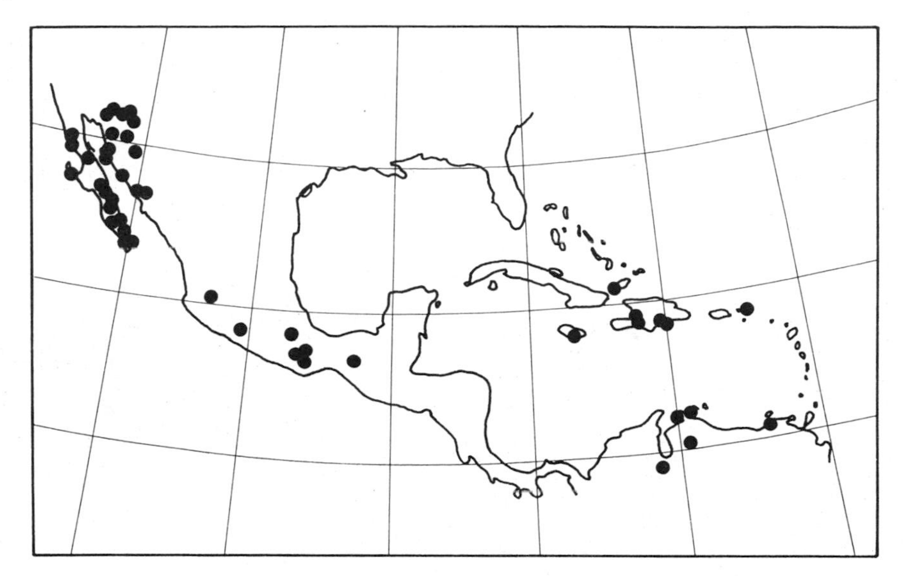

FIG. 9. Geographic distribution of *Candida ingens.*

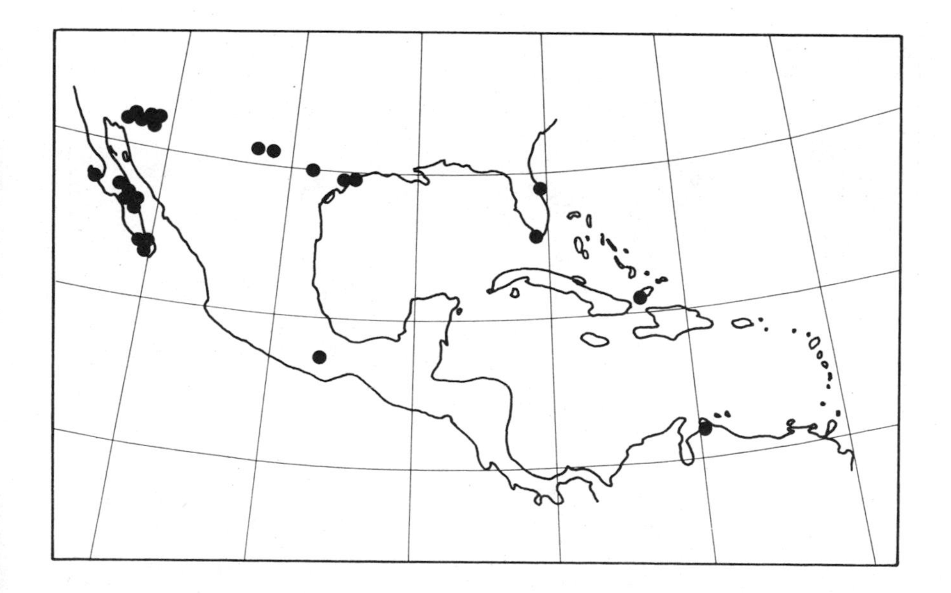

FIG. 10. Geographic distribution of *Candida mucilagina.*

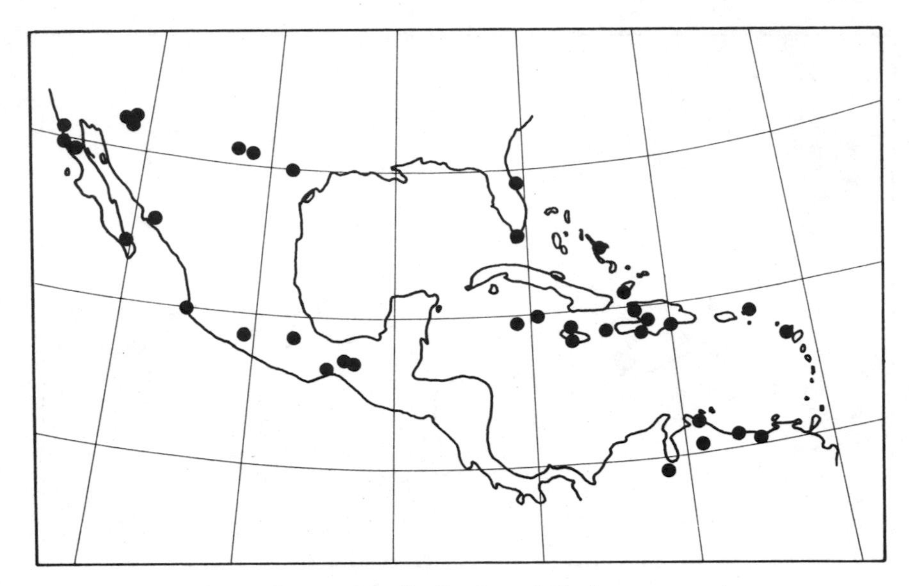

FIG. 11. Geographic distribution of *Clavispora opuntiae.*

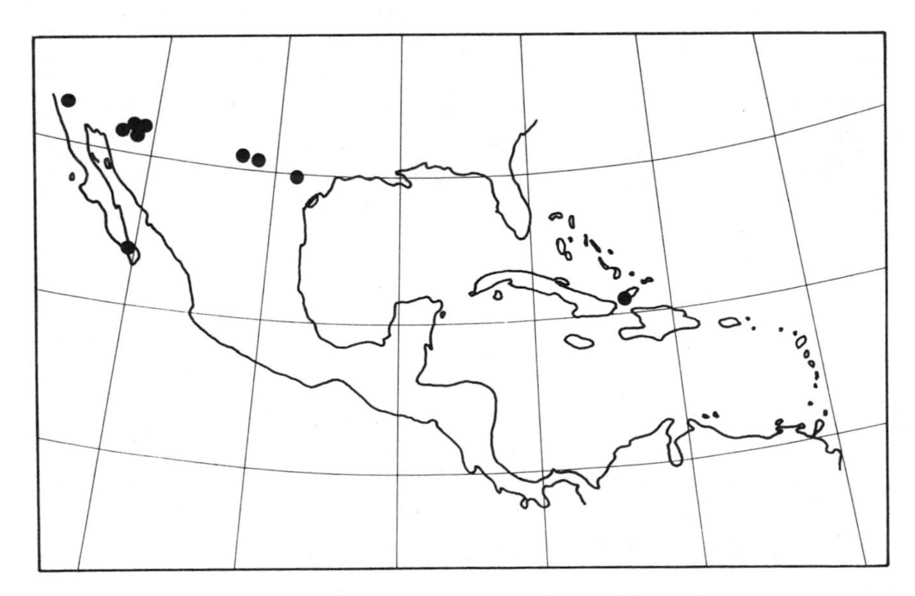

FIG. 12. Geographic distribution of *Hansenula polymorpha.*

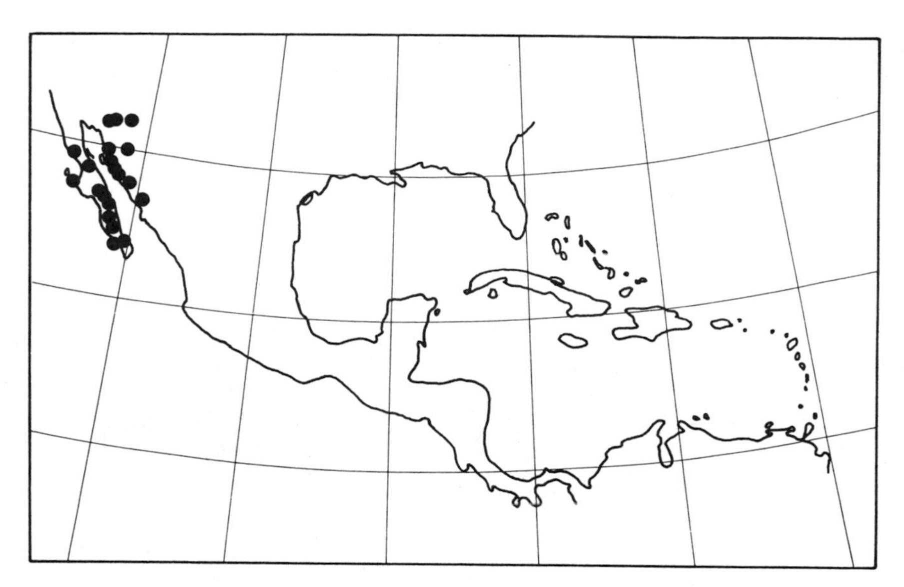

FIG. 13. Geographic distribution of *Pichia heedii.*

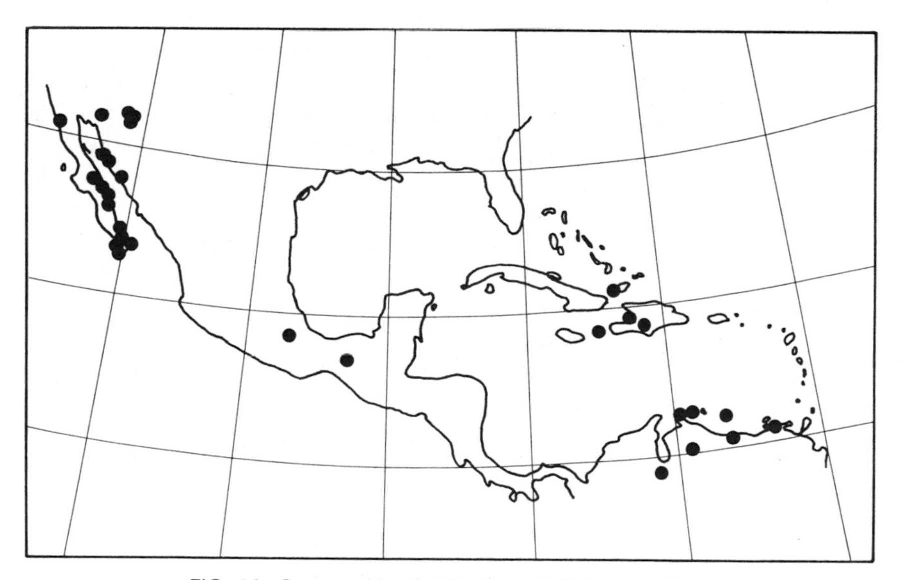

FIG. 14. Geographic distribution of *Pichia mexicana.*

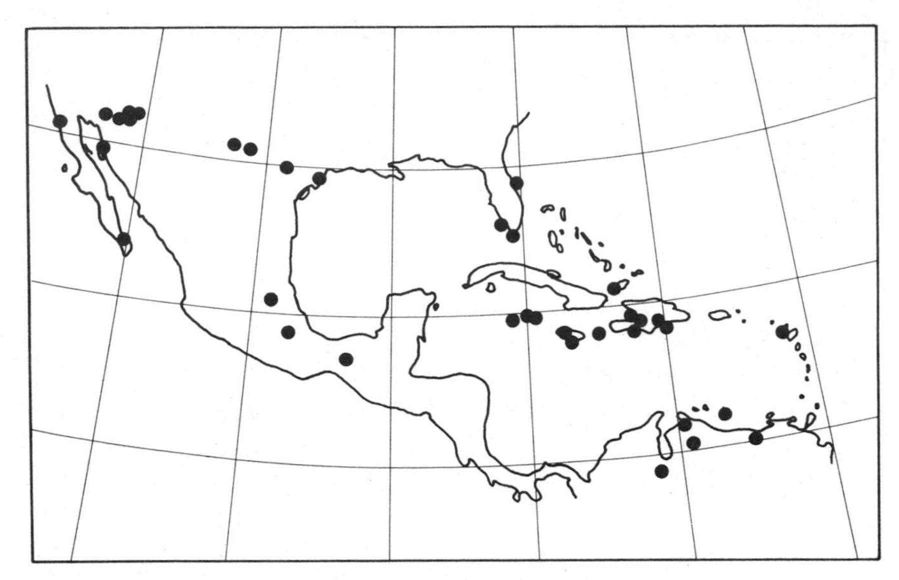

FIG. 15. Geographic distribution of *Prototheca* sp.

Opuntieae). *Candida ingens* has never been found in the fruit habitat, but accounts for 7–8% of the yeasts in the rots of columnar cacti (Stenocereinae and Pachycereinae habitats). This yeast is found in all regions, but is not commonly found in Australia or Hawaii. It is infrequently encountered in Opuntieae habitats (0.5%) and this might explain why it is not present in the introduced *Opuntia* communities studied in Australia and Hawaii. *Candida mucilagina* is relatively common in Opuntieae habitats (5.3% of the isolates), has never been found in cactus fruit rots, and is frequently found in all samples from the southwestern United States (17% of the samples contained this species and 7.4% of the regional yeast community was *C. mucilagina*). *Clavispora opuntiae* also is frequently found in Opuntieae yeast communities (5.1% of the isolates), but unlike *C. mucilagina,* it is found frequently in the fruit yeast community as well (3.7% of the isolates). *Clavispora opuntiae* is one of the most commonly encountered yeast species in eastern Australia (Barker *et al.*, 1983, 1984; Starmer *et al.*, 1987*a*), where it lives in cladode and fruit rots of Opuntieae species. *Hansenula polymorpha* is restricted to Opuntieae yeast communities, where it accounts for 3.8% of the isolates. It is also almost entirely restricted to the southwestern United States. It is found outside of the cactus habitat and may invade the Opuntieae habitat when other habitats it uses, such as tree fluxes, are present. *Pichia heedii* is highly host and region specific. This species accounts for 14.2% of the isolates of Pachycereinae yeast communities, and is restricted to the Sonoran Desert and the southwestern United States (when *Carnegiea gigantea* or

Lophocereus schottii is present). It has never been isolated from the Opuntieae or fruit habitats and is rarely isolated from Stenocereinae rots. *Pichia mexicana* probably represents a complex of species. It has been recovered from Stenocereinae, Opuntieae, and "other cactus" habitats (2.6–4.7% of the isolates of these communities), but is rarely found in the Pachycereinae or fruit habitats (0.4–0.7% of the isolates of these communities). It is most commonly isolated in Venezuela, where 14.7% of the samples contained this species, and where it accounted for 7.4% of the regional yeast community (Table VI). It is frequent in the Sonoran Desert (5.5% of the samples, 3.2% of the community), the southwestern United States (3% of the samples, 1.3% of the community), and the Caribbean (2.2% of the samples, 1.3% of the community), but it is relatively rare in southern Mexico, where it occurred in less than 1% of the samples and accounted for less than 1% of the regional yeast community. *Prototheca* (a yeastlike colorless *Chlorella*) is frequent in all host categories except for the fruit habitat. It is common in Opuntieae rots, where it accounts for 7.8% of the community. This species has high geographic diversity, but is relatively infrequent in Mexico (Sonoran Desert and southern Mexico).

COMMUNITY COMPOSITION

Yeast communities for the three cactus types (Stenocereinae, Pachycereinae, and Opuntieae; Tables VII–IX, respectively) were constructed by retaining yeast species that accounted for at least 6% of the yeast isolates of the habitat type for any one region. This criterion ensured that at least the top five members of any one community would be represented and accounted for 78–94% of the isolates of the communities. *Pichia cactophila, Candida sonorensis,* and *Cryptococcus cereanus* are listed for all three habitat types and are often ranked first, second, or third in frequency of occurrence in the regional comparison of the three community types. These three species form the nucleus of most of the cactus communities other than the fruit community.

Yeast Communities of Species in the Subtribe Stenocereinae

Studies on the yeast communities of *Stenocereus gummosus* (Starmer, 1982*a*), and *S. thurberi* (Fogleman and Starmer, 1985) have shown that the variation in species composition and relative abundance in these communities resides mostly at the level of individual plants within a locality. This variation is thought to be a consequence of having sampled plants whose rots had reached different stage of a successional process. Rots may differ also from one another in the history of their colonization. Species composition or relative proportions exhibited very few differences when the yeasts of the two cactus species or the communities from different localities were compared.

The Stenocereinae yeast community (Table VII) is characterized by the uniform distribution of *Candida ingens* across regions, and more specific distributions of the other species. *Pichia mexicana* was found in all regions, but was most prevalent in the communities sampled in the southwestern United States. *Pichia kluyveri* var. *eremophila* was not found in Venezuela or the Caribbean and was most frequently isolated in the southwestern United States. In a similar manner, *P. amethionina* var. *amethionina* was primarily found in the Sonoran Desert, with some isolates occurring in the southwestern United States. *Candida deserticola* was absent in Stenocereinae yeast communities sampled in Venezuela and the southwestern United States, while *Prototheca* was not found in the Sonoran Desert. The general pattern that emerges from the distribution of yeast communities of Stenocereinae species across regions is that when changes occur, they involve replacements or additions of relatively unrelated species.

Yeast Communities of Species in the Subtribe Pachycereinae

The number of yeast species in Pachycereinae yeast communities is similar to that in Stenocereinae yeast communities when the 6% criterion is applied (11 and 12 species, respectively). Venezuela is not included in Table VIII because no Pachycereinae samples were obtained from that region. Although these communities share species with the Stenocereinae yeast communities (*Pichia cactophila, Candida sonorensis, Cryptococcus cereanus, Candida ingens,* and to some extent *Pichia kluyveri* var. *eremophila* in southern Mexico), many species are almost unique to the Pachycereinae habitat. *Pichia pseudocactophila,* which is closely related to *P. cactophila* (Holzschu *et al.*, 1983), was present alongside *P. cactophila* in Mexico and the United States, but not in the Caribbean. *Pichia heedii* is dominant in Pachycereinae yeast communities in the Sonoran Desert and almost as frequent as *P. cactophila* in the southwestern United States, but it has not been isolated from other regions. *Pichia thermotolerans* and *P. antillensis* are related species (about 25% DNA relatedness) that occupy Pachycereinae species in different regions (United States and Sonoran Desert versus Caribbean). In a similar manner the two varieties of *P. amethionina* (var. *pachycereana* and "*fermentans*") are divided along the same geographic boundaries (Table VIII). *Clavispora lusitaniae* was found frequently in Pachycereinae habitats in southern Mexico and is also known from other habitats (Lachance *et al.*, 1986). The general pattern that is apparent in the Pachycereinae yeast communities is different from that seen in the Stenocereinae yeast communities. In the Pachycereinae habitat, yeasts appear to replace one another along lines of genetic relationship, and thus geography has been important in fostering species splitting

TABLE VII. Dominant Yeasts of Stenocereinae Yeast Communities across Regions

	Region[a]				
	US	SD	SM	CA	VZ
Pichia cactophila	**16.4**	**36.1**	**32.2**	**33.8**	**32.7**
Candida sonorensis	**13.1**	**22.7**	**12.1**	**7.8**	**13.3**
Cryptococcus cereanus	**11.5**	4.5	**17.2**	**19.6**	5.3
Candida ingens	**11.5**	**6.5**	**10.3**	6.4	**12.4**
C. deserticola	0.0	**5.7**	**8.0**	**17.4**	0.0
Prototheca sp.	3.3	0.0	1.7	**8.2**	**9.7**
Pichia kluyveri var. *eremophila*	**11.5**	5.1	4.0	0.0	0.0
P. amethionina var. *amethionina*	3.3	**8.0**	0.0	0.0	0.0
P. mexicana	6.6	3.7	0.6	0.9	3.5
Candida valida	0.0	0.0	0.0	0.0	**11.5**
"C. caseinolytica"	8.2	0.0	0.0	0.0	0.0
Total percent of community	85.4	92.3	86.1	94.1	88.4
Total isolates in community	61	352	174	219	113

[a]Percent of total yeast community of Stenocereinae species in each region (*Geotrichum* is not included). The top five ranking species are shown in boldface. Region: US, southwestern United States; SD, Sonoran Desert; SM, southern Mexico; CA, islands in the Caribbean; VZ, Venezuela. (See yeast collections for regional definitions.)

TABLE VIII. Dominant Yeasts of Pachycereinae Yeast Communities across Regions

	Region[a]			
	US	SD	SM	CA
Pichia cactophila	**18.7**	**9.5**	**13.0**	**41.5**
P. heedii	**16.8**	**33.6**	0.0	0.0
Cryptococcus cereanus	**9.3**	**8.8**	**13.0**	**8.7**
Candida ingens	7.5	**13.9**	**8.7**	1.6
C. sonorensis	4.7	5.1	**13.0**	**14.2**
P. thermotolerans	**12.1**	6.6	0.0	0.0
P. pseudocactophila	4.7	**8.8**	4.3	0.0
P. amethionina var. *pachycereana*	**11.2**	3.6	0.0	0.0
Clavispora lusitaniae	0.0	1.5	**17.4**	0.0
P. antillensis	0.0	0.0	0.0	10.4
"P. amethionina var. *fermentans"*	0.0	0.0	0.0	**9.8**
P. kluyveri var. *eremophila*	0.0	0.0	8.7	0.0
Total percent of community	85.0	91.4	78.1	86.2
Total isolates in community	107	137	23	183

[a]Percent of total yeast community of Pachycereinae species in each region (*Geotrichum* is not included). The top five ranking species are shown in boldface type. Region: US, southwestern United States; SD, Sonoran Desert; SM, southern Mexico; CA, islands in the Caribbean; VZ, Venezuela. (See yeast collections for regional definitions.)

and thus speciation of lineages specific to Pachycereinae cactus species. The role of host plants within the Pachycereinae category cannot be regarded as unimportant because the species of cactus do differ across regions (i.e., *Cephalocereus royenii* in the Caribbean and *Carnegiea gigantea* in the southwestern United States).

Yeast Communities of Species of the Tribe Opuntieae

The yeast communities associated with rotting Opuntieae cladodes exhibit about the same amount of variation when communities collected in different seasons in the same region are compared as when communities collected in different regions are contrasted (Starmer *et al.*, 1987*a*). A study of *Opuntia stricta* yeast communities from the same plants sampled over time has strong successional and historical components (Barker *et al.*, 1987). Starmer and Phaff (1983) found Opuntieae yeast communities of the same host species to show as much variation from locality to locality as Stenocereinae yeast communities show from host species to host species. Taken together, these results indicate that the Opuntieae yeast community is somewhat more variable over space than are Stenocereinae communities.

The Opuntieae yeast communities (Table IX) are dominated by *Pichia cactophila, Candida sonorensis,* and *Cryptococcus cereanus,* and share to some extent the two varieties of *P. amethionina* (var. *pachycereana* and var. "*fermentans*") with the yeast communities of Pachycereinae species. *Clavispora opuntiae* was found in all regions and appears to be Opuntieae specific. Differences from region to region are due to replacement by related forms (i.e., the varieties of *P. amethionina*) or by regional specificity of relatively unrelated species as is apparent for *C. mucilagina, H. polymorpha,* and *P. kluyveri.* Previous analysis has shown the general composition of yeast communities of Opuntieae to be similar to that of Stenocereinae yeast communities (Starmer and Phaff, 1983). This is mainly due to the three dominant species and their relatively high incidence in both communities.

Yeast Communities of Cactus Fruit

An analysis of the yeasts associated with *Opuntia stricta* fruit in the Caribbean and eastern Australia shows that fruit yeast communities are probably influenced to a great extent by other local and regional yeast communities (Starmer *et al.*, 1987*a*) and thus there are great differences in the species composition when these communities are compared across regions. There is no doubt that the yeasts found in decaying cactus fruit differ qualitatively from the yeasts found in neighboring cactus pads (Starmer *et al.*, 1988). The reasons for these differences

TABLE IX. Dominant Yeasts of Opuntieae Yeast Communities across Regions

	Region[a]				
	US	SD	SM	CA	VZ
Pichia cactophila	**15.3**	**21.9**	**25.7**	**24.4**	**24.0**
Cryptococcus cereanus	**23.4**	**8.2**	**22.9**	**12.6**	**15.6**
Candida sonorensis	**12.1**	**17.8**	**8.6**	**21.6**	**14.5**
Prototheca sp.	**7.4**	5.5	**2.9**	**8.6**	**10.6**
Candida mucilagina	**8.5**	0.0	0.0	2.6	1.1
Clavispora opuntiae	3.0	**6.8**	**15.7**	**6.3**	7.3
Hansenula polymorpha	7.1	1.4	0.0	0.2	0.0
Pichia mexicana	1.1	5.5	0.0	2.2	**11.2**
"*P. amethionina* var. *fermentans*"	0.7	0.0	1.4	6.1	0.0
P. kluyveri var. *kluyveri*	0.1	6.8	1.4	2.2	0.0
P. amethionina var. *pachycereana*	0.8	**8.2**	0.0	0.0	0.0
Total percent of community	79.5	82.1	78.6	86.8	84.3
Total isolates in community	903	73	70	509	179

[a]Percent of total yeast community of Opuntieae species in each region (*Geotrichum* is not included). The top five ranking species are shown in boldface type. Region: US, southwestern United States; SD, Sonoran Desert; SM, southern Mexico; CA, islands in the Caribbean; VZ, Venezuela. (See yeast collections for regional definitions).

TABLE X. Dominant Yeasts of Cactus Fruit Yeast Communities across Regions

	Region[a]		
	SD	SM	CA
Kloeckera apis	0.0	**35.6**	**17.6**
Kl. apiculata	**59.4**	0.0	4.7
Issatchenkia terricola	**40.6**	0.0	5.2
Pichia barkeri	0.0	0.0	**9.8**
Candida krusei	0.0	**4.4**	**6.7**
P. cactophila	0.0	0.0	**6.7**
P. norvegensis	0.0	0.0	**5.7**
P. kluyveri	0.0	**13.3**	1.6
P. membranaefaciens	0.0	**6.7**	3.1
Candida sp.	0.0	**20.0**	3.6
Total percent of community	100.0	80.0	64.7
Total isolates in community	32	45	193

[a]Percent of total yeast community of rotting fruit in each region (*Geotrichum* is not included). The top five ranking species are shown in boldface type. Region: SD, Sonoran Desert; SM, southern Mexico; CA, islands in the Caribbean. (See yeast collections for regional definitions).

are unclear, but they could be due to a combination of factors such as vectors, competition, the production of killer toxins (Starmer *et al.*, 1987*c*), and host plant chemistry. Our compilation of fruit community yeasts (Table X) shows that there are sharp differences among regions and that these differences are generally replacements (i.e., *Kloeckera apis* for *K. apiculata*) and perhaps additions (the Caribbean region has many more species of yeast found in fruits). The comparison of fruit communities across regions has some faults in this case. First, the samples from the Sonoran Desert consisting of 19 fruits came from a single locality in the cape region of Baja California, Mexico. The data from the Caribbean and southern Mexico are probably more representative of the fruit yeast community spectrum and serve to illustrate the difference between regions and the qualitative differences as compared to the yeasts found in stems of cactus (Tables VII–IX).

CACTOPHILIC *DROSOPHILA* AND YEASTS

In general, yeast communities associated with different species groups and species groupings of the family Drosophilidae are distinct (Begon, 1982; Starmer, 1981). The physiological abilities of these yeast communities show similarities that parallel the evolutionary history of several separate *Drosophila* radiations (Starmer, 1981). The correlation of the physiological and metabolic abilities of the yeast communities with the phylogenetic history of the *Drosophila* suggests that yeasts and *Drosophila* may be coadapted. Coadaptation between *Drosophila* and yeasts probably did not occur in terms of strict nutritional requirements. On the contrary, it may have involved a combination of diffuse mutualisms among the yeasts in the community (Starmer and Fogleman, 1986) and benefits to the flies from the yeasts, such as substrate detoxification (Starmer *et al.*, 1986*b;* Starmer, 1982*b*), and complementary culture conditioning by mixed cultures of yeasts (Starmer and Fogleman, 1986; Starmer and Barker, 1986).

Species of *Drosophila* are thought to be one of the primary vectors of yeasts (Gilbert, 1980). Drosophilids, along with other cactus-breeding and feeding insects, such as dipterans (nereids, syrphids, dolichopodids), beetles (staphylinids), bugs (*Dactylopius*), and moths (*Cactoblastis, Melitara*), all probably serve to disperse yeasts to new cactus habitats. When these insects expand their distribution by shifting host plants or by moving into new regions, they are presumed to spread and expand the distribution of the yeasts that they carry. The cactophilic drosophilids show a great affinity for yeasts living in cactus, and studies have demonstrated that the dispersal of yeasts by adults is probably not passive (Starmer *et al.*, 1987*b;* Ganter *et al.*, 1986; Ganter, 1988).

The geographic distribution and host plant utilization patterns of *Drosophila* (Heed, 1977, 1982; Heed and Mangan, 1986; Fogleman and Heed, 1989) coupled with information on the cytology (Wasserman, 1982*a, b*) and morphology (Throckmorton, 1982) of known species indicates that a single lineage of *Drosophila* radiated in the New World about 30 million years ago. This radiation, known as the New World Virilis Repleta Radiation (Throckmorton, 1975, 1982), gave rise to the repleta group, which consists of 71–77 species, most of which are known to use cactus as a breeding and feeding habitat. Most of the cactophilic species of *Drosophila* are well known in southwestern United States and reasonably well known in Mexico (Patterson and Stone, 1952). The origins of species in the Sonoran Desert have been studied by Heed (1977, 1982). The species occupying the columnar cacti of the Sonoran Desert each have a separate origin from lineages outside of the desert, and all have relatives that use either *Opuntia* sp. cactus or columnar cacti as host plants. Furthermore, the majority of species of columnar cacti also have independent origins (Gibson, 1982). These unique invasions of *Drosophila* and cacti explain how many of the species of yeasts were initially partitioned on the basis of host plants in many regions. For instance, a similar sequence of events probably occurred in the West Indies, as discussed below. However, the maintenance of intraregional partitioning of the yeasts over evolutionary time is more difficult to assess, but it is useful to address the problem with examples from *Drosophila.* First, however, it is important to make a comparison of the present distribution of the flies and the yeasts across regions in order to ferret out any pattern that may be meaningful on this level.

With the inclusion of the information of the West Indian cactophilic *Drosophila* fauna currently under study (Wasserman, 1982*a, b;* Heed, 1989), a broad-scale comparison of the distribution of the yeasts and the *Drosophila* may be made across the regions of the southwestern United States, the Sonoran Desert, and the Caribbean islands. A total of 24 yeast species and varieties were selected from Table III on the criterion that they were represented by at least 20 isolates and there was more than one isolate from each of the three regions. No isolates from fruit and no *Geotrichum* or *Prototheca* species were scored. In regards to the *Drosophila,* a total of 21 species were selected from the same three regions (Table XI). The list is made up of members of the mulleri subgroup of the repleta species group mentioned above and as defined by Wasserman (1982*a*). All the species are cactus breeders.

The differences in distribution and endemicity between the yeasts and the
Drosophila fauna currently under study (Wasserman, 1982*a, b;* Heed, 1989), a
regions, while none of the 21 *Drosophila* species has as wide a distribution. By comparison, 48% of the species of *Drosophila* are found in only one or another of these regions, while 12.5% of the yeast species and varieties are so restricted. Thus, the yeasts are not as geographically differentiated as the *Drosophila,* probably because of a high rate of transport between regions by a variety of

TABLE XI. Distribution of Species in the Cactophilic Mulleri Subgroup of *Drosophila* in Three Regions

Drosophila species	Region[a]		
	US	SD	CA
Drosophila aldrichi	+	+	
D. arizonensis	+	+	
D. eremophila		+	
D. hamatofila	+	+	
D. longicornis	+	+	
D. mainlandi		+	
D. mayaguana			+
D. mettleri	+	+	
D. meridiana	+	+	
D. mojavensis	+	+	
D. mulleri	+		+
D. nigrospiracula	+	+	
D. richardsoni			+
D. spenceri		+	
D. stalkeri			+
D. wheeleri	+	+	
Undescribed species			
A			+
B			+
C			+
D		+	
E		+	

[a]US, Southwestern United States; SD, Sonoran Desert; CA, islands of the Caribbean.

insects and other vectors and also because of the asexuality and/or lack of outbreeding in the most common yeasts, as discussed previously. Thus, speciation events in the *Drosophila* have little bearing on speciation events in the yeasts. The two systems are evidently evolving at different rates for the most part.

In regard to the host plant specificity of specific yeasts, the first example pertains to "*Pichia amethionina* var. *fermentans*" in the Caribbean region. Three separate lineages in the mulleri subgroup of *Drosophila* have invaded the islands, giving rise to six endemic species (W. B. Heed and D. Grimaldi, unpublished). One of these lineages consists of *D. stalkeri* and *D. richardsoni,* both of which have been bred only from *Cephalocereus royenii* and *Opuntia stricta*. "*Pichia amethionina* var. *fermentans*" is found chiefly in the Caribbean region (82% of 72 isolates) and 73% of these came from the rots of the two cacti mentioned above. This close association is very suggestive and also the inference that this yeast is derived from South America is a strong one.

The second example of yeast host plant specificity involves three of the five species and varieties of yeasts that are strongly associated only with *Pachycereus* hosts. It is most significant that three of them are restricted to the Sonoran Desert and extreme southwestern United States. They are *P. thermotolerans, P. amethionina* var. *pachycereana,* and *P. heedii.* Two species of Sonoran Desert *Drosophila* are almost completely restricted to the *Pachycereus* hosts in this region. They are *D. pachea,* which breeds in senita cactus (*Lophocereus schottii*), and *D. nigrospiracula,* which breeds in saguaro (*Carnegiea gigantea*) and cardon (*Pachycereus pringlei*). These cases of a strong association of flies to cactus in the West Indies and in the Sonoran Desert definitely aid in understanding the host plant specificity of certain yeasts.

In contrast to the *Pachycereus* hosts, *Stenocereus* shares most of its yeast species and varieties with the *Opuntia* hosts, as discussed previously. One may ask whether or not *Drosophila* host plant use has any bearing on this relationship. There is a relationship in that there are three cases of closely related species of *Drosophila* using *Opuntia* and *Stenocereus* species. One case is in Mexico, one is in the West Indies, and one is in Venezuela. The most thoroughly documented case is the evolution of the two *Stenocereus* species in western Mexico, *D. arizonensis* and *D. mojavensis,* from a cytologically more ancetral species (*D.* sp. *N*) that lives in *Opuntia* (Heed, 1982). This shift in host plant type from *Opuntia* to *Stenocereus* is believed to be more difficult to achieve than the reverse occurrence because of the substantial concentrations of secondary plant products in the *Stenocereus* types that reduce the fitness of an uninitiated invading species (Kircher, 1982; Ruiz and Heed, 1989). Thus, the shift must take place in isolation. The speciation event accompanying the host shift necessarily would protect the newly founded population from future hybridizations with the ancestral *Opuntia* populations, permitting selection to proceed uninhibited. Speciation in the *Drosophila* once again has no influence on speciation in the yeasts.

In summary, then, four points of intersection have been described where *Drosophila* behavior may have influenced the geographic distribution and host plant specificity of the yeasts. The first emphasizes the independent origin of the fly species and the columnar cactus species now inhabiting the Sonoran Desert and also the independent origin of many of the fly species in the Caribbean region. By implication, the major lineages of the cacti also arrived independently. Also by implication, different types of yeasts arrived in association with the different flies and cacti. Therefore, it is not too surprising that the plant host has a larger influence on the composition of the yeast community than does the geography of these organisms if they are closely adapted to their particular host.

The second point illustrates the same phenomenon in a different way. The yeasts show only slight geographic differentiation compared to the *Drosophila* and thus, for whatever reason, they must be evolving at different rates. The slower rate in the yeasts emphasizes the importance of the host plant over geographic isolation for community structure. The third point gives examples of host plant specificity in the flies in various regions which might explain the specificity of several of the yeasts. Lastly, major host plant shifts in the *Drosophila* do occur by speciation in the flies, but usually not in the yeasts.

SPECULATIONS ON ORIGINS, EVOLUTION, AND SPECIATION

Several features of the geographic and host plant distribution of the common cactophilic yeasts (*Pichia cactophila, Candida sonorensis,* and *Cryptococcus cereanus*) argue for a relatively old association with cactus. These include the following points: (1) the common yeasts are found wherever cacti are found and although their relative frequencies change from plant type to plant type and region to region, they are almost always ranked among the top three or four species of the community; (2) the phenotypes (physiological) of *P. cactophila* and *Candida sonorensis* are remarkably stable and consistent; *Cryptococcus cereanus,* on the other hand, may represent a complex of species, the details of which have yet to be worked out; (3) discernible relatives (other than distant relatives known from the same genera) are not found outside of the cactus habitat,; and (4) other than being isolated from insects that are associated with cactus and occasionally from rotting fruit of cactus (*P. cactophila* and *Candida sonorensis*), these species are completely restricted to the cactus habitat. They are not found in other yeast habitats that are physically near the cactus habitat, such as tree fluxes of mesquite, cottonwood, and oak (Ganter *et al.*, 1986) and are not found in other regions where cactus does not occur [e.g., the Sierra Nevada mountains or the Great Lakes region (Lachance *et al.*, 1982)]. Many of these points are true for complexes of cactophilic yeasts (i.e., cactophila, amethionina, opuntiae, and deserticola complexes) where divergence, variety formation, and speciation probably has occurred only within the cactus habitat. This is not true, however, for the kluyveri complex, which appears to have origins from outside the cactus habitat. The relationships among the complexes are not known, but two possibilities are (1) that they have definitive connections to one another with no closer lineages living outside the cactus habitat, or (2) they each have different origins from outside the cactus niche and thus represent separate invasions.

Cactophila Complex

The three members of the cactophila complex show a typical pattern of host plant utilization. *Pichia cactophila* is widespread and nonspecific, while the other two members occupy either Opuntieae (*P. norvegensis*) or columnar cactus (*P. pseudocactophila*). Although the geographic distributions of *P. norvegensis* and *P. pseudocactophila* are more or less mutually exclusive, they completely overlap with the distribution of *P. cactophila* (compare Figs. 1 and 2). This pattern is difficult to interpret in terms of speciation by either host plant or geographic separation because *P. cactophila* occupies almost all hosts in every region. The sexuality of these species may help in understanding their present distributions. *Pichia cactophila* is homothallic, while the other two species are heterothallic. By being homothallic and as a result facultatively self-fertile, *P. cactophila* is isolated genetically, a condition which favors the development of new species free from the effects of recombination and gene flow from the parent species. Heterothallism would have been the initiating factor in the divergence of *P. norvegensis* and *P. pseudocactophila,* by facilitating recombination and adaptation, and ultimately resulting in complete speciation. If *P. cactophila* also had been heterothallic, then divergence into the present geographic and host distributions may not have been possible. It is interesting in this context to consider the widespread nature of *C. sonorensis,* which is ameiotic (until evidence to the contrary) and does not have any known close relatives outside or within the cactus habitat.

Amethionina Complex

Members of this complex have fairly distinct host and geographic distribution, with "*P. amethionina* var. *fermentans*" being the most widespread and having the greatest host diversity. This complex shows a stepwise replacement in taxa as it moves from the Caribbean ("*P. amethionina* var. *fermentans*") to the southwestern United States (both *P. amethionina* var. *pachycereana* and the variety *amethionina*) to the Sonoran Desert, where *P. amethionina* var. *amethionina* dominates (Fig. 3). This pattern is not free from host plant influences, as host shifts accompany the replacement, especially when considering occupation in the Sonoran Desert. "*Pichia amethionina* var. *australensis*" has been found only in eastern Australia and is closely related to *P. amethionina* var. *amethionina* (H. J. Phaff, unpublished data). Given that the taxonomic separation of the varieties is still somewhat problematic, the patterns we give here are only considered approximate. As they are presently understood, they indicate speciation by both geographic isolation and host plant separation. If the fruit habitat is considered as ancestral to the cactus stem and cladode habitat, then "*P. amethionina* var. *fermentans*" would represent the living representative of the lineage that gave rise to the complex and would point to either the Caribbean or southern Mexico as the likely point of origin. However, until more is known about the yeasts from other regions of South America, this suggestion is considered preliminary.

Opuntiae Complex

The analysis of this group is complicated by the fact that two of its members are known only from collections in eastern Australia, where cacti and the yeasts associated with them were introduced. Thus, the origin and native host use patterns of *P. opuntiae* and "*P. opuntiae* var. hem" are unknown. Both species do use *Opuntia stricta* in eastern Australia, where *P. opuntiae* is widespread and "*P. opuntiae* var. hem" is locally restricted to a region near Brisbane, Queensland (W. T. Starmer, unpublished). The other two species (*P. antillensis* and *P. thermotolerans*) share the same host plant category (Pachycereinae), but occupy different regions (Caribbean and Sonoran Desert, respectively, Fig. 4). A small number of isolates of related forms have been found in Venezuela, southern Mexico, the Sonoran Desert, and South Africa, but until further information is obtained from regions of South America other than Venezuela, we can only speculate that *P. antillensis* and *P. thermotolerans* are relatively old lineages that have remained in similar hosts. *Pichia antillensis* could be a relict, especially considering that it lives on islands in the Lesser Antilles of the relatively old Caribbean islands (Pindell and Dewey, 1982) and that its host is *Cephalocereus royenii,* which belongs to an early lineage of the Pachycereinae (Gibson and Horak, 1978; Gibson, 1982; Gibson *et al.,* 1986).

Kluyveri Complex

Pichia kluyveri var. *kluyveri* is not frequent in cactus and when it has been found it is mainly in fruit or Opuntieae cladodes. The variety *P. k.* var. *eremophila* is recovered from both Stenocereinae and Opuntieae rots, but the occupation of these plant types appears to divide the variety into sexual and asexual forms, with the sexual form in Opuntieae rots and the asexual form in Stenocereinae necroses (Phaff *et al.,* 1987). The variety *P. k.* var. *cephalocereana* is found on the island of Monserrat in the Caribbean and more recently on the island of St. Martin (H. J. Phaff, unpublished). It probably represents a unique, isolated lineage that has not spread to other localities or regions. The kluyveri complex is the only group which shows evolutionary connections to habitats other than cactus. The complex probably originated in fruits, subsequently spreading to Opuntieae (*P. kluyveri* var. *kluyveri* and *P. kluyveri* var. *eremophila*), and then to Stenocereinae (*Candida eremophila*). More information on this complex will be obtained when more isolates from fruits and similar habitats are studied. The geographic distribution of members of the kluyveri complex is shown in Fig. 5.

Deserticola Complex

The two members of this complex are separated by host plants and geographic boundaries (Fig. 6). The asexual form (*C. deserticola*) is found mainly in Stenocereinae species and is geographically widespread. The sexual form is found in low numbers in both Pachycereinae and Opuntieae habitats and is geographically restricted. This pattern is somewhat difficult to understand, as the origin of the sexual form from the asexual one is not considered likely. It would be important to understand the nature of asexuality in *C. deserticola* in this context. Several complexes show asexuality when their members are associated with Stenocereinae [i.e., *C. eremophila, P. amethionina* var. *amethionina,* which is often asexual when isolated from Stenocereinae (Starmer *et al.*, 1978), *C. deserticola,* and some strains of *P. cactophila*]. The asexual condition may arise as an adaptation to new host chemistry (Stenocereinae tissue is known to contain abundant quantities of surface-active triterpene glycosides that can inhibit microbial growth). Thus, we speculate that *P. deserticola* gave rise to *C. deserticola.* Alternatively, both forms may have a common ancestor that lived (or has yet to be detected) in the cactus habitat, having given rise to both forms.

Common and Frequently Isolated Cactus Yeasts

Cryptococcus cereanus is the anamorph of *Sporopachydermia cereana.* The genus *Sporopachydermia* has two other members, one of which, *S. quercuum,* has been found associated with tree fluxes (Bowles and Lachance, 1983). It is thus possible that the cactus-inhabiting members of the *Cryptococcus cereanus* complex are derived from tree flux inhabitants. This is a parallel to the situation in the cactus-associated *Pichia* species complexes, where many distantly related forms are found associated with tree fluxes (Phaff and Starmer, 1987). Until more is known about the interrelationships of the members of the genus *Sporopachydermia* and the *Cryptococcus cereanus* complex, speculation on the origins and evolution of the group is difficult.

The phylogenetic position of *Candida sonorensis* with respect to other yeasts, within or outside cactus-inhabiting taxa, is not at all clear. Its metabolic ability, especially the ability to utilize methanol as a sole source of carbon, may link it with free-living *Hansenula* or *Pichia* species that live in tree fluxes and also use methanol.

Most of the other frequently isolated yeast species (Tables V and VI) have no close relatives within or outside of the cactus habitat. Exceptions to this are *Hansenula polymorpha* and *Clavispora opuntiae*. *Hansenula polymorpha* occurs in soils as well as tree fluxes and comparisons of the nucleotide composition of DNA between cactus isolates and soil isolates have shown that the two types of isolates differ and have diverged genetically to a small extent (Starmer *et al.*, 1986*a*). This situation is similar to that for the *P. kluyveri* complex, where it is thought that the member taxa originate from the fruit habitat. In the case of *H. polymorpha* it is possible that cactus (Opuntieae)-inhabiting strains have originated from close relatives that live in tree fluxes. The geographic distribution of the cactus-inhabiting strains (Fig. 12) implicates a possible origin from northern regions of North America because strains have not been recovered in southern Mexico or Venezuela and isolates are rare in the Caribbean. Phaff *et al.* (1956) *and Shehata et al.* (1955) have shown that *H. polymorpha* (syn. *H. angusta*) is a common species in the intestinal tract of various wild species of noncactophilic *Drosophila* in mountainous regions of central and southern California.

Clavispora opuntiae is closely related to *Clavispora lusitaniae*. *Clavispora lusitaniae* is found associated with Pachycereinae habitats in southern Mexico (Table VIII), but it is also found in other habitats, such as fruit, agave, effluents, and warm-blooded animals (Lachance *et al.*, 1986). The wide spectrum of habitat use together with analysis of restriction maps of ribosomal DNA in the two *Clavispora* species suggest that *Clavispora lusitaniae* may be more representative of the ancestral lineage of the group (Lachance *et al.*, 1986). This is another case where the origin of a cactus-restricted species (*Clavispora opuntiae*) may have come from outside of the cactus yeast community.

OVERALL VIEW OF HOST AND GEOGRAPHIC DISTRIBUTION

One of the central questions we have posed is: What is the role of host plant and geography in determining the distribution of cactus-specific yeasts and yeast communities? One method of analysis is to compare the similarities of communities from the same or different host plant categories across regions to evaluate the effect of host plant and to compare the similarities among communities within a region (necessarily from different host plants) to the similarities among communities from different regions. This comparison can be made by calculating a measure of similarity such as the Euclidean distance or Pearson's correlation coefficient. We have calculated the correlation coefficient using the frequency of isolation of each taxon from a particular community and region as one observation. Comparison of the four community types (Stenocereinae, Pachycereinae,

Opuntieae, and fruit) across the five regions shows that yeast communities from the same host plant type are more similar to one another than are communities from the same region but different host plants. This conclusion is based on a 2 × 2 analysis of variance that compared communities categorized as coming from different or the same hosts and from different or the same regions. This analysis only includes main effects of host and region, but no interaction component, because a comparison of host effects within a region is not possible (i.e., the data are structured such that the host is not replicated in each region). Host means showed a significant effect ($F = 12.44$, d.f. $= 1, 207$; $P < 0.001$), with communities from the same host being more similar ($r = 0.63$) than communities from different hosts and the same region ($r = 0.41$). Region means were not significantly different ($F = 0.47$, d.f. $= 1, 207$; $P > 0.05$), with communities from the same region having the same similarity as communities from different regions ($r = 0.45$). The same analysis was conducted with the fruit community removed. In this comparison the same result was obtained. Host means were significantly different ($F = 5.17$, d.f. $= 1, 88$; $P < 0.05$; $r = 0.72$ for the same host, $r = 0.63$ for different hosts) and region means were not significantly different ($F = 0.17$, d.f. $= 1, 88$; $P > 0.05$; $r = 0.66$ for the same region, $r = 0.65$ for different regions). It thus appears that the host has a larger influence on the composition of the community than does geography.

Acknowledgments

We are grateful for the assistance and collaboration of Francesc Peris and Antonio Fontdevila. We are especially thankful to them for allowing us access to their collection records on yeasts from Venezuela. The field assistance of many collaborators has been instrumental in obtaining our samples for analysis. We appreciate the assistance of William Etges, James C. Fogleman, Philip Ganter, William R. Johnson, Martin M. Miller, Jean Russell, and Richard Thomas. Collections in the Caribbean were facilitated by cruises CF-8205 and CF-8314 of the research vessel Cape Florida. The laboratory assistance of Virginia Aberdeen and Mary Miranda is gratefully appreciated. This work was supported by NSF grants (W. T. Starmer, H. J. Phaff, and W. B. Heed) and Natural Science and Engineering Research Council of Canada Operating Grants (M.-A. Lachance).

REFERENCES

Barker, J. S. F., and Starmer, W. T., eds., 1982, *Ecological Genetics and Evolution: the Cactus–Yeast–Drosophila Model System,* Academic Press, Sydney, Australia.

Barker, J. S. F., Toll, G. L., and East, P. D., 1983, Heterogeneity of the yeast flora in the breeding sites of cactophilic *Drosophila, Can. J. Microbiol.* **29:**6–14.

Barker, J. S. F., East, P. D., Phaff, H. J., and Miranda, M., 1984, The ecology of the yeast flora in necrotic *Opuntia* cacti and of associated *Drosophila* in Australia, *Microb. Ecol.* **10:**379–399.

Barker, J. S. F., Starmer, W. T., and Vacek, D. C., 1987, Analysis of spatial and temporal variation in the community structure of yeasts associated with decaying *Opuntia* cactus, *Microb. Ecol.* **14:**267–276.

Begon, M., 1982. Yeasts and *Drosophila,* in: *The Genetics and Biology of Drosophila,* Vol. 3B (M. Ashburner, H. L. Carson, and J. N. Thompson, Jr., eds.), pp. 345–384, Academic Press, New York, New York.

Bowles, J. M., and Lachance, M. A., 1983, Patterns of variation in the yeast florae of exudates in an oak community, *Can. J. Bot.* **61:**2984–2995.

De Camargo, R., and Phaff, H. J., 1957, Yeast occurring in *Drosophila* flies and in fermenting tomato fruits in northern California, *Food Res.* **22:**367–372.

Fogleman, J. C., 1982, The role of volatiles in the ecology of cactophilic *Drosophila,* in: *Ecological Genetics and Evolution: The Cactus–Yeast–Drosophila Model System* (J. S. F. Barker and W. T. Starmer, eds.), pp. 191–206, Academic Press, Sydney, Australia.

Fogleman, J. C., and Heed, W. B., 1989, Columnar cacti and desert *Drosophila:* The chemistry of host plant specificity, in: *Special Biotic Relationships in the Arid Southwest* (J. O. Schmidt, ed.), pp. 1–24. University of New Mexico Press, Albuquerque, New Mexico.

Fogleman, J. C., and Starmer, W. T., 1985, Analysis of the community structure of yeasts associated with the decaying stems of cactus. III. *Stenocereus thurberi, Microb. Ecol.* **11:**165–173.

Fogleman, J. C., Starmer, W. T., and Heed, W. B., 1981, Larval selectivity for yeast species by *Drosophila mojavensis* in natural substrates, *Proc. Natl. Acad. Sci. USA* **78:**4435–4439.

Fogleman, J. C., Starmer, W. T., and Heed, W. B., 1982, Comparisons of yeast florae from natural substrates and larval guts of southwestern *Drosophila, Oecologia* **52:**187–191.

Ganter, P. F., 1987, The vectoring of cactophilic yeasts by *Drosophila, Oecologia* **75:**400–404.

Ganter, P. F., Starmer, W. T., Lachance, M. A., and Phaff, H. J., 1986, Yeast communities from host plants and associated *Drosophila* in southern Arizona: New isolations and analysis of the relative importance of hosts and vectors on community composition, *Oecologia* **70:**386–392.

Gibson, A. C., 1982, Phylogenetic relationships of Pachycereeae, in: *Ecological Genetics and Evolution: The Cactus–Yeast–Drosophila Model System* (J. S. F. Barker and W. T. Starmer, eds.), pp. 3–13, Academic Press, Sydney, Australia.

Gibson, A. C., and Horak, K. E., 1978, Systematic anatomy and phylogeny of Mexican columnar cacti, *Ann. Mo. Bot. Gard.* **65:**999–1057.

Gibson, A. C., Spencer, K. C., Bajaj, R., and McLaughlin, J. L., 1986, The ever-changing landscape of cactus systematics, *Ann. Mo. Bot. Gard.* **73:**532–555.

Gilbert, D. G., 1980, Dispersal of yeasts and bacteria by *Drosophila* in a temperate forest, *Oecologia* **46:**135–137.

Heed, W. B., 1977, Ecology and genetics of Sonoran Desert *Drosophila,* in: *Ecological Genetics: The Interface* (P. F. Brussard, ed.), pp. 109–126, Springer-Verlag, New York.

Heed, W. B., 1982, The origin of *Drosophila* in the Sonoran Desert, in: *Ecological Genetics and Evolution: The Cactus–Yeast–Drosophila Model System* (J. S. F. Barker and W. T. Starmer, eds.), pp. 65–80, Academic Press, Sydney, Australia.

Heed, W. B., 1989, Origin of *Drosophila* of the Sonoran Desert revisited. In search for a founder event and the description of a new species in the Eremophila Complex, in: *Genetics, Speciation and the Founder Principle* (L. V. Giddings, K. Y. Kaneshiro, and W. W. Anderson, eds.), pp. 253–278, Oxford University Press, Oxford.

Heed, W. B., and Mangan, R. L., 1986, Community ecology of the Sonoran Desert *Drosophila,* in: *The Genetics and Biology of Drosophila* (M. Ashburner, H. L. Carson, and J. N. Thompson, Jr., eds.), Vol. 3e, pp. 311–345, Academic Press, New York.

Holzschu, D. L., and Phaff, H. J., 1982, Taxonomy and evolution of some ascomycetous cactophilic yeasts, in: *Ecological Genetics and Evolution: The Cactus–Yeast–Drosophila Model System* (J. S. F. Barker and W. T. Starmer, eds.), pp. 127–141, Academic Press, Sydney, Australia.

Holzschu, D. L., Phaff, H. J., Tredick, J., and Hedgecock, D., 1983, *Pichia pseudocactophila,* a new species of yeast occurring in necrotic tissue of columnar cacti in the North American Sonoran Desert, *Can. J. Microbiol.* **29:**1314–1322.

Kircher, H. W., 1982, Chemical composition of cacti and its relationship to Sonoran Desert *Drosophila,* in: *Ecological Genetics and Evolution: The Cactus–Yeast–Drosophila Model System* (J. S. F. Barker and W. J. Starmer, eds.), pp. 143–158, Academic Press, Sydney, Australia.

Lachance, M. A., Metcalf, B. J., and Starmer, W. T., 1982, Yeasts from exudates of *Quercus, Ulmus, Populus,* and *Pseudotsuga:* New isolation and elucidation of some factors affecting ecological specificity, *Microb. Ecol.* **8:**191–198.

Lachance, M. A., Phaff, H. J., Starmer, W. T., Moffitt, A., and Olson, L. G., 1986, Interspecific discontinuity in the genus *Clavispora* Rodrigues de Miranda by phenetic analysis, genomic deoxyribonucleic acid reassociation, and restriction mapping of ribosomal deoxyribonucleic acid, *Int. J. Syst. Bacteriol.* **36:**524–530.

Lachance, M. A., Starmer, W. T., and Phaff, H. J., 1988, Identification of yeasts found in decaying cactus tissue, *Can. J. Microbiol.* **34:**1025–1036.

Lodder, J., 1970, ed., *The Yeasts—A Taxonomic Study,* North-Holland, Amsterdam.

Patterson, J. T., and Stone, W. S., 1952, *Evolution in the Genus Drosophila,* Macmillan.

Phaff, H. J., and Starmer, W. T., 1987, Yeasts associated with plants, insects and soil, in: *The Yeasts,*2nd ed. (A. H. Rose and J. S. Harrison, eds.), Vol. 1, pp. 123–179, Academic Press, New York.

Phaff, H. J., Miller, M. W., Recca, J. A., Shifrine, M., and Mrak, E. M., 1956, Studies on the ecology of *Drosophila* in the Yosemite area of California. II. Yeasts found in the alimentary canal of *Drosophila, Ecology* **37:**533–538.

Phaff, H. J., Starmer, W. T., Miranda, M., and Miller, M. W., 1978, *Pichia heedii,* a new species of yeast indigenous to necrotic cacti in the North American Sonoran Desert, *Int. J. Syst. Bacteriol.* **28:**326–331.

Phaff, H. J., Starmer, W. T., Tredick, J., and Miranda, M., 1985, *Pichia deserticola* and *Candida deserticola,* two new species of yeasts associated with necrotic stems of cacti, *Int. J. Syst. Bacteriol.* **35:**211–216.

Phaff, H. J., Starmer, W. T., and Tredick-Kline, J., 1987, *Pichia kluyveri sensu lato.* A proposal for two new varieties and a new anamorph, in: *Proceedings of the Symposium on the Expanding Realm of Yeast-Like Fungi, Amersfoort, The Netherlands. Studies in Mycology,* pp. 403–414, Centraalbureau voor Schimmelcultures, Baarn.

Pindell, J., and Dewey, J. F., 1982, Permo-Triassic reconstruction of western Pangea and the evolution of the Gulf of Mexico/Caribbean region, *Tectonics* **1:**179–211.

Ruiz, A., and Heed, W. B., 1988, Host-plant specificity in the cactophilic *Drosophila mulleri* species complex, *J. Anim. Ecol.* **57:**237–249.

Sang, J. H., 1978, The nutritional requirements of *Drosophila,* in: *The Genetics and Biology of Drosophila* (M. Ashburner and T. R. F. Wright, eds.), Vol. 2, pp. 159–192, Academic Press, New York.

Shehata, A. M. E. T., Mrak, E. M., and Phaff, H. J., 1955, Yeasts isolated from *Drosophila* and from their suspected feeding places in southern and central California, *Mycologia* **47:**799–811.

Starmer, W. T., 1981, A comparison of *Drosophila* habitats according to the physiological attributes of the associated yeast communities, *Evolution* **35:**35–52.

Starmer, W. T., 1982*a,* Analysis of the community structure of yeasts associated with the decaying stems of cactus. I. *Stenocereus gummosus, Microb. Ecol.* **8:**71–81.

Starmer, W. T., 1982*b,* Associations and interactions among yeasts, *Drosophila* and their habitats, in: *Ecological Genetics and Evolution: The Cactus–Yeast–Drosophila Model System* (J. S. F. Barker and W. T. Starmer, eds.), pp. 159–174, Academic Press, Sydney, Australia.

Starmer, W. T., and Barker, J. S. F., 1986, Ecological genetics of the *Adh-1* locus of *Drosophila buzzatii*, *Biol. J. Linn. Soc.* **28:**373–385.

Starmer, W. T., and Fogleman, J. C., 1986, Coadaptation of *Drosophila* and yeasts in their natural habitat, *J. Chem. Ecol.* **12:**1037–1055.

Starmer, W. T., and Phaff, H. J., 1983, Analysis of the community structure of yeasts associated with the decaying stems of cactus. II. *Opuntia* species, *Microb. Ecol.* **9:**247–259.

Starmer, W. T., Phaff, H. J., Miranda, M., and Miller, M. W., 1978, *Pichia amethionina,* a new yeast associated with the decaying stems of cereoid cacti, *Int. J. Syst. Bacteriol.* **28:**433–441.

Starmer, W. T., Phaff, H. J., Tredick, J., Miranda, M., and Aberdeen, V., 1984, *Pichia antillensis,* a new species of yeast associated with necrotic stems of cactus in the Lesser Antilles, *Int. J. Syst. Bacteriol.* **34:**350–354.

Starmer, W. T., Ganter, P. F., and Phaff, H. J., 1986*a,* Quantum and continuous evolution of DNA base composition in the yeast genus *Pichia, Evolution* **40:**1263–1274.

Starmer, W. T., Barker, J. S. F., Phaff, H. J., and Fogleman, J. C., 1986*b,* Adaptations of *Drosophila* and yeasts: Their interactions with the volatile 2-propanol in the cactus–microorganism–*Drosophila* model system, *Aust. J. Biol. Sci.* **39:**69–77.

Starmer, W. T., Lachance, M. A., and Phaff, H. J., 1987*a,* A comparison of yeast communities found in necrotic tissue of cladodes and fruits of *Opuntia stricta* on islands in the Caribbean Sea and where introduced into Australia, *Microb. Ecol.* **14:**179–192.

Starmer, W. T., Phaff, H. J., Bowles, J. M., and Lachance, M. A., 1987*b,* Yeasts vectored by insects feeding on decayed saguaro cactus, *Southw. Nat.* **33:**362–363.

Starmer, W. T., Ganter, P. F., Aberdeen, M. A., Lachance, M. A., and Phaff, H. J., 1987*c,* The ecological role of killer yeasts in natural communities of yeasts, *Can. J. Microbiol.* **33:**783–796.

Starmer, W. T., Aberdeen, V., and Lachance, M. A., 1988, The yeast community associated with *Opuntia stricta* (Haworth) in Florida, with regard to the moth *Cactoblastis cactorum* (Berg), *Fl. Sci.* **51:**7–11.

Throckmorton, L. H., 1975, The phylogeny, ecology, and geography of *Drosophila,* in: *Handbook of Genetics* (R. C. King, ed.), Vol. 3, pp. 421–469, Plenum Press, New York.

Throckmorton, L. H., 1982, Pathways of evolution in the genus *Drosophila* and the founding of the Repleta group, in: *Ecological Genetics and Evolution: The Cactus–Yeast–Drosophila Model System* (J. S. F. Barker and W. T. Starmer, eds.), pp. 33–47, Academic Press, Sydney, Australia.

Vacek, D. C., Starmer, W. T., and Heed, W. B., 1979, The relevance of the ecology of *Citrus* yeasts to the diet of *Drosophila, Microb. Ecol.* **5:**43–49.

Wasserman, M., 1982*a,* Cytological evolution in the *Drosophila repleta* species group, in: *Ecological Genetics and Evolution: The Cactus–Yeast–Drosophila Model System* (J. S. F. Barker and W. T. Starmer, eds.), pp. 49–64, Academic Press, Sydney, Australia.

Wasserman, M., 1982*b,* Evolution of the *repleta* group, in: *The Genetics and Biology of Drosophila* (M. Ashburner, L. Carson, and J. N. Thompson, Jr., eds.), Vol. 3b, pp. 61–139, Academic Press, New York.

8

Mosaic Pattern of Heterochronies

Variation and Diversity in Pourtalesiidae (Deep-Sea Echinoids)

B. DAVID

INTRODUCTION

The Roman mosaics of the Villa Casale in Sicily are beautiful and fascinating. The majority depict wild animals with great realism, but one of them figures naiads accompanied by marine creatures showing imaginary morphologies. However, other very uncommon morphological patterns, not so different from these mosaics, are sometimes recorded in our living world. Such extreme morphologies, situated at the boundaries of variation, are often encountered in extreme conditions, as illustrated by the strange shapes assumed by life in the deep sea. In this respect, the Recent deep-sea forms of the Order Holasteroida are among the most curious irregular sea urchins: ''In the transformation of the Echinoid body . . . *Echinosigra paradoxa* may well deserve the prize; its body being so highly specialized that its obvious to describe it as having a head, neck, body, and tail [. . .]. Other very peculiar forms are the triangular *Ceratophysa* and the pyramidal *Echinocrepis*'' (Mortensen, 1950, pp. 132–133).

The evolutionary meaning of oddness has been elegantly stressed by Alberch, who has demonstrated the discontinuous nature of morphological space, using ontogenetic processes and ''monsters'' (Alberch, 1980, 1989). Similarly, morphologically bizarre deep-sea echinoids form an ideal field of research to retrace ontogenetic trajectories and to observe how diversity develops. Such research goes directly to the heart of recent evolutionary studies dealing with the relationship between ontogeny and phylogeny, a subject which has attracted a

B. DAVID • CNRS Associate Unit 157, Earth Sciences Center, F-21100 Dijon, France.

renewed interest in recent years (e.g., Gould, 1968, 1977, 1984; Alberch, 1982; Bonner, 1982; McNamara, 1983, 1985). These studies focused on the explanation of diversity, or of evolution, through ontogenetic processes, and due to their didactic purpose, they remained generalized. The organisms were often globally visualized and their variations ascribed to a single type of heterochrony. But in reality, organisms are complex entities comprising numerous traits and components whose ontogenetic fluctuations are not necessarily congruent. More detailed study reveals that the observed differences between two organisms (or two morphs, two species, two groups, etc.) often result in fact in the juxtaposition of many heterochronies, certain features being accelerated, whereas others are delayed [see Gould (1977) on human evolution; Alberch and Alberch (1981) on tropical salamanders; Guerrant (1982) on flowers, and McKinney (1984) on echinoids].

The main purpose of this chapter is to illustrate this complexity by reconstructing the heterochronic mosaic which has led to the diversification of the deep-sea echinoid family Pourtalesiidae. The differences between genera will be described in terms of ontogenetic variations in order to understand which heterochronies are involved and which are the most dominant in this family.

Pourtalesiidae are sea urchins which do not exhibit the usual pentaradiate symmetry of echinoids. Moreover, they often have strange shapes (Fig. 1). The two original features shown in Fig. 1 for *Ceratophysa ceratopyga* (Agassiz) are not expressed in juveniles, but they become progressively more and more obvious during ontogeny. The same general ontogenetic pattern is followed to varying degrees by all the Pourtalesiidae, which associate a more or less strange shape to a more or less altered symmetry. Differences among adult structures might correspond to variations in a unique developmental process.

The Pourtalesiidae thus have three advantages for the current study: (1) they make up a homogeneous group; (2) the different genera have long ontogenetic trajectories where extreme transformations are achieved; (3) the variation among genera seems to relate to variation in ontogeny.

MATERIAL AND METHODS

The Pourtalesiidae belong to the Holasteroida. Holasteroids correspond to an order of heart-shaped sea urchins which arose during the early Cretaceous and the living representatives of which are mostly deep-sea dwellers. Among the seven families belonging to this order, the Pourtalesiidae appear strongly differentiated from other holasteroids. It corresponds to a marginal and well-delimited group (Fig. 2).

The Pourtalesiidae include seven genera: *Ceratophysa, Cystocrepis, Echi-*

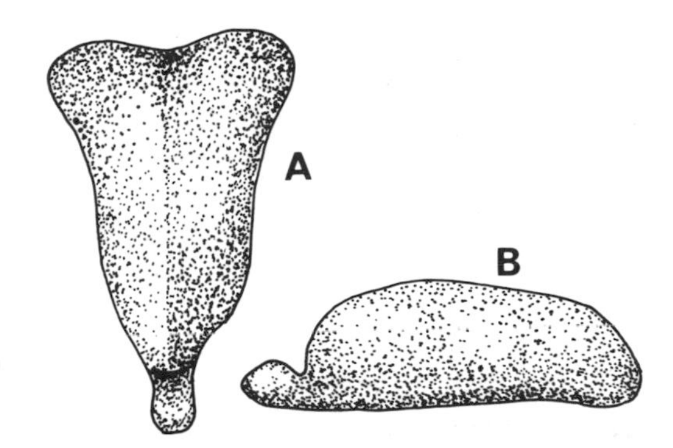

FIG. 1. *Ceratophysa ceratopyga* (Agassiz). (A) Apical and (B) lateral views. [Redrawn from Agassiz (1881).]

nosigra, Echinocrepis, Helgocystis, Pourtalesia, and *Spatagocystis* [following the classification by Mortensen (1950, 1951).] All are extant forms, but *Pourtalesia* has also been recorded from the Miocene of Japan (Kikuchi and Nikaido, 1985). These genera share several apomorphies and constitute a monophyletic group of strong cohesion (David, 1985). They are thus the result of a unique history, and the comparison of their ontogenesis is fully justifiable.

The bulk of the material of this study is made up of *Pourtalesia miranda* Agassiz and *Echinosigra phiale* (Wyv. Thomson). Several hundred specimens of these species were dredged from three localities in the northeast Atlantic: (1) the Rockall trough, where the two genera are sympatric; (2) the Porcupinc scabight, where *E. phiale* were collected; and (3) the Bay of Biscay, where the two genera colonize independently the northern margin (*Echinosigra*) and southern margin

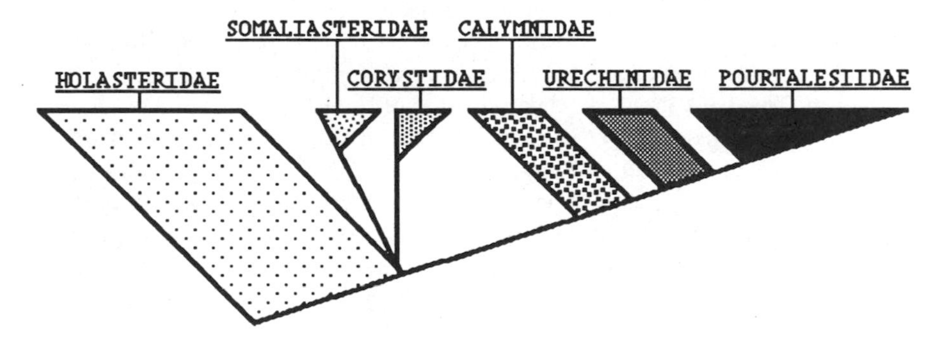

FIG. 2. Simplified cladogram of major subdivisions of Holasteroida. The general pattern is adapted from David (1988); the taxonomic units refer to classical family names taken from Smith (1984), amended by Foster and Philip (1978). The apomorphic groupings are represented by triangles, and the plesiomorphic sets (not still fully resolved) by trapezoidal areas. (The Stenonasteridae are not included in the cladogram, because they are insufficiently known.)

(*Pourtalesia*). Juveniles are always abundant in the samples, and were collected at the same stations as the adults. Precise bibliographic or museum data provide complementary information about *Cystocrepis, Echinocrepis,* and *Spatagocystis*. Both *Ceratophysa* and *Helgocystis* are poorly known (Table I).

A two-part investigation has been undertaken.

1. A detailed comparison of the ontogenetic trajectories of *Pourtalesia* and *Echinosigra* in order to understand the possible relationship between ontogenetic development and the resultant adult morphologies. This is done using statistical analyses and involves numerous features.

2. Extrapolation of the model to all genera within the family and to other types of characters. This is accomplished in a less formal manner, without statistical analysis and mainly considering adult forms, due to the scarcity of the juvenile material.

A MODEL OF ONTOGENETIC COMPARISON: *POURTALESIA* AND *ECHINOSIGRA*

Methods of Comparison

Preliminary note. Ontogenetic comparisons are widely used, and have given rise to a rich, but sometimes idiosyncratic vocabulary. The terms used in this chapter agree with the general literature [for definitions, see Gould (1977), McNamara (1982, 1986), and Dommergues *et al.* (1986)].

The development of an organism can be represented in an ($n+2$)-dimensional space: age + size + n shape parameters. This type of representation, proposed by Alberch *et al.* (1979), allows the description, comparison, and eventual quantification of ontogenetic trajectories for different components of an organism. That is, it allows the determination of heterochrony from the analysis of a few factors: mainly the coefficients of size and shape changes, as well as the age or the size at the onset and cessation of growth. Practically the analysis can dissociate all the parameters. The ontogenetic trajectory of a structure or of a component of the organism is thus projected onto a size–shape plane or an age–shape plane. This leads to classic bivariate analyses. The shape parameter can be defined by a ratio between two dimensions or by other more convenient indexes. The size parameter can be given by a preferential dimension chosen as reference. The question of the age parameter is trickier because age data are generally absent in purely morphological approaches, unless a reliable growth zonation provides a regular function of time. When age data are unavailable, the size can eventually be used to replace the age parameter. But the equating of size and age has to be performed with great care because the linearity of their correlation admits numerous exceptions [for more information on this equating of size and

TABLE I. Genera of the Family Pourtalesiidae: Species Diversity and Geographic Distribution

Genus	Number of Species	Specimens collected (since 1896)	Geographic distribution	Basis of the study
Ceratophysa	1	1+ fragments	Southern Indian Ocean	Bibliography: Agassiz (1881), Lovén (1883) Museography: British Museum
Cystocrepsis	1	1+ fragments	Eastern Pacific Ocean (Panama)	Bibliography: Agassiz (1904), Mortensen (1907)
Echinocrepis	2	2+ fragments	Southern Indian Ocean, northern Pacific Ocean	Bibliography: Agassiz (1881), Lovén (1883), Mironov (1973) Museography: British Museum
Echinosigra	3	10^2 (often broken)	Worldwide (except Arctic)	Oceanographic cruises: *E. phiale*, northeastern Atlantic Ocean Bibliography: Gage (1984)
Helgocystis	1	≈5 (incomplete)	Antarctic Ocean, southern Pacific Ocean	Bibliography: Agassiz (1881), Lovén (1883), Mortensen (1907, 1950, 1951) Museography: British Museum
Pourtalesia	≈10	10^2	Worldwide	Oceanographic cruises: *P. miranda*, northeastern Atlantic Ocean
Spatagocystis	1	≈10	Southern Indian Ocean	Bibliography: Agassiz (1881, 1904) Museography: British Museum

age see Dommergues *et al.* (1986) and McKinney (1988)]. All heterochronic processes discussed below will be obtained using a size parameter as the basis of ontogenetic comparison. In a strict sense, they do not refer to heterochronies (i.e., timing processes), and thus they should be qualified as "allometric" as recently suggested by McKinney (1988). However, in the following sections, this "allometric" qualification will be made only implicitly, in order to avoid cumbersomeness in the text.

The main outlines of the test of *Pourtalesia* and *Echinosigra* were characterized by 12 principal measurements and four complementary calculated variables which represent sectors of the adoral side. These 16 parameters refer to the whole morphology as well as the architecture (plate arrangement) of the test (Fig. 3). Twelve indexes were also calculated. They correspond to ratios which allow changes in proportion to be taken into account. These are the parameters of shape supporting the ontogenetic analysis.

The measurements were made on 65 intact specimens, divided into three groups: 23 juveniles not recognizable (Rockall trough and Bay of Biscay); 15 determined *P. miranda* (Rockall trough and Bay of Biscay); and 27 determined *E. phiale* (Rockall trough and Porcupine seabight), including 9 specimens which were directly measured on the figures from Gage (1984).

Ontogenetic Trajectories: General Tendencies

It is possible to observe the plate arrangement of a young sea urchin at 2 mm size. But at this length, and up to 3 or 4 mm, young *Echinosigra* and *Pourtalesia* are differentiated neither by body shape nor by architecture. The general outline of the test is oval. The arrangement of the ambulacra and interambulacra is regular, and the odd interambulacrum 5 is continuous (Fig. 4A). In fact, an uncertainty exists, and it is only the coexistence of adult types together with young types in the same samples which allows one to suppose that the very young specimens belong to the two respective genera.

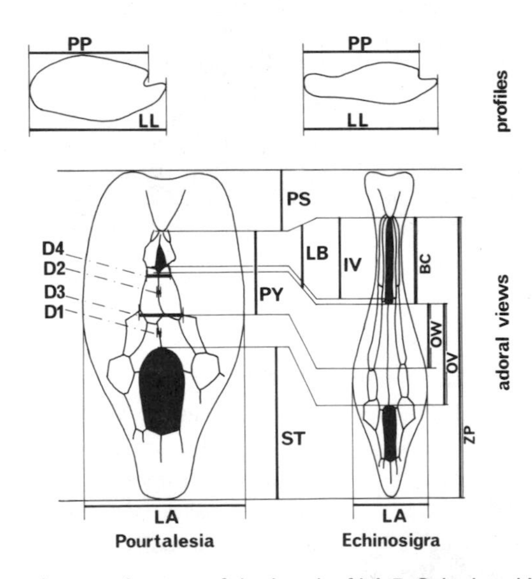

FIG. 3. Morphological parameters. Measurements (bold lines): LL, length of test; PP, periproct to anterior end of test; LA, width of test; PS, peristome to posterior end of test; LB, length of labrum; IV, length of plates I.a.1 or V.b.1; PV, oral invagination to particular points (triple points), defined as meeting points of plates I.a.2 + I.b.2 + I.b.3 and V.b.2 + V.a.2 + V.a.3 respectively; ST, anterior edge of sternum to posterior end of test; D1, width I.a.2–V.b.2; D2, width I.b.2–V.a.2; D3, width between triple points; D4, width I.ab.2–4.ab.2. Calculated variables (double lines): ZP = LL − PS, posterior part of test; BC = maximum between IV and LB; OV = ZP − (BC+ST), whole length of the break of I.A.5; OW = PY − BC, length of the anterior part of the break of I.A.5. Calculated indexes: AMB = LA/LL; RPP = PP/LL; RPS = PS/LL; RPY = PY/LL; RST = ST/LL; RBC = BC/ZP; ROV = OV/ZP; ROW = OW/ZP; RD1 = D1/LA; RD2 = D2/LA; RD3 = D3/LA; RD4 = D4/LA. In the odd interambulacrum 5, labrum and sternum are in black. The lettering of plates follows the classic Lovén rules, and the architectural interpretation is from David (1987).

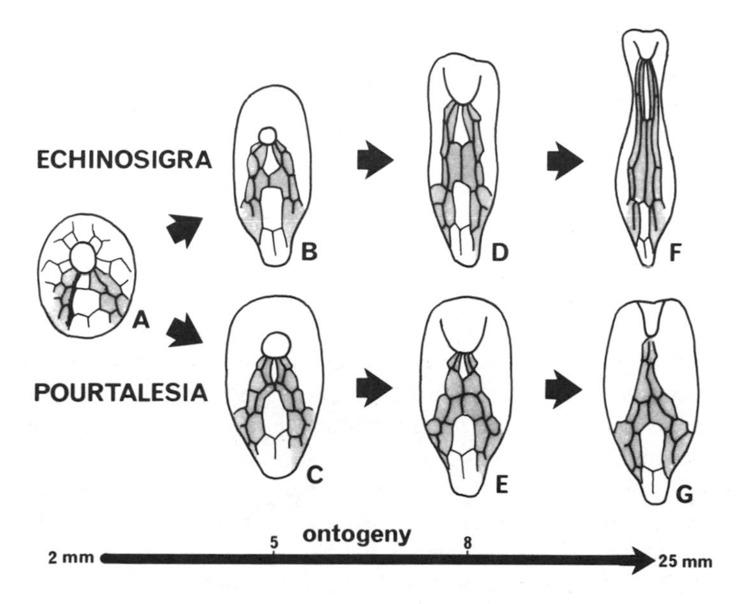

FIG. 4. Main ontogenetic stages of *Echinosigra* and *Pourtalesia* (oral sides). (A) Undeterminable juvenile (2 mm in length); (B,C) juveniles showing a single break in I.A.5 (5 mm in length); (D,E) juveniles with a single (*Echinosigra*) or double break of I.A.5 (*Pourtalesia*) (8 mm in length); (F,G) adults (25 mm in length). Posterior ambulacra are stippled.

At 4–5 mm, young *Echinosigra* elongates, whereas *Pourtalesia* stays more shortened. However, their architectural development remains identical: members of the two genera simultaneously reach the stage where the labrum and sternum (the two first plates of interambulacrum 5) are separated by the meeting of the adjacent ambulacral plates (Figs. 4B and 4C). This relative plate movement can be referred to the process of plate translocation described by McNamara (1987) with regard to some spatangoid genera.

After 7–8 mm, the differences are greater. *Echinosigra* retains the same architectural pattern with a single break of interambulacrum 5, but the test stretches considerably and the plates, while keeping a relatively stable array, elongate a great deal (Fig. 4D). On the other hand, *Pourtalesia* changes little in body shape, but achieves an architectural pattern with a twofold break of interambulacrum 5 (Fig. 4E).

Further development shows the continuation of these initial trends, and more complex breaks are sometimes achieved by *Pourtalesia* (Figs. 4F and 4G) [for a complete description of the architectural growth of *Pourtalesia* see David (1987)].

Using other holasteroids as an outgroup (for example, the holasterian genus *Holaster*, which could be considered as the paragon of the common Holasteroid, or the urechinid genera *Plexechinus* and *Urechinus*, which are closely related to

Pourtalesiidae; see Fig. 2), one can orient the comparison between *Pourtalesia* and *Echinosigra*. It is then possible to understand their differences as resulting from an architectural peramorphosis in *Pourtalesia* and from a peramorphosis of the general body shape in *Echinosigra*.

Ontogenetic Trajectories: Detailed Analyses

The preceding section has shown the general frame of the divergences between *Pourtalesia* and *Echinosigra* with a twofold peramorphosis (both architecture and shape). Here, the processes of divergence between the two genera are analyzed by studying detailed ontogenetic trajectories of particular features. For this, bivariate analyses conforming to the model of Alberch *et al.* (1979) were carried out on 12 indexes.

At this point one has to address the problem of the choice of a size parameter. Such a choice is crucial since it partly influences the nature of the results. Indeed, a size parameter which would not have the same significance in the two genera could introduce a distortion between the ontogenetic trajectories and biased results could follow. Bearing this in mind, the choice of length, which is usually used as a proxy for size in irregular echinoids, is inappropriate in the case of *Echinosigra* and *Pourtalesia* species, which have, for this parameter, largely different growth dynamics. Consequently, when the comparison is made between heterogeneous genera, the main risks of distortion can be avoided by the choice of the linear combination of all measured variables as a size parameter. In the present example, each variable is modified by its coefficient on the first eigenvector of the principal components factorial analysis. This results in (see Fig. 3 for definition of parameters) SIZE = 0.98LL + 0.77LA + 0.91PS + 0.98PP + 0.88LB + 0.98ST + 0.98IV + 0.99PY + 0.56D1 + 0.57D2 + 0.97D3 + 0.83D4.

The following sections will describe and illustrate the main heterochronic tendencies observed between *Pourtalesia* and *Echinosigra*.

Position of the Peristome

During development, the peristome shows an important forward migration. This migration occurs at the same rate in the two genera: the growth curves can almost be superimposed (index RPS, Fig. 5A). The only difference relates to the extension of the trend in *Echinosigra,* which attains a larger size than *Pourtalesia* (at least than *P. miranda*). Thus, in adult *Echinosigra* the mouth occupies a slightly more anterior position as a result of the extension of the growth allometry. This arrangement of the two curves could be interpreted in two ways: either *Echinosigra* is hypermorphic or *Pourtalesia* is progenetic. In a strict sense, each

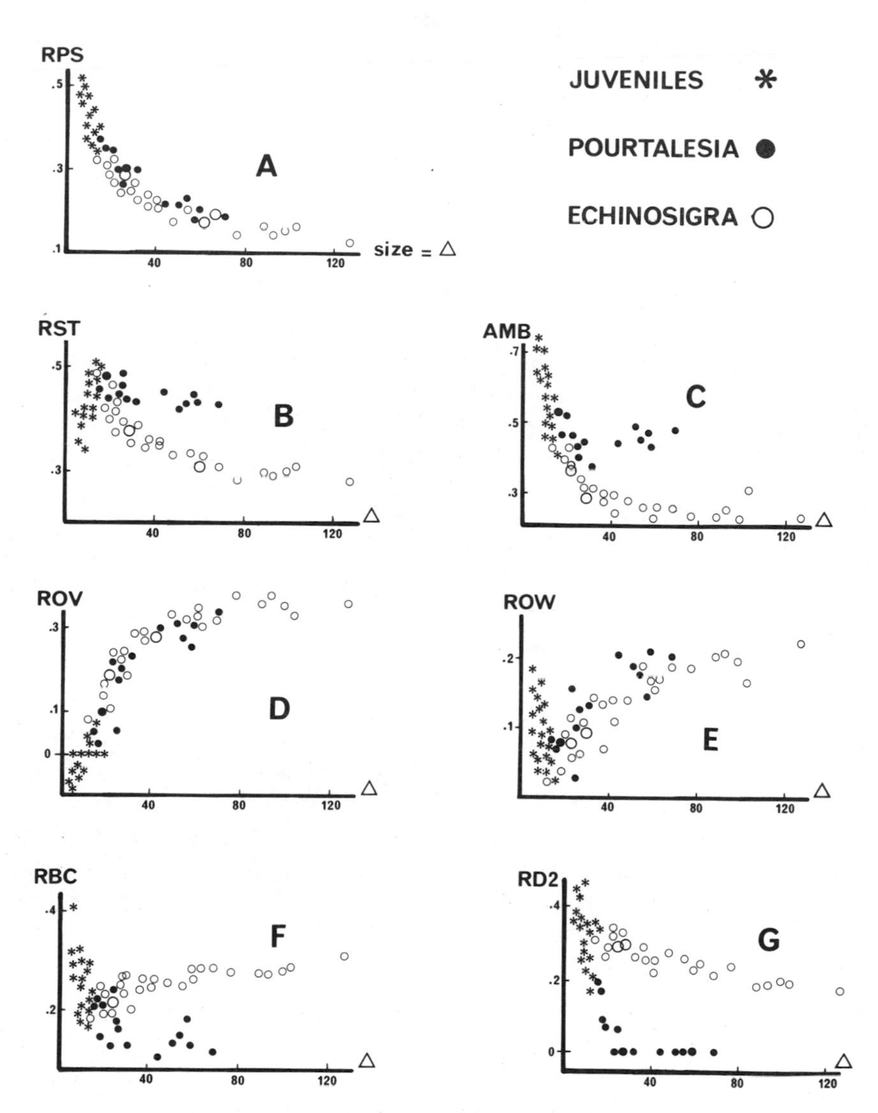

FIG. 5. Ontogenetic trajectories. Scatter diagrams of seven shape indexes versus a size parameter Δ (for the definition of this parameter, see text). Different types of divergence between trajectories of *Echinosigra* and *Pourtalesia* are presented: (A,D,E) *superimposed,* (B,C,G) *homologous,* and (F) *distinct* trajectories. The meaning of parameter abbreviations is given in the caption of Fig. 3.

of these heterochronies involves the maturation of the organisms. In the absence of knowledge on the biological age, and in the absence of a reliable stratigraphic polarity of the two genera (namely a certitude about their relative ancestry), it is not possible to determine which process has operated. The comparison remains unpolarized.

Sternum

In *Pourtalesia* and *Echinosigra,* the sternum and the episternals are large plates which bear strong locomotory spines and which have an important functional use. For this character (variable ST and index RST) the two genera have homologous ontogenetic trajectories (Fig. 5B). At the beginning of development, the anterior edge of the sternum moves toward the front: the growth of the back of the test is predominant. After this first stage, the trend is reversed and the sternum returns slightly backward. The phenomenon is shared by the two genera, but it is more rapid and persists longer in *Echinosigra.* The gap between the two curves expresses either an acceleration in *Echinosigra* or a retardation in *Pourtalesia.* Each of these heterochronies affects the morphological development. They concern the rates of change in shape, and are visualized on the diagram by the slope of the curves.

Reference to outgroup comparisons allows one to interpret the fraction of the difference resulting in the divergence of *Echinosigra* and *Pourtalesia* (difference in the slopes of ontogenetic trajectories). An interpretation of the divergence could thus be advanced by referring to the morphology of other genera which serve as outgroups. A comparison with the closely related *Plexechinus* (sister group Urechinidae; see Fig. 2), where the sternum occupies a very anterior position, suggests that *Echinosigra* is accelerated in comparison to *Pourtalesia.* But reference to the outgroups provides no help in trying to interpret the fraction of the difference resulting in the longer extension of the trend in *Echinosigra* (difference in the length of ontogenetic trajectories) (see preceding section).

Outline of the Ambitus

The problem is similar for the outline of the ambitus (index AMB, Fig. 5C). *Echinosigra* and *Pourtalesia* share the same lengthening trend. But this trend is lower in *Pourtalesia,* where it stops and even reverses at the end of growth, whereas *Echinosigra* continues to stretch until it attains an extremely elongate shape. *Echinosigra* is accelerated in comparison to *Pourtalesia.* This parameter (AMB) expresses quantitatively the general tendency described above qualitatively.

Break of Interambulacrum 5

One of the most characteristic features of the Pourtalesiidae is the break of interambulacrum 5, represented here by variables OV and OW. *Echinosigra* and *Pourtalesia* share the same ontogenetic trajectory (the same allometry) for the relative width of the break (index ROV, Fig. 5D). At the beginning of the development, the sternum is very close to the plates surrounding the peristome; then it separates from these plates and moves away steadily. This widening of the break occurs at the same rate in the two genera, but it attains a greater relative width in large adult *Echinosigra* by prolongation of the allometry.

Estimation of the anterior part of the break (given by variable OW) does not have the same architectural significance in the two genera. In *Pourtalesia* it can be compared to external ambulacral plates I.b.2 and V.a.2, whereas in *Echinosigra* it corresponds to the adoral part of the internal ambulacral plates I.a.2 and V.b.2 (Fig. 3). However, an analysis of growth proves that it has the same geometric importance. Figure 5E shows that, despite an important variability, *Echinosigra* and *Pourtalesia* retain, along a complex curve, very similar values. As with the preceding feature, the prolongation of the allometry leads in large *Echinosigra* to a slight amplification of the character.

The double architectural break of interambulacrum 5 of *Pourtalesia* thus appears quantitatively equivalent to the single break of *Echinosigra.* This relation is true considering the whole break (index ROV) or only its adoral section (index ROW). The ontogenetic trajectories of the two genera are similar at the morphological level despite an architectural disparity. Such an antagonism indicates a mosaic pattern of the morphological and architectural characteristics in relation to heterochronies.

Peristomial Group

The peristomial group (variable BC and index RBC) is made up of plates which border the posterior margin of the peristome: labrum and surrounding ambulacral plates. It represents anatomical and functional entity, even though its architecture may be variable (David, 1983). In juveniles of both genera, the relative size of the peristomial group decreases while being highly variable (Fig. 5F). This decreasing trend continues in *Pourtalesia,* but slows down. Conversely, in *Echinosigra,* the decreasing trend is stopped and even reversed, the peristomial group becoming involved in the growth in length. This leads to very strong divergences between the curve and leads to the question of whether the two genera really share common ontogenetic pathways. If they do, this raises another question: With what parameter should the comparison be established? In the case of the attained morphologies (raw values on the curves) *Pourtalesia* transcends *Echinosigra* and is accelerated. On the other hand, in the case of the growth dynamic (slopes of the curves), *Echinosigra* reaches adult growth rates at an earlier stage and becomes in turn accelerated. There remains a last question, specific to these sea urchins: How should the comparison be oriented? Indeed, the related genera which could be used as outgroups have rather different growth patterns and cannot serve as reference. These three questions raised about the peristomial group indicate that the method proposed by Alberch *et al.* (1979), as attractive it may be in theory, can come up against some difficulties in practical application.

Transversal Distances

Four variables (D1, D2, D3, and D4) represent the distance between four pairs of homologous plates on both sides of the plane of symmetry (Fig. 3).

Echinosigra and *Pourtalesia* have the same ontogenetic trajectory for the first parameter (index RD1). The diagram of RD2 (Fig. 5G) shows that plates I.b.2 and V.a.2 are drawing close together. This trend is shared by the two genera, but appears more intense in *Pourtalesia*. In comparison to the urechinids *Urechinus* or *Plexechinus,* the observed divergence can be interpreted as an acceleration in *Pourtalesia*. A closely similar process has been observed in RD3 and RD4.

Despite its apparent simplicity, this group of variables raises the problem of the intercorrelation of features apparently well differentiated. Thus, the indexes RD2, RD3, and RD4 have mutual coefficients of partial correlation (excluding the size effect) which vary from 0.80 to 0.87. With such correlations, the heterochrony determined on one index largely controls those observed in the other two. More generally, when one considers simultaneously a large set of characters, they are never fully independent, even after being released from size influence. Such hidden correlations are very difficult to predict *a priori,* but they are equally difficult to evaluate, particularly in the case of complex curves. There is thus a risk of introducing an element of redundancy into the matrix of the heterochronies, which forbids all quantitative assessments and requires accurate analysis.

Summary of the Divergence

Table II recapitulates feature by feature the comparison between *Echinosigra* and *Pourtalesia,* and takes into account the reservations which have been formulated above. Three categories of ontogenetic transformations allow an explanation of the observed differences between the two genera.

TABLE II. Detailed Ontogenetic Comparisons between *Echinosigra* and *Pourtalesia*: Morphological Synthesis[a]

Feature	General growth trend	Trajectory	Result	Impact		Heterochronic processes
				Range	Rhythm	
RPS	Forward migration	Superimposed >>	E>P	3	0	E hypermorphic or P progenetic
ROV	Increase	Superimposed >>	E>P	1	0	E hypermorphic or P progenetic
ROW	Complex	Superimposed >>	E#P	0–1	0	E hypermorphic or P progenetic
RD1	Decrease, then stability	Superimposed >>	E=P	0	0	E hypermorphic or P progenetic
AMB	Lengthening	Homologous >··>	E>P	0	3–4	E hypermorphic or P progenetic and E accelerated
RPP	Backward migration	Homologous >··>	E>P	2	1	E hypermorphic or P progenetic and E accelerated
RST	Complex	Homologous >··>	E>P	3	2–3	E hypermorphic or P progenetic and E accelerated
RPY	Complex	Homologous >··>	E>P	5	2–3	E hypermorphic or P progenetic and E accelerated
RD2	Decrease	Homologous >··>	P>E	1	5	E hypermorphic or P progenetic and P accelerated
RD3	Irregular decrease	Homologous >··>	P>E	1	2–4	E hypermorphic or P progenetic and P accelerated
RD4	Complex	Homologous >··>	P>E	3	3–4	E hypermorphic or P progenetic and P accelerated
RBC	No shared trend	Distincts ><	E≠P	—	—	Innovation

[a]The table recapitulates, for each feature listed in column 1 (see Fig. 3 for the calculation of these indexes), the morphological trend observed during the growth (column 2), the type of relationship between the ontogenetic trajectories of the two genera, with differences focusing on the range (>>), the rhythm (>··>), or the nature (><) of trajectories (column 3), the result of the divergence that corresponds to the achieved difference between the two genera in relation to the dynamic of the growth trend (=, identical; #, slighly different; >, more advanced than; E, *Echinosigra*; P, *Pourtalesia*) (column 4), the morphological impact of the divergence, with a distinction between the impact of range differences (column 5) and rhythm differences (column 6) (relative scales varying from 0 to 5), and the ontogenetic processes involved (column 7).

Amplitude Difference

Pourtalesia and *Echinosigra* differ only in the amplitude of their ontogenetic trajectories, which otherwise stay *superimposed*. The morphological implications of such a difference depend on the degree of allometry of adult growth. At one extreme, if the end of the growth is harmonious, the difference has no morphological effect and the two genera stay identical for the feature considered.

Rate Difference

Pourtalesia and *Echinosigra* also differ in the rate of their ontogenetic trajectories, which otherwise are the same. Their ontogenetic trajectories are similar, but not superimposed, and could be qualified as *homologous*. One of the genera (either *Echinosigra* or *Pourtalesia*) is thus accelerated and transcends the

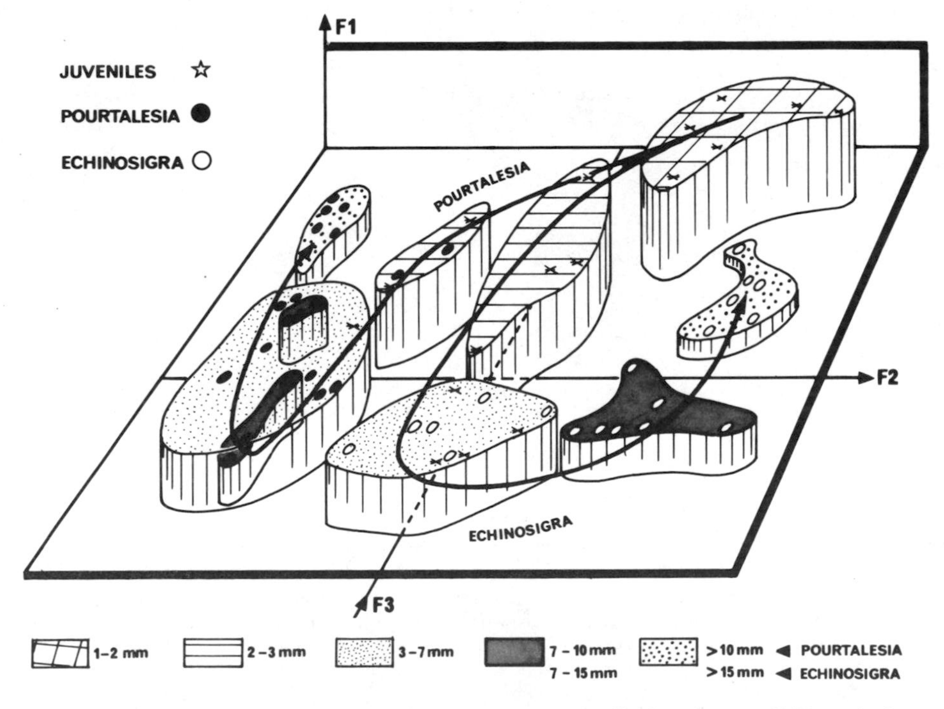

FIG. 6. Synthetic picture of the divergence between *Echinosigra* and *Pourtalesia.* Principal components analysis realized on calculated indexes; plotting of 56 specimens on the F1–F2–F3 factorial space (involving 91.7% of the total variance). The specimens have been gathered in five size ranges, and the two arrows represent the ontogenetic dynamics of each genus.

other by reaching a more advanced morphology. This type of heterochrony has been observed for most of the features. It results in generally pronounced adult morphological differences, especially since they can add to those induced by the preceding heterochrony.

Complex Difference

Finally, *Pourtalesia* and *Echinosigra* can differ by the nature of their ontogenetic trajectories, which appear to be *distinct*. Comparisons thus become more difficult and in this case the concept of heterochrony itself can be challenged.

To summarize, the whole divergence observed between the two genera is due to a mosaic distribution of these three categories of ontogenetic transformations shared among the features of the test. This mosaic pattern also has a quantitative dimension: the same heterochrony can vary in intensity, according to the features.

Multivariate analysis provides a synthetic picture of such a divergence which cannot be reduced to a single process (Fig. 6). Principal components analysis, calculated on the indexes, shows that the divergence between *Echinosigra* and *Pourtalesia* increases progressively along a double loop running on the factorial plan F2–F3, while decreasing along the axis F1. This synthetic picture shows that it is impossible to explain the deviation of the two branches by a global heterochrony.

GENERAL ONTOGENETIC COMPARISON: THE SEVEN GENERA

The foregoing statements underline the fact that the divergence between two taxa does not necessarily result from a single process, but can be seen as resulting from many heterochronies associated in a complex combination. On the basis of this result, the purpose of this section is to expand the field of comparison to all genera of the family (with the exceptions of *Ceratophysa* and *Helgocystis,* which are not sufficiently known). The comparisons extend to five major characters, or groups of correlated features, for which the ontogenetic interpretation is clear. The study will proceed in three steps.

1. To define a morphocline for each of the selected characters, i.e., to arrange sequentially the different states taken by this character in the family. This morphocline could be identified by reference to the ontogenetic data of *Echinosigra* and *Pourtalesia.*

2. If possible, to orient the morphocline by reference to other genera (by

applying the criterion of outgroup comparison), and thus to determine whether the sequence of states is pedomorphic or peramorphic. This step should allow relative evolutionary values to be attributed to the different states recognized in the first step.

3. To recapitulate the observed tendencies for the five characters in the seven genera.

The two first steps will be made simultaneously, character by character; the third will take the form of a synthesis.

Architecture of the Plastron

In Pourtalesiidae, the main alteration of the pentaradiate symmetry affects the plastronal area situated behind the peristome on the oral side. This alteration corresponds to a break of interambulacrum 5, the labrum being separated from the sternum by plates of posterior ambulacra joining together on the axis of symmetry (Fig. 7). This break is of variable length and corresponds to a more or less important architectural repatterning. Absent in the juveniles (Fig. 8, state A), it could be limited at the junction of one pair of ambulacral plates (Fig. 8, state B, and Fig. 7B), expand to two pairs of ambulacral plates (Fig. 8, state C), even to a third pair of ambulacral plates (Fig. 8, state D, and Fig. 7A), or as far as the interambulacral plates (Fig. 8, state E). States C, D, and E are often accompanied by an inhibition of growth of the labrum. This induces a supplementary disjunction due to the meeting of the peristomial ambulacral plates. Another characteristic is that, in all the observed cases of state D, the junction occurs between nonhomologous plates, because only one of the external ambulacral plates I.b.3 or V.a.3 reaches the axis of symmetry (Fig. 7A).

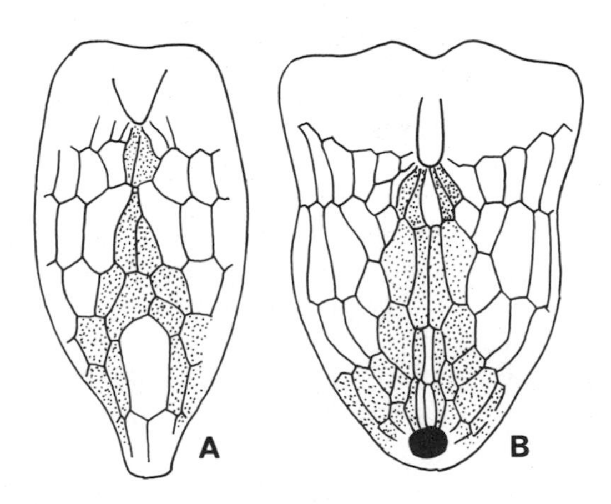

FIG. 7. Architectural diagrams of the adoral side of (A) *Pourtalesia alcocki* and (B) *Echinocrepis cuneata,* showing two kinds of break of I.A.5. Posterior ambulacra are stippled.

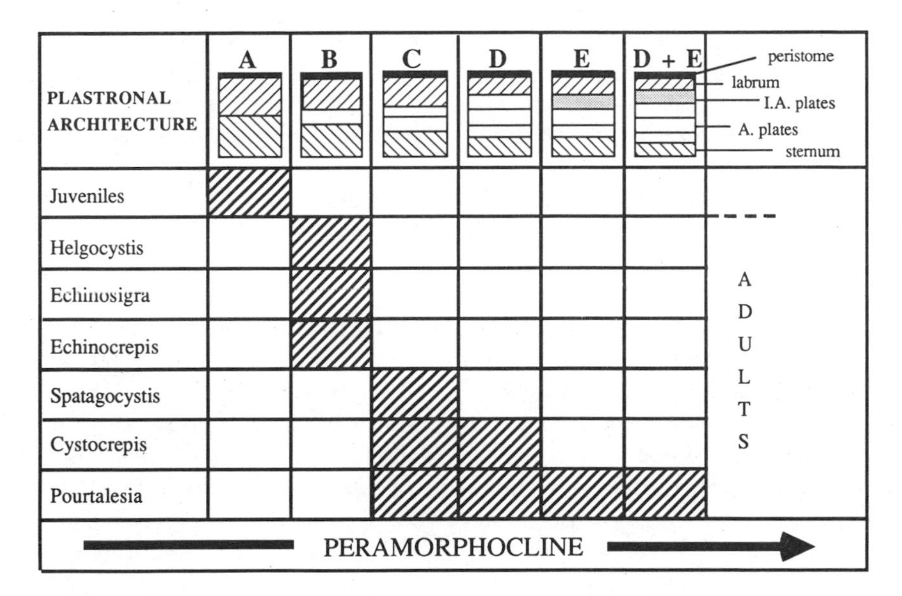

FIG. 8. Morphocline of plastronal architecture in Pourtalesiidae. Distinction among six discrete states (upper row) sequentially arranged, and lettered from A to D + E (states B and D are also illustrated in Figs. 7B and 7A, respectively). The ordinary states reached by adults are noted for each genus (hatched areas). The orientation of the morphocline, as inferred by outgroup comparison, is given (lower row). The figure includes only those genera whose character is sufficiently known.

This succession defines a morphocline which is going from continuous interambulacrum 5 to an interambulacrum 5 interrupted by a three- or four-plate series. This tendency appears peramorphic by comparison with other holasteroids showing the ancestral condition (architecture less or not altered). The ontogenetic process involved into this peramorphic pattern is probably an acceleration. Indeed, it cannot be a hypermorphosis, because of the relative small size of *Pourtalesia,* and it cannot be a predisplacement, because of the similarity between the young stages of *Echinosigra* and *Pourtalesia.* The adult forms of the different genera of the Pourtalesiidae family occupy the states B to (D+E) on this peramorphocline.

Posterior Profile of the Test

During the ontogeny of *Echinosigra* and *Pourtalesia,* the periproct rotates and becomes more and more vertical. The adjacent areas of the test begin to differentiate into particular structures: the hood above the periproct, and the rostrum underneath (Fig. 9, states A and B). At the adult stage of these two genera, the periproct is completely vertical, the hood slightly overhanging, and the rostrum bears a sharply delimited fasciole (Fig. 9, state C). In *Spatagocystis,* the periproct reaches a reverse position by rotating beyond the vertical. This rotating process is linked with a hyperdevelopment of the hood. Correlatively, the rostrum is reduced and shifts downward; laterally it bears a diffuse fasciole (Fig. 9, states D and E). *Echinocrepis rostrata* has a marginal periproct, a very prominent hood, and a small rostrum (Fig. 9, state F). *Echinocrepis cuneata* and *Cystocrepis* illustrate the achievement of the trend (Fig. 9, state G). The periproct is completely inframarginal, the extended hood makes up the whole back of the test, and the rostrum is reduced to a slight projection of the adoral side.

These few states define a morphocline which, by direct comparison with the sister group urechinids (genus *Plexechinus*), can be interpreted as being peramorphic. Figure 9 summarizes the repartition of the seven Pourtalesiidae along this peramorphocline. But this peramorphocline does not express real complete ontogenetic sequences, because some genera, such as *Echinocrepis* or *Cystocrepis* (reaching the extreme states of F and G), could not pass during their

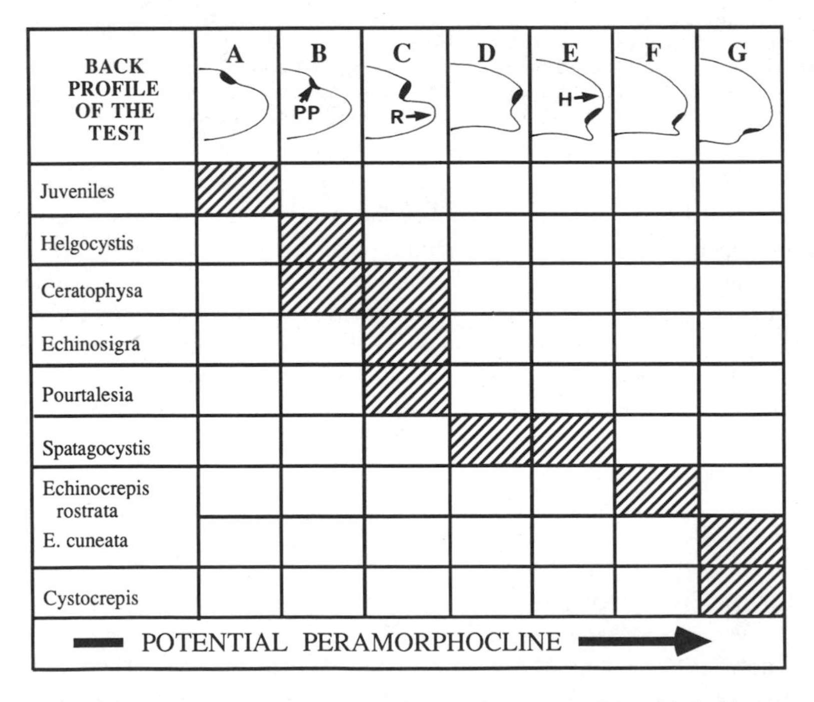

FIG. 9. Morphocline of the back profile of the test in Pourtalesiidae. Distinction among seven states (same symbols as in Fig. 8). PP, periproct; R, rostrum; H, hood.

development by all intermediary states of B–E. Indeed, these states could have been lost through a deletion or condensation process equivalent to a predisplacement. For this reason, the peramorphocline illustrated by Fig. 9 has to be considered as virtual. Early variations in the growth parameters of the rostrum and hood could explain such condensed developments: the common juvenile stage differentiates directly in states B–C, D–E, or F–G according to the value of these initial parameters.

This type of variation illustrates some limits of heterochrony (Alberch, 1985; Wake, 1989), and it might correspond to a process of early bifurcation

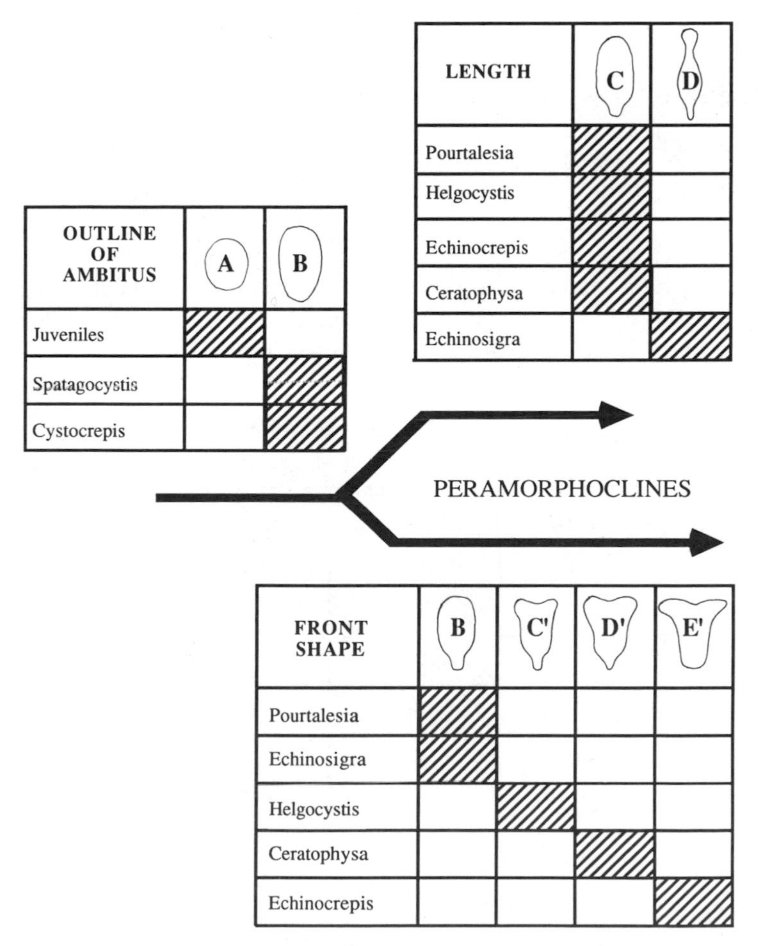

FIG. 10. Morphoclines of the outline of the ambitus in Pourtalesiidae. Distinction among seven states belonging to two trends. Same conventions as in Fig. 8.

inducing some directionality in further ontogenetic transformations (in the sense of Oster and Alberch, 1982).

Outline of the Ambitus (Excluding the Rostrum)

The trend leading to the very elongated test of *Echinosigra* (see above, analysis of the growth) could be taken as a reference to establish a morphocline of the outline of the ambitus (Fig. 10, states A–D). But genera *Helgocystis, Ceratophysa,* and *Echinocrepis* acquire a triangular ambitus by enlargement of the anterior part of the test, and do not fit this morphocline. It is therefore necessary to propose a second morphocline, to encompass all the encountered shapes (Fig. 10, state C′–E′). As for the preceding character, these clines can be interpreted as being peramorphic.

Sternum

The relative size of the sternum is very variable among genera: from "normal" to tiny. It is possible to construct a quantitative morphocline for such a character. Indeed, the sternum of the Pourtalesiidae is a more or less quadrangular plate the size of which can be estimated by the surface area of the ex-inscribed rectangle. Figure 11 gives a five-class assessment of the observations,

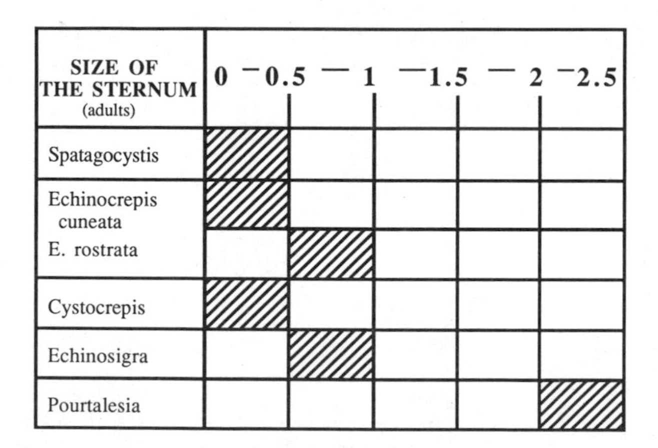

SIZE OF THE STERNUM (adults)
0 — 0.5 — 1 — 1.5 — 2 — 2.5
Spatagocystis
Echinocrepis cuneata
E. rostrata
Cystocrepis
Echinosigra
Pourtalesia

FIG. 11. Morphocline of the size of the sternum in Pourtalesiidae. Distinction among five states. The coefficient approximates the area of the sternum, each genus being equated with a standard length; it thus takes a relative value. Same conventions as in Fig. 8.

using a relative, unitless index. However, the trend expressed through these five classes cannot be oriented, because the outgroup comparisons do not provide sufficient reliable information. The different genera could thus be classified in one or another sense. But beyond this trivial problem of orientation arises a basic question: Is this type of classification really significant? Indeed, it is possible that the different genera may have completely independent ontogenetic trajectories for this character, any morphocline becoming then a purely formal and artificial construction. In general terms, the question must be asked regarding all non-orientable morphoclines, and convincing arguments have to be found to justify each of them.

In the current example, the high correlation between the size of the sternum and the downward shift of the periproct or the shape of the posterior profile of the test suggests that the constructed morphocline has a biological reality. It can thus be retained.

Orientation of the Peristome

During the development of *Pourtalesia,* the peristome rotates progressively within the anterior groove forming in ambulacrum III. It finishes, at the adult stage, by occupying a vertical or even a reverse position at the bottom of a deep invagination. This position is not achieved by all the Pourtalesiidae, and the degree of verticality of the peristome forms a morphocline, which, for convenience, has been divided into four states (Fig. 12, states A–D). This cline can be

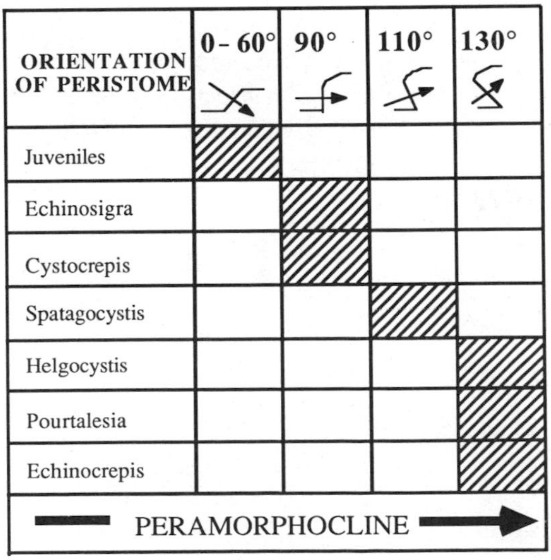

ORIENTATION OF PERISTOME	0 - 60°	90°	110°	130°
Juveniles				
Echinosigra				
Cystocrepis				
Spatagocystis				
Helgocystis				
Pourtalesia				
Echinocrepis				

PERAMORPHOCLINE

FIG. 12. Morphocline of the orientation of the peristome in Pourtalesiidae. Distinction among four states. The small arrows of the upper row indicate the outside of the peristomial opening. Same conventions as in Fig. 8.

viewed as peramorphic, the peristome being flush with the test in related genera *Plexechinus* and *Urechinus*.

Synthesis

The foregoing observations have been gathered together in Table III, where the different genera are arranged sequentially. In each morphocline, an increasing value has been attributed to each of them according to their position along the morphocline (the character "sternum" has been omitted from the table because it is insufficiently known). Globally, there is little consistency among genera: the order of the genera varies greatly according to the clines. This reflects the nature of the mosaic pattern of heterochronies in the whole family.

DISCUSSION AND CONCLUSION

Methodological Remarks

A number of lessons can be gained from the example of the Pourtalesiidae. They range from precise remarks to a general examination of the overall proceedings.

The Size Reference

It is always difficult to choose a size reference, particularly when ontogenetic comparisons are being made among different types of shape. Yet this is a crucial choice, since it affects the type of heterochrony identified. It is therefore necessary to use a size parameter (single feature, or as done here, a combination of features) which reflects a homology of growth among all the types of shape compared.

The Correlation between Characters

Organisms are complex entities, and comparing them is never an easy process. In order to obtain faithful restitution of differences, all attempts at comparison should be carried out by means of analyses dealing with a large number of characters. However, except in rare cases where general trends are evident, a heterochronic synthesis is often difficult to achieve. Indeed, the resultant trend cannot be regarded as a simple addition of elementary trends, since one never fully controls all the correlations between characters. It could sometimes become necessary to take in account partial correlation coefficients between

TABLE III. Summary of Trends in the Studied Morphoclines (Figs. 8–12)[a]

	Helgocystis	*Ceratophysa*	*Echinosigra*	*Echinocrepis*	*Spatagocystis*	*Cystocrepis*	*Pourtalesia*	
Cline 1, plastron	1	Unknown	1	1	2	3	4	Peramorphocline
Cline 2, back profile	1	1–2	2	4	3	5	2	Peramorphocline (potential)
Cline 3, ambitus	2, 2′	2, 3′	3, 1′	2, 4′	1	1	2, 1′	Peramorphocline (twice)
Cline 4, sternum	—	—	—	—	—	—	—	?
Cline 5, peristome	3	unknown	1	3	2	1	3	Peramorphocline

[a]For each cline, the different genera are numbered according to their relative position (numbering excluding the juvenile states, and cline number 4, which is insufficiently known). The unconformity between the different clines testifies to the mosaic pattern of characters in relation to evolution (see text).

features, and thus to introduce a quantitative aspect in the analysis of divergences. But that will increase the difficulties of interpretation and must be used with a great care.

The Reference of Comparison

Once the divergences are known, it remains to orient them. Such an orientation requires that two problems be resolved: first, the choice of a reference of comparison (outgroup); then the comparison of the ontogenetic trajectories with this reference. The choice of the outgroup must be done accurately, regarding the pattern of relationships between the studied group and the other taxa. Concerning the second point, there are few problems when the heterochronies affect only the rates of change in shape (acceleration and neoteny). But the solution becomes more hazardous when size or age must also be considered (hypermorphosis and progenesis; pre- and postdisplacement), because size is a rather unstable parameter, especially sensitive to the influence of local factors.

The General Framework

The purpose of this chapter has been to provide an assessment of the variation and diversity in ontogeny. But this type of analysis is part of a larger framework which leads to the study of relationships and cladistic analyses. The overall proceedings are organized according to three steps.

1. First, the study relies on a postulate which presupposes that the analysis is dealing with a set of related taxa. The relationships between these taxa could be supported by a character analysis (in the example of Pourtalesiidae, the seven genera share many apomorphies), or be simply assumed. But in both cases (facts or hypotheses), they allow the following comparisons to be made.

2. Second, the study includes the analysis of complete ontogenetic trajectories as well as the analysis of adult characteristics (analyses of diversity leading to the identification of morphoclines and to their assessment as pedomorphic or peramorphic). This second step, more than being just of descriptive interest, allows the interpretation of divergences in terms of alteration of the ontogeny. It therefore provides evidence for evolutionary processes generating variation and diversity. Furthermore, the degree of conformity between the compared ontogenies (estimated by the relative proportion of similar trajectories, namely superimposed and homogous trajectories) can provide a reliable proof of the validity of relationships admitted in the first step.

3. Finally, this methodology could lead to the testing or the determination of the relationships within the group under study. Such an attempt to introduce ontogeny into phylogenetic systematics has already been done by numerous

authors. Some of them, the so-called "pattern cladists," advocate that ontogenetic data should be primary for deducing homology and inferring sister-group relationships (Nelson, 1978; Nelson and Platnick, 1981). Others hold that the ontogenetic criterion, although basically reliable, is not always absolute and sufficient. It must be used critically (Fink, 1982; Kluge, 1985). The above example of a mosaic pattern of heterochronies strongly supports this latter claim in attesting that "ontogeny does not always provide unambiguous evidence of the hierarchy of genealogy." (Kluge, 1985, p. 24). Notably, other arguments must be used to clear up uncertainties. These arguments are of three types. (1) Use the principle of parsimony or a weighting of characters; this kind of argument gives rise to a major methodological problem which has been already largely discussed (Patterson, 1982; Novacek, 1987). (2) Increase the number of characters taken into account, especially by joining characters which are rather independent of ontogeny [namely, characters which do not change or change little during ontogeny, such as nonmetrical traits, which have been largely used in mammals (Berry, 1968, 1979; Sjovold, 1977)]. (3) Integrate palaeontological data, although the fossil record might be sometimes regarded as incomplete.

These three arguments can all provide additional data, sufficient to undertake the determination of relationships, and to obtain some evidence for phylogenetic transformation. Because the last two are not related to ontogeny, any congruence that they could share with characters of the mosaic would provide more powerful data about phylogeny. Moreover, they can be combined.

Remarks about Evolution

The wide diversity observed among the various extant adult forms of a taxon is a manifestation of their evolution, and every analysis, even partial, dealing with this diversity could provide significant evolutionary data. Within this framework, the analysis of ontogenetic trajectories, or more indirectly, adult comparisons using ontogeny, appear very helpful guides in identifying the processes underlying the evolution of these forms.

In the example of the Pourtalesiidae, the results show that the evolution and diversification of the family proceeded by means of a generalized peramorphosis. Moreover, such an evolutionary trend leads to extrapolated shapes, and supports well the idea of the morphological originality of the deep-sea organisms. This peramorphic tendency is clearly apparent with regard to the ontogenetic comparison between *Echinosigra* and *Pourtalesia,* where the differences between ontogenetic trajectories (homologous itineraries) always came from an acceleration (see Table II). This process manifests itself again in the comparison of the seven genera of the family, since all the orientable morphoclines turn out to be peramorphic. It is noteworthy that almost all the characters retained in this paper are peramorphic, and I do not know any other character capable of supporting a pedomorphic trend. Moreover, all the genera are to various degrees involved in peramorphosis (Table III). Given these results, all leading to recapitulative trends, one can readily talk about a generalized peramorphosis.

But this overall uniformity hides a more complex reality. Indeed, the mosaic pattern of characters supposes that each genus can be simultaneously peramorphic and aperamorphic according to the character under consideration (in the sense used by McNamara, 1982). That is, each genus could be viewed as a heterogeneous combination of more or less contradictory trends. None presents a global uniformity, with a majority of characters sharing the same degree of peramorphosis.

To summarize, two points must be emphasized. (1) The general history of the Pourtalesiidae appears globally peramorphic. This apparent uniformity is emphasized from a retrospective point of view. (2) The internal diversification of the family results from a complex evolution involving a differential distribution of the peramorphosis, depending both on the genera and on the characters. In view of these statements, I would like to highlight two subjects.

Challenge to the Direct Adaptive Significance of Clines

The mosaic pattern of characters allows the acquisition of analogous structures in different developmental ways. For example, in *Echinosigra* and *Pourtalesia,* the break of interambulacrum 5 reaches the same geometric size, even while having a different architecture. In other words, the same functional result, which is mainly the forward migration of the peristome, is achieved either by the acceleration of the position of the plates (architecture) in *Pourtalesia,* or by the acceleration of the shape of the plates in *Echinosigra*. This leads to a reconsideration of the intrinsic adaptive significance of a character such as architecture. The observed architectural patterns are probably by-products of ontogenetic pathways which lead primarily to an anterior, well-fitted, position of the mouth. Within the Pourtalesiidae, one may expect that there is a wide range of pathways to achieve similar adaptive structures. The differential degree of completion of architectural peramorphosis among genera of the family could only have been a possible way to explore this wide range of ways. From this assumption, the direct adaptive significance of pedomorphoclines and peramorphoclines could more generally be challenged. It could only be a secondary effect of developmental constraints. Such a reconsideration of adaptation is well supported by the critique of the "Panglossian paradigm" elegantly explained by Gould and Lewontin (1979).

Challenge to the Direct Use of Ontogeny in Cladistics

Put back in a cladistic context, each observed peramorphocline corresponds to a sequential spectrum of apomorphies, where the more advanced the state of a character, the higher will be the level of the apomorphy. Cladistic analyses of such multistate characters are all the harder to do, since characters are arranged in a mosaic pattern. This pattern is responsible for the weak congruence between peramorphoclines, thus inducing much homoplasy (reversion and convergence). To a certain extent, if the mosaic increases too much in complexity, "the signal (synapomorphy) could be swamped out by the noise (homoplasy)" (Novacek, 1987). Consequently, it will be very difficult to infer a hierarchy among genera, and thus to obtain some information about relationships. Yet, ontogenetic analyses provide theoretically a good line of evidence for relationship assumptions. But starting from a complex network of progressive apomorphies, it is really impossible to construct cladograms and to achieve phylogenies without using any other characters, namely characters not related to the ontogeny (see above, methodological remarks).

CONCLUSION

Nothing is easy! The current case, dealing with deep-sea echinoids, reveals that ontogenetic analyses of variation and diversity exemplifies this adage.

When a large set of characters is analyzed, it frequently appears that the heterochronic trends are arranged in a complex mosaic pattern. More generally, this suggests that the processes involved in the history of groups are complex, and one must avoid the failure of an oversimplification. That is, heterochronies are not a universal means to understand evolution. They have intrinsic limitations: some aspects of variation and diversity cannot be referred to heterochronies (such as the characters of the peristomial area, or the area of the sternum in Pourtalesiidae). They also have extrinsic limitations: the precise determination of heterochronies related to size is never easy because this parameter is very fluctuating (the choice between progenesis and hypermorphosis cannot be made in the comparison between *Echinosigra* and *Pourtalesia*). Particular limitations of the use of heterochronies occur when the taxa concerned share several different ontogenetic patterns. Dealing with a mosaic pattern, ontogeny fails to provide a reliable synthesis of divergences, and relationships have to be established using external parameters. In other words, ontogeny, if it remains necessary, is rarely sufficient to investigate thoroughly the diversity and to infer the phylogeny of related taxa.

These considerations emphasize several limitations of ontogenetic analyses.

They agree with some recent work stressing the nonuniversality of heterochronies (Oster and Alberch, 1982; Alberch, 1985; McKinney, 1988). On the other hand, they are in conflict with some dogmas leading to disputable generalizations. Nevertheless, I wish that these remarks might be understood not as rejecting ontogenetic analyses, but more as a genuine appeal in favor of their critical use.

Summary

This study is an attempt to test the limitations of heterochronies. It is done through the analysis of variation and diversity in deep-sea echinoids of the family Pourtalesiidae. The complexity of the processes involved in the diversification of the family is emphasized, and the limitation of use of ontogenetic studies stressed. In the first part, statistical analyses of ontogenetic trajectories of *Echinosigra* and *Pourtalesia* (the two main genera of the family) are carried out on 12 shape parameters. The differences recorded between these genera belong to different types, depending on whether they focus on range, rate, or mode of ontogenetic trajectories. Their synthesis provides some evidence for a mosaic pattern of heterochronies. In the second part, the seven genera of the family are simultaneously compared using five multistate characters. Morphoclines are characterized, and their pedo- or peramorphic nature is determined by outgroup comparisons. The whole comparison suggests a mosaic pattern of heterochronies. The evolution of Pourtalesiidae is complex. It has resulted from a global peramorphic trend, including a differential distribution of peramorphosis among genera and characters. The direct adaptative significance of some features is challenged, and the importance of homoplasy emphasized. In the light of this example, important methodological points are discussed: choice of a size parameter, orientation of heterochronic clines, integration of ontogenetic data into a cladistic frame. The conclusion stresses that ontogenetic analysis of heterochronies is limited, and cannot be used as a universal way of investigating diversity.

Acknowledgments

I am grateful to my colleagues J. L. Dommergues and B. Laurin for some valuable discussions about ontogeny, and to M. K. Hecht and K. J. McNamara for their generous help in preparing the English version of this manuscript. I am very obliged to D. S. M. Billett for his information on Pourtalesiidae. I wish to thank G. Paterson (British Museum) and the staff of the Centre National de Tri d'Océanographie Biologique (CENTOB) for making specimens available to me. I am also obliged to C. Fourcault and K. Sebedio for technical assistance. Oceanographic cruises were from IFREMER.

This research was supported by CNRS: it is a contribution to the program on ontogenesis and evolution of the Action Spécifique Programmée (ASP) Évolution.

REFERENCES

Agassiz, A., 1881, Report on the Echinoidea dredged by H.M.S. Challenger, during the years 1873–1876, *Rep. Sci. Results Voyage HMS Challenger Zool.* **3**:1–321.

Agassiz, A., 1904, The panamic deep sea echini, *Mem. Mus. Comp. Zool. Harv. Coll.* **31**:1–243.

Alberch, P., 1980, Ontogenesis and morphological diversification, *Am. Zool.* **20**:653–657.

Alberch, P., 1982, The generative and regulatory roles of development in evolution, in: *Environmental Adaptation and Evolution* (S. Mossakowski and G. Roth, eds.), pp. 19–36, Gustav Fischer, Stuttgart.

Alberch, P., 1985, Problems with the interpretation of developmental sequences, *Syst. Zool.* **34**:46–58.

Alberch, P., 1989, Orderly monsters—Evidence for internal constraint in development and evolution, in: *Ontogenese et évolution* (B. David, J. L. Dommegues, J. Chaline, and B. Laurin, eds.), *Géobios* (Mémoire spécial 12), in press.

Alberch, P., and Alberch, J., 1981, Heterochronic mechanisms of morphological diversification and evolutionary change in the neotropical Salamander, *Bolitoglossa occidentalis* (Amphibia: Plethodontidae), *J. Morphol.* **167**:249–264.

Alberch, P., Gould, S. J., Oster, G. F., and Wake, D. B., 1979, Size and shape in ontogeny and phylogeny, *Paleobiology* **5**:296–317.

Berry, R. J., 1968, The biology of non-metrical variation in mice and men, in: *The Skeletal Biology of Earlier Human Populations* (D. R. Brothwell, ed.), pp. 103–133, Pergamon Press, London.

Berry, R. J., 1979, Genes and skeletons, ancient and modern, *J. Hum. Evol.* **8**:669–677.

Bonner, J. T., ed., 1982, *Evolution and Development,* Berlin.

David, B., 1983, Isolement géographique de populations benthiques abyssales: Les *Pourtalesia jeffreysi* (Echinoidea Holasteroida) en Mer de Norvège, *Oceanol. Acta.* **6**:13–20.

David, B., 1985, Significance of architectural patterns in the deep-sea echinoids Pourtalesiidae, in: *Echinodermata, Proceedings of the Fifth International Echinoderm Conference, Galway, 1984,* (B. F. Keegan and B. D. S. O'Connor, eds.), pp. 237–243, Balkema, Rotterdam.

David, B., 1987, Dynamics of plate growth in the deep-sea echinoid *Pourtalesia miranda* Agassiz: A new architectural interpretation, *Bull. Mar. Sci.* **40**:29–47.

David, B., 1988, Origins of the deep-sea holasteroid fauna, in: *Echinoderm Phylogeny and Evolutionary Biology* (C. R. C. Paul and A. B. Smith, eds.), pp. 331–346, Clarendon Press, Oxford.

Dommergues, J. L., David, B., and Marchand, D., 1986, Les relations ontogenèse–phylogenèse: Applications paléontologiques, *Géobios* **19**:335–356.

Fink, W., 1982, The conceptual relationship between ontogeny and phylogeny, *Paleobiology* **8**:254–264.

Foster, R. J., and Philip, G. M., 1978, Tertiary holasteroid echinoids from Australia and New Zealand, *Palaeontology* **21**:791–822.

Gage, J. D., 1984, On the status of the deep-sea echinoids *Echinosigra phiale* and *E. paradoxa*, *J. Mar. Biol. Assoc. UK* **64:**157–170.
Gould, S. J., 1968, Ontogeny and the explanation of form: An allometric analysis, *J. Paleontol.* **42** (Suppl.):81–98.
Gould, S. J., 1977, *Ontogeny and Phylogeny,* Harvard University Press, Cambridge, Massachusetts.
Gould, S. J., 1984, Morphological channeling by structural constraint: Convergence in styles of dwarfing and gigantism in *Cerion,* with a description of two new fossil species and a report on the discovery of the largest Cerion, *Paleobiology* **10:**172–194.
Gould, S. J., and Lewontin, R. C., 1979, The spandrels of San Marco and the Panglossian paradigm: A critique of the adaptationist programme, *Proc. R. Soc. Lond. B* **205:**581–598.
Guerrant, E. O., 1982, Neotenic evolution of *Delphinium nudicaule* (Ranunculacea): A hummingbird-pollinated larkspur, *Evolution* **36:**699–712.
Kikuchi, Y., and Nikaido, A., 1985, The first occurrence of abyssal echinoid *Pourtalesia* from the middle Miocene Tatsukuroiso Mudstone in Ibaraki Prefecture, northeastern Honshu, Japan, *Annu. Rep. Geosci. Univ. Tsukuba* **11:**32–34.
Kluge, A. G., 1985, Ontogeny and phylogenetic systematics, *Cladistics* **1:**13–27.
Lovén, S., 1883, On *Pourtalesia* a genus of Echinoidea, *Kgl. Sven. Vet. Akad. Handl.* **19**(1881)**:**1–95.
McKinney, M. L., 1984, Allometry and heterochrony in an Eocene echinoid lineage: Morphological change as a by-product of size selection, *Paleobiology* **10:**407–419.
McKinney, M. L., 1988, Classifying heterochrony: Allometry, size, and time, in: *Heterochrony in Evolution* (M. L. McKinney, ed.), pp. 17–34, Plenum Press, New York.
McNamara, K. J., 1982, Heterochrony and phylogenetic trends, *Paleobiology* **8:**131–142.
McNamara, K. J., 1983, The earliest *Tegulorynchia* (Brachiopoda: Rhynchonellida) and its evolutionary significance, *J. Paleontol.* **57:**461–473.
McNamara, K. J., 1985, Taxonomy and evolution of the cainozoic spatangoid echinoid *Protenaster, Palaeontology* **28:**311–330.
McNamara, K. J., 1986, A guide to the nomenclature of heterochrony, *J. Paleontol.* **60:**4–13.
McNamara, K. J., 1987, Plate translocation in spatangoid echinoids: Its morphological, functional and phylogenetic significance, *Paleobiology* **13:**312–325.
Mironov, A. N., 1973, Nouvelle espèce abyssale d'oursins du genre *Echinocrepis* et diffusion de la famille des Pourtalesiidae (Echinoidea, Meridosternina), *Tr. P. P. Shirshov Inst. Okeanol.* **91:** 239–247 [in Russian].
Mortensen, T., 1907, Echinoidea (part 2), *Danish Ingolf-Expedition* **4:**1–200.
Mortensen, T., 1950, *A Monograph of the Echinoidea—Spatangoida,* Vol. 5, Part 1, Reitzel, Copenhagen.
Mortensen, T., 1951, *A Monograph of the Echinoidea—Spatangoida,* Vol. 5, Part 2, Reitzel, Copenhagen.
Nelson, G., 1978, Ontogeny, phylogeny, palaeontology and the biogenetic law, *Syst. Zool.* **27:**324–345.
Nelson, G., and Platnick, N., 1981, *Systematics and Biogeography. Cladistics and Vicariance,* Columbia University Press, New York.
Novacek, M. J., 1987, Characters and cladograms: Examples from zoological systematics, in: *Biological Metaphor and Cladistic classification* (H. M. Hoenigswald and L. F. Wiener, eds.), pp. 181–191, University of Pennsylvania Press, Philadelphia, Pennsylvania.
Oster, G., and Alberch, P., 1982, Evolution and bifurcation of developmental programs, *Evolution* **36:**444–459.
Patterson, C., 1982, Morphological characters and homology, in: *Problems of Phylogenetic Reconstruction* (K. A. Joysey and A. E. Friday, eds.), pp. 21–74, Academic Press, New York.
Sjovold, T., 1977, Non-metrical divergence between skeletal populations. The theoretical foundation

and biological importance of C. A. B. Smith's mean measure of divergence, *Ossa* **4**(Suppl.):1–133.

Smith, A. B., 1984, *Echinoid Palaeobiology,* Allen and Unwin, London.

Wake, D. B., 1989, Phylogenetic implications of ontogenetic data, in: *Ontogenese et évolution* (B. David, J. L. Dommergues, J. Chaline, and B. Laurin, eds.), Géobios, Memoire Spécial 12, in press.

9

Tempo and Mode of Morphological Evolution in Two Neogene Diatom Lineages

ULF SORHANNUS

INTRODUCTION

Ever since Simpson published *Tempo and Mode in Evolution* (Simpson, 1944) and *The Major Features of Evolution* (Simpson, 1953), there has been a substantial amount of work, especially during the past 10 years, on rates and patterns of morphological evolution (e.g., Eldredge and Gould, 1972; Kellogg, 1975, 1983; Kellogg and Hays, 1975; Bookstein *et al.*, 1978; Gingerich, 1976; Gould and Eldredge, 1977; Malmgren and Kennett, 1981; Hoffman, 1982; Hecht, 1983; Levinton, 1983; Malmgren *et al.*, 1983; Lazarus *et al.*, 1985; Hecht and Hoffman, 1986; Lande, 1986; Lazarus, 1986; Sorhannus *et al.*, 1988; Fenster *et al.*, 1989). Many of the studies that followed Simpson's pioneering work in evolutionary paleontology directly utilized as their basis his methods, observations, and ideas for further research. One of the major contentions of Simpson's work was that morphological transformations within lineages proceed gradually through time (phyletic evolution), eventually giving rise to new species. His findings were both a contribution to and in agreement with the modern synthesis, which states that "all evolution is simply an accumulation of small genetic changes guided by natural selection and that transpecific evolution is nothing but an extrapolation of events that take place within populations and species" (Mayr, 1963).

The concept of punctuated equilibrium proposed by Eldredge and Gould

---

ULF SORHANNUS • Department of Biology, Queens College of the City University of New York, Flushing, New York 11367.

(1972) claims that evolutionary change is associated with the origin of new species through genetic revolution (Mayr, 1963) in small allopatric populations and that stasis within lineages is prevalent. This pattern of distribution of evolutionary change within lineages is also asserted to predominate in the fossil record (Gould and Eldredge, 1977), thus refuting the importance of phyletic evolution in the speciation process as conceived by Simpson (1944, 1953). Consequently, the gaps in the fossil record are real and informative about rates of evolution. Another claim made by the proponents of punctuated equilibrium (Gould and Eldredge, 1977) is that speciation, the source of macroevolutionary variation, is qualitatively different from local adaptation within populations, thus requiring special macroevolutionary mechanisms; that is, microevolution is decoupled from macroevolution (e.g., Stanley, 1975, 1979; Gould and Eldredge, 1977).

As a result of the reconsideration of the tempo and mode of morphological evolution, a debate, which is still unresolved and actually based on few case studies, over the two extreme positions, phyletic gradualism and punctuated equilibrium, was triggered (e.g., Eldredge and Gould, 1972; Kellogg, 1975; Bookstein *et al.*, 1978; Gingerich, 1976; Gould and Eldredge, 1977; Malmgren and Kennett, 1981; Williamson, 1981; Hecht, 1983; Hecht and Hoffman, 1986; Kellogg, 1988; Hoffman, 1988; Levinton, 1988). Phyletic gradualism, which was Eldredge and Gould's (1972) conception of phyletic evolution, is a pattern of steady directional change in morphology resulting in the origination of new species. In this controversy punctuated equilibrium, which is a pattern of evolution where species stasis is punctuated by a geologically short period of time through which new species originate, and phyletic gradualism have usually been considered as exclusive alternatives without recognizing the possibility of various temporal combinations of the two extreme viewpoints. If punctuated equilibrium and phyletic gradualism represent two extreme patterns of evolution which are interconnected by a wide spectrum of transitional patterns and rates, it may be suggestive that these are merely a description of "fast" and "slow" phyletic change. On the other hand, if patterns of evolution are found to be explicitly separated into punctuated equilibrium and phyletic gradualism, that is, without any transitional patterns, or dominated by punctuated equilibrium, it may present a challenge to the modern synthesis in terms of it being at the most incomplete.

This work investigates whether the patterns of morphological evolution in two planktic Neogene diatom lineages are dominated by punctuated equilibria, phyletic gradualism, or by patterns and rates intermediate to these two extremes. This is done by fitting a set of hierarchical linear models, which represent a whole spectrum of patterns, to the data through a linear regression technique. The quantification of the morphological sequences through time are considered in the light of the null hypothesis of a random walk (Bookstein, 1987).

SOME LIMITATIONS OF THIS STUDY

The fossil record, especially the late Neogene and Quaternary microfossil record, is unique in that it provides data on evolutionary and ecological processes over greater time scales than those available to neontologists. On the other hand, the paleontological record also imposes restrictions on the types of problems that can be investigated with great confidence; this is largely due to the discontinuous nature of the preservation process and the difficulty of precise time correlation among localities (Jablonski *et al.,* 1986). Even though the deep-sea record of shelled microplankton has turned out to be rather promising in documenting the speciation process (Berggren and Casey, 1983), it still presents some difficulties, which I shall discuss in more detail.

Species Recognition

One of the major problems in this study on diatom evolution is that of identifying species; this problem obviously depends on the species definition used. Since the hypothesis of punctuated equilibrium and much of evolutionary theory are phrased in terms of the origination of biological species (Schopf, 1982), defined by Mayr (1963) "as groups of actually or potentially interbreeding populations in nature which are reproductively isolated from other such groups," it is imperative to examine some of the difficulties associated with using such a concept for this study of the tempo and mode of evolution.

Mayr's (1963) nondimensional species concept can most easily be applied to sympatric populations at a particular instant in time. The multidimensional species notion, which is a practical concept, must be based on the theoretical nondimensional species concept (Bock, 1986) when testing the concept of punctuated equilibrium.

Species among Bacillariophyceae are primarily recognized by structural features of the silica shell, and it is not known to what extent these features reflect the genetic makeup of the populations and thus biological species (Guillard and Kilham, 1977). There are two major problems with recognizing biological species in diatoms based on phenotypic information: first, the existence of cryptic sibling species, which may result in lumping of phenotypically similar, but reproductively isolated groups; second, to determine when intergroup morphological variability reflects reproductive isolation (Wood *et al.,* 1987). Moreover, it has been discovered that among some present planktonic diatom genera the cells of one species produce the characteristic valve structures of more than one species (Guillard and Kilham, 1977; Wood, 1959). The biological species concept due to Mayr (1963) applied to diatoms is also difficult due to the fact that some "species" reproduce only vegetatively—under such circumstances there exist no biological species (Bock, 1986)—as opposed to the predominant mode of alternating between asexual and sexual reproduction. The extant *Rhizosolenia bergonii* Peragallo, which has been included in this study, is known to reproduce sexually, since it forms auxospores (e.g., Cupp, 1943).

In this investigation I have recognized phyletic lineages in terms of temporal morphological sequences that were established through "vertical" comparisons. The morphospecies, which is recognized based on morphological distinctness or "gaps" validated through horizontal comparisons, has been employed to distinguish among morphological entities on a particular time level. Such distinct morphotypes may or may not reflect biological species.

Preservational Problems with *Rhizosolenia*

In all cases the middle part of the frustule of the investigated *Rhizosolenia* lineages has been broken off and a complete shell has never been recovered, except for the extant *Rhizosolenia bergonii* Peragallo. Consequently, the preserved condition of the valve largely determined the number of investigated variables as well as the arbitrary choice of measurements, which in this case was restricted to three (Fig. 1). This represents a serious limitation to this study, since such a few variables may not reflect biospecies evolution in these two *Rhizosolenia* lineages.

Geographic/Ecophenotypic Variation

Both Gould and Eldredge (1977) and Bookstein *et al.* (1978) emphasize geographic variation in studies of the tempo and mode of evolution because the presence of spatial variation in combination with directional migration will potentially obscure local patterns of temporal change in morphology. Consequently, such a phenomenon may give rise to false stasis, gradualism, and punctuations in the investigated traits and must be considered to be a substantial problem. Many workers have drawn their conclusions about tempo and mode based on data taken from a restricted area within the geographic range of a particular lineage (e.g., Kellogg, 1975; Malmgren *et al.*, 1983; Gingerich, 1976; Williamson, 1981). Even though migration of species, morphological clines, and geographic variants present an uncertainty in distinguishing "real" evolutionary patterns from spurious ones, the paleomagnetic reversal record, which forms worldwide synchronous time "slices" of the deep-sea cores, enables us to get an estimate of geographic variation at different time levels.

If there is a specific pattern of geographic variation present, one has to be concerned about whether directional migration took place and the effects it might have had on temporal patterns of change in morphology. Directional migration can be inferred if there is a stepwise directional pattern of first appearances which can be determined to be significantly different within the limits of our resolution in at least three widely separated sites. After satisfactorily demonstrating directional migration, the impacts of geographic variants on temporal evolution may or may not be substantial; this is virtually impossible to show and will thus remain a problem when drawing conclusions about tempo and mode of evolution. However, Bookstein *et al.* (1978) have pointed out that a relative increase in the character variances at consecutive time levels, in addition to bimodal/multimodal frequency distributions, are expected in sites affected by directional movement of geographic variants, assuming no change in morphology during migration. This approach suggested by Bookstein *et al.* (1978) is not sufficiently conclusive because a temporal increment in character variances and the presence of bimodal/multimodal frequency distributions may be a result of other phenomena, such as the origination of polymorphism, relaxation of the selection pressure, and/or aggregation in conjunction with directional evolution or any combination of the above.

The Stratigraphic Record

Studies on tempo and mode in evolution utilize data extracted from the fossil record, which is generally incomplete with respect to fine time scales; this presents problems with interpreting morphological sequences and may in some cases lead to erroneous conclusions. Consequently, the presence of gaps may obscure gradual evolution and give rise to a false impression of a punctuated event (e.g., Bookstein *et al.*, 1978; Levinton and Simon, 1980). For instance, Raup and Crick (1982) discovered in their analysis of character evolution in *Kosmoceras* that gaps in morphology are associated with sedimentary breaks more often than expected by chance, although the majority of discontinuous sediments are not linked to gaps in morphology. The existence of gaps in the fossil record may be a result of missing sediments (Schindel, 1980, 1982; Dingus and Sadler, 1982) which is the outcome of the discontinuous nature of the sedimentation process or nonpreservation or nondeposition of fossil material. Additionally, Schindel (1982) pointed out that "most profound breaks" should be interpreted as periods during which the habitat deteriorated, resulting in migration and subsequent recolonization at the return of hospitable conditions. Methods for assessing the completeness of stratigraphic sections at a particular level of resolution have been developed in order to evaluate results of studies on the tempo and mode of evolution (Sadler, 1981; Dingus and Sadler, 1982;

Schindel, 1982). Even though these methods are rather crude, McKinney and Schoch (1983) have demonstrated that failure to use them may give rise to misinterpretations of data. Due to the discontinuous nature of fossiliferous sediments, many workers (e.g., van Andel, 1981) have questioned the utility of evolutionary patterns that are derived from the record; this is without doubt one of the most serious limitations imposed by the stratigraphic record on studies concerned with tempo and mode. However, Dingus and Sadler (1982) pointed out that testing of neontological concepts in the biostratigraphic record requires organisms with a dense fossil record, long generation times, and sections which are fairly recent and have high rates of sediment accumulation. Unfortunately, diatoms have a relatively short generation time, but otherwise they fulfill the above requirements. Sampling and resampling of the same population in different places, if possible, gives confidence to the data if patterns are "similar" in all the localities.

Another major problem encountered when using stratigraphic data in evolutionary studies is time averaging, that is, aggregation of fossils resulting in the accumulation of thousands of generations into a particular stratum. Levinton and Simon (1980) indicated that the phenomenon of aggregation tends to obscure short-term trends as well as evolutionary patterns in general. Similarly, Bookstein *et al.* (1978) discussed the problem of inferring a gradual, punctuated mode of change and stasis from data pooled over several thousand generations. According to Bookstein *et al.* (1978), hidden randomness in data increases as a function of aggregation, and the hypothesis of punctuated equilibrium should not be applied to highly aggregated data, while gradualism may be demonstrated in situations where a series of samples with overlapping frequency distributions and with rather constant variance show directed change. Likewise, stasis cannot be inferred unambiguously based on highly accumulated fossil data. For instance, a morphological sequence that first appears to be in equilibrium may be taken for a gradual trend as a longer time span is sampled (Bookstein *et al.*, 1978). Mixing of marine sediments by burrowing organisms as well as by diagenetic (compaction) and environmental factors (rate of sedimentation and the turbulence level) influence the vertical position of fossils and represents one of the primary causes of time-averaged fossiliferous sediments, especially in shallow shelf deposits (Fursich, 1978). Diatoms generally have a short generation time, ranging from 4.4 hr in small forms to 24 hr or more in very large species (Werner, 1977); this will give rise to highly aggregated assemblages, since the fossils are not preserved at such a fine scale of resolution due to the previously mentioned factors. Consequently, if the rate of evolution is measured as change from one generation to another (Bookstein *et al.*, 1978), inferences made about mode of evolution in diatoms may be unreliable. An additional difficulty in conjunction with aggregation, which specifically pertains to diatoms, is ontogenetic variation. Since most diatom species go through size diminution as they divide asexually, also called the Macdonald–Pfitzer rule, it could lead to false patterns of change, especially in a situation of differential preservation of cell sizes at separate points in time.

The paleomagnetic reversal record, which has been dated independent of the organisms through radiometric techniques, forms the basis for constructing the time axis in my study. The major drawback with this procedure is that the temporal horizon of samples located between the magnetic reversals is calculated through linear interpolation without considering variation in the sedimentation rate. Consequently, the exact temporal position of such samples must be interpreted with caution.

Quantification of Morphological Sequences

Bookstein (1987; personal communication) points out that in studies of tempo and mode in evolution the null hypothesis of a random walk, which is the "simplest" explanation for a temporal sequence of means, must be rejected before "rates" of evolution can be "measured." The reason for this is that there exists no underlying forcing function if the null hypothesis is accepted, but only a step variance. From a purely mathematical point of view "an evolutionary rate is a derivative of some quantitative feature with respect to time and such derivatives don't exist for random walks" (Bookstein, 1987). Since a random walk may imitate trends, punctuations, and stasis, we are always faced with the null hypothesis as an alternative explanation for a temporal sequence of morphological changes. Bookstein (1987, 1988) has developed a statistical test (Range Test) which can distinguish trends, punctuations, and stasis due to random walks from those that are nonrandom. The question that arises when the null hypothesis is accepted is whether the observed time sequence is a result of a random walk or a deterministic process. The reason for this is that a random walk does not always imply random morphological changes in a population; in fact, the changes may be deterministic, but the "causal" factor(s) may vary randomly (Fisher, 1986). Moreover, Bookstein's suggestions must be interpreted within a biological framework, since populations of organisms must be "functional" in their environment and are thus ultimately restricted as to what is morphologically possible at a particular point in time. In other words, a biological system cannot be truly random in the sense of a sequence of coin flips, but it may be random within "restrictive boundaries." For instance, Raup and Crick (1981) discussed a model in which a random walk occurs within selective boundaries and M. Hecht (personal communication) suggested that the same phenomenon may take place within developmental boundaries. Since inferences about random morphological variation present in temporal and spatial fossiliferous sequences seem to be a controversial issue, it will present a problem in interpreting morphological sequences in the fossil record. Nevertheless, in this work I have used the Range Test developed by Bookstein (1987) for evaluating the validity of the null hypothesis of random walk for explaining the time series.

Many investigations on tempo and mode of morphological evolution have been performed without any definition of punctuated equilibrium, stasis, and phyletic gradualism (e.g., Gingerich, 1974, 1976; Kellogg, 1975; Ozawa, 1975; Williamson, 1981). Consequently, one worker's punctuated equilibrium has been another's phyletic gradualism [e.g., Gould and Eldredge (1977) and their interpretation of Kellogg's data] and vice versa. To test for the presence of punctuated equilibrium, phyletic gradualism, and stasis as patterns we need a "rule of thumb" to follow in order to avoid chaos. For instance, Gould (1982) defined "geologically instantaneous" as 1% or less of later existence in stasis. The problem with this is that lineages characterized by long duration will be more prone to show punctuated patterns than the ones with relatively short temporal distributions.

An alternative approach was taken by Bookstein *et al.* (1978), who defined punctuated equilibrium, phyletic gradualism, and stasis from a statistical point of view. Thus, instead of testing either model, they proposed to estimate jointly the comparative contribution of punctuated, phyletic change and stasis to explaining the variance in the data. This can be done by fitting a set of hierarchical linear models which describe phyletic gradualism or punctuated equilibrium or any temporal combination of these to the raw data through piecewise linear regression. The most parsimonious model or the one that explains the most variance with the fewest number of parameters is considered to be the "best" fitting linear representation of evolution. This technique has been employed in this investigation of the tempo and mode in evolution, since the advantage of this procedure is that it makes it possible to classify rather objectively morphological sequences as being either static, gradual, or punctuated. However, these models simply ignore nonlinearities, thus assuming that nonzero rates are constant. According to Bookstein (1975), nonlinear data may be fitted through polynomial regression. Such a technique is not preferable, since it brings about drawbacks, such as "susceptibility to systematic regional biases, faulty behavior at the limits of the predictor range and the creation of artificial mathematical features (extrema, points of inflection)" (Bookstein, 1975). To the contrary, piecewise linear regression requires that the worker detects "breaks" in the data where the trend appears to bend. The position of these "breaks" may be inferred by visually inspecting the distribution of the data (Bookstein, 1975). Thus, a morphological time series characterized by sections with a distinct direction of change justify a subdivision of a time sequence into subseries (Bookstein, 1975).

MATERIAL AND METHODS

Temporal and Geographic Distribution of *Rhizosolenia*

One of the lineages I have examined in this investigation of tempo and mode is the fossil form *Rhizosolenia praebergonii* Mukhina—a centric diatom, which belongs to the family Rhizosoleniaceae—which is widely used for biostratigraphic correlation in upper Pliocene and lower Pleistocene sediments of the equatorial Pacific Ocean and Indian Ocean (Mukhina, 1965, 1969; Burckle and Opdyke, 1977; Burckle, 1972, 1978; Sancetta, 1982; Barron, 1985). The Neogene geographic distribution range also incorporated the northwest Pacific Ocean (Koizumi, 1968; Koizumi and Tanimura, 1985), the north Atlantic Ocean (Schrader, 1977; Baldauf, 1984), and the equatorial and subtropical Atlantic Ocean (E. Fourtanier and C. Sancetta, personal communication). The first appearance of *R. praebergonii* in the equatorial Pacific Ocean was in the middle of the Gauss Chron at 3.1 million years ago (Mya) and it disappeared in the early Pleistocene near the middle of the Olduvai Subchron or 1.6 Mya (Burckle and Trainer, 1979). In higher latitudes of the Pacific Ocean the temporal distribution range becomes progressively shorter, the first appearance datum levels ranging from upper Gauss to lower Matuyama (approximately 2.7 to 2.3 Mya) and the last appearance from lower Matuyama Chron to just above the Olduvai event (approximately 2.2 to 1.7 Mya) (Burckle *et al.*, 1985). In the equatorial Indian Ocean *R. praebergonii* appeared later than in the equatorial Pacific Ocean in upper Kaena Subchron (approximately 2.9 Mya) (Burckle and Opdyke, 1977).

Rhizosolenia bergonii Peragallo, which is an extant diatom species among *Rhizosoleniaceae* with a relatively good fossil record, is included in this analysis because it is thought to be the ancestral form of the previously discussed lineages. This is a relatively large celled diatom species which rarely shows high abundances in modern diatom assemblages (Guillard and Kilham, 1977), and it also seems to fluctuate in abundance in the fossil record. Both at present and in the past *R. bergonii* has been characterized by having a geographic distribution incorporating the tropical and subtropical regions of the Pacific, Indian, and Atlantic Oceans (e.g., Guillard and Kilham, 1977; Cupp, 1943). The temporal range extends from early Miocene to present (L. H. Burckle, personal communication).

Morphological Variables Studied

Permanent slides were prepared after the method of Schrader (1974). The sampling interval in this study ranged from 10 cm to 1 m. Due to the discontinuous nature of the sedimentation process, the samples are not evenly distributed in time; where possible, the minimum sample spacing is on the order of 0.03 million years, while the maximum spacing is approximately 0.3 million years. Core samples, which ranged from 1 to 2 cm^3 in volume, thus averaging 0.003 million years of sediment, were broken down in 10% HCl to remove $CaCO_3$ and organic material. Each sample went through a series of six washes with distilled water at 90 min intervals, after which 3–5 drops of the residue was pipetted onto a coverslip. When the water had evaporated, a process that was speeded up by putting the coverslip onto a moderately heated plate, the coverslip was fixed onto a microscope slide by using Permount. Before the samples were studied the compound was allowed to dry for a couple of days.

At each investigated site approximately 30 specimens, which were chosen randomly, per sample were measured and on each specimen the following measurements were taken with an ocular micrometer at a magnification of 1000×: (1) length of apical process (AL), (2) width of the valve 8 μm below its apex (WC), and (3) height of the hyaline area at the apex of the valve (HH) (Fig. 1). Since one micrometer unit equals 0.4 μm in this study, the variables were measured to the closest 0.4 μm. The choice of variables was arbitrary, although it was largely dictated by the preserved condition of the valves of the two *Rhizosolenia* lineages (Fig. 1). In all cases, the middle part of the frustule has been broken off; the complete shell has never been recovered, except for the extant *R. bergonii*.

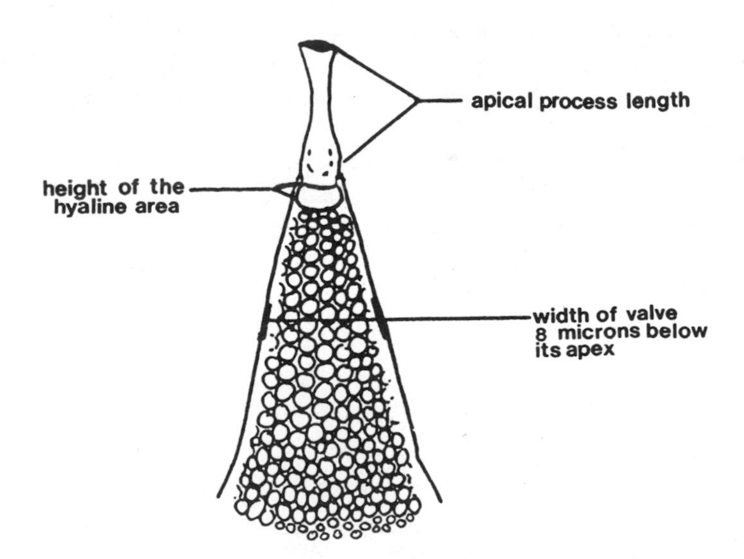

FIG. 1. Schematic diagram of the genus *Rhizosolenia* showing the measurements that were taken on the valve: apical process length (AL), width of the valve (WC), height of the hyaline area (HH).

The Cores

The investigated lineages were recovered from Deep Sea Drilling Project (DSDP) and Lamont–Doherty Geological Observatory (L-DGO) cores, which have been taken from all parts of the ocean. The coverage of the world ocean by deep-sea cores is complete in the sense that all biotic provinces are represented. In piston cores as opposed to rotary cores the earth's paleomagnetic reversal record can be identified and tied to an absolute time scale; this is done by correlating the polarity sequence of the deep-sea core to terrestrial volcanic rocks in which the reversal events have been dated through the potassium–argon method. Sites that produced rotary cores that lack a reliable magnetic record can be tied to cores for which absolute dates are available, through oxygen isotope and calcium carbonate data as well as biostratigraphic markers. Each magnetic reversal is a globally synchronous event which makes correlation between sites far away from each other possible and more reliable. The temporal position for samples located between magnetic reversals was calculated through linear interpolation using the revised geomagnetic time scale of Berggren *et al.* (1983). Thus, the construction of the temporal axis in this investigation assumes a constant sedimentation rate. Marine cores generally have a continuous sedimentation, allowing samples to be taken from a continuous but time-averaged fossil record.

The deep-sea cores used in this study (Fig. 2) were chosen based on several criteria, such as the presence of a good paleomagnetic reversal and stable isotope record as well as the abundance of the investigated lineages throughout their known geographic and temporal distribution range. Moreover, several sites were deliberately selected within the same water masses rather close to each other; this facilitates comparisons of patterns of evolution among locations (Fig. 2). The Atlantic cores were omitted in this study since the investigated lineages have a poor fossil record there. Consequently, all the deep-sea sections considered are located in the equatorial upwelling zone of the Indian and Pacific Oceans, a zone characterized by having a relatively good Neogene and Quaternary microfossil record and time control (Fig. 2).

DSDP site 157 is located within the Peru water mass (1° 45′ S; 85° 54′ W). Since this core was obtained by rotary drilling from the southern edge of the Panama Basin on the Carnegie Ridge, there is no paleomagnetic reversal record available. Consequently, the temporal position of the samples from site 157 is inferred using correlation of this core with the central Pacific sites (RC12-66 and DSDP site 573, which have good magnetostratigraphy) by means of chemostratigraphy (calcium carbonate concentration) and various biostratigraphic indicators (Burckle, 1978). The average sedimentation rate of this site throughout the temporal scope considered was approximately 76 m/million years, which is relatively high in comparison to the central Pacific sites.

A core that is also closely associated with the Peru water mass even though

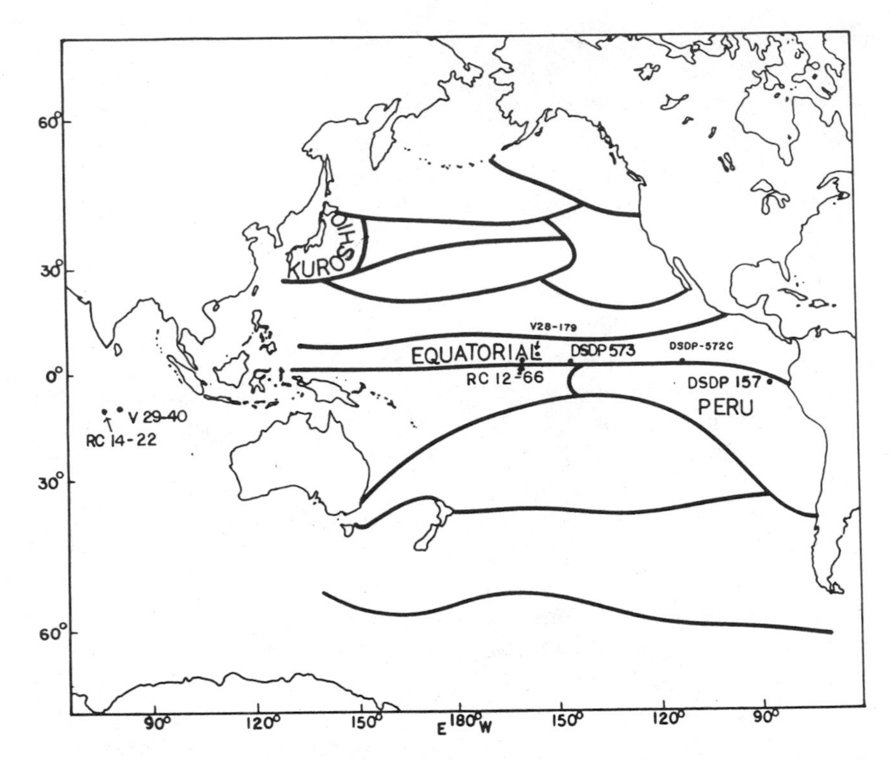

FIG. 2. Locations of the sites sampled and their position with respect to the water masses.

it is within of the central equatorial Pacific upwelling zone is DSDP site 572C; this site is an intermediate (1° 26′ N, 83° 43.9′ W) on an east–west transect between the central equatorial and eastern equatorial cores (Fig. 2). Since the chronology of magnetic reversals is not directly available in this deep-sea section, the dating at site 572C has been done through correlation with the paleomagnetics in L-DGO core V28-179 and DSDP site 573 in the central Pacific using calcium carbonate concentration (Prell, 1985). This site has a moderate sedimentation rate for the time duration considered in this study, an average rate of approximately 16 m/million years.

DSDP site 573 was taken from beneath the central equatorial water mass (0° 29′ N, 133° 18′ W) of the Pacific upwelling system (Fig. 2). Paleomagnetic data are directly available because site 573 is a hydraulic piston core; thus, the dates are generated directly from the revised geomagnetic time scale of Berggren *et al.* (1983). The average sedimentation rate of the late Pliocene and early Pleistocene sections is 14.5 million years, which is lower than the rates in the easternmost sites.

Both V28-179 (4° 37′ N: 139° 36′ W) and RC12-66 (2° 36′ N, 148° 12′ W) are L-DGO cores, which are located to the northwest of DSDP site 573 in the equatorial water mass, not far apart from each other (Fig. 2). In these hydraulic piston cores a good isotope and magnetic record has been recovered and has been worked out by Shackleton and Opdyke (1977) for V28-179 and Foster and Opdyke (1970) for RC12-66; the absolute times of the reversals were adopted from the Berggren *et al.* (1983) time scale. The average sedimentation rate was rather low for both V28-179 and RC12-66 throughout the sampled section, approximately 3 m/million years for both cores.

The westernmost sites considered in my investigation were the L-DGO cores V29-40 (10° 29′ S, 78° 3′ E) and RC14-22 (11° 27′ S, 75° 9′ E) from the equatorial upwelling region of the Indian Ocean (Fig. 2). The paleomagnetic and biostratigraphic data for both V29-40 and RC14-22 sites have been reported by Burckle and Opdyke (1977). The time axis was generated from the Berggren *et al.* (1983) revised geomagnetic time scale. Since a large portion of V29-40 is missing across the Pliocene/Pleistocene boundary, RC14-22 was chosen to cover this part of the temporal distribution range of *R. praebergonii* and *R. bergonii*. Due to the fact that these cores are located close to each other (Fig. 2), the data have been combined into a single section as if they were collected from the same site; thus, in V29-40, *R. praebergonii* and *R. bergonii* were sampled from the Upper Mammoth Subchron to the Matuyama/Gauss boundary and in RC14-22 from the Matuyama/Gauss boundary up to the Lower Olduvai Subchron. The average sedimentation rate for both cores was of the same magnitude as in the central Pacific sites, or 4 m/million years for both V29-40 and RC14-22.

ANALYSIS

Ancestor–Descendant Relationships

Before one addresses questions about tempo and mode of evolution, one has to reconstruct phylogenetic relationships among the lineages that are being investigated, since punctuated equilibrium and phyletic gradualism are hypotheses about modes of lineage origination. The use and importance of paleontological versus neontological data as well as the kind of methodology used in reconstructing phylogenies have been, and still are, celebrated issues in systematics (e.g., Schaeffer *et al.*, 1972; Cracraft and Eldredge, 1979; Lazarus and Prothero, 1984). The debate has primarily been in reference to the vertebrate and macroinvertebrate fossil records. According to Lazarus and Prothero (1984), most deep-sea microfossils lack a sufficient number of hierarchically nested sets of characters and show frequent convergence as well as iterative evolution, thus making complex phylogenetic inferential methods ineffective. Since iterative evolution occurred in *R. praebergonii* (U. Sorhannus, E. Fenster, A. Hoffman, and L. Burckle, unpublished) as well as due to the limited number of characters used in this study, a cladistic approach to inferring the evolutionary relationships would be unreliable according to the principles outlined by Lazarus and Prothero (1984, and references therein). To the contrary, our stratigraphic data seem to fulfill the criteria for reliability (Lazarus and Prothero, 1984), since the morphological sequence of *Rhizosolenia* is found in the same temporal order in all the investigated localities, as well as due to good geographic control and preservation of the examined lineages. Therefore, the potential ancestor–descendant relationships between *R. praebergonii* and *R. bergonii* should be interpreted by close sampling, rather than through a cladistic approach, of temporally separated populations. Such sampling is made possible by the high quality of the deep-sea stratigraphic record. However, this approach requires three assumptions: (1) the direct ancestor has been preserved, (2) the direct ancestor will be included in the samples, and (3) morphological data, which delimit phena, are appropriate for phylogenetic analysis. As pointed out by Prothero and Lazarus (1980), given these assumptions, ancestors can be recognized when the hypothetical ancestor can be inferred to be older than its descendant and when all possible ancestral populations are sampled. This converts the problem of the recognition of ancestors to a problem of sampling. Through the efforts of various deep-sea drilling programs, there is now a large number of sites available that cover a broad geographic area. Thus, it is possible to sample most of the potential ancestral populations as well as descendant lineages through time over their geographic range. Moreover, the chronostratigraphy of many of these sites can be reliably determined by three independent correlative methods: biostratigraphy, chemostratigraphy, and magnetostratigraphy. In addition, radiometric dating of magnetostratigraphy makes possible the calibration of evolutionary events in absolute time as well as positioning of samples in time.

Given this reasoning, the potential ancestor–descendant relationships between *R. praebergonii* and *R. bergonii* were investigated through sampling rather closely spaced temporal populations. *R. praebergonii* primarily occurred in the equatorial region of the Pacific, Atlantic, and Indian Oceans and it first appeared in the equatorial Pacific in the middle of the Gauss Magnetic Chron as a transitional form (Burckle and Trainer, 1979). However, 200,000 years later, *R. praebergonii* appeared in the Indian Ocean. Consequently, the ancestor–descendant analysis was restricted to the equatorial Pacific region of its geographic, range ensuring that all possible ancestral populations were sampled. *R. praebergonii* was morphologically closer to *R. bergonii* than to any other *Rhizosolenia* lineage present at its first appearance in the five examined locations (Figs. 5–9); this supports the existence of phylogenetic affinity between the two lineages. Moreover, an evolutionary relationship between these two lineages is supported on their relative temporal positions in the stratigraphic record; that is, *R. bergonii* occurred prior to, during, and after the origination of *R. praebergonii*. In addition, after the first appearance, AL and HH of *R. praebergonii* started to diverge substantially from the presumed ancestral form, while WC did not indicate such a sharp separation from *R. bergonii* (Figs. 5–9). Based on the above arguments, it can be concluded that *R. praebergonii* originated from *R. bergonii* in the equatorial region of the Pacific Ocean and, more importantly, that the criteria presented in Prothero and Lazarus (1980) for testing the hypothesis of an ancestor–descendant relationship have been satisfied. Based on similar criteria as above, *R. praebergonii* may also have evolved from *R. bergonii* in the Indian Ocean.

Random Walk

In studies of tempo and mode in morphological evolution the appropriate null hypothesis is a random walk, since it is the "simplest" explanation for a temporal sequence of mean values (Bookstein, 1987, and personal communication). Consequently, this null hypothesis should be tested and if it is not rejected, the results are very difficult to interpret. This is because failure to reject the null model does not necessarily imply random morphological change in a population. In fact, shifts in the relative frequencies of morphotypes in a population may very well be deterministic, since the causal factor(s) may very randomly. However, if the null model is rejected, stasis and anagenesis are deviations from the central random walk model in "opposite directions" (Bookstein, 1987). Bookstein (1987) has developed a statistic from the mathematical literature of random walk which can be applied to empirical data, such as mine. Consequently, this Range Test will be used to investigate the presence of temporal random walks in the mean values of AL, HH, and WC of the three *Rhizosolenia* forms. The "reduced speeds" in the equation that follows form the basis for calculating the test statistic x. According to Bookstein (1987), the reduced speeds (ratio), which is the expected deviation from the starting size (in μm) in the examined character under a random walk over a fixed time interval of 1 million years is calculated based on the following equation:

$$\frac{|S_{n_k} - S_{n_{k-1}}|}{(n_k - n_{k-1})^{0.5}}$$

$S_{nk} - S_{nk+1}$ is the absolute difference in morphology between two subsamples, which is scaled by the square-root of the elapsed time, or $(n_k - n_{k+1})^{0.5}$, between the two subsamples.

The upper 5% of the distribution of x, that is, values above 2.25, is in this study considered to represent change improbably large for being due to a random walk alone, and consequently an evolutionary interpretation of anagenesis. Likewise, the lowest 5% of the distribution of the test statistic x, or 0.62 represents temporal variation which is improbably constrained, thus corresponding to stasis. The null model of random walk separates these two findings by an interval incorporating values between 2.25 and 0.62. The procedures for calculating the test statistic x and the corresponding probabilities are taken from Bookstein (1987).

In every one of the three investigated regions, (the Indian Ocean core, core RC12-66, in the central Pacific Ocean, and DSDP site 157 in the eastern Pacific Ocean), AL in *R. praebergonii* was found to have x values ranging from 2.49 to 5.227 with corresponding probabilities between 0.976 and 0.99999 (Table I). Hence, the pattern of evolution in AL in these sites can be interpreted as being anagenetic. To the contrary, the apparent changes in the length of the apical process of *R. praebergonii* in cores V28-179, 573, and 572C of the central Pacific Ocean were all consistent with a random walk. The temporal sequences of mean HH in *R. praebergonii* in the Indian Ocean core, core RC12-66, site 573, and site 157 are all consistent with anagenesis (Table II). For instance, in RC12-66, one has $x = 4.089$, which corresponds to a probability of 0.9999; this is strongly suggestive of anagenesis. The time series of HH in V28-179 and site 572C conform to the null hypothesis (Table II). As far as WC is concerned, cores RC12-66, V28-179, 573, and 157 indicate temporal variability which is quite

TABLE I. Random Walk Analysis of the Length of the Apical Process (AL)

Site	Lineage[a]	Mean reduced speed[b]	Expected STD[c]	Observed "range"[d]	x	$1-F(x^{-2})$[e]	Result[f]
Indian Ocean	R.p.	2.67	3.35	10.26	3.15	0.997	Anag
RC12-66	R.p.	1.88	2.35	10.26	5.23	0.9999	Anag
V28-179	R.p.	4.42	5.54	11.57	1.59	0.774	RW
DSDP573	R.p.	3.84	4.81	6.94	1.68	0.813	RW
DSDP572C	R.p.	4.14	5.19	7.30	1.49	0.731	RW
DSDP157	R.p.	4.71	5.90	14.08	2.49	0.976	Anag
Indian Ocean	R.b.	3.07	3.84	5.3	1.42	0.685	RW
RC12-66	R.b.	2.60	3.26	6.14	2.46	0.971	Anag
DSDP573	R.b.	2.49	3.12	5.22	2.13	0.931	RW/Anag
DSDP157	R.b.	7.13	8.93	23.85	3.14	0.997	Anag

[a]R.p., *Rhizosolenia praebergonii*; R.b., *R. bergonii*.
[b]Expected deviations from starting size under a random walk of 1 million years.
[c]The expected standard deviations (STD) in 1 million years under a random walk.
[d]The average range of variability of the investigated time interval.
[e]Probability values.
[f]Anag, anagenesis; RW, random walk.

TABLE II. Random Walk Analysis of the Height of the Hyaline Area (HH)[a]

Site	Lineage	Mean reduced speed	Expected STD	Observed "range"	x	$1-F(x^{-2})$	Result
Indian Ocean	R.p.	0.91	1.13	3.33	3.02	0.994	Anag
RC12-66	R.p.	1.15	1.44	4.91	4.09	0.9999	Anag
V28-179	R.p.	2.34	2.93	7.29	1.90	0.881	RW
DSDP573	R.p.	1.11	1.39	3.34	2.80	0.990	Anag
DSDP572C	R.p.	1.42	1.78	2.64	1.58	0.774	RW
DSDP157	R.p.	1.24	1.56	5.80	3.88	0.9998	Anag
Indian Ocean	R.b.	1.68	2.10	1.08	0.53	0.015	Stas
RC12-66	R.b.	1.89	2.37	5.96	3.28	0.998	Anag
DSDP573	R.b.	1.94	2.43	4.04	2.12	0.931	RW/Anag
DSDP157	R.b.	3.22	4.04	12.19	3.56	0.9992	Anag

[a]See Table I for abbreviations and explanation.

typical of a random walk (Table III). Nevertheless, two temporal sequences, one in site 572C and the other in the Indian Ocean core, show significant anagenetic change in WC (Table III).

The evolution of AL in *R. bergonii* is compatible with anagenesis in core RC12-66 and site 157 (Table I). However, this is not the case in the Indian Ocean core and site 573, where the x values are 1.4186 and 2.1299, respectively, with corresponding probabilities of 0.695 and 0.93. Moreover, in all the sites where *R. bergonii* was sampled, the time series of HH was found to be incompatible with random walks (Table II). For instance, the change of HH in site 157 and

TABLE III. Random Walk Analysis of the Valve Width (WC)[a]

Site	Lineage	Mean reduced speed	Expected STD	Observed "range"	x	$1-F(x^{-2})$	Result
Indian Ocean	R.p.	2.80	3.51	8.70	2.55	0.98	Anag
RC12-66	R.p.	1.85	2.31	3.18	1.65	0.83	RW
V28-179	R.p.	2.39	2.99	4.22	1.08	0.437	RW
DSDP573	R.p.	2.23	2.80	3.78	1.57	0.774	RW
DSDP572C	R.p.	3.23	4.04	12.28	3.23	0.9977	Anag
DSDP157	R.p.	3.26	4.08	6.80	1.74	0.834	RW
Indian Ocean	R.b.	1.67	2.09	2.25	1.11	0.478	RW
RC12-66	R.b.	1.42	1.77	2.65	1.94	0.895	RW
DSDP573	R.b.	2.14	2.68	8.36	3.97	0.9998	Anag
DSDP157	R.b.	3.54	4.44	8.93	2.37	0.964	Anag

[a]See Table I for abbreviations and explanation.

core RC12-66 tend very strongly against anagenesis, as evidenced by two improbably high x values, 3.556 and 3.283, respectively (Table II). In the Indian Ocean core, the variable is found to be significantly in stasis. The values for WC of *R. bergonii* in the Indian Ocean core and in RC12-66 follow a random walk through time (Table III). However, in sites 157 and 573 the temporal sequence of WC are significantly in anagenesis (Table III).

Hierarchical Linear Models

Bookstein *et al.* (1978) proposed a methodology for analyzing temporally ordered data from the fossil record for punctuated equilibrium, phyletic gradualism, and stasis by estimating the comparative contributions of the respective models or an admixture of them to explaining the variance in the raw data. This is done by fitting a set of hierarchical linear models to the data through linear regression techniques (see the appendix for instructions for the models). The method is hierarchical in the sense that the total variance explained by the most general (most inclusive) model in its class, that is, the one having the largest number of parameters, can be further explained by more specialized (simpler) models with fewer parameters. A "step-down F-test" (Bookstein *et al.*, 1978, Appendix I; Zar, 1984) is used to choose the best fitting model, that is, to find the simplest model in the hierarchy which is not significantly different from the most inclusive model in terms of explaining the variance in the data (parsimony criterion) (Bookstein *et al.,* 1978).

In this work the linear representations were constructed with the intention to quantify morphological change not just during and prior to the divergence of the descendant from its ancestor, but also during the time the descendant had become an established lineage, which is inferred from an apparent "break" in the data (Figs. 3a–3c). Such a class of models are of great interest in testing the relative frequency of phyletic gradualism, punctuated equilibrium, and various temporal combinations of these in the data.

The following hierarchical linear models (illustrated in Figs. 3a–3c) have been fitted to the data in this study using the REG procedure in SAS Institute (1985):

I. One lineage without "break."
   1. Gradualism in one lineage.

II. One lineage and one "break."
   2. Stage 1: gradualism; stage 2: stasis.
   3. Stage 1: stasis; stage 2: gradualism.
   4. Stage 1: gradualism; stage 2: gradualism.

III. Two lineages and one "break."

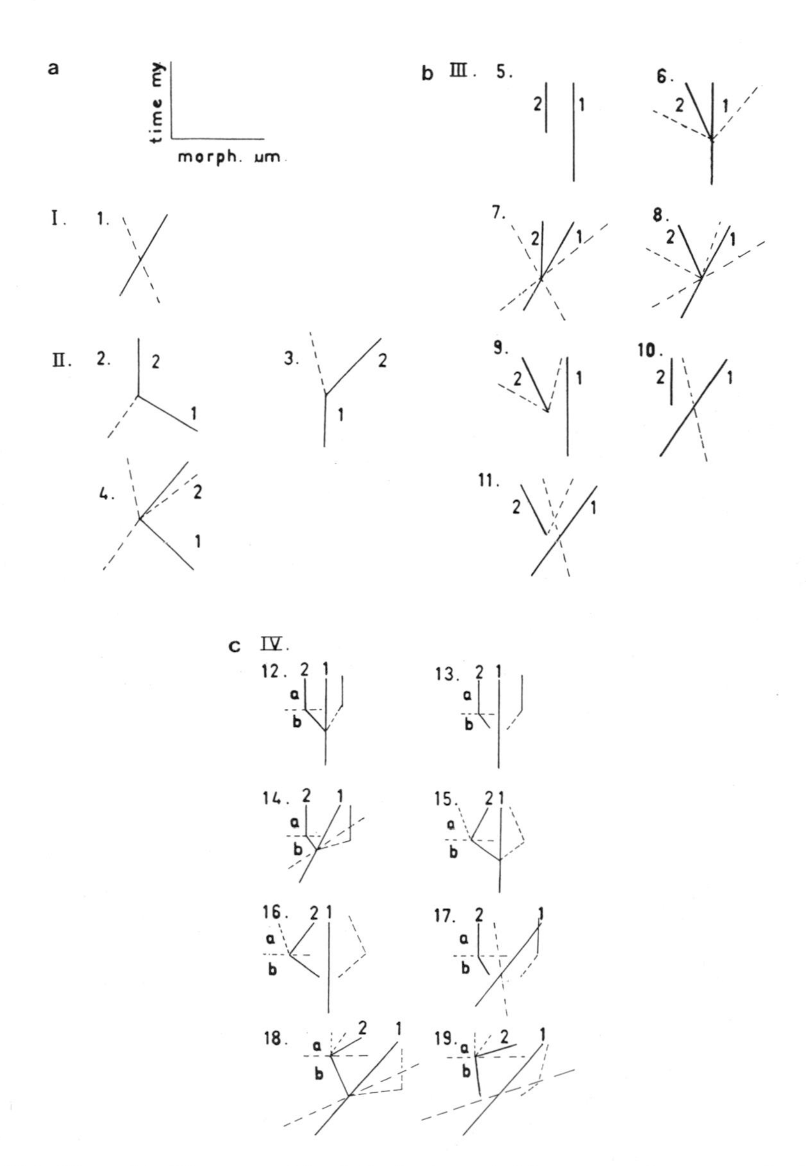

FIG. 3. Hierarchical linear models plotted as bounded line segments on a graph whose ordinate is time (million years) and whose abscissa is the metric morphological variable (μm). Dashed lines indicate that the direction of change is not fixed. (a) Models of classes I and II. The numbers within each model designate substages (1 and 2). (c) (b) Models of class III. The numbers within each model designate different lineages (1 and 2). Models of class IV. The numbers within each model designate different lineages (1 and 2). There are substages A and B recognized within lineage 2.

TABLE IV. The Best "Fitting" Linear Representations of Evolution of the Length of the Apical Process (AL), Height of the Hyaline Area (HH), and the Width of the Valve (WC) in *R. praebergonii* and *R. bergonii*[a]

Site	Variable	N	Model	Sum of squares	r^2
Indian Ocean	AL	965	10	8761.9	0.7361
RC12-66	AL	1123	14	8506.3	0.8012
V28-179	AL	370	2	1548.4	0.6581
DSDP573	AL	1191	13	10127.7	0.7781
DSDP572C	AL	509	4	1629.6	0.6216
DSDP157	AL	1422	9	13896.1	0.7201
Indian Ocean	HH	952	11	3710.5	0.8029
RC12-66	HH	1095	12	2086.4	0.7973
V28-179	HH	347	2	194.1	0.5819
DSDP573	HH	1066	16	1909.4	0.7564
DSDP572C	HH	506	4	346.3	0.6952
DSDP157	HH	1384	13	2690.2	0.8039
Indian Ocean	WC	604	11	645.2	0.4859
RC12-66	WC	997	16	500.7	0.3585
V28-179	WC	365	1	28.57	0.0659
DSDP573	WC	1061	9	510.0	0.3071
DSDP572C	WC	489	1	92.75	0.1136
DSDP157	WC	1203	11	1083.1	0.3904

[a]N is the number of specimens measured for AL, HH, and WC. r^2 is the coefficient of determination or the proportion of variance explained by the best fitting model.

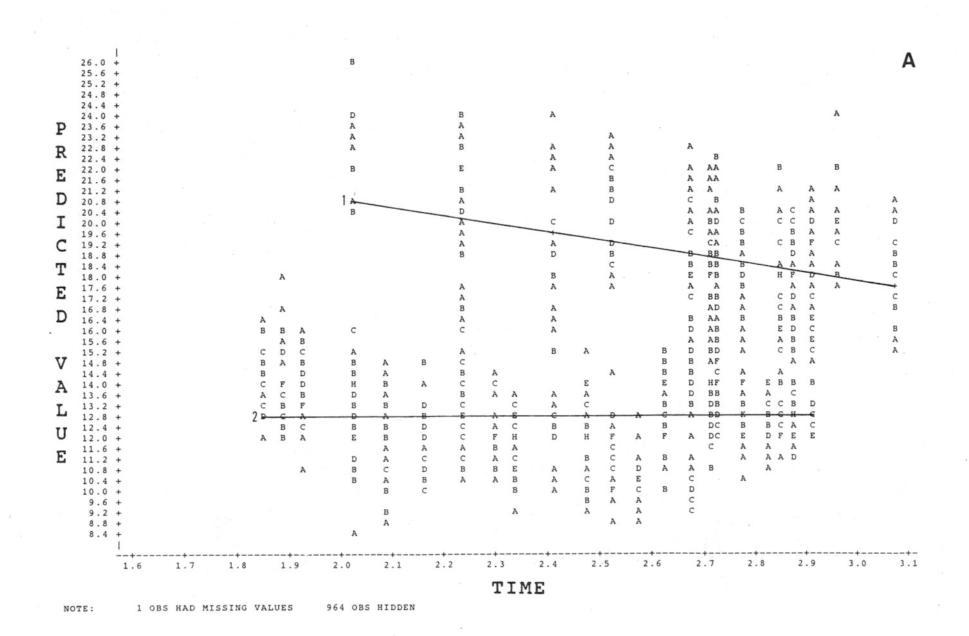

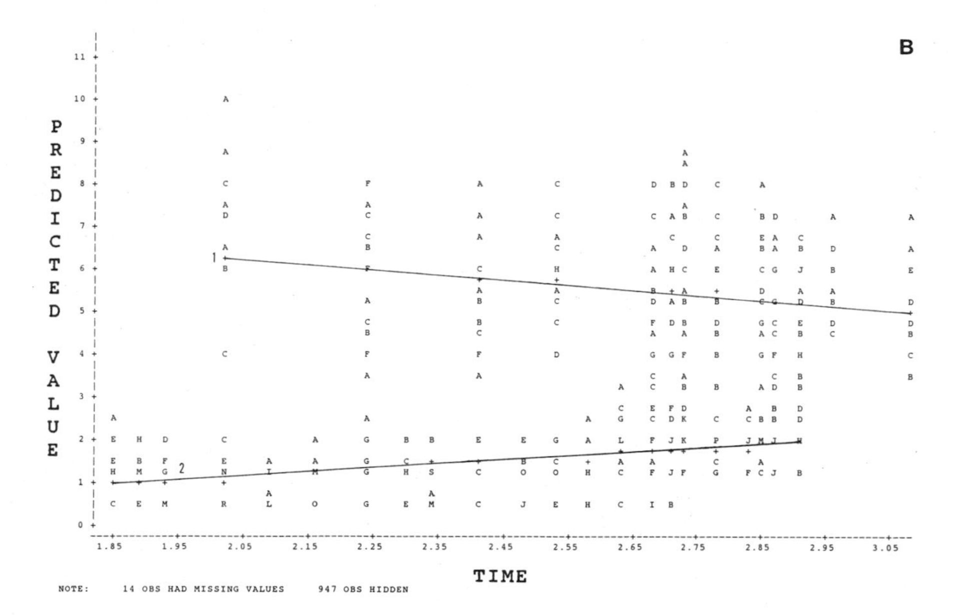

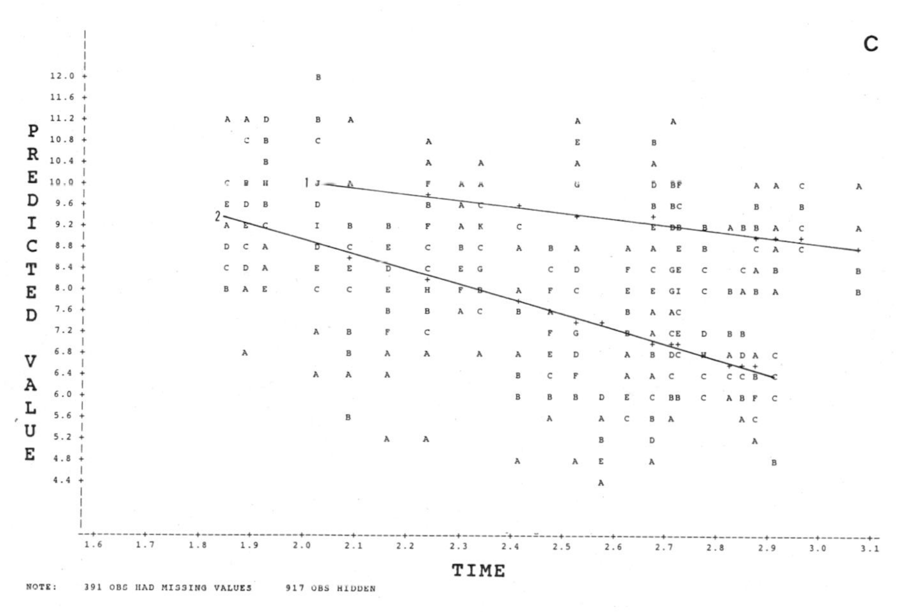

FIG. 4. Changes through time (million years) in (a) the length (μm) of the apical process (AL), in *R. praebergonii* (lineage 2) and *R.bergonii* (lineage 1) in the Indian Ocean sites as predicted by the "best fitting" linear model 10 when tested against the sample scatters; in (b) the height (μm) of the hyaline area (HH) in *R. praebergonii* (lineage 2) and *R. bergonii* (lineage 1) in the Indian Ocean sites as predicted by the "best fitting" linear model 11 when tested against the sample scatters; and in (c) the width (μm) of the valve (WC) in *R. praebergonii* (lineage 2) and *R. bergonii* (lineage 1) in the Indian Ocean sites as predicted by the "best fitting" linear model 11 when tested against the sample scatters. Legend: A = 1 observation, B = 2 observations, etc.

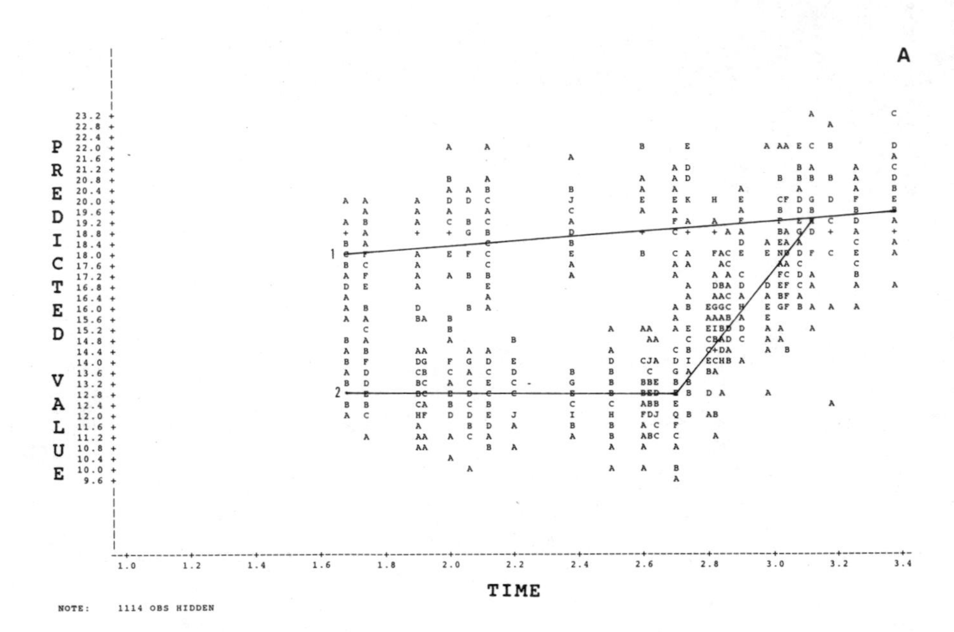
A
PREDICTED VALUE
TIME
NOTE: 1114 OBS HIDDEN

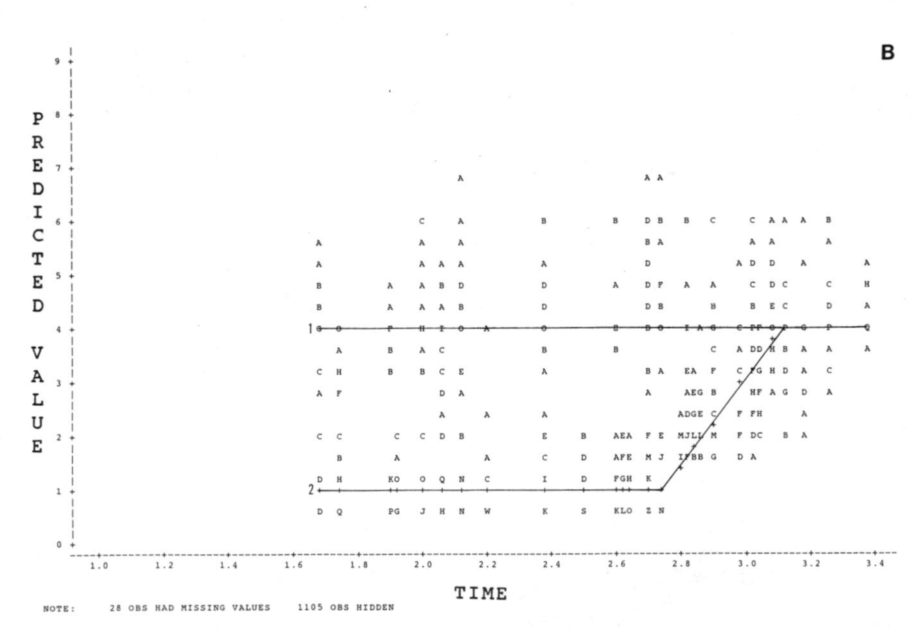
B
PREDICTED VALUE
TIME
NOTE: 28 OBS HAD MISSING VALUES 1105 OBS HIDDEN

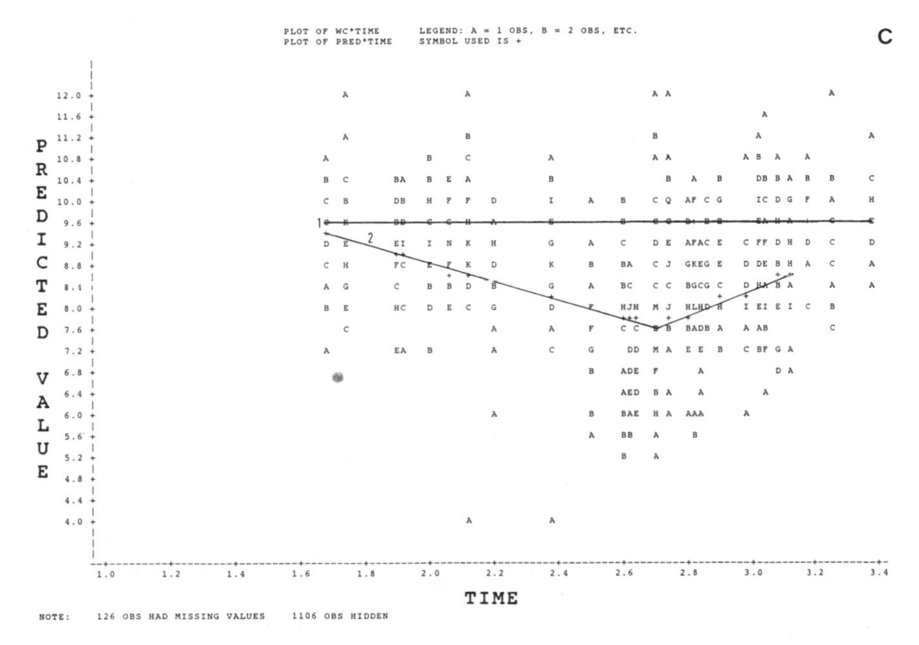

FIG. 5. Changes through time (million years) in (a) the length (μm) of the apical process (AL) in *R. praebergonii* (lineage 2) and *R. bergonii* (lineage 1) in site RC12-66 as predicted by the "best fitting" linear model 14 when tested against the sample scatters; in (b) the height (μm) of the hyaline area (HH) in *R. praebergonii* (lineage 2) and *R. bergonii* (lineage 1) in site RC12-66 as predicted by the "best fitting" linear model 12 when tested against the sample scatters; and in (c) the width (μm) of the valve (WC) in *R. praebergonii* (lineage 2) and *R. bergonii* (lineage 1) in site RC12-66 as predicted by the "best fitting" linear model 16 when tested against the sample scatters. Legend: A = 1 observation, B = 2 observations, etc.

5. Stasis in lineages 1 and 2; lineage 2 originated from lineage 1 through a punctuation.
6. Stasis in lineage 1 and gradualism in lineage 2; lineage 2 originated from lineage 1 through gradualism.
7. Gradualism in lineage 1 and stasis in lineage 2; lineage 2 originated from lineage 1 by remaining static while lineage 1 changed gradually.
8. Gradualism in lineages 1 and 2; lineage 2 originated from lineage 1 through gradualism while lineage 1 changed gradually.
9. Gradualism in lineage 2 and stasis in lineage 1; lineage 2 originated from lineage 1 by a punctuation, after which it continued to change gradually.
10. Gradualism in lineage 1 and stasis in lineage 2; lineage 2 originated from lineage 1 as it changed gradually through a punctuation, after which it remained static.

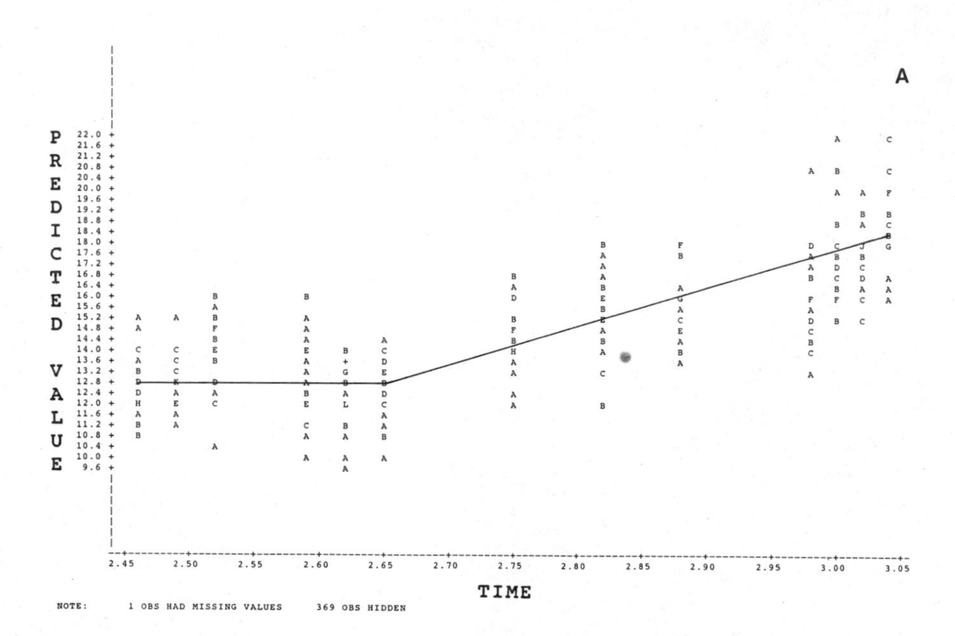
A
PREDICTED VALUE
22.0
21.6
21.2
20.8
20.4
20.0
19.6
19.2
18.8
18.4
18.0
17.6
17.2
16.8
16.4
16.0
15.6
15.2
14.8
14.4
14.0
13.6
13.2
12.8
12.4
12.0
11.6
11.2
10.8
10.4
10.0
9.6
2.45 2.50 2.55 2.60 2.65 2.70 2.75 2.80 2.85 2.90 2.95 3.00 3.05
TIME
NOTE: 1 OBS HAD MISSING VALUES 369 OBS HIDDEN

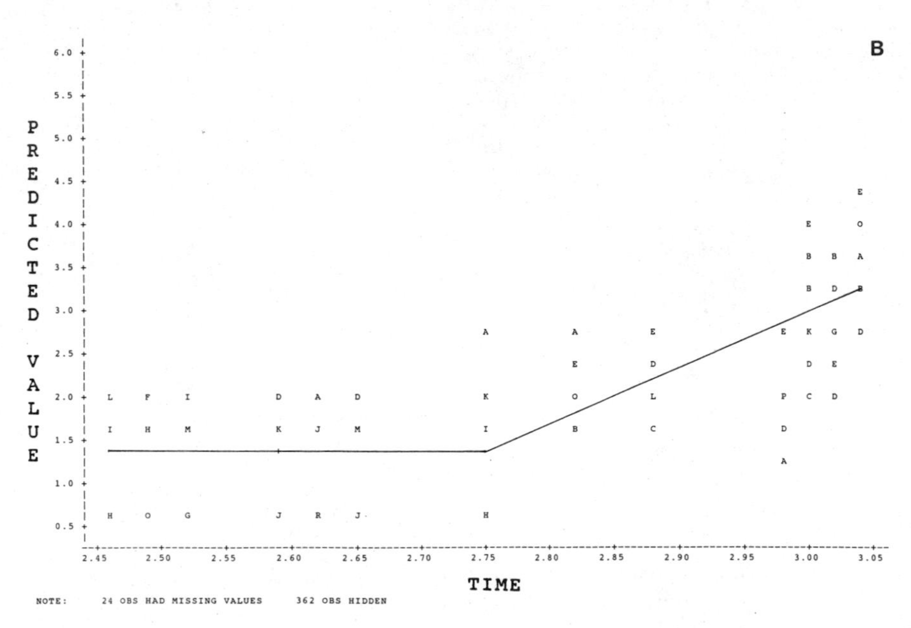
B
PREDICTED VALUE
6.0
5.5
5.0
4.5
4.0
3.5
3.0
2.5
2.0
1.5
1.0
0.5
2.45 2.50 2.55 2.60 2.65 2.70 2.75 2.80 2.85 2.90 2.95 3.00 3.05
TIME
NOTE: 24 OBS HAD MISSING VALUES 362 OBS HIDDEN

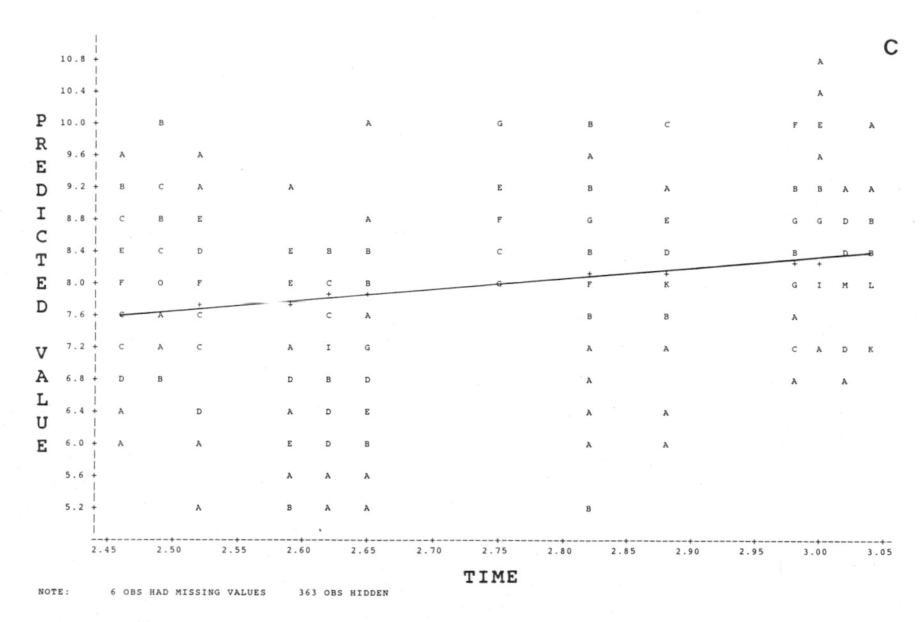

FIG. 6. Changes through time (million years) in (a) the length (μm) of the apical process (AL) in *R. praebergonii* in site V28-179 as predicted by the "best fitting" linear model 2 when tested against the sample scatters; in (b) the height (μm) of the hyaline area (HH) in *R. praebergonii* in site V28-179 as predicted by the "best fitting" linear model 2 when tested against the sample scatters; and in (c) the width (μm) of the valve (WC) in *R. praebergonii* in site V28-179 as predicted by the "best fitting" linear model 1 when tested against the sample scatters. Legend: A = 1 observation, B = 2 observations, etc.

11. Gradualism in lineages 1 and 2; lineage 2 originated from lineage 1 through a punctuation, after which both lineages continued to change gradually.

IV. Two lineages and two "breaks."

12. Stasis in lineage 1 and gradualism (b) and subsequent stasis (a) in lineage 2; lineage 2 originated from lineage 1 through gradualism.
13. Stasis in lineage 1 and gradualism (b) and subsequent stasis (a) in lineage 2; lineage 2 originated from lineage 1 through a punctuation.
14. Gradualism in lineage 1 and gradualism (b) and subsequent stasis (a) in lineage 2; lineage 2 originated from lineage 1 through gradualism while it changed gradually.
15. Stasis in lineage 1 and successive gradualism (a,b) in different directions in lineage 2; lineage 2 originated from lineage 1 through gradualism.
16. Stasis in lineage 1 and successive gradualism (a,b) in different

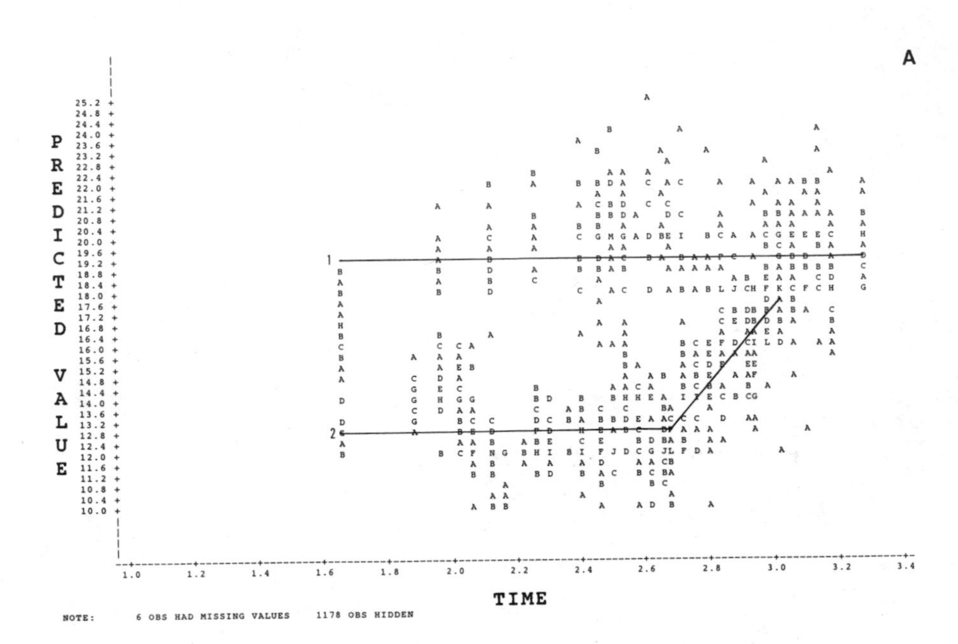
A
PREDICTED VALUE
25.2 24.8 24.4 24.0 23.6 23.2 22.8 22.4 22.0 21.6 21.2 20.8 20.4 20.0 19.6 19.2 18.8 18.4 18.0 17.6 17.2 16.8 16.4 16.0 15.6 15.2 14.8 14.4 14.0 13.6 13.2 12.8 12.4 12.0 11.6 11.2 10.8 10.4 10.0
1.0 1.2 1.4 1.6 1.8 2.0 2.2 2.4 2.6 2.8 3.0 3.2 3.4
TIME
NOTE: 6 OBS HAD MISSING VALUES 1178 OBS HIDDEN

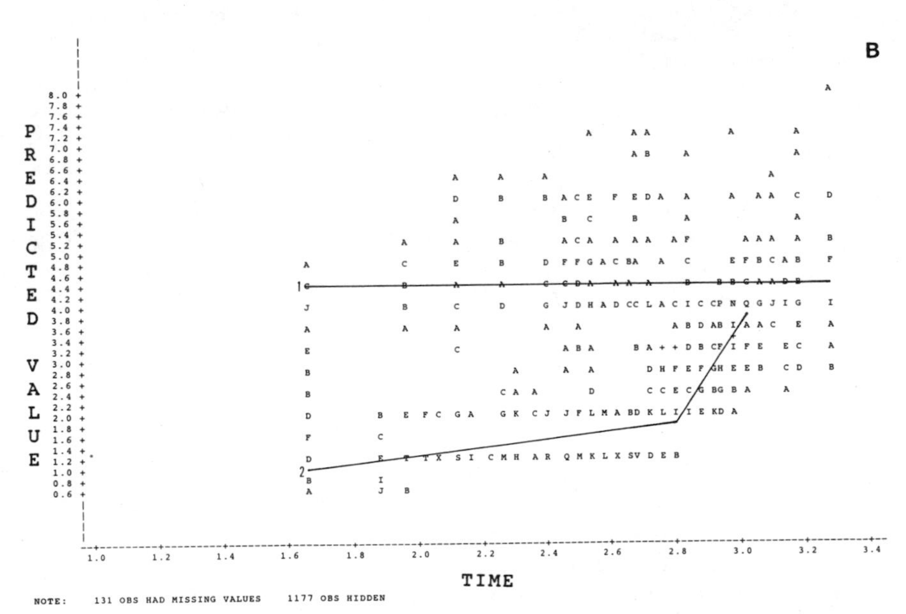
B
PREDICTED VALUE
8.0 7.8 7.6 7.4 7.2 7.0 6.8 6.6 6.4 6.2 6.0 5.8 5.6 5.4 5.2 5.0 4.8 4.6 4.4 4.2 4.0 3.8 3.6 3.4 3.2 3.0 2.8 2.6 2.4 2.2 2.0 1.8 1.6 1.4 1.2 1.0 0.8 0.6
1.0 1.2 1.4 1.6 1.8 2.0 2.2 2.4 2.6 2.8 3.0 3.2 3.4
TIME
NOTE: 131 OBS HAD MISSING VALUES 1177 OBS HIDDEN

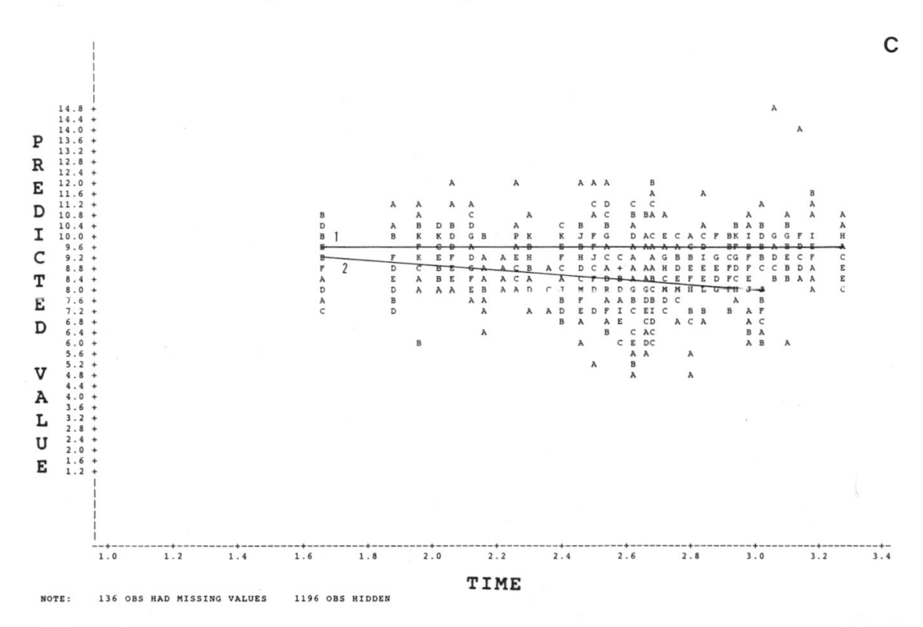

FIG. 7. Changes through time (million years) in (a) the length (μm) of the apical process (AL) in *R. praebergonii* (lineage 2) and *R. bergonii* (lineage 1) in DSDP site 573 as predicted by the "best fitting" linear model 13 when tested against the sample scatters; in (b) the height (μm) of the hyaline area (HH) in *R. praebergonii* (lineage 2) and *R. bergonii* (lineage 1) in DSDP site 573 as predicted by the "best fitting" linear model 16 when tested against the sample scatters; and (c) in the width (μm) of the valve (WC) in *R. praebergonii* (lineage 2) and *R. bergonii* (lineage 1) in DSDP site 573 as predicted by the "best fitting" linear model 9 when tested against the sample scatters. Legend: A = 1 observation, B = 2 observations, etc.

directions in lineage 2; lineage 2 originated from lineage 1 through a punctuation.

17. Gradualism in lineage 1 and gradualism (b) and subsequent stasis (a) in lineage 2; lineage 2 originated from lineage 1 through punctuation.
18. Gradualism in lineage 1 and successive gradualism in different directions (a,b) in lineage 2; lineage 2 originated from lineage 1 through gradualism.
19. Gradualism in lineage 1 and successive gradualism (a,b) in different directions in lineage 2; lineage 2 originated from lineage 1 through a punctuation.

In this study, model 5 represents the extreme version of punctuated equilibrium, while model 8 illustrates phyletic gradualism or the most drastic version of phyletic evolution (Figs. 3a–3c). The other models describe evolutionary

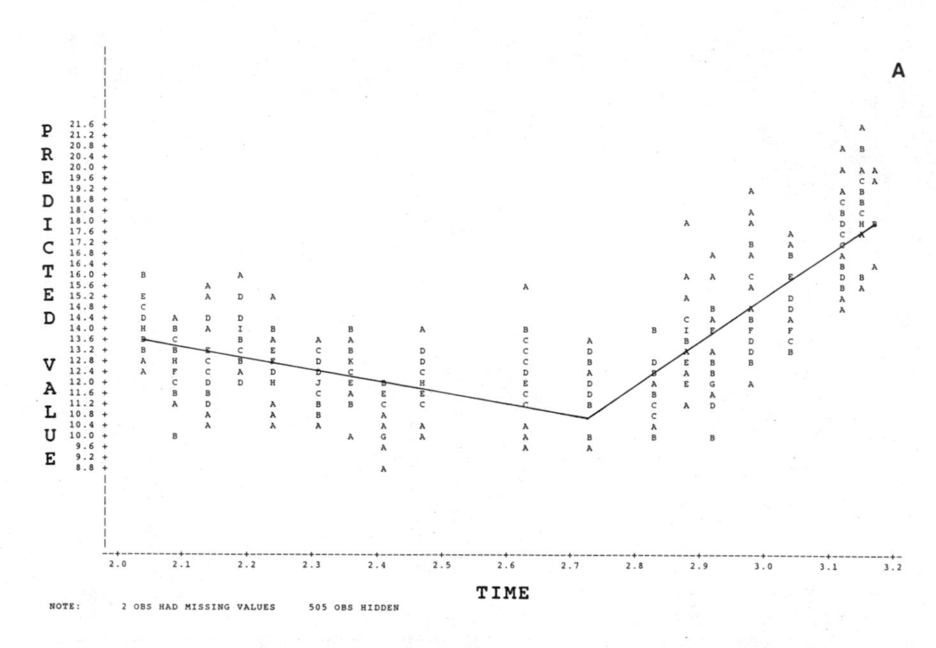
A
PREDICTED VALUE
21.6
21.2
20.8
20.4
20.0
19.6
19.2
18.8
18.4
18.0
17.6
17.2
16.8
16.4
16.0
15.6
15.2
14.8
14.4
14.0
13.6
13.2
12.8
12.4
12.0
11.6
11.2
10.8
10.4
10.0
9.6
9.2
8.8
2.0 2.1 2.2 2.3 2.4 2.5 2.6 2.7 2.8 2.9 3.0 3.1 3.2
TIME
NOTE: 2 OBS HAD MISSING VALUES 505 OBS HIDDEN

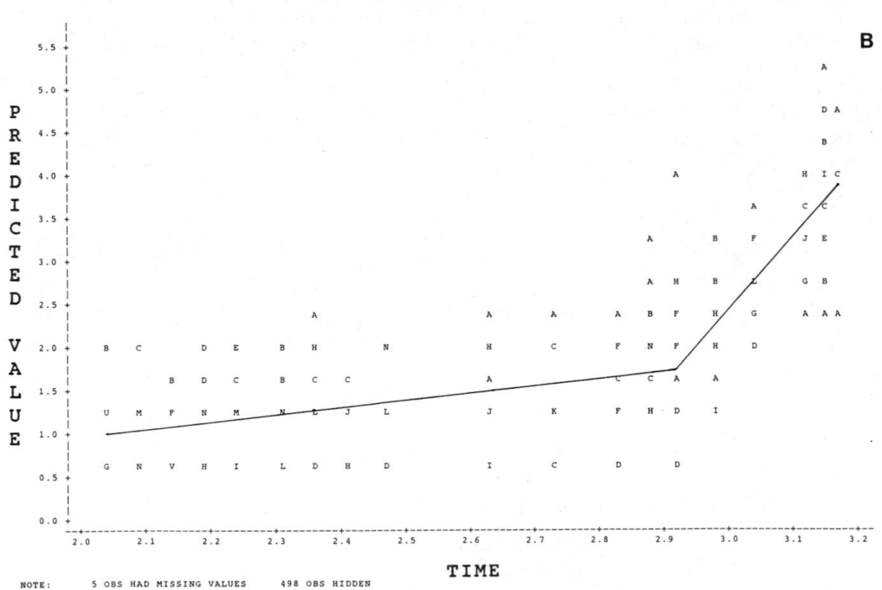
B
PREDICTED VALUE
5.5
5.0
4.5
4.0
3.5
3.0
2.5
2.0
1.5
1.0
0.5
0.0
2.0 2.1 2.2 2.3 2.4 2.5 2.6 2.7 2.8 2.9 3.0 3.1 3.2
TIME
NOTE: 5 OBS HAD MISSING VALUES 498 OBS HIDDEN

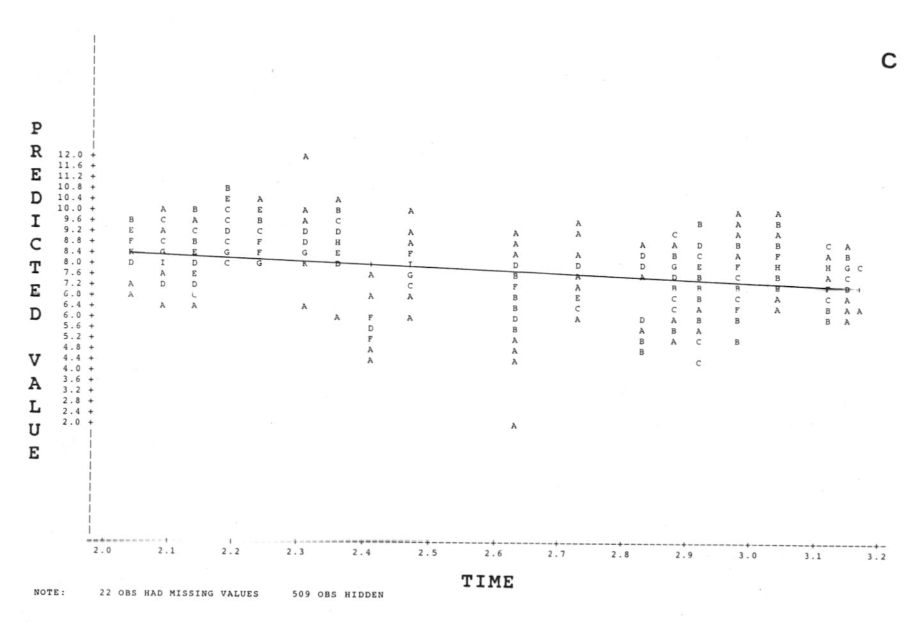

FIG. 8. Change through time (million years) in (a) the length (μm) of the apical process (AL) in *R. praebergonii* in DSDP site 572C as predicted by the "best fitting" linear model 4 when tested against the sample scatters; in (b) the height (μm) of the hyaline area (HH) in *R. praebergonii* in DSDP site 572C as predicted by the "best fitting" linear model 4 when tested against the sample scatters; and in (c) the width (μm) of the valve (WC) in *R. praebergonii* in DSDP site 572C as predicted by the "best fitting" linear model 1 when tested against the sample scatters. Legend: A = 1 observation, B = 2 observations, etc.

patterns that are transitional with respect to punctuated equilibrium and phyletic gradualism.

The category of models fitted, that is, with *R. praebergonii* and *R. bergonii a priori* defined, were determined by the presence of a "break" between the ancestral and descendant lineages as well as a "breaks" or a shift in the direction of change within the descendant lineage.

The initial "break" at 3.1 Mya between *R. praebergonii* and *R. bergonii* for all three variables took place through a significant punctuation as opposed to gradual change in 10 out of 12 possible cases (Table IV; Figs. 4–9). AL, HH, and WC of *R. praebergonii* continued to evolve in a gradual manner after the incipient first appearance, except after the first appearance in the Indian Ocean, where AL remained static (Table IV; Fig. 4a). As *R. praebergonii* became established as a distinct lineage (2.7 Mya), it remained static as opposed to showing gradual change for all three characters in 7 cases out of 18 (includes the cores where just *R. praebergonii* was sampled). *R. bergonii* exhibited character

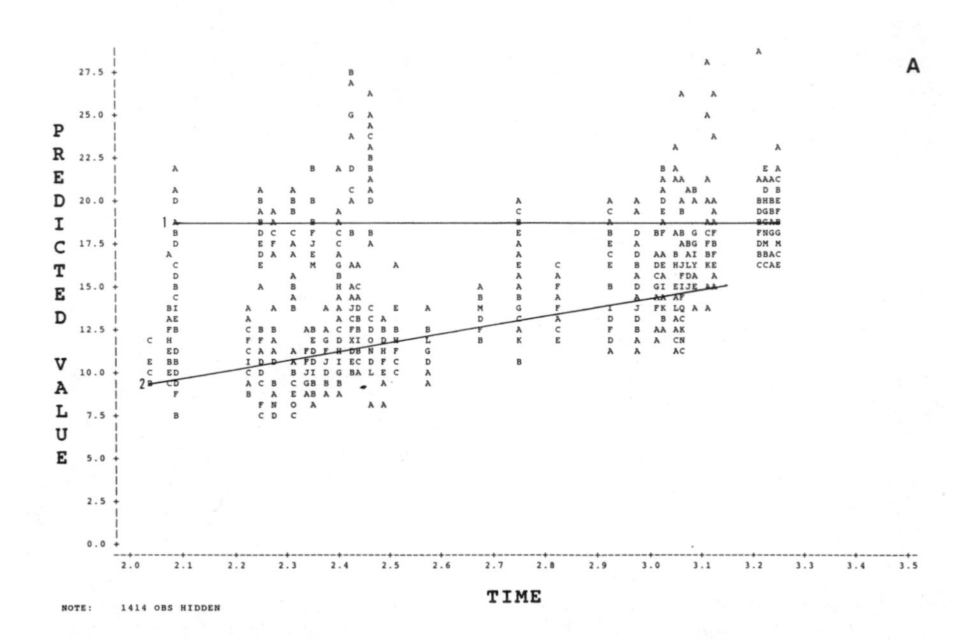
A
PREDICTED VALUE
27.5
25.0
22.5
20.0
17.5
15.0
12.5
10.0
7.5
5.0
2.5
0.0
2.0 2.1 2.2 2.3 2.4 2.5 2.6 2.7 2.8 2.9 3.0 3.1 3.2 3.3 3.4 3.5
TIME
NOTE: 1414 OBS HIDDEN

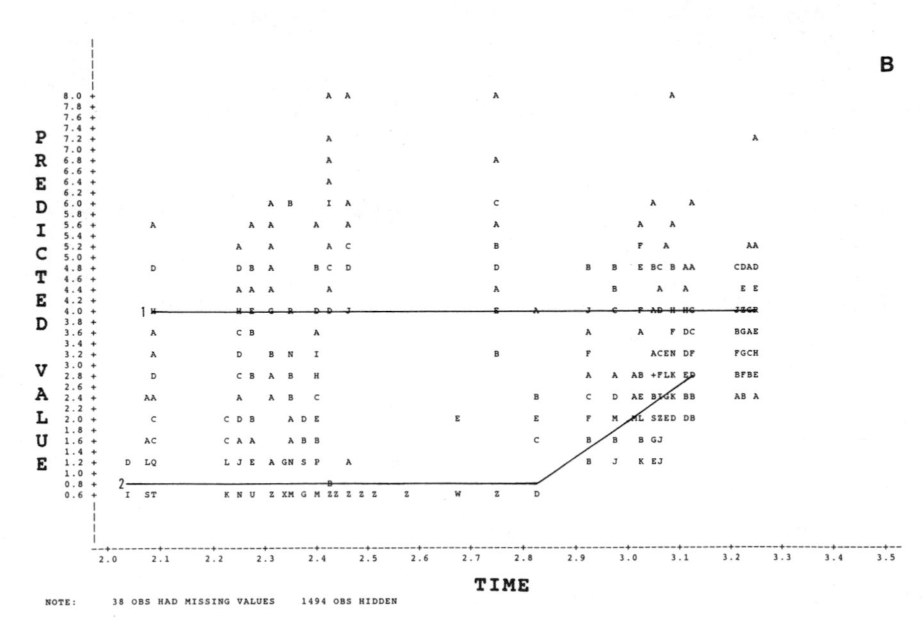
B
PREDICTED VALUE
8.0
7.0
6.0
5.0
4.0
3.0
2.0
1.0
0.6
2.0 2.1 2.2 2.3 2.4 2.5 2.6 2.7 2.8 2.9 3.0 3.1 3.2 3.3 3.4 3.5
TIME
NOTE: 38 OBS HAD MISSING VALUES 1494 OBS HIDDEN

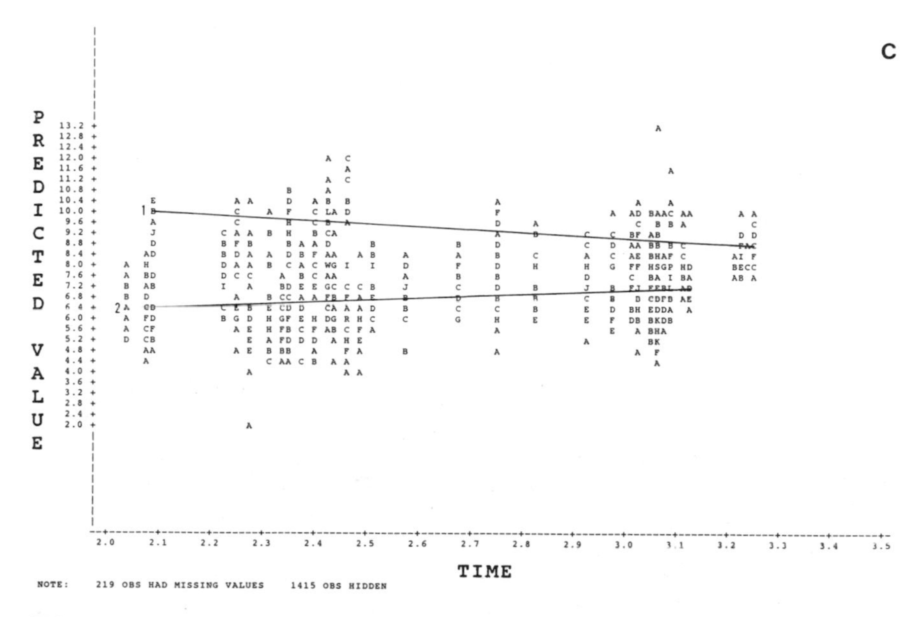

FIG. 9. Changes through time (million years) in (a) the length (μm) of the apical process (AL) in *R. praebergonii* (lineage 2) and *R. bergonii* (lineage 1) in DSDP site 157 as predicted by the "best fitting" linear model 9 when tested against the sample scatters; in (b) the height (μm) of the hyaline area (HH) in *R. praebergonii* (lineage 2) and *R. bergonii* (lineage 1) in DSDP site 157 as predicted by the "best fitting" linear model 13 when tested against the sample scatters; and (c) in the width (μm) of the valve (WC) in *R. praebergonii* (lineage 2) and *R. bergonii* (lineage 1) in DSDP site 157 as predicted by the "best fitting" linear model 11 when tested against the sample scatters. Legend: A = 1 observation, B = 2 observations, etc.

stasis 7 times out of a possible 12, that is, gradual change appears to have been less frequent.

The best fitting linear representation of the evolution of AL and HH in the two lineages accounts for between 0.65 and 0.80 of the sum of squares, while WC indicates a much lower range, between 0.06 and 0.61 (Table IV).

DISCUSSION

Random Walk

The results of the Range Test are very difficult to interpret, since the failure to reject the null hypothesis of a random walk does not necessarily imply random morphological change in a population; this has, for instance, been pointed out by Fisher (1986). In fact, shifts in the relative frequencies of morphotypes in a population may be deterministic even though the test statistics indicates a random walk, since the causal factor(s), which are tracked by the deme, may vary randomly.

The mean length of the apical process indicates an anagenetic pattern of change in *R. praebergonii* in the Indian Ocean core, core RC12-66 and site 157, while the same time series is random in core V28-179, site 573, and site 572C even though these are located in the same water mass as core RC12-66 (Fig. 2; Tables I–III). Moreover, AL shows a very significant trend in core RC12-66, since the x value is well past the 0.9999 tail of its distribution (Table I). How should these discrepancies be interpreted? Since the number of laterally distinct "panmictic" populations of oceanic plankton tend to fall between water masses or major oceanographic boundaries (Lazarus, 1983; Prothero and Lazarus, 1980), *R. praebergonii* in the central Pacific cores was most likely part of the same population in the central equatorial region and thus should have evolved in a similar fashion. This interpretation also applies to *R. praebergonii* in site 572C, which, in addition, could have been affected by migration of forms from the population in the Peru water mass (Fig. 2), thus giving rise to a spuriously random temporal sequence of AL. Consequently, the pattern of evolution in all the central equatorial samples should be interpreted to follow the same anagenetic pattern that was observed in RC12-66.

Anagenesis was also found to dominate the pattern of change of HH in *R. praebergonii* in the Indian Ocean core as well as in RC12-66, site 573, and site 157 (Table II). In this case, HH in core RC12-66 also exhibits x values which end up past the 0.9999 tail of the distribution of this statistic. Such high x values are very likely an indication of a directional pattern of evolution. The temporal variation of HH in *R. praebergonii* in site 572C may also be given the same interpretation as for AL, that is, spurious randomness due to migration of unique morphotypes from the eastern population. In this case the eastern equatorial population, which is restricted to the Peru water mass, also exhibit "highly significant" anagenesis in HH. In this case, it may be justified to conclude that HH in *R. praebergonii* also exhibited largely nonrandom variation through time in the three water masses.

The change in the mean width of the valve in *R. praebergonii* is consistent with a random walk in the eastern and central equatorial cores, except in Site 572C (Table III). This core has a unique position in the Pacific Ocean in that it was drilled in an area where the central equatorial and the Peru water mass meet. Thus, the evolutionary patterns in this site could have been affected by environmental parameters unique to each location as well as by migration of distinct morphotypes between the eastern and central regions. The temporal sequences of

AL and HH in *R. praebergonii* in site 572C also deviated from the findings in the more centrally located cores as well as the sampled location (site 157) in the east. Despite the fact that WC in sites RC12-66, V28-179, and 573 varied through time in a random fashion, it cannot be determined whether this is due to a deterministic tracking of a randomly varying causal factor(s) or not. A more likely explanation is that developmental constraints (M. Hecht, personal communication) and selection operated on this character at the extremes of an adaptive peak within which WC was free to vary randomly (cf. Raup and Crick, 1981). The same explanation can also be applied to the sequence in site 157. The width of the valve in *R. praebergonii* in the Indian Ocean exhibited a significant upward trend.

Both AL and HH in *R. bergonii* indicate a nonrandom pattern of change in almost all the cores in which it was sampled (Tables I and II). The exception is the temporal sequence of AL in the Indian Ocean, in which the variable conforms to a random walk. This result in the Indian Ocean site is very hard to explain, since the same trait in *R. praebergonii* indicates anagenesis. The anagenetic pattern in 573 is a borderline case: the x value is just above the critical value of 2.25. This is also a case where the change in AL should be interpreted as being anagenetic, since *R. bergonii* in RC12-66 and 573 probably belonged to the same "panmictic population." The same explanation also pertains to the change in HH, since the situation is similar; that is, *R. bergonii* in RC12-66 is "significantly" in anagenesis for this character, while in 573 it is a borderline case.

The width of the valve in *R. bergonii* changed in a nonrandom fashion in site 157. I similarly, interpret the pattern of change in WC in the central equatorial population even though this trait is consistent with a random walk in core RC12-66, since the temporal sequence in site 573 is anagenetic (Table III). In the Indian Ocean core the WC conforms to random walk, which is also hard to interpret, since no other sites have been sampled within this population.

Aside from the statistical/mathematical arguments of Bookstein (1987), the time sequences of the three variables are most likely nonrandom in *R. praebergonii* and *R. bergonii* because the "same event" is found to have taken place in possibly two semi-isolated populations at the same time; this is unlikely to have been a result of random morphological variation, but rather due to some common cause(s). The above argument is further supported by the iterative pattern of evolution in *R. praebergonii* with respect to *Rhizosolenia sigmoida* Sorhannus in site 157. Such a repetative change in morphology may have been triggered by some cyclic environmental factor(s) (U. Sorhannus, E. Fenster, A. Hoffman, and L. Burckle, unpublished). The probability of generating two such temporally consecutive cladogenetic events in AL, HH, and WC purely by chance must be very low. Based on the above considerations, there is good basis for believing that variability in the investigated morphological traits was largely nonrandom;

this would justify the quantification of the temporal sequences through piecewise linear regression.

Mode of Evolution

The kinds of linear representations that were fitted to this data set will lead to certain conclusions about the relative frequency of punctuations, stasis, and gradual change in morphology as well as how these relate to each other temporally (Figs. 3a–3c).

The significant incipient punctuations (''breaks'') in morphology, which are very common in the investigated temporal sequences in the central and eastern Pacific Ocean, are not at all the kinds of ''breaks'' predicted by the punctuated equilibrium hypothesis (Eldredge and Gould, 1972; Gould and Eldredge, 1977), since these did not give rise to distinct morphological entities in a geologically short period of time. Consequently, these types of punctuations cannot be taken as corroborative evidence for the extreme punctuated model of evolution (e.g., Hecht and Hoffman, 1986) (model 5). The initial shifts in morphology of *R. praebergonii* in the eastern and central Pacific Ocean may reflect a very short period of time when evolution was accelerated, but not punctuated in the way Eldredge and Gould (1972) claimed it would. Another possible explanation for the significant ''breaks'' between *R. praebergonii* and *R. bergonii* is that they are artifacts of aggregation. This phenomenon, according to Bookstein *et al.* (1978), increases hidden randomness in temporal data, and thus punctuated equilibria should be interpreted with great caution.

The fact that *R. praebergonii* was nearly indistinguishable from *R. bergonii* at 3.1 Mya and the presence of all the intermediate forms between the ancestor and the descendant support phyletic evolution as the most appropriate evolutionary mode during the morphological divergence between *R. praebergonii* and *R. bergonii* in the Pacific Ocean (Figs. 4–9). This argument is further favored by the results of fitting the hierarchical linear models to the data (Table IV). The only predictions of the extreme punctuated model of evolution (model 5) that hold for the data in the Pacific Ocean are the cases of stasis in *R. bergonii* (3.1 to 1.7 Mya) and in *R. praebergonii* after (2.7 to 1.7 Mya) it became established as a distinct lineage.

An exception to the observations made in the Pacific Ocean are the significant ''breaks'' between *R. praebergonii* and *R. bergonii* in the Indian Ocean, which gave rise to the former as a distinct morphospecies (Figs. 4a–4c). However, there is evidence for these punctuations being both an artifact of migration and an independent origination followed by parallel evolution. The migration hypothesis in conjunction with evolution is supported by the fact that the first appearance of *R. praebergonii* took place 200,000 years later in the Indian

Ocean, while the independent origination hypothesis is strengthened by the existence of *R. praebergonii* in the Indian Ocean as a *distinct* morphospecies 200,000 years earlier than in the Pacific Ocean. If such an abrupt "break" in morphology was "real," that is, as a result of local evolution, it would support punctuated equilibrium, but not in its extreme version (model 5), since gradual change followed the punctuation in either the descendant or the ancestor or both (Table IV; Figs. 4a–4c).

Stasis and rapid evolutionary change in the fossil record is in no way inconsistent with neo-Darwinian evolutionary mechanisms (e.g., Charlesworth *et al.,* 1982; Lande, 1986). Consequently, there is no reason to rethink neo-Darwinian evolution—at least not as far as this data set is concerned.

The results of this analysis indicate that the patterns of evolution in both *R. praebergonii* and *R. bergonii* represent different degrees of combinations of punctuated equilibrium and phyletic gradualism rather than being dominated by either one of the extreme patterns (model 5 or 8). Thus, neither model 5 nor model 8 was found to be the best representation of the morphological evolution in any of the investigated sites. Malmgren *et al.* (1983) also found a pattern of evolution which exhibits a long period of morphological stasis punctuated by relatively rapid gradual phyletic change, however, without resulting in lineage splitting, in the late Neogene planktonic foraminifera lineage *Globoratalia tumida*. Furthermore, these workers suggested that this pattern, which they call punctuated gradualism, may be a common form of evolution in planktonic foraminifera (Malmgren *et al.*, 1983). This conclusion also seems to be confirmed in diatoms.

SUMMARY

Based on criteria for identifying ancestor–descendant relationships (Prothero and Lazarus, 1980), it can be concluded that *R. praebergonii* originated from *R. bergonii* in the equatorial Pacific Ocean via a cladogenetic event at 3.1 Mya. The ancestor–descendant relationships are substantiated by the fact that at the first appearance *R. praebergonii* was nearly indistinguishable from *R. bergonii* as far as the investigated characters are concerned. Through time, differences in morphology were enhanced due to a relatively rapid decrease in the length of the apical process, height of the hyaline area, and the width of the valve in *R. praebergonii,* whereas *R. bergonii* underwent a slight increase in size as far as the investigated characters are concerned. Moreover, *R. bergonii* is stratigraphically positioned prior to the first appearance of *R. praebergonii* and it is also found both during and after the cladogenetic event in all the investigated sites; this lends further support for the ancestor–descendant relationships. As *R.*

praebergonii became established as distinct lineages at around 2.7 Mya, the morphology continued to change at a relatively slower rate until it disappeared at approximately 1.7 Mya. The first appearance of *R. praebergonii* in the Indian Ocean took place 200,000 years later than in the Pacific Ocean. Based on such a stepwise pattern of first appearances in conjunction with the absence of intermediate morphologies between *R. praebergonii* and *R. bergonii* in the Indian Ocean, the possibility of *R. praebergonii* migrating into the Indian Ocean after its evolution in the Pacific Ocean arises. An alternative interpretation of the data is that *R. praebergonii* evolved independently from *R. bergonii* in the Indian Ocean, after which it underwent parallel evolution; this would imply that "*R. praebergonii*" in the Indian and Pacific Oceans were actually two different species.

With some exceptions, both *R. praebergonii* and *R. bergonii* exhibited a nonrandom pattern of change in AL and HH within each hypothesized population, while the result with respect to WC is rather inconsistent. In addition to the results of the Range Test (Bookstein, 1987), the fact that the divergence event between *R. praebergonii* and *R. bergonii* took place simultaneously in two supposedly semi-isolated populations and the fact that *R. praebergonii* underwent iterative evolution in the eastern Pacific Ocean (U. Sorhannus, E. Fenster, A. Hoffman, and L. Burckle, unpublished) lend support to the patterns of evolution being largely nonrandom. This is because the probability of such events taking place at the same time in two semi-isolated populations or in the same place at two different times purely by chance is very low. Consequently, it can be argued that some common causal factor(s) may have been involved.

Based on fitting a set of hierarchical linear models to the temporal scatter of the data, it can be concluded that the evolution of *R. praebergonii* from *R. bergonii* (3.1 to 2.7 Mya) and it subsequent establishment as a distinct morphospecies (2.7 to 1.7 Mya) in the Pacific Ocean indicate a spectrum of temporal combinations of the two extreme patterns, punctuated equilibrium (model 5) and phyletic gradualism (model 8). However, the relative frequency of stasis which is predicted by the extreme punctuated equilibrium model (model 5) was rather high in *R. bergonii* but also in *R. praebergonii* as it became established as a distinct morphospecies. The significant punctuation in the morphology of *R. praebergonii* which was observed in the Indian Ocean is of the nature predicted by the extreme punctuated model (5) of evolution. However, in all instances this event was associated with gradual phyletic change in either the ancestor or the descendant or both. There is also evidence for *R. praebergonii* migrating into the Indian Ocean as it originated in the Pacific Ocean as well as for independent evolution. Thus, the pattern observed in the Indian Ocean may be an evolutionary punctuation (in the sense of model 5) which was closely connected with phyletic change or it may just be an artifact of migration in conjunction with evolution.

CONCLUSION

The temporal patterns of morphological evolution in *R. praebergonii* and *R. bergonii* reflect a spectrum of combinations of the two extreme patterns of evolution, phyletic gradualism and punctuated equilibrium, rather than being dominated by either one. Likewise, the geographic variability in the patterns of evolution among Indian, central Pacific, and eastern Pacific Oceans indicate temporal combinations of gradual and punctuated change in morphology as well as morphological stasis. Consequently, the claims made about the history of life primarily being dominated by either the extreme form of punctuated equilibrium (model 5) or by the most drastic mode of phyletic evolution, that is, phyletic gradualism (model 8), are in this case refuted. In addition, the findings in this investigation suggest that there is no need to rethink neo-Darwinian evolution. The reason for this is that in this study punctuated equilibrium and phyletic gradualism appear to be two extreme patterns which are interconnected by a whole spectrum of transitional patterns and tempos—describing slow and fast phyletic evolution—rather than being mutually exclusive with distinct underlying evolutionary mechanisms.

APPENDIX

The statistical procedures for fitting the hierarchical linear models presented in Figs. 3a–3c are described in Table V. All the models in this study were fitted through linear regression techniques using the REG procedure in SAS (1985). Each dependent variable (AL, HH, and WC) was regressed upon time T and/or time levels shifted S (except for the model describing pure stasis or model 5) as well as other "dummy variables" (also see Bookstein *et al.* 1978, Appendix I). The intercept is automatically included in all the regressions.

TABLE V. Statistical Procedures for Fitting Hierarchical Linear Models of Fig. 3

Model	Variable(s) on which AL, HH, or WC was regressed	Definition of variable(s) (Fig.3)
1	T	T = time level of sampling points for a lineage
2	S	Stage 1: S = time levels shifted $(T_1-T_1, T_1-T_2, T_1-T_n)$; stage 2: S = value of stage 1 at "bend"
3	S	Stage 1: S = value of stage 2 at "bend"; stage 2: S = time levels shifted
4	T, S	Stage 1: T = time level, S = 0; stage 2: T = time level, S = time levels shifted
5	P	Lineage 1: P = 0; lineage 2: P = 1; P = punctuation

(continued)

TABLE V. *(Continued)*

Model	Variable(s) on which AL, HH, or WC was regressed	Definition of variable(s) (Fig.3)
6	S	Lineage 1: S = time levels shifted; lineage 2: $S = 0$
7	S	Lineage 1: S = time levels shifted; lineage 2: S = value of lineage 1 at furcation
8	T, S	Lineage 1: T = time level, $S = 0$; lineage 2: T = time level, S = time level shifted
9	S, P	Lineage 1: $S = 0$, $P = 0$; lineage 2: S = time levels shifted, $P = 1$
10	S,P	Lineage 1: S = time levels shifted, $P = 0$; lineage 2: S = value of lineage 1 at furcation, $P = 1$
11	T, S, P	Lineage 1: T = time level, $S = 0$, $P = 0$; lineage 2: T = time level, S = time levels shifted, $P = 1$
12	S	Lineage 1: $S = 0$; lineage 2, stage b: S = time levels shifted stage a: S = value of stage b at the "bend"
13	S, P	Lineage 1: $S = 0$, $P = 0$; lineage 2: $P = 0$; stage b, S = time levels shifted; stage 1, S = value of stage b at the "bend"
14	S_1, S_2	Lineage 1: S_1 = time levels shifted, $S_2 = 0$; lineage 2, stage b: $S_1 = S_1$ for lineage 1, S_2 = time levels shifted; lineage 2, stage a: $S_1 = S_1$ of lineage 1 at "bend," S_2 = value of stage b at bend
15	S_1, S_2	Lineage 1: $S_1 = 0$, $S_2 = 0$; lineage 2, stage b: S_1 = time levels shifted, $S_2 = 0$; stage a: S_1 = value of stage b at "bend," S_2 = time levels shifted
16	S_1, S_2, P	Lineage 1: $S_1 = 0$, $S_2 = 0$, $P = 0$; lineage 2, stage b: S_1 = time levels shifted, $S_2 = 0$, $P = 1$; stage a: S_1 = value of stage b at "bend," S_2 = time levels shifted, $P = 1$
17	S_1, S_2, P	Lineage 1: S_1 = time levels shifted, $S_2 = 0$, $P = 0$; lineage 2, stage b: $S_1 = S_1$ for lineage a, S_2 = time levels shifted, $P = 1$; stage a: $S_1 = S_1$ of lineage 1 at "bend," S_2 = value of stage b at "bend," $P = 1$
18	T, S_1, S_2	Lineage 1: T = time levels, $S_1 = 0$; $S_2 = 0$; lineage 2, stage b: T = time levels, S_1 = time levels shifted, $S_2 = 0$; stage a: T = time levels, S_1 = value of stage ba at "bend," S_2 = time level shifted
19	T, S_1, S_2, P	Lineage 1: T = time levels, $S_1 = 0$, $S_2 = 0$, $P = 0$; lineage 2, stage b: T = time levels, S_1 = time levels shifted, $S_2 = 0$, $P = 1$; stage a: T = time levels, S_1 = value of stage b at "bend," S_2 = time level shifted, $P = 1$

Acknowledgments

I am very grateful to Walter Bock, Fred Bookstein, Lloyd Burckle, Eugene Fenster, Max Hecht, Antoni Hoffman, Les Marcus, Stanely Salthe, and Constance Sancetta for constructive criticism and discussions of the material in this chapter. Moreover, I thank Davida Kellogg and Norman Gilinsky for their criticism, which markedly improved this chapter. The Lamont–Doherty Geological Observatory and the Deep Sea Drilling Project (DSDP) provided the material for this study, while CUNY–UCC provided the computing facilities.

REFERENCES

Baldauf, J., 1984, Cenozoic diatom biostratigraphy and paleoceanography of Rockall Plateau region. North Atlantic Deep Sea Drilling Project 81, in: *Initial Reports DSDP 81* (D. G. Roberts *et al.*, eds.), pp. 439–478, U. S. Government Printing Office, Washington, D.C.

Barron, J. A., 1985, late Eocene to Holocene diatom biostratigraphy of the equatorial Pacific Ocean, Deep Sea Drilling Project Leg 85, in: *Initial Reports DSDP 85* (L. Mayer *et al.*, eds.), pp. 413–456, U. S. Government Printing Office, Washington, D.C.

Berggren, W. A., and Casey, R. E., eds., 1983, Symposium on tempo and mode of evolution from micropaleontological data, *Paleobiology* **9**:326–428.

Berggren, W. A., Hamilton, N., Johnson, D. A., Pujal, C., Weiss, W., Cepek, P., and Gombos, V., 1983, Magnetostratigraphy of Deep Sea Drilling Project Leg 72, Sites 515–518, Rio Grande Rise (South Atlantic), in: *Initial Reports DSDP 72* (P. F. Barker *et al.*, eds.), pp. 939–948, U. S. Government Printing Office, Washington, D.C.

Bock, W. J., 1986, Species concepts, speciation, and macroevolution, in: *Modern Aspects of Species* (K. Iwatsuki, P. H. Raven, and W. J. Bock, eds.), pp. 31–57, University of Tokyo Press, Tokyo.

Bookstein, F. L., 1975, On a form of piecewise linear regression, *Am. Stat.* **29**:116–117.

Bookstein, F. L., 1987, Random walk and the existence of evolutionary rates, *Paleobiology* **13**:446–464.

Bookstein, F. L., 1988, Random walk and the biometrics of morphological characters, in: *Evolutionary Biology,* Vol. 23 (M. K. Hecht and B. Wallace, eds.), pp. 369–398, Plenum Press, New York.

Bookstein, F. L., Gingerich, P. D., and Kluge, A. G., 1978, Hierarchical linear modeling of the tempo and mode of evolution, *Paleobiology* **4**:120–130.

Burckle, L. H., 1972, Late Cenozoic planktonic diatom zones from the eastern equatorial Pacific, *Nova Hedwigia* **39**:217–246.

Burckle, L. H., 1978, Early Miocene to Pliocene diatom datum levels for the equatorial Pacific, in: *Proceedings of the 2nd Working Group Meeting, Biostratigraphic Datum Planes of the Pacific Neogene Project 114* (S. Wiryosujono and E. Marks, eds.), pp. 25–44, Bandung.

Burckle, L. H., and Opdyke, N. D., 1977, Late Neogene diatom correlations in the circum-Pacific, in: *Proceedings of the International Congress on Pacific Neogene Stratigraphy,* pp. 255–284, Tokyo.

Burckle, L. H., and Trainer, J., 1979, Lake Pliocene diatom levels in the central Pacific, *Micropaleontology* **25**:281–293.

Burckle, L. H., Morley, J. J., Koizumi, I., and Bleil, U., 1985, Assessment of diatom and radiolarian high- and low-latitude zonations in northwest Pacific sediments: Comparison based upon magnetstratigraphy, in: *Initial Reports DSDP 86* (G. R. Heath *et al.*, eds.), pp. 781–785, U. S. Government Printing Office, Washington, D.C.

Charlesworth, B., Lande, R., and Slatkin, M., 1982, A Neo-Darwinian commentary on macroevolution, *Evolution* **36**(3):474–498.

Cracraft, J., and Eldredge, N. E., eds., 1979, *Phylogenetic Analysis and Paleontology,* Columbia University Press.

Cupp, E. E., 1943, *Marine Plankton Diatoms of the West Coast,* University of California Press, Berkeley, California.

Dingus, L., and Sadler, P. M., 1982, The effects of stratigraphic completeness on estimates of evolutionary rates, *Syst. Zool.* **31**(4):400–412.

Eldredge, N., and Gould, S. J., 1972, Punctuated equilibria: An alternative to phyletic gradualism, in: *Models of Paleobiology* (T. J. M. Schopf, ed.), pp. 82–115, Freeman, Cooper and Co., San Francisco, California.

Fenster, E. J., Sorhannus, U., Burckle, L., and Hoffman, A., 1989, Morphological evolution in the Neogene diatom *Nitzschia jouseae* Burckle, *Hist. Biol.* **2**:In press.

Fisher, D. C., 1986, Progress in organismal design, in: *Patterns and Processes in the History of Life* (D. M. Raup and D. Jablonski, eds.), pp. 99–117, Springer-Verlag, Berlin.

Foster, J. H., and Opdyke, N. D., 1970, Upper Miocene to recent magnetic stratigraphy in deep-sea sediments, *J. Geophys. Res.* **75**:4465–4473.

Fursich, F. T., 1978, The influence of faunal condensation and mixing on the preservation of fossil benthic communities, *Lethia* **11**:243–250.

Gingerich, P. D., 1974, Stratigraphic record of early Eocene Hyopsodus and the geometry of mammalian phylogeny, *Nature* **248**:107–109.

Gingerich, P. D., 1976, Paleontology and phylogeny: Patterns of evolution at the species level in Early Tertiary mammals, *Am. J. Sci.* **276**:1–28.

Gould, S. J., 1982, The meaning of punctuated equilibrium and its role in validating a hierarchical approach to macroevolution, in: *Perspectives on Evolution* (R. Milkman, ed.), pp. 83–104, Sinauer, Sunderland, Massachusetts.

Gould, S. J., and Eldredge, N., 1977, Punctuated equilibria: The tempo and mode of evolution reconsidered, *Paleobiology* **3**:115–151.

Guillard, R. R. L., and Kilham, P., 1977, The ecology of marine planktonic diatoms, in: *The Biology of Diatoms* (D. Werner, ed.), pp. 372–469, University of California Press, Berkeley, California.

Hecht, M. K., 1983, Microevolution, developmental processes, paleontology and the origin of vertebrate higher categories, in: *Modalities, Rhythmes et Mecanismes* (J. Chaline, ed.), pp. 289–294, CNRS, Paris.

Hecht, M. K., and Hoffman, A., 1986, Why not Neo-Darwinism? A critique of paleobiological challenges, *Oxf. Surv. Evol. Biol.* **3**:1–46.

Hoffman, A., 1982, Punctuated versus gradual mode of evolution: A reconsideration, in: *Evolutionary Biology,* Vol. 15 (M. K. Hecht, B. Wallace, and G. T. Prance, eds.), pp. 411–436, Plenum Press, New York.

Hoffman, A., 1988, *Arguments on Evolution: A Paleontologist's Perspective,* Oxford University Press, Oxford.

Jablonski, D., Gould, S.J., and Raup, D. M., 1986, The nature of the fossil record: A biological perspective, in: *Patterns and Processes in the History of Life* (D. M. Raup and D. Jablonski, eds.), pp. 7–21, Springer-Verlag, Berlin.

Kellogg, D. E., 1975, The role phyletic change in the evolution of *Pseudocubus vema* (*Radiolaria*), *Paleobiology* **1**:359–370.

Kellogg, D. E., 1983, Phenology of morphologic change in radiolarian lineages from-deep cores: Implications for macroevolution, *Paleobiology* **9**:355–362.

Kellogg, D. E., 1988, "And then a miracle occurs"—Weak links in the chain of argument from punctuation to hierarchy, *Biol. Philos.* **3**:3–28.

Kellogg, D. E., and Hays, J., 1975, Microevolutionary patterns in Late Cenozoic *Radiolaria, Paleobiology* **1**:150–160.

Koizumi, I., 1968, Tertiary diatom flora of Oga Peninsula, Akita Prefecture, northeast Japan, *Tohoku Univ. Sci. Rep.* **40**:171–240.

Koizumi, I., and Tanimura, Y., 1985, Neogene diatom biostratigraphy of the middle latitude western north Pacific, Deep Sea Drilling Project Leg 86, in: *Initial Reports DSDP 89* (G. R. Heath *et al.*, eds.), pp. 269–300, U. S. Government Printing Office, Washington, D.C.

Lande, R., 1986, The dynamics of peak shifts and pattern of morphological evolution, *Paleobiology* **12:**343–354.

Lazarus, D., 1983, Speciation in pelagic Protista and its study in the planktonic microfossil record: A review, *Paleobiology* **9:**327–340.

Lazarus, D., 1986, Tempo and mode of morphological evolution near the origin of the radiolarian lineage *Pterocanium prismatium, Paleobiology* **12:**175–189.

Lazarus, D., and Prothero, D. R., 1984, The role of stratigraphic and morphologic data in phylogeny, *J. Paleontol.* **58:**163–172.

Lazarus, D., Scherer, R. P., and Prothero, D. R., 1985, Evolution of the radiolarian species-complex *Pterocanium:* A preliminary survey, *J. Paleontol.* **59:**183 220.

Levinton, J. S., 1983, Stasis in progress: The empirical basis of macroevolution, *Annu. Rev. Ecol. Syst.* **14:**103–137.

Levinton, J. S., 1988, *Genetics, Paleontology, and Macroevolution,* Cambridge University Press, Cambridge.

Levinton, J. S., and Simon, C. M., 1980, A critique of the punctuated equilibria model and implications for the detection of speciation in the fossil record, *Syst. Zool.* **29:**130–142.

Malmgren, B. A., and Kennett, J. P., 1981, Phyletic gradualism in a Late Cenozoic planktonic foraminiferal lineage; DSDP Site 284, southwest Pacific, *Paleobiology* **7:**230–240.

Malmgren, B. A., Berggren, W. A., and Lohmann, G. P., 1983, Evidence of punctuated gradualism in the late Neogene *Globoratalia tumida* lineage of planktonic foraminifera, *Paleobiology* **9:** 377–384.

Mayr, E., 1963, *Animals, Species and Evolution,* Harvard University Press, Cambridge, Massachusetts.

McKinney, M. L., and Schoh, R. M., 1983, A composite terrestrial Paleocene section with completeness estimates, based upon magnetostratigraphy, *Am. J. Sci.* **283:**801–814.

Mukhina, V. V., 1965, New species of diatoms from bottom sediments of the equatorial region of the Pacific, in: *Novum Systematica Plantae non Vascularium,* pp. 22–25, Botanical Institute and Academy of Sciences of the USSR, Moscow.

Mukhina, V. V., 1969, Biostratigraphy of sediments and some questions of paleogeography of the tropical region of the Pacific and Indian Oceans, in: *Micropaleontology and Organigenous Sedimetation in the Oceans* (A. P. Jouse, ed.), pp. 5284, Nauka, Moscow.

Ozawa, T., 1975, Evolution of *Lepidolina multiseptata* (Permian foraminifer) in East Asia, *Mem. Fac. Sci. Kyushu Univ. Ser. D Geol.* **23:**117–164.

Prell, W. L., 1985, Pliocene stable isotopes and carbonate stratigraphy (Holes 572C and 573A): Paleoceanography data bearing on the question of Pliocene glaciation, in: *Initial Reports DSDP 85* (L. Mayer *et al.*, eds.), pp. 723–734, U. S. Government Printing Office, Washington, D.C.

Prothero, D. R., and Lazarus, D., 1980, Planktonic microfossils and the recognition of ancestors, *Syst. Zool.* **29:**119–129.

Raup, D. M., and Crick, R. E., 1981, Evolution of single characters in the Jurassic ammonite *Kosmoceras, Paleobiology* **7**(2)**:**90–100.

Raup, D. M., and Crick, R. E., 1982, *Kosmoceras:* evolutionary jumps and sedimentary breaks, *Paleobiology* **8:**90–100.

Sadler, P. M., 1981, Sediment accumulation and the completeness of stratigraphic sections, *J. Geol.* **89:**569–584.

Sancetta, C., 1982, Biostratigraphic and paleoceanographic events in the eastern equatorial Pacific; Results of Deep Sea Drilling Project Leg 69, in: *Initial Reports DSDP 69* (W. L. Prell *et al.*, eds.), pp. 311–342, U. S. Government Printing Office, Washington, D.C.

SAS Institute, 1985, *SAS User's Guide: Statistics,* Version 5, SAS Institute, Cary, North Carolina.

Schaeffer, B., Hecht, M. K., and Eldredge, N., 1972, Phylogeny and paleontology, in *Evolutionary Biology,* Vol. 6 (T. Dobzhansky, M. K. Hecht, and W. C. Steere, eds.), pp. 31–46, Appleton-Century-Crofts, New York.

Schindel, D. E., 1980, Microstratigraphic sampling and the limits of paleontological resolution, *Paleobiology* **6:**408–426.
Schindel, D. E., 1982, Resolution analysis: A new approach to gaps in the fossil record, *Paleobiology* **8:**340–353.
Schopf, T. J. M., 1982, A critical assessment of punctuated equilibria. I. Duration of taxa, *Evolution* **36:**1144–1157.
Schrader, H. J., 1974, Cenozoic marine plankton diatom stratigraphy of the tropical Indian Ocean, in: *Initial Reports DSDP 24* (R. L. Fisher *et al.*, eds.), pp. 887–968, U. S. Government Printing Office, Washington, D.C.
Schrader, H. J., 1977, Diatom biostratigraphy DSDP Project Leg 37, in: *Initial Reports DSDP 37* (F. Aumento *et al.*, eds.), pp. 967–975, U. S. Government Printing Office, Washington, D.C.
Shackleton, N. J., and Opdyke, N. D., 1977, Oxygen isotope and paleomagnetic evidence for early Northern Hemisphere glaciation, *Nature* **270:**216–219.
Simpson, G. G., 1944, *Tempo and Mode in Evolution,* Columbia University Press, New York.
Simpson, G. G., 1953, *The Major Features of Evolution,* Columbia University Press, New York.
Sorhannus, U., Fenster, E.J., Burckle, L. H., and Hoffman, A., 1988, Cladogenetic and anagenetic changes in the morphology of *Rhizosolenia praebergonii,* Muhkina, *Hist. Biol.* **1:**185–205.
Stanley, S. M., 1975, A theory of evolution above the species level, *Proc. Natl. Acad. Sci. USA* **72:** 646–650.
Stanley, S. M., 1979, *Macroevolution—Pattern and Process,* Freeman, San Francisco.
Van Andel, T. H., 1981, Consider the incompleteness of the geological record, *Nature* **294:**397–398.
Werner, D., 1977, Silicate metabolism, in: *The Biology of Diatoms* (D. Werner, ed.), pp. 110–149, University of California Press, Berkeley, California.
Williamson, P. G., 1981, Paleontological documentation of speciation in Cenozoic molluscs from Turkana Basin, *Nature* **293:**437–443.
Wood, E. J. F., 1959, An unusual diatom from the Antaarctic, *Nature* **184:**962–963.
Wood, A. M., Lande, R., and Fryxell, G. A., 1987, Quantitative genetic analysis of morphological variation in an Antarctic diatom grown at two light intensities, *J. Phycol.* **23:**42–54.
Zar, J. H., 1984, *Biostatistical Analysis,* 2nd ed., Prentice-Hall, Englewood Cliffs, New Jersey.

10

Scientific Methodologies in Collision

The History of the Study of the Extinction of the Dinosaurs

MICHAEL J. BENTON

INTRODUCTION

The extinction of the dinosaurs is a major topic in modern paleontology, and indeed in a variety of disciplines that touch on the history of the earth and the history of life. It has also become a major theme in popular accounts, in newspaper and TV reports of science, and in museums. However, this has not always been the case.

About 70 years ago, the extinction of the dinosaurs was regarded as a minor hiccup in the progression of life, no more significant than, say, the extinction of the labyrinthodont amphibians, or the origin of the mammal-like reptiles. It acquired a certain notoriety in the 1950s and 1960s because of its popular appeal, and a vast array of hypotheses was presented, many of them rather bizarre in retrospect. The methods of research and the criteria of hypothesis testing during these years were often very loose.

Finally, in the 1970s and 1980s, major advances in paleobiological methodology and in geochemical and astrophysical research focused strong attention on the question, both from scientists and from the press. The three phases of study of dinosaurian extinction this century may be designated very broadly as follows: (1) the nonquestion phase (up to 1920); (2) the dilettante phase (1920–1970); (3) the professional phase (1970 onward).

MICHAEL J. BENTON • Department of Geology, The Queen's University of Belfast, Belfast BT7 1NN, Northern Ireland, United Kingdom; *present address:* Department of Geology, University of Bristol, Bristol BS8 1RJ, England, United Kingdom.

These three phases are not temporally exclusive, but are based on the majority of publications on the subject each year. General opinions and approaches to the question of the extinction of the dinosaurs have changed in a progressive way from phase 1 to phase 2 to phase 3, and clearly the majority of scientists have followed the broad trends. However, a number of individuals approached this subject in new ways long before the majority of researchers did. For example, Nopcsa (1911, 1917) was considering reasons for dinosaurian extinction in a "dilettante" way long before this became the common approach, in the 1930s and 1940s. At the other end of the spectrum are those individuals who persisted in the old way long after the main body of their science had moved on. So, for example, several paleontologists in the 1950s and 1960s regarded the extinction of the dinosaurs as a nonquestion, while even today many regard this as a subject not capable, not worthy, of serious scientific analysis. On the other hand, several highly significant serious studies were carried out in the 1960s. Hence, the dates given above are only very approximate indications.

The aims of this chapter are to review these stages in the study of the extinction of the dinosaurs and to illustrate the key aspects of each with published examples. Most attention is given to the present, professional phase, the roles of biology and physics in the debate, and the role of the press.

THE NONQUESTION PHASE (1825–1920)

Early Nineteenth Century Views of Extinction

The extinction of the dinosaurs was not regarded as a major event by most 19th century scientists for two reasons. Before 1840, very few dinosaurs had been described, and their uniqueness had not been recognized, so that the early 19th century catastrophists did not focus their attention on them. After 1840, and the naming of the Dinosauria, catastrophism was replaced by uniformitarianism in geology, and Darwinian natural selection emphasized the gradual and continuous nature of extinction. It was assumed then that the dinosaurs had dwindled in a progressive fashion during the Cretaceous. Thus, most 19th century and early 20th century paleontologists did not regard the extinction of the dinosaurs as a particular problem.

The first dinosaurs were described in 1824 and 1825, but they were initially interpreted simply as curious giant lizards. The idea of extinction at the species or the genus level was generally accepted by 1825, and there seemed to be little unusual in finding a few such curious extinct reptiles. The theological arguments of the 18th century, that to admit the possibility of extinction was to accuse God of having made a mistake in Creation, were no longer made (Mayr, 1982;

Bowler, 1984; Buffetaut, 1987), since pragmatic arguments against extinction had been effectively rebuffed. Discoveries of mammoths and mastodons from 1750 onward convinced many that there were extinct species. Nevertheless, as late as 1771, Thomas Pennant could write about these fossil elephants, "as yet the living animal has evaded our search; it is more than possible that it yet exists in some of those remote parts of the vast new continent, unpenetrated as yet by Europeans" (Pennant, 1771). New discoveries of the remains of large fossil animals that had never been found alive—the giant ground sloth, the giant Irish deer, giant cattle, and the much older plesiosaurs and ichthyosaurs—allowed Cuvier and others to demonstrate the reality of extinction to the satisfaction of most savants by 1825.

Cuvier's work on fossil vertebrates from 1799 onward convinced him that many large mammals had become extinct in the not too distant past. He was also able to demonstrate successive earlier faunas of vertebrates and invertebrates in the Tertiary sediments of the Paris Basin. Cuvier invoked a catastrophic explanation for the extinction of these earlier faunas:

> Life in those times was often disturbed by these frightful events. Numberless living things were victims of such catastrophes: some, inhabitants of the dry land, were engulfed in deluges; others, living in the heart of the sea, were left stranded when the ocean floor was suddenly raised up again; and whole races were destroyed forever, leaving only a few relics which the naturalist can scarcely recognise. (Cuvier, 1825, Vol. 1, p. 9.)

Cuvier did not, however, extrapolate from his theory of localized catastrophic extinctions to the global scale. He imagined that physical catastrophes acted within basinal areas only, since life must have survived elsewhere in order to provide the continuity of the fossil record.

Cuvier also expressed views on the vertebrates of the Mesozoic and Cenozoic. At that time (Cuvier, 1825), the dinosaurs were barely known, and the only Mesozoic reptiles were aquatic forms—Jurassic crocodilians, ichthyosaurs, and plesiosaurs, Cretaceous turtles, and older Permian lizardlike animals. Cuvier, then, argued that, until the end of the Mesozoic, vertebrates were essentially marine, or at most swamp-dwellers. He regarded the Tertiary mammals as the first fully dry-land forms. Cuvier recognized a major faunal change, then, at the Cretaceous–Tertiary (K–T) boundary, and he related it to a supposed major phase of regression when the great Mesozoic seas retreated.

The idea of a major shift in vertebrate evolution at the K–T boundary from the sea to the land was rejected by 1830 because of increasing evidence of terrestrial reptiles and mammals in the Mesozoic. Mantell (1831) heralded this in a popular paper entitled, "The geological age of reptiles," and this transition from an Age of Reptiles to an Age of Mammals has held sway ever since.

Buckland (1823) and others more explicitly extrapolated Cuvier's ideas to develop the fully catastrophist geology in which truly global calamities had occurred. Buckland developed his catastrophist hypotheses from his studies of mammal bones in Pleistocene caves. He argued that these "relics of the Flood" had been wiped out by a universal deluge, and extrapolated this back to cover earlier major faunal replacements. Critics of catastrophism, such as Fleming (1826), pointed out that most of the mammals in the Pleistocene caves could have been hunted to death by early humans, and he also indicated that earlier extinctions might also have been caused by predators. Increasing knowledge of the fossil record led to the view by 1849 that at least 29 catastrophes were needed (Bourdier, 1969), each corresponding to a major stratigraphic boundary.

The causes of these catastrophes were not clear, but Buckland wrote, in reference to the sudden change in climate after the last catastrophe (the Flood), "What this cause was, whether a change in the inclination of the earth's axis, or the near approach of a comet, or any other cause or combination of causes purely astronomical, is a question the discussion of which is foreign to the object of the present memoir" (Buckland, 1823, pp. 47–48). These remarkably modern-sounding speculations gave way later to the proposal that the cooling earth had induced the major catastrophes (Buckland, 1836).

The noncatastrophist view was presented by Lyell (1832). He also regarded extinction as important, but more at the species level than necessarily globally. His view was that the species present at any time depended on the environmental conditions, and when such conditions became hostile, extinction inevitably followed.

In the 1820s and 1830s, extinction was viewed broadly in five ways. One view was that extinctions had occurred catastrophically a number of times when most species were wiped out (Buckland). Agassiz took this view one step further, and considered that all species were killed off at times of extinction, to be replaced by a fresh divine Creation. The third view was that extinctions occurred essentially at the local level owing to catastrophic physical effects (Cuvier). The fourth and fifth views were that extinctions occurred at the species level: because conditions became unsuitable (Lyell), or because the species had outlived its natural span.

The Dinosauria

By 1842, several genera of dinosaurs had been described, including *Megalosaurus, Iguanodon,* and *Hylaeosaurus,* and in that year, Richard Owen recognized them as a distinctive order for which he introduced the term Dinosauria (Owen, 1842). Desmond (1979) has argued that Owen "invented" the Dinosauria for a very specific purpose that had little to do with descriptive paleontology, but a great deal to do with then-current views of the history of life. Between 1830 and 1855, there was an active debate about the idea of progression

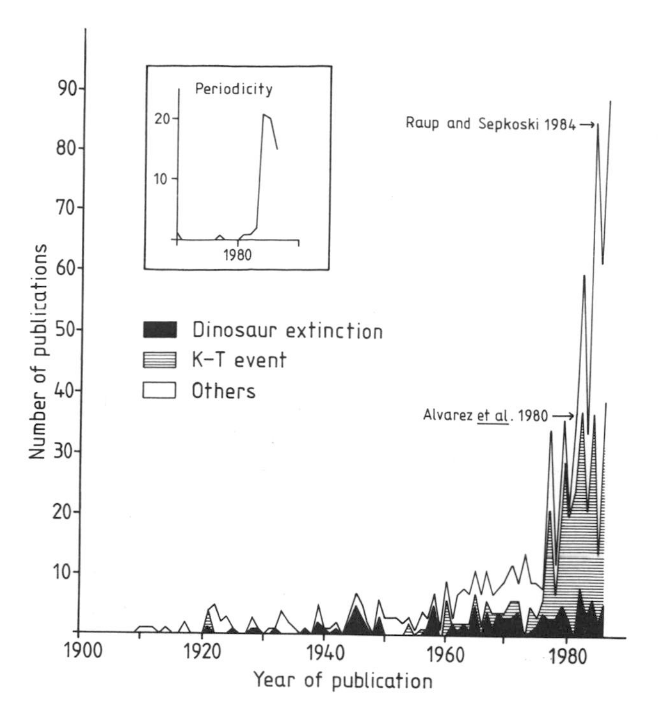

FIG. 1 The increasing rate of publication on dinosaur extinction, and the K-T event, from 1900 to 1986. Figures are plotted year by year, based on a compilation of publications that are concerned solely or mainly with dinosaur extinction, with the K-T event (including stratigraphic accounts), and with broad questions of extinction that relate to the K-T event (shown as "others"). Since 1980, most of the publications that fall in the last category concern the periodicity debate, and these are also plotted separately in the inset. Two seminal papers, in 1980 and 1984, are noted in the diagram.

tion of the dinosaurs alone has remained fairly constant (Fig. 1) at a figure between 0 and 8 per annum (mean 2.0 per annum in the 1960s; 2.8 per annum in the 1970s; 3.7 per annum in the 1980s). The total number of papers has been boosted by general interest in the K–T boundary events, by the rise of catastrophic extraterrestrial theories [particularly in the 1970s, and after the Alvarez *et al.* (1980) paper], and by the more recent interest in the periodicity of mass extinctions following the publication of Raup and Sepkoski (1984). Papers devoted essentially to the K–T event, but mentioning dinosaurian extinction, and papers about periodicity of mass extinctions are indicated separately in Fig. 1. It can be seen that these two categories constitute most of the overall increase in number of publications over the past few years. Overlaid on the general rising pattern are considerable annual fluctuations in number of papers; the peaks generally correspond to the publication of symposium volumes on the K–T event

in the history of life (Bowler, 1976). One view was that life had progressed from simple to more and more complex forms through time, as argued by Robert E. Grant and others, a view emanating from Buffon and Lamarck in the 18th century. Richard Owen opposed this simple notion, and sought to argue that the fossil record showed that degeneration was really the order of the day. Owen's new Dinosauria were described by him (Owen, 1842) as animals "which in structure most nearly approach Mammalia," and which "from their superior adaptation to terrestrial life, [may be concluded] to have enjoyed the function of such a highly organised centre of circulation in a degree more nearly approaching that which now characterizes the warm-blooded Vertebrata." Owen then used these "mammal-like" dinosaurs as direct evidence against the progressionist doctrines of Lamarck, Grant, and others, arguing that these earliest reptiles were the most advanced of all, and that the stock had degenerated to leave only the poor crocodiles, lizards, snakes, and turtles of today.

Owen (1842) argued that the Creator had chosen the Mesozoic Era as suitable for the dinosaurs because of its different atmospheric conditions. He believed that the air then was deficient in oxygen, and that this suited the dinosaurs. As reptiles, they had lower metabolic rates than the birds and mammals and could survive on less energy. He argued that oxygen levels rose during the Mesozoic and that the atmosphere became more "invigorating." The world then became uninhabitable for the huge saurians, and they died out, together with the giant marine reptiles and the flying pterosaurs. This argument was essentially circular, since Owen's evidence for low oxygen levels in the Mesozoic was simply the presence of dinosaurs and other prehistoric reptiles and the virtual absence of mammals and birds. He was explaining the presence of these early reptiles in Mesozoic rocks, very different from the dominant mammals and birds of today, in terms of precise matching by a benevolent Creator of living things and physical environments. He was not primarily trying to explain why the dinosaurs died out when they did—for him, that would have been a preordained event in the Creator's plan.

The progressionists and the supporters of degeneration models for the history of life did not regard the extinction of the dinosaurs as a particular issue, since there were so few of them known, and since they had far more pressing controversial issues to consider: the role of the Creator, the possibility of progression, transmutation (evolution), and so on. The progressionists would probably have interpreted the disappearance of the dinosaurs in terms of the isolated death and replacement of each of the three species then known. The nonprogressionists would have had a number of attitudes. Owen's view, that living reptiles are degenerate forms compared to the dinosaurs and other extinct reptiles, is an example of a theory of "discontinuous progression" (Bowler, 1976), in which a group is created and subsequently remains constant or declines. Another nonprogressionist view, espoused by Charles Lyell, was that living things matched the available habitats and they changed in response to physical changes. Thus, if conditions were right at some time in the future, the dinosaurs would return to inhabit the earth. In none of these views did suprageneric extinction really come into question.

Post-Darwinian Interpretations

The advent of the Darwinian theory of evolution by natural selection in 1859 and subsequent studies of phylogeny based on the fossil record by Haeckel, Huxley, Cope, Marsh, and others did not lead to any real discussion of mass extinctions, nor of the extinction of the dinosaurs in particular, as issues worthy of explanation. Indeed, Darwin (1859) saw the rapid disappearance of the ammonites at the end of the Cretaceous as probably as much to do with gaps in the fossil record as with any real phenomenon. In his paper "On the classification of the Dinosauria . . . ," Huxley (1870) noted the 16 genera of dinosaurs that were known at that date, spanning most of the Mesozoic Era, and based on discoveries in Europe, North America, Africa, and Asia. However, neither here nor elsewhere did he have any statements to make regarding their extinction. Marsh (1882) listed 46 dinosaur genera, and simply noted that they "continued in diminishing numbers to the end of the Cretaceous period, when they became extinct." In a later work, Marsh (1895) listed 68 genera of Dinosauria, but made no mention of their extinction at all.

In the late 19th and early 20th centuries, many paleontologists and biologists moved to non-Darwinian viewpoints, such as orthogenesis and finalism (Bowler, 1983). These models assumed that evolution was directed in some way and that there are regular patterns laid out along which evolution proceeds. The dinosaurs could be viewed as primitive lumbering beasts that *had* to give way to the more advanced mammals.

This then led to views of *racial senility*—the belief that certain long-lived groups of animals became old and their store of evolutionary novelty dried up. There was a parallel here between the life-span of an individual plant or animal and that of an evolutionary stock. Youth and early racial vigor were equated and seen to be just as inevitable as the old age and death of an individual and the racial senescence and eventual extinction of a major phylogenetic group. According to this view, the dinosaurs had been around for a long time, and they simply ran out of the genetic variability that was necessary to survive. The remarkable horns, frills, and spines of some late Cretaceous dinosaurs were occasionally cited as evidence for this racial senility, but in general the extinction of the dinosaurs remained a nonquestion.

Standard textbooks of general paleontology and vertebrate paleontology of the latter half of the 19th century and the first decades of the 20th, like the technical

papers, barely mention the extinction of the dinosaurs. Woodwar notes that "toward the close of the Mesozoic period [sic], . . gradually became extinct," and later (p. 418), he states that the c Cretaceous "became more specialised and almost fantastic ju disappear." Von Zittel (1902, 1932), Hutchinson (1910), J Williston (1925), and Kuhn (1937) barely refer to this topic at all. 1950s and 1960s, several authors on vertebrate palaeontology contin to ignore the extinction of the dinosaurs as a topic (e.g., von Huene, 1 1964; A. H. Müller, 1968).

THE RISE OF INTEREST IN DINOSAURIAN EXTINCTIO

I hope to show below that the study of the extinction of dinosau through several phases in the 20th century. There was a long time, fro 1920 to 1970, when it was seen generally as a moderately interestin moderately amusing) topic, attracting only a few pages of scientific comm each year. Then, after 1970, the level of interest increased, as did the ra publication.

An attempt has been made to survey everything that has been publishe the extinction of the dinosaurs in order to provide a basis for this surve attitudes to this topic. I searched through my reprint collection, noting all te nical papers, books, and popular works that dealt solely with the extinction of dinosaurs, or that covered the Cretaceous–Tertiary (K–T) boundary event, that devoted at least a few pages to the question. I supplemented these figure with references given in the various *Bibliographies of Vertebrate Palaeontolog* (indexed 1928–1973). The total number of publications is 597. When these are plotted according to the year of publication (Fig. 1), it is clear that the rate of publication has been erratic, but has risen in the 1960s and again, dramatically, in the late 1970s and 1980s.

The total number of publications per annum range from 0 to 7 (mean 1.8) for 1910–1959, from 3 to 11 (mean 7.8) for 1960–1969, from 8 to 36 (mean 15.2) for 1970–1979, and from 21 to 85 (mean 50.1) for 1980–1986. Until 1960, most of the papers on the subject of the extinction of the dinosaurs were isolated publications that did not give rise to any real debate, but some, such as the work of Cowles and Bogert in the 1940s on reptilian thermoregulation (e.g., Cowles, 1945, 1949; Colbert *et al.*, 1946), gave rise to brief flurries of papers. In the 1960s and 1970s, more sustained work on dinosaurian biology and on paleoclimates led to a steady number of four or five papers every year.

The main boosts in rates of publication seem to have come from non-dinosaurian quarters. Indeed, the annual number of papers devoted to the extinc-

(1960, 1979), on catastrophic models for mass extinction (1977, 1982 [two]), and on general aspects of mass extinction and diversity in the fossil record (1977, 1981, 1984 [three], 1985 [two], 1986 [three]).

The pattern of publication about the extinction of the dinosaurs may be divided, rather artificially, as has already been noted, into two main phases, the first running from about 1920 to 1970, and the second from 1970 to the present day.

THE DILETTANTE PHASE (1920–1970)

Racial Senility

During the early part of the 20th century, a small number of authors began to treat the extinction of the dinosaurs as an event, and as an event that was worth explaining. A typical early account was given by Arthur Smith Woodward (1910) in his address to the British Association for the Advancement of Science. He argued that the end of the dinosaurs was largely the result of racial senility. He pointed to the great spinescence, excess growth, and loss of teeth of the later dinosaurs as evidence. Loomis (1905, p. 842) wrote about the dorsal bony plates of the stegosaurs: "with such an excessive load of bony weight entailing a drain on vitality, it is little wonder that the family was short-lived." Racial senility as a theory of extinction held sway in some quarters for a long time. Schuchert (1924, p. 12) wrote: "when races are senile, or overspecialized, or are giants of their stocks, they are apt to disappear with the great physiographic and climatic changes that periodically appear in the history of the earth." These kinds of views were also expressed by Beurlen (1933) and Swinton (1939), for example, although they had been earlier rejected by Stromer von Reichenbach (1912), Hennig (1922), and Audova (1929).

Ideas of orthogenesis and racial senility of the dinosaurs were effectively demolished by the advent of the modern synthesis or neo-Darwinian model of evolution in the 1930s and 1940s. There was no longer any place for preordained patterns in this view of evolution. However, the notions of racial senility and of dinosaurs as inferior, lumbering monsters still live on, even if rather subconsciously, in the minds of scientists and the public alike. There seem to be two aspects of this lingering "crypto-orthogenesis": (1) we would like to believe in some form of progress as an optimistic view of the meaning and aim of life; and (2) we would like to believe that *Homo sapiens,* and mammals in general, are at the top of the tree, and that they somehow proved it by overwhelming the dinosaurs. Many evolutionists have wrestled with the idea of progress in neo-Darwinism (e.g., Julian Huxley, George Gaylord Simpson, Francisco Ayala,

Theodosius Dobzhansky), or in punctuated macroevolution (e.g., Steven Stanley), but they have met with difficulties (Gould, 1985; Benton, 1987). It has proved virtually impossible to define ''progress'' in a modern view of evolution in anything other than a Whig manner [i.e., we can only see it from our present standpoint, and judge it with present values (Schopf, 1979)]. The lumbering dinosaur doomed to extinction is an insistent metaphor that refuses to die.

Biotic and Physical Factors

A number of authors, however, eschewed racial senility, and concentrated on biotic and physical factors that might have caused the extinction of the dinosaurs. Baron Franz Nopcsa was one of the first. He noted, for example (Nopcsa, 1911, p. 148), that the great amount of cartilage that he believed was necessary for growth to huge size ''perhaps . . . was one of the causes for the rapid extinction of the Sauropoda.'' Later Nopcsa (1917, p. 345) summarized a number of views on the extinction of the dinosaurs: their ''low power of resistance,'' their huge size, a shortage of food, or ''a reduction in their sexual functions.'' He hints that a key factor may be the supposed ''increase in function of the hypophysis'' (pituitary gland) which was related to their very large body size. Secretions from the pituitary caused giantism, partly by the production of large masses of cartilage as precursors of bone, and partly by a form of acromegaly, or pathological excess thickening and overgrowth of limb bones and facial bones. He notes that ''the increase in weight of the limbs in the dinosaurs recalls the eunuch condition.''

William Diller Matthew (1921) presented early evidence for a model of dinosaurian extinction that involved gradual topographic change and progressive replacement by mammals. His study of the late Cretaceous and Paleocene in North America suggested that there was extensive mountain building and continental uplift. The dinosaur faunas, which were adapted to lowland and marsh situations, were displaced, and the placental mammals, which were adapted to upland zones, moved in. The new mammal faunas apparently had their origins in Asia. Nopcsa (1922, pp. 110–113) presented a modified version of this model, although he laid more stress on floral changes and the loss of marsh-type vegetation.

Other suggestions of these times were that climatic cooling was the cause (Jakovlev, 1922), that disease levels had risen markedly in the late Cretaceous dinosaurs (Moodie, 1923), that early mammals ate all of the dinosaur eggs (Wieland, 1925), or that volcanic eruptions were responsible (L. Müller, 1928).

Audova (1929) reviewed the whole question of the extinction of the dinosaurs in some detail in a remarkable 61-page paper in the short-lived German journal *Palaeobiologica*. He rejected racial senility and simple natural selection as explanations, and focused on environmental change. His favored view, after surveying geological evidence on paleotemperatures and physiological evidence on the thermoregulation of modern reptiles, was that temperature had declined gradually, and that this acted directly on the dinosaurs and other Mesozoic reptiles by preventing proper embryonic development.

This approach typified much of the literature on dinosaurian extinction from the 1930s to the 1960s. Only a small number of papers was published on this topic, ranging from detailed investigations of dinosaurian physiology (e.g., Colbert *et al.*, 1946) and of patterns of diversity and extinction (e.g., Newell, 1952; Simpson, 1952), to brief suggestions and musings on the subject [e.g., de Laubenfels (1956): "Dinosaur extinction: One more hypothesis"].

A Survey of Theories for the Death of the Dinosaurs

It is not possible to discuss all of the hypotheses for dinosaurian extinction that were proposed during the dilettante phase. Jepsen (1964) listed 40 separate hypotheses, and there are doubtless many more than that. I summarize these hypotheses, together with key references where possible, and add numerous others that I have tracked down. In order to bring some semblance of order into this list, I have tried to classify the hypotheses under major headings. The list is probably more comprehensive than any survey yet published, but it is probably still far from complete.

In addition, I have attempted to distinguish the theories that were presented as deliberate jokes (marked with an asterisk) from those that were intended seriously (no asterisk), and those that, in retrospect at least, would appear to be testable at least in part (T) from those that seem to offer no hope of refutation (no T). Hypotheses marked (T?) were presented as speculative ideas which would require some assumptions and extensive testing, while those marked (T) were supported by some evidence and had been tested to some extent.

I. *Biotic causes*

A. "Medical problems"

1. Metabolic disorders
   a. Slipped vertebral discs
   b. Malfunction or imbalance of hormone systems
      i. Overactivity of pituitary gland and excessive (acromegalous?) growth of bones and cartilage (Nopcsa, 1917)
      ii. Malfunction of pituitary gland leads to excess growth of unnecessary and debilitating horns, spines, and frills
      iii. Imbalances of vasotocin and estrogen levels leading to pathological thinning of egg shells (Erben *et al.*, 1979)

c. Diminution of sexual activity (Nopcsa, 1917)
d. Cataract blindness (Croft, 1982)
e. Disease: caries, arthritis, fractures, and infections reached a maximum in late Cretaceous reptiles (Moodie, 1923) (T?)
f. Epidemics
g. Parasites
h. AIDS caused by increasing promiscuity (F. Hoyle and N. C. Wickramasinghe, press reports, 1986–1987) (*)
i. Change in ratio of DNA to cell nucleus

2. Mental disorders
   a. Dwindling brain and consequent stupidity (Raymond, 1939, pp. 148–150)
   b. Absence of consciousness, and absence of the ability to modify behavior (Fremlin, 1979)
   c. Development of psychotic suicidal factors (*)
   d. *Paleoweltschmerz* (*)
3. Genetic disorders: excessive mutation rate induced by high levels of cosmic rays and/or ultraviolet light, leading to small population size burdened by a high genetic load, and consequent vulnerability to environmental shock (Tsakas and David, 1987)

B. Racial senility (''phylogeronty'')
   1. Evolutionary drift into senescent overspecialization, as evinced in gigantism, spinescence (e.g., loss of teeth, and ''degenerate form'') (e.g., Woodward, 1910; Schuchert, 1924; Beurlen, 1933; Swinton, 1939)
   2. Racial old age [Will Cuppy (1964): ''the Age of Reptiles ended because it had gone on long enough and it was all a mistake in the first place''] (*)
   3. Increasing levels of hormone imbalance leading to ever-increasing growth of unnecessary horns and frills (see above)

C. Biotic interactions
   1. Competition with other animals
      a. Competition with the mammals—invasion of North America by Asian mammals (Nopcsa, 1922) (T?)
      b. Competition with caterpillars, which ate all of the plants (Flanders, 1962)
   2. Predation
      a. Overkill capacity by predators (the carnosaurs ate themselves out of existence)
      b. Egg-eating by mammals, which reduced hatching success of the young, and drained gene pools (Wieland, 1925; Thaler, 1965)
   3. Floral changes
      a. Spread of angiosperms and reduction in availability of gymnosperms, ferns, etc. This led to a reduction of fern oils in dinosaur

in the history of life (Bowler, 1976). One view was that life had progressed from simple to more and more complex forms through time, as argued by Robert E. Grant and others, a view emanating from Buffon and Lamarck in the 18th century. Richard Owen opposed this simple notion, and sought to argue that the fossil record showed that degeneration was really the order of the day. Owen's new Dinosauria were described by him (Owen, 1842) as animals "which in structure most nearly approach Mammalia," and which "from their superior adaptation to terrestrial life, [may be concluded] to have enjoyed the function of such a highly organised centre of circulation in a degree more nearly approaching that which now characterizes the warm-blooded Vertebrata." Owen then used these "mammal-like" dinosaurs as direct evidence against the progressionist doctrines of Lamarck, Grant, and others, arguing that these earliest reptiles were the most advanced of all, and that the stock had degenerated to leave only the poor crocodiles, lizards, snakes, and turtles of today.

Owen (1842) argued that the Creator had chosen the Mesozoic Era as suitable for the dinosaurs because of its different atmospheric conditions. He believed that the air then was deficient in oxygen, and that this suited the dinosaurs. As reptiles, they had lower metabolic rates than the birds and mammals and could survive on less energy. He argued that oxygen levels rose during the Mesozoic and that the atmosphere became more "invigorating." The world then became uninhabitable for the huge saurians, and they died out, together with the giant marine reptiles and the flying pterosaurs. This argument was essentially circular, since Owen's evidence for low oxygen levels in the Mesozoic was simply the presence of dinosaurs and other prehistoric reptiles and the virtual absence of mammals and birds. He was explaining the presence of these early reptiles in Mesozoic rocks, very different from the dominant mammals and birds of today, in terms of precise matching by a benevolent Creator of living things and physical environments. He was not primarily trying to explain why the dinosaurs died out when they did—for him, that would have been a preordained event in the Creator's plan.

The progressionists and the supporters of degeneration models for the history of life did not regard the extinction of the dinosaurs as a particular issue, since there were so few of them known, and since they had far more pressing controversial issues to consider: the role of the Creator, the possibility of progression, transmutation (evolution), and so on. The progressionists would probably have interpreted the disappearance of the dinosaurs in terms of the isolated death and replacement of each of the three species then known. The nonprogressionists would have had a number of attitudes. Owen's view, that living reptiles are degenerate forms compared to the dinosaurs and other extinct reptiles, is an example of a theory of "discontinuous progression" (Bowler, 1976), in which a group is created and subsequently remains constant or declines. Another nonprogressionist view, espoused by Charles Lyell, was that living things matched the available habitats and they changed in response to physical changes. Thus, if conditions were right at some time in the future, the dinosaurs would return to inhabit the earth. In none of these views did suprageneric extinction really come into question.

Post-Darwinian Interpretations

The advent of the Darwinian theory of evolution by natural selection in 1859 and subsequent studies of phylogeny based on the fossil record by Haeckel, Huxley, Cope, Marsh, and others did not lead to any real discussion of mass extinctions, nor of the extinction of the dinosaurs in particular, as issues worthy of explanation. Indeed, Darwin (1859) saw the rapid disappearance of the ammonites at the end of the Cretaceous as probably as much to do with gaps in the fossil record as with any real phenomenon. In his paper "On the classification of the Dinosauria . . . ," Huxley (1870) noted the 16 genera of dinosaurs that were known at that date, spanning most of the Mesozoic Era, and based on discoveries in Europe, North America, Africa, and Asia. However, neither here nor elsewhere did he have any statements to make regarding their extinction. Marsh (1882) listed 46 dinosaur genera, and simply noted that they "continued in diminishing numbers to the end of the Cretaceous period, when they became extinct." In a later work, Marsh (1895) listed 68 genera of Dinosauria, but made no mention of their extinction at all.

In the late 19th and early 20th centuries, many paleontologists and biologists moved to non-Darwinian viewpoints, such as orthogenesis and finalism (Bowler, 1983). These models assumed that evolution was directed in some way and that there are regular patterns laid out along which evolution proceeds. The dinosaurs could be viewed as primitive lumbering beasts that *had* to give way to the more advanced mammals.

This then led to views of *racial senility*—the belief that certain long-lived groups of animals became old and their store of evolutionary novelty dried up. There was a parallel here between the life-span of an individual plant or animal and that of an evolutionary stock. Youth and early racial vigor were equated and seen to be just as inevitable as the old age and death of an individual and the racial senescence and eventual extinction of a major phylogenetic group. According to this view, the dinosaurs had been around for a long time, and they simply ran out of the genetic variability that was necessary to survive. The remarkable horns, frills, and spines of some late Cretaceous dinosaurs were occasionally cited as evidence for this racial senility, but in general the extinction of the dinosaurs remained a nonquestion.

Standard textbooks of general paleontology and vertebrate paleontology of the latter half of the 19th century and the first decades of the 20th, like the technical papers, barely mention the extinction of the dinosaurs. Woodward (1898, p. 213) notes that "toward the close of the Mesozoic period [sic], . . . the Dinosaurs gradually became extinct," and later (p. 418), he states that the dinosaurs of the Cretaceous "became more specialised and almost fantastic just before they disappear." Von Zittel (1902, 1932), Hutchinson (1910), Jaekel (1911), Williston (1925), and Kuhn (1937) barely refer to this topic at all. Even into the 1950s and 1960s, several authors on vertebrate palaeontology continued virtually to ignore the extinction of the dinosaurs as a topic (e.g., von Huene, 1956; Orlov, 1964; A. H. Müller, 1968).

THE RISE OF INTEREST IN DINOSAURIAN EXTINCTION

I hope to show below that the study of the extinction of dinosaurs went through several phases in the 20th century. There was a long time, from about 1920 to 1970, when it was seen generally as a moderately interesting (and moderately amusing) topic, attracting only a few pages of scientific commentary each year. Then, after 1970, the level of interest increased, as did the rates of publication.

An attempt has been made to survey everything that has been published on the extinction of the dinosaurs in order to provide a basis for this survey of attitudes to this topic. I searched through my reprint collection, noting all technical papers, books, and popular works that dealt solely with the extinction of the dinosaurs, or that covered the Cretaceous–Tertiary (K–T) boundary event, or that devoted at least a few pages to the question. I supplemented these figures with references given in the various *Bibliographies of Vertebrate Palaeontology* (indexed 1928–1973). The total number of publications is 597. When these are plotted according to the year of publication (Fig. 1), it is clear that the rate of publication has been erratic, but has risen in the 1960s and again, dramatically, in the late 1970s and 1980s.

The total number of publications per annum range from 0 to 7 (mean 1.8) for 1910–1959, from 3 to 11 (mean 7.8) for 1960–1969, from 8 to 36 (mean 15.2) for 1970–1979, and from 21 to 85 (mean 50.1) for 1980–1986. Until 1960, most of the papers on the subject of the extinction of the dinosaurs were isolated publications that did not give rise to any real debate, but some, such as the work of Cowles and Bogert in the 1940s on reptilian thermoregulation (e.g., Cowles, 1945, 1949; Colbert *et al.*, 1946), gave rise to brief flurries of papers. In the 1960s and 1970s, more sustained work on dinosaurian biology and on paleoclimates led to a steady number of four or five papers every year.

The main boosts in rates of publication seem to have come from non-dinosaurian quarters. Indeed, the annual number of papers devoted to the extinc-

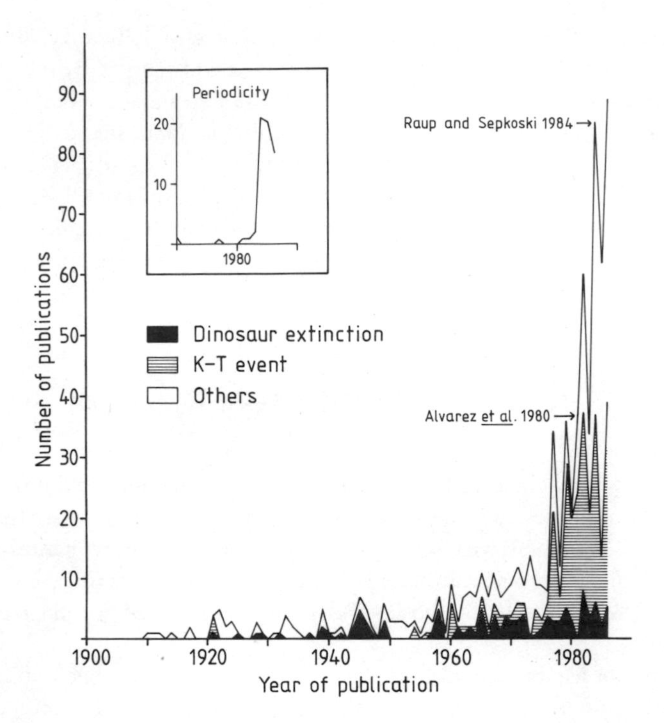

FIG. 1 The increasing rate of publication on dinosaur extinction, and the K-T event, from 1900 to 1986. Figures are plotted year by year, based on a compilation of publications that are concerned solely or mainly with dinosaur extinction, with the K-T event (including stratigraphic accounts), and with broad questions of extinction that relate to the K-T event (shown as "others"). Since 1980, most of the publications that fall in the last category concern the periodicity debate, and these are also plotted separately in the inset. Two seminal papers, in 1980 and 1984, are noted in the diagram.

tion of the dinosaurs alone has remained fairly constant (Fig. 1) at a figure between 0 and 8 per annum (mean 2.0 per annum in the 1960s; 2.8 per annum in the 1970s; 3.7 per annum in the 1980s). The total number of papers has been boosted by general interest in the K–T boundary events, by the rise of catastrophic extraterrestrial theories [particularly in the 1970s, and after the Alvarez *et al.* (1980) paper], and by the more recent interest in the periodicity of mass extinctions following the publication of Raup and Sepkoski (1984). Papers devoted essentially to the K–T event, but mentioning dinosaurian extinction, and papers about periodicity of mass extinctions are indicated separately in Fig. 1. It can be seen that these two categories constitute most of the overall increase in number of publications over the past few years. Overlaid on the general rising pattern are considerable annual fluctuations in number of papers; the peaks generally correspond to the publication of symposium volumes on the K–T event

(1960, 1979), on catastrophic models for mass extinction (1977, 1982 [two]), and on general aspects of mass extinction and diversity in the fossil record (1977, 1981, 1984 [three], 1985 [two], 1986 [three]).

The pattern of publication about the extinction of the dinosaurs may be divided, rather artificially, as has already been noted, into two main phases, the first running from about 1920 to 1970, and the second from 1970 to the present day.

THE DILETTANTE PHASE (1920–1970)

Racial Senility

During the early part of the 20th century, a small number of authors began to treat the extinction of the dinosaurs as an event, and as an event that was worth explaining. A typical early account was given by Arthur Smith Woodward (1910) in his address to the British Association for the Advancement of Science. He argued that the end of the dinosaurs was largely the result of racial senility. He pointed to the great spinescence, excess growth, and loss of teeth of the later dinosaurs as evidence. Loomis (1905, p. 842) wrote about the dorsal bony plates of the stegosaurs: "with such an excessive load of bony weight entailing a drain on vitality, it is little wonder that the family was short-lived." Racial senility as a theory of extinction held sway in some quarters for a long time. Schuchert (1924, p. 12) wrote: "when races are senile, or overspecialized, or are giants of their stocks, they are apt to disappear with the great physiographic and climatic changes that periodically appear in the history of the earth." These kinds of views were also expressed by Beurlen (1933) and Swinton (1939), for example, although they had been earlier rejected by Stromer von Reichenbach (1912), Hennig (1922), and Audova (1929).

Ideas of orthogenesis and racial senility of the dinosaurs were effectively demolished by the advent of the modern synthesis or neo-Darwinian model of evolution in the 1930s and 1940s. There was no longer any place for preordained patterns in this view of evolution. However, the notions of racial senility and of dinosaurs as inferior, lumbering monsters still live on, even if rather subconsciously, in the minds of scientists and the public alike. There seem to be two aspects of this lingering "crypto-orthogenesis": (1) we would like to believe in some form of progress as an optimistic view of the meaning and aim of life; and (2) we would like to believe that *Homo sapiens,* and mammals in general, are at the top of the tree, and that they somehow proved it by overwhelming the dinosaurs. Many evolutionists have wrestled with the idea of progress in neo-Darwinism (e.g., Julian Huxley, George Gaylord Simpson, Francisco Ayala,

Theodosius Dobzhansky), or in punctuated macroevolution (e.g., Steven Stanley), but they have met with difficulties (Gould, 1985; Benton, 1987). It has proved virtually impossible to define "progress" in a modern view of evolution in anything other than a Whig manner [i.e., we can only see it from our present standpoint, and judge it with present values (Schopf, 1979)]. The lumbering dinosaur doomed to extinction is an insistent metaphor that refuses to die.

Biotic and Physical Factors

A number of authors, however, eschewed racial senility, and concentrated on biotic and physical factors that might have caused the extinction of the dinosaurs. Baron Franz Nopcsa was one of the first. He noted, for example (Nopcsa, 1911, p. 148), that the great amount of cartilage that he believed was necessary for growth to huge size "perhaps . . . was one of the causes for the rapid extinction of the Sauropoda." Later Nopcsa (1917, p. 345) summarized a number of views on the extinction of the dinosaurs: their "low power of resistance," their huge size, a shortage of food, or "a reduction in their sexual functions." He hints that a key factor may be the supposed "increase in function of the hypophysis" (pituitary gland) which was related to their very large body size. Secretions from the pituitary caused giantism, partly by the production of large masses of cartilage as precursors of bone, and partly by a form of acromegaly, or pathological excess thickening and overgrowth of limb bones and facial bones. He notes that "the increase in weight of the limbs in the dinosaurs recalls the eunuch condition."

William Diller Matthew (1921) presented early evidence for a model of dinosaurian extinction that involved gradual topographic change and progressive replacement by mammals. His study of the late Cretaceous and Paleocene in North America suggested that there was extensive mountain building and continental uplift. The dinosaur faunas, which were adapted to lowland and marsh situations, were displaced, and the placental mammals, which were adapted to upland zones, moved in. The new mammal faunas apparently had their origins in Asia. Nopcsa (1922, pp. 110–113) presented a modified version of this model, although he laid more stress on floral changes and the loss of marsh-type vegetation.

Other suggestions of these times were that climatic cooling was the cause (Jakovlev, 1922), that disease levels had risen markedly in the late Cretaceous dinosaurs (Moodie, 1923), that early mammals ate all of the dinosaur eggs (Wieland, 1925), or that volcanic eruptions were responsible (L. Müller, 1928).

Audova (1929) reviewed the whole question of the extinction of the dinosaurs in some detail in a remarkable 61-page paper in the short-lived German journal *Palaeobiologica*. He rejected racial senility and simple natural selection as explanations, and focused on environmental change. His favored view, after surveying geological evidence on paleotemperatures and physiological evidence on the thermoregulation of modern reptiles, was that temperature had declined gradually, and that this acted directly on the dinosaurs and other Mesozoic reptiles by preventing proper embryonic development.

This approach typified much of the literature on dinosaurian extinction from the 1930s to the 1960s. Only a small number of papers was published on this topic, ranging from detailed investigations of dinosaurian physiology (e.g., Colbert *et al.*, 1946) and of patterns of diversity and extinction (e.g., Newell, 1952; Simpson, 1952), to brief suggestions and musings on the subject [e.g., de Laubenfels (1956): ''Dinosaur extinction: One more hypothesis''].

A Survey of Theories for the Death of the Dinosaurs

It is not possible to discuss all of the hypotheses for dinosaurian extinction that were proposed during the dilettante phase. Jepsen (1964) listed 40 separate hypotheses, and there are doubtless many more than that. I summarize these hypotheses, together with key references where possible, and add numerous others that I have tracked down. In order to bring some semblance of order into this list, I have tried to classify the hypotheses under major headings. The list is probably more comprehensive than any survey yet published, but it is probably still far from complete.

In addition, I have attempted to distinguish the theories that were presented as deliberate jokes (marked with an asterisk) from those that were intended seriously (no asterisk), and those that, in retrospect at least, would appear to be testable at least in part (T) from those that seem to offer no hope of refutation (no T). Hypotheses marked (T?) were presented as speculative ideas which would require some assumptions and extensive testing, while those marked (T) were supported by some evidence and had been tested to some extent.

I. *Biotic causes*

A. ''Medical problems''

1. Metabolic disorders
    - a. Slipped vertebral discs
    - b. Malfunction or imbalance of hormone systems
        - i. Overactivity of pituitary gland and excessive (acromegalous?) growth of bones and cartilage (Nopcsa, 1917)
        - ii. Malfunction of pituitary gland leads to excess growth of unnecessary and debilitating horns, spines, and frills
        - iii. Imbalances of vasotocin and estrogen levels leading to pathological thinning of egg shells (Erben *et al.*, 1979)

c. Diminution of sexual activity (Nopcsa, 1917)
d. Cataract blindness (Croft, 1982)
e. Disease: caries, arthritis, fractures, and infections reached a maximum in late Cretaceous reptiles (Moodie, 1923) (T?)
f. Epidemics
g. Parasites
h. AIDS caused by increasing promiscuity (F. Hoyle and N. C. Wickramasinghe, press reports, 1986–1987) (*)
i. Change in ratio of DNA to cell nucleus

2. Mental disorders
   a. Dwindling brain and consequent stupidity (Raymond, 1939, pp. 148–150)
   b. Absence of consciousness, and absence of the ability to modify behavior (Fremlin, 1979)
   c. Development of psychotic suicidal factors (*)
   d. *Paleoweltschmerz* (*)
3. Genetic disorders: excessive mutation rate induced by high levels of cosmic rays and/or ultraviolet light, leading to small population size burdened by a high genetic load, and consequent vulnerability to environmental shock (Tsakas and David, 1987)

B. Racial senility (''phylogeronty'')
   1. Evolutionary drift into senescent overspecialization, as evinced in gigantism, spinescence (e.g., loss of teeth, and ''degenerate form'') (e.g., Woodward, 1910; Schuchert, 1924; Beurlen, 1933; Swinton, 1939)
   2. Racial old age [Will Cuppy (1964): ''the Age of Reptiles ended because it had gone on long enough and it was all a mistake in the first place''] (*)
   3. Increasing levels of hormone imbalance leading to ever-increasing growth of unnecessary horns and frills (see above)

C. Biotic interactions
   1. Competition with other animals
      a. Competition with the mammals—invasion of North America by Asian mammals (Nopcsa, 1922) (T?)
      b. Competition with caterpillars, which ate all of the plants (Flanders, 1962)
   2. Predation
      a. Overkill capacity by predators (the carnosaurs ate themselves out of existence)
      b. Egg-eating by mammals, which reduced hatching success of the young, and drained gene pools (Wieland, 1925; Thaler, 1965)
   3. Floral changes
      a. Spread of angiosperms and reduction in availability of gymnosperms, ferns, etc. This led to a reduction of fern oils in dinosaur

diets, and to lingering death by terminal constipation (Baldwin, 1964) (*)

b. Floral change and loss of marsh vegetation (Nopcsa, 1922) (T?)
c. Floral change and increase in forestation, leading to a loss of habitat (Krassilov, 1981) (T?)
d. Reduction in availability of plant food as a whole
e. Presence of poisonous tannins and alkaloids in the angiosperms (Swain, 1976) (T?)
f. Presence of other poisons in plants (T?)
g. Lack of calcium and other necessary minerals in plants (T?)
h. Rise of angiosperms, and of their pollen, led to extinction of dinosaurs by terminal hay fever (Dott, 1983) (*)

II. *Abiotic (physical) causes*

A. Terrestrial explanations

1. Climatic change
   a. Climate became too hot as a result of high levels of carbon dioxide in the atmosphere, and the "greenhouse effect" (McLean, 1978); extinction was caused by the high temperature and increased aridity (Colbert *et al.*, 1946), which either inhibited spermatogenesis (Cowles, 1945), unbalanced the male : female ratio of hatchlings (Ferguson and Joanen, 1982), killed off juveniles (Cowles, 1949), or led to overheating in summer, especially if the dinosaurs were endothermic (Cloudsley-Thompson, 1978) (T?)
   b. Climate became too cold (Jakovlev, 1922; Nopcsa, 1922), and this led to extinction because it was too cold for embryonic development (Audova, 1929), because the endothermic dinosaurs lacked insulation and could not maintain a constant body temperature (L. S. Russell, 1965; Bakker, 1972, 1975), and they were also too large to hibernate (Cys, 1967), or, even if they were inertial homeotherms (i.e., not endotherms), the cold winter temperatures finished them off (Spotila *et al.*, 1973) (T?)
   c. Climate became too dry (Colbert *et al.*, 1946) (T?)
   d. Climate became too wet (T?)
   e. Reduction in climatic equability and increase in seasonality (Axelrod and Bailey, 1968) (T)
2. Atmospheric change
   a. Changes in the pressure or composition of the atmosphere [e.g., excessive amounts of oxygen from photosynthesis (Schatz, 1957)] (T?)
   b. High levels of atmospheric oxygen, leading to fires following an impact (Anderson, 1987) (T)

   c. Low levels of carbon dioxide removed the "breathing stimulus" of endothermic dinosaurs (Wieland, 1942) (T?)
   d. Excessively high levels of carbon dioxide in the atmosphere and asphyxiation of dinosaur embryos in the eggs (Oelofson, 1978) (T?)
   e. Extensive vulcanism and the production of volcanic dust (T?)
   f. Poisoning by selenium from volcanic lava and dust (Koch, 1967) (T?)
   g. Toxic substances in the air, possibly produced from volcanoes, which caused thinning of dinosaur egg shells (Erben, 1972) (T?)
3. Oceanic and topographic change
   a. Marine regression (Ginsburg, 1964; Newell, 1967; Hallam, 1984) (T)
   b. Lowering of global sea level leading to dinosaur extinction, on the assumption that they were underwater organisms (Wilfarth, 1949) (T?)
   c. Floods (T?)
   d. Mountain building, for example, the Laramide Revolution (Matthew, 1921) (T?)
   e. Drainage of swamp and lake habitats (Swinton, 1939) (T?)
   f. Stagnant oceans caused by high levels of carbon dioxide (Keith, 1983) (T)
   g. Bottom-water anoxia at start of transgression (Hallam, 1984) (T)
   h. Spillover of Arctic water (fresh) from its formerly enclosed condition into the oceans, which led to reduced temperatures worldwide, reduced precipitation, and a 10-year drought (Gartner and Keany, 1978; Gartner and McGuirk, 1979) (T?)
   i. Reduced topographic relief, and reduction in terrestrial habitats (Tappan, 1968; Bakker, 1977) (T?)
4. Other terrestrial catastrophes
   a. Sudden vulcanism (L. Müller, 1928; Vogt, 1972; McLean, 1982) (T?)
   b. Fluctuation of gravitational constants (T?)
   c. Shift of the earth's rotational poles (T?)
   d. Extraction of the moon from the Pacific Basin (T?)
   e. Poisoning by uranium sucked up from the soil (Neruchev, 1984) (T?)

B. Extraterrestrial explanations
1. Entropy; increasing chaos in the Universe and hence loss of large organized life forms
2. Sunspots
3. Cosmic radiation and high levels of ultraviolet radiation (Marshall, 1928; Schindewolf, 1958) (T?)
4. Destruction by solar flares of the ozone layer, and letting in ultraviolet radiation (Stechow, 1954; Reid *et al.*, 1976) (T?)

5. Ionizing radiation (Terry and Tucker, 1968; Ruderman, 1974; Yayanos, 1983) (T?)
6. Electromagnetic radiation and cosmic rays from the explosion of a nearby supernova (D. A. Russell and Tucker, 1971; D. A. Russell, 1971; Tucker, 1977) (T?)
7. Interstellar dust cloud (Renard and Rocchia, 1984) (T?)
8. Flash heating of atmosphere by entry of meteorite (de Laubenfels, 1956) (T?)
9. Oscillations about the galactic plane (Hatfield and Camp, 1970) (T?)
10. Impact of an asteroid (Alvarez *et al.*, 1980), a comet (Hsü, 1980), or comet showers (Hut *et al.*, 1987), which caused extinction by a number of postulated mechanisms (see below) (T)

Problems with the "Dilettante" Approach

Certain of these suggestions are perfectly reasonable ideas on the basis of present knowledge, but the obviously ludicrous nature of many has had two consequences. First, many paleontologists were led to believe that this approach was the only one to a study of mass extinction, and therefore that mass extinctions were of little importance to a serious paleontologist. Second, the whole approach was apparently so easy and such fun that everyone felt that they had the opportunity, if not the duty, to solve the question of why the dinosaurs died out. Many of the ideas listed above were presented by nonpaleontologists, and certainly most of the authors had little first-hand knowledge of the late Cretaceous fossil record of dinosaurs—hence the ''dilettante'' soubriquet. A large number of the theories, most of which were published in standard scientific journals by scientists who were no doubt expert in their own fields, show a remarkable relaxation of scientific standards. It was as if, at the mere mention of ''dinosaur extinction,'' scientists breathed a sigh of relief and felt freed from the straitjacket of normal scientific hypothesis-testing.

I believe that there are four main arguments in support of this view.

1. Many of the authors demonstrated an ignorance of the basic paleontological data. For example, the hypotheses were often restricted to explaining why the dinosaurs alone died out, and no mention was made of the marine plankton, invertebrates, and other vertebrates that also disappeared. The question of the survivors of the K–T event was often not tackled: some scenarios were so extreme or catastrophic that it is hard to understand how the land plants, insects, frogs, lizards, snakes, crocodiles, turtles, birds, placental mammals, and so on were not detectably affected. Assumptions have also been made about the suddenness and synchroneity of the K–T event, facts that are not yet established in detail. In other cases, the timing of evolutionary events is wrong: for example, the flowering plants appeared 40–50 million years before the K–T event, the

mammals 150 million years before. Neither group could have caused the demise of the dinosaurs unless some other major evolutionary innovation in one or the other is called in.

2. A number of the theories apparently ignored basic biological principles. Could caterpillars really compete with herbivorous dinosaurs and eat all of the plants (Flanders, 1962)? Could dinosaurs really have been like automata, and unable to modify their behavior (Fremlin, 1979)? Is it possible to model a terrestrial biosphere in which a single biotic factor—epidemics, parasites, glandular malfunction, competition, predation—would lead to a complete ecological breakdown?

3. The mode of argumentation in many papers was by strong advocacy. "If it is assumed that dinosaurs were endothermic/that UV radiation was increasing during the Cretaceous/that caterpillars competed for food with plant-eating dinosaurs, then it follows that . . . . If it is further assumed that climates were becoming warmer, or colder, or drier, or wetter, then it follows that . . . ." It is rare to find careful weighing of evidence both for and against particular hypotheses.

4. There is the overall assumption by some authors that the whole subject is really just a parlor game, and not terribly serious. If a vertebrate paleontologist were to write an account of his or her theory of the origin of the universe or of a cure for cancer or of why caterpillars turn into butterflies, he or she would probably fail to get into print in a reputable scientific journal. However, most of the dilettante theories of the extinction of the dinosaurs were published in very reputable journals: *Science, Nature, American Naturalist, Journal of Paleontology, Evolution,* and so on.

Fortunately, scientific approaches to the question of the extinction of the dinosaurs, the K–T event, and extinctions in general have improved markedly over the past 20 years or so.

THE PROFESSIONAL PHASE (1970 ONWARD)

Background

The professional phase, in which investigators attempted to study the *pattern* of events at the K–T boundary in detail and to present testable hypotheses for the extinction of the dinosaurs, began in the mid 1960s. Sloan and Van Valen (Sloan, 1964, 1970; Van Valen and Sloan, 1972, 1977) attempted to analyze in detail the changes in vertebrate faunas across the K–T boundary in Montana. L. S. Russell (1965), Ostrom (1970), and Bakker (1971, 1972) began detailed studies of dinosaurian physiology and its relation to extinction, while

Axelrod and Bailey (1968) and others surveyed paleobotanical evidence of climatic change. Newell (1962, 1967), Valentine (1969, 1974), and Raup (1972) began general studies of the fossil record and the identification of mass extinctions. Catastrophic terrestrial and extraterrestrial scenarios for extinction were explored by Terry and Tucker (1968), Hatfield and Camp (1970), Crain (1967), Hays (1971), D. A. Russell and Tucker (1971), and Urey (1973), while explanations involving sea-level change were presented by Ginsburg (1964), Newell (1967), and others. Studies in the 1970s and 1980s have generalized from tackling the question of dinosaur extinction alone to the problem of the whole K–T event.

None of the proposed scenarios for K–T extinction, or the extinction of dinosaurs alone, is conducive to a single test. However, each entails a number of hypotheses, some of which are testable. For example, tests can be made of the pattern of a particular event [what was its duration, synchroneity worldwide, synchroneity for all taxa, what did and did not go extinct (any size, habitat, geographic correlations?), absolute and relative taxic effects] as well as many particular predictions made by the different scenarios (e.g., synchroneity of geochemical anomalies worldwide and with the extinction events, terrestrial and extraterrestrial sources of geochemical spikes, matching of timing of extinction events with other physical changes—topography, sea level, climate, chemical composition of seawater, temperature, atmospheric composition).

The Current Scenarios

The two main current scenarios to explain the K–T event (including the extinction of the dinosaurs) are the "gradualist" ecological succession model of Van Valen and others and the "catastrophist" extraterrestrial impact model of Luis Alvarez and others. Recent reviews include Van Valen (1984), Officer *et al.* (1987), and Hallam (1987) on the one side, and L. Alvarez (1983, 1987) and W. Alvarez (1986) on the other side of the issue. There is a considerable amount of evidence of different kinds for both scenarios: mainly paleontological and stratigraphic for the ecological succession model, and mainly geochemical and astrophysical for the extraterrestrial impact model. A "catastrophist" would envisage that the main extinction event lasted less than, say, 1 year, while a gradualist would regard the time span as somewhat more than, say, 1000 years (Van Valen, 1984). At these levels of distinction, the stratigraphy is not good enough to distinguish these time spans: although very different on a biological time scale, they are both geologically "instantaneous." Gradualists typically view the events as very long term; Sloan *et al.* (1986) suggests 7 million years for the extinction of the dinosaurs, with an acceleration in the rate in the last 0.3 million years. On the other hand, catastrophists have offered a range of timings.

Alvarez *et al.* (1980, p. 1099) showed durations of 1–5000 years for the whole event, with the main extinctions occurring in 1–10 years. However, Hsü (1984) has indicated that the knock-on effects of an impact could have lasted for more than 1000 years. Other analyses of the geochemical evidence have given durations of 0.1–0.25 million years (Rocchia *et al.*, 1984; Officer and Drake, 1985), figures that overlap the timings of many gradualist models.

The recent proposal of "stepwise extinction" patterns (Kauffman, 1986; Hut *et al.*, 1987) expands the time scale for extraterrestrially induced mass extinction to 3.5 million years or more. The proposal is that showers of comets arrive on the surface of the earth over intervals of typically 1–3 million years and that pulses of cometary impact cause a major global extinction event in three or four steps. This notion could be seen as a compromise between the instant catastrophe models and the gradualist models (Raup, 1986), a kind of additive long-term catastrophe. However, the stepwise extinction model derives from the extraterrestrial catastrophe models, and it explains mass extinctions by bolide impact, even if by several impacts spaced over 1–3 million years. This may be a necessary modification to the single-impact scenario, as a result of more detailed paleontological evidence, but it is in no way concordant with the earthbound gradualist models as regards the ultimate causation of extinction.

The "Gradualist" Models

The gradual ecosystem evolution model has been largely based on the progressive appearance of a mammal community (the *Protungulatum* Community) of distinctive Paleocene aspect in the last 300,000 years of the Cretaceous in Montana. As the mammals increase in abundance, the dinosaurs apparently decline until they disappear altogether (Van Valen and Sloan, 1977; Sloan *et al.*, 1986). This gradual replacement is explained in terms of diffuse competition between dinosaurs and mammals set against a major change in habitats. The lush subtropical dinosaur habitats were apparently giving way to cooler temperate forests which favored the mammals. It has been argued, however (Fastovsky and Dott, 1986), that the occurrence of dinosaurs and mammals in these particular cases are in channel sediments that cannot be dated as either Cretaceous or Paleocene.

The gradualist scenario has been extended to cover all aspects of the K–T events. Thus, Perch-Neilsen *et al.* (1982) note that the planktonic extinctions took 10,000 years, and various groups were already declining well before the boundary (Signor and Lipps, 1982). A variety of sea-bottom livers and filter-feeders died out, but sea-bottom predators and detritus-feeders were little affected. Extinction patterns of many marine groups show gradual declines throughout the late Cretaceous (Kauffman, 1984). Many gradualists also link the extinctions to marine regressions, as well as to declining temperatures (e.g., Ginsburg, 1964; Hallam, 1984, 1987). Gradualists also argue that the fact that many groups did not go extinct at the K–T boundary is hard to understand in the face of some of the devastating catastrophist scenarios (Buffetaut, 1984). On land, placental mammals, birds, lizards, snakes, crocodiles, champsosaurs, and tortoises and other freshwater organisms were little affected, and the plant record shows only modest and gradual changes (Hickey, 1981).

The "Catastrophist" Models

There is a huge literature now on the catastrophist extraterrestrial scenarios [for review see L. Alvarez (1983, 1987), Hsü (1984), W. Alvarez (1986)]. Basically, these postulate the impact of an asteroid or a comet on the earth. The impact caused mass extinctions either by throwing up a vast dust cloud which blocked out the sun and prevented photosynthesis (Alvarez *et al.*, 1980), by releasing cyanide (Hsü, 1980), by flash heating of the atmosphere on entry (Hsü, 1980), by excessive cooling of the atmosphere (Turco *et al.*, 1983), by releasing poisonous arsenic and osmium (Hsü *et al.*, 1982), by global wildfires (Wolbach *et al.*, 1985), or by a combination of darkness, extreme cold, and acid rain (Prinn and Fegley, 1987). These chemical and darkness models are postulated to explain most of the extinctions, while the extinction of the dinosaurs is often ascribed to thermal stress (Hsü, 1984).

The best evidence for the impact hypothesis is said by some to be the iridium (Ir) enhancement at the boundary, now recorded from nearly 60 locations worldwide (Alvarez, 1983), while others emphasize the importance of shocked quartz (Bohor *et al.*, 1987). Further evidence includes the occurrence of spherules in a few sections (Smit and Kyte, 1984), the similarity of the ratios of elements in boundary clays to those of chondrites, isotopic changes in O and C, and the actual presence of clays at the boundary in so many sections. A catastrophic extinction is also indicated by abrupt shifts in pollen ratios at some K–T boundaries (Tschudy *et al.*, 1984; Wolfe and Upchurch, 1986), as well as abrupt plankton and other extinctions in certain sections (Alvarez *et al.*, 1984*a;* Surlyk and Johansen, 1984).

In opposition to the catastrophist explanations, Van Valen (1984) and Hallam (1987) note the following criticisms: supposed extraterrestrial material below the K–T boundary, absence of the effects of elimination of stratospheric ozone, and problems with the darkness, cooling, acid rain, and other predicted results of impact. Officer and Drake (1985) and Hallam (1987) have further argued that the Ir could be of terrestrial volcanic origin, and that the Ir spikes lasted from 10,000 to 100,000 years. This idea of a relatively long Ir spike and the evidence for long-term decline of dinosaurs and certain marine groups could suggest a combination of the gradualist and catastrophist views in which certain groups were already declining when a bolide impact, or intense vulcanism, finished them off—a kind of "last-strawist" view point (cf. Buffetaut, 1984).

A CONFLICT OF STYLES

Physics versus Paleontology

The two main models for dinosaurian extinction are based on rather different kinds of data, as mentioned above: essentially paleontological and stratigraphic for the gradualist models, and mainly geochemical and astrophysical for the catastrophic models. This has inevitably meant that it has been hard for the proponents of one view to assess the evidence that supposedly favors the other view, as noted by Van Valen (1984). There is, however, apparently a more fundamental source of potential conflict between certain biologists and physicists, or "soft" scientists and "hard" scientists, as Raup (1986, p. 212) describes the pecking order.

The initial publications by Alvarez *et al.* (1980) were greeted skeptically by many paleontologists and geologists with long-term expertise on aspects of the K–T boundary. No doubt, they resented the intrusion into their subject by a group of physicists, and Luis Alvarez's (1983, p. 632) lengthy catalogue of his team's credentials (a physicist, two nuclear chemists, and a geologist) may not seem so unusual in view of this resentment: "suddenly I realised that we combined in one group a wide range of scientific capabilities, and that we could use these to shed some light on what was really one of the greatest mysteries in science—the sudden extinction of the dinosaurs." Walter Alvarez (1986) characterizes the two main geological prejudices that the impacters faced: a deeply-held belief in uniformitarianism and gradualism among many geologists, and detailed objections from paleontologists who saw no general large-scale instant catastrophe in the fossil record.

The crux of the dispute was outlined by Jastrow (1983, p. 152): "Professor Alvarez was pulling rank on the palaeontologists. Physicists sometimes do that; they feel they have a monopoly on clear thinking. There is a power in their use of math and the precision of their measurements that transcends the power of the softer sciences." The very titles of their papers could be seen to exemplify this: Alvarez *et al.* (1980) is "Extraterrestrial cause for the Cretaceous–Tertiary extinction—Experimental results and theoretical implications," while Alvarez (1983) is "Experimental evidence that an asteroid impact led to the extinction of many species 65 million years ago." Van Valen (1984, p. 122) commented that "to call [the Alvarez] evidence "experimental" is misleading propaganda; it refers merely to the fact that some observations were made in the laboratory rather than in the field, not to an active experimental test.''

Luis Alvarez is surprisingly revealing throughout his 1983 paper. He is dismissive of his critics:

> I think the first two points—that the asteroid hit, and that the impact triggered the extinction of much of the life in the sea—are no longer debatable points. Nearly everybody now believes them. But there are always dissenters. I understand that there is even one famous American geologist who does not yet believe in plate tectonics. . . . People have telephoned with facts and figures to throw the theory into disarray, and written articles with the same intent, but in every case the theory has withstood these challenges. (Alvarez, 1983, p. 67.)

He later outlines the advantages of physics in comparison with paleontology: ''The field of data analysis is one in which I have had a lot of experience'' (Alvarez, 1983, p. 638), and ''In physics, we do not treat seriously theories with such low *a priori* probabilities'' (p. 640). He writes, ''That is something that made me very proud to be a physicist, because a physicist can react instantaneously when you give him some evidence that destroys a theory that he previously had believed. . . . But that is not true in all branches of science, as I am finding out'' (p. 629). Public utterances from Luis Alvarez about his ''opponents'' have frequently been more critical than these examples to the point of being libellous (e.g., Browne, 1988). Van Valen (1984, pp. 136–137) complains about how Alvarez ''makes fun'' of paleontologists, while Halstead (1986) and Archibald (1987) argue that it is wrong to argue that physics is necessarily better than paleontology—it is only different.

On the other hand, much of the distrust of the physicists by certain paleontologists has surely been unfounded, as Raup (1986, pp. 104–105) notes. He quotes at length a statement by Robert Bakker, a dinosaur paleobiologist, originally published in the *New York Times:*

> The arrogance of those people is simply unbelievable. They know next to nothing about how real animals evolve, live and become extinct. But despite their ignorance, the geochemists feel that all you have to do is crank up some fancy machine and you've revolutionized science. The real reasons for the dinosaur extinctions have to do with temperature and sea level changes, the spread of diseases by migration and other complex events. But the catastrophe people don't seem to think such things matter. In effect, they're saying this: ''We high-tech people have all the answers, and you paleontologists are just primitive rock hounds.''

The impact hypothesis, being a new idea, was initially at a disadvantage, as Raup (1986, pp. 195–197) argues. In order to displace the established wisdom, the impacters would have to present overwhelmingly strong evidence which would be much more critically scrutinized than the gradualists' evidence. Indeed, the catastrophist hypothesis was on trial in a way that the gradualist hypothesis was not. As Clemens *et al.* (1981) wrote of the proposed asteroid impact, ''analyses of the paleobiological data suggest that such an event is not required to explain the biotic changes during the Cretaceous–Tertiary transition.''

Styles of Argumentation

Elsabeth Clemens (1986) has analyzed the nature of the debate about ''asteroids and dinosaurs,'' and she argues that there are many nonscientific undercurrents, such as styles of argumentation and the role of professional and popular publication.

First, the broadly-based research enterprise that has developed around the question of the extinction of the dinosaurs—geologists, paleontologists, chemists, physicists, astronomers—is not a single community. It is a body consisting of several factions, each going in different directions, and with very little communication, a point noted also by Van Valen (1984, p. 121). Clemens goes on to suggest that the Alvarez theory gained rapid notice and acceptance in many quarters because catastrophism in geology was becoming intellectually fashionable. Catastrophic theories for mass extinction had been made for years (e.g., de Laubenfels, 1956; Terry and Tucker, 1968; McLaren, 1970; D. A. Russell and Tucker, 1971; Urey, 1973; Reid *et al.*, 1976), but they contradicted the strictly gradualistic ''geological dogma'' of the day. However, the supernova theories (Terry and Tucker, 1968; D. A. Russell and Tucker, 1971; Béland *et al.*, 1977) probably did prepare the way for the asteroid theory.

Clemens (1986) argues that it was the mode of presentation of the Alvarez hypothesis that won it such wide attention and acceptance: ''In a sense, the problem of the K–T boundary was framed so as to be amenable to the methods of particle physics.'' The bulk of the long 1980 paper (14 pages in all) was confined to the geological and physical evidence for an impact, and the physical results of the impact. The discussion of the biological results of the impact occupies only half a page. The paper was restricted then to a rather simple astrophysical hypothesis which could be tested in many ways, and the more complex aspects of stratigraphic imprecision and complexity of the evolution of biological communities were largely omitted. These issues had to be taken on board later, however. Alvarez *et al.* (1984*b*) allowed from 10^4 to 10^5 years for the overall length of time involved in the extinctions, while Alvarez *et al.* (1984*a,* p. 1135) note that ''the paleontological record thus bears witness to terminal-Cretaceous extinctions on two time scales: a slow decline unrelated to the impact and a sharp truncation synchronous with and probably caused by the impact.'' The recent proposal of a stepwise extinction model involving comet showers (Kauffman, 1986; Hut *et al.*, 1987) extends the catastrophists' time scale to 3.5 million years or more. However, by 1984, the simplicity of the ''instant-extinction'' model of

1980 had ensured its general acceptance by many scientists. The later modifications noted here are seen by Clemens (1986, pp. 434, 441) as rather *ad hoc* qualifiers that tend to protect the impact theory from refutation by stratigraphic or paleontological evidence. The extension of the impact scenario to "nuclear winter" models in 1981, to theories of extraterrestrially induced periodic mass extinctions in 1983, and to the comet shower model of 1987 further helped to cement its professional and popular appeal.

The Role of the Professional and Popular Press

Clemens (1986) then goes on to argue that the nature of the professional and popular press has largely shaped the development of models of dinosaurian extinction since 1980. She points out that the 1980 *Science* article (Alvarez *et al.*, 1980) was twice as long as such articles usually are, and it was published in a prominent position, at the start of the issue. This one article gained a very wide readership, particularly in the United States, whereas articles that presented similar theories at the same time (Smit and Hertogen, 1980; Hsü, 1980; Ganapathy, 1980) were much less widely read (Hoffman and Nitecki, 1985). Since 1980, it has been alleged, pro-impact papers have been much favored by the editorial board of *Science,* and the argument has spilled over into the commentary and review sections of leading journals and into the newspapers (e.g., Lewin, 1983, 1985*a,b;* Maddox, 1985; Hoffman, 1985; Browne, 1985, 1988). Clemens (1986) suggests that the very format of publication has had a restrictive effect, since most of the debate has been carried on so far in the pages of *Science* and *Nature,* both of which normally publish only very short papers of three or four pages in length, and both of which require papers to be readily understandable to a wide audience. It is easier to present a simple hard view, such as the impact, she argues, than to argue about the imprecision of present methods for dating rocks, or the complexity of biological communities.

CONCLUSION

There are clearly a number of layers to the "catastrophists versus gradualists" controversy, ranging from purely scientific aspects, to matters of style, modes of argumentation, and the nature of publication. These all add spice to the controversy, but do not necessarily lead to progress toward its resolution. At present, it is hard to see how the two viewpoints will be fused, since there can only be one correct explanation for the extinction of the dinosaurs, whether "gradualistic," "catastrophic," or a bit of both. From its rather modest image only 20 or so years ago, the extinction of the dinosaurs has now become one of the most studied unique events in the history of life. Indeed, there is now a sort of "bandwagon" effect, as new topics are spawned—nuclear winter, periodicity, comet showers—which keep public interest alive, and which keep levels of funding for research at record levels. In general, the controversy has been good for the historical sciences (geology, paleontology), and it can only be hoped that the uncomfortable grating between physicists and paleontologists will eventually lead to a more satisfactory and fully cooperative research effort.

Acknowledgments

I thank Peter Bowler, Steve Yearley, Max Hecht, and anonymous reviewers for helpful remarks.

REFERENCES

Alvarez, L. W., 1983, Experimental evidence that an asteroid impact led to the extinction of many species 65 million years ago, *Proc. Natl. Acad. Sci. USA* **80:**627–642.

Alvarez, L. W., 1987, Mass extinctions caused by large bolide impacts, *Phys. Today* **1987**(7)**:**24–33.

Alvarez, L. W., Alvarez, W., Asaro, F., and Michel, H. V., 1980, Extraterrestrial cause for the Cretaceous–Tertiary extinction—Experimental results and theoretical implications, *Science* **208:**1095–1108.

Alvarez, W., 1986, Toward a theory of impact crises, *Eos* **67**(35)**:**649, 653–655, 658.

Alvarez, W., Kauffman, E. G., Surlyk, F., Alvarez, L. W., Asaro, F., and Michel, H. V., 1984*a*, Impact theory of mass extinctions and the invertebrate fossil record, *Science* **223:**1135–1141.

Alvarez, W., Alvarez, L. W., Asaro, F., and Michel, H. V., 1984*b*, The end of the Cretaceous: Sharp boundary or gradual transition?, *Science* **223:**1183–1186.

Anderson, I., 1987, Dinosaurs breathed air rich in oxygen, *New Sci.* **116**(1585)**:**25.

Archibald, J. D., 1987, Stepwise and non-catastrophic late Cretaceous terrestrial extinctions in the Western Interior of North America: Testing observations in the context of an historical science, *Mém. Soc. Géol. Fr.* **150:**45–52.

Audova, A., 1929, Aussterben der Mesozoischen Reptilien, *Palaeobiologica* **2:**222–245, 365–401.

Axelrod, D. I., and Bailey, H. P., 1968, Cretaceous dinosaur extinction, *Evolution* **22:**595–611.

Bakker, R. T., 1971, Dinosaur physiology and the origin of mammals, *Evolution* **25:**636–658.

Bakker, R. T., 1972, Anatomical and physiological evidence of endothermy in dinosaurs, *Nature* **238:**81–85.

Bakker, R. T., 1975, Dinosaur renaissance, *Sci. Am.* **232**(4)**:**58–78.

Bakker, R. T., 1977, Tetrapod mass extinctions—A model of the regulation of speciation rates and immigration by cycles of topographic diversity, in: *Patterns of Evolution as Illustrated by the Fossil Record* (A. Hallam, ed.), pp. 439–468, Elsevier, Amsterdam.

Baldwin, E., 1964, *An Introduction to Comparative Biochemistry,* 4th ed., Cambridge University Press, Cambridge.

Béland, P., Feldman, P., Foster, J., Jarzen, D., Norris, G., Pirozynski, K., Reid, G., Roy, J. R., Russell, D., and Tucker, W., 1977, Cretaceous–Tertiary extinctions and possible terrestrial and extraterrestrial causes, *Syllogeus* **1977**(12):1–162.

Benton, M. J., 1987, Progress and competition in macroevolution, *Biol. Rev.* **62:**305–338.

Beurlen, K., 1933, Vom Aussterben der Tiere, *Nat. Mus.* **63:**1–8, 55–63, 102–106.

Bohor, B. F., Modreski, P. J., and Foord, E. E., 1987, Shocked quartz in the Cretaceous–Tertiary boundary clays: Evidence for a global distribution, *Science* **236:**705–709.

Bourdier, F., 1969, Geoffroy Saint-Hilaire versus Cuvier: The campaign for paleontological evolution (1825–1838), in: *Toward a History of Geology* (C. J. Schneer, ed.), pp. 36–63, Harvard University Press, Cambridge, Massachusetts.

Bowler, P. J., 1976, *Fossils and Progress: Paleontology and the Idea of Progressive Evolution in the Nineteenth Century,* Science History Publications, New York.

Bowler, P. J., 1983, *The Eclipse of Darwinism: Anti-Darwinian Evolution Theories in the Decades around 1900,* Johns Hopkins University Press, Baltimore, Maryland.

Bowler, P. J., 1984, *Evolution: The History of an Idea,* University of California Press, Berkeley, California.

Browne, M. W., 1985, Dinosaur experts resist meteor extinction idea: Paleontologists say dissenters risk harm to their careers, *N. Y. Times* **1985**(29 October)**:**21–22.

Browne, M. W., 1988, Debate over dinosaur extinction takes an unusually rancorous turn, *N. Y. Times* **1988**(19 January)**:**19, 23.

Buckland, W., 1823, *Reliquiae Diluvianae,* Murray, London.

Buckland, W., 1836, *Geology and Mineralogy Considered with Reference to Natural Theology,* Murray, London.

Buffetaut, E., 1984, Selective extinctions and terminal Cretaceous events, *Nature* **310:**276.

Buffetaut, E., 1987, *A Short History of Vertebrate Palaeontology,* Croom Helm, London.

Clemens, E. S., 1986, Of asteroids and dinosaurs: The role of the press in the shaping of scientific debate, *Soc. Stud. Sci.* **16:**421–456.

Clemens, W. A., Archibald, J. D., and Hickey, L. J., 1981, Out with a whimper not a bang, *Paleobiology* **7:**293–298.

Cloudsley-Thompson, J. L., 1978, *Why the Dinosaurs Became Extinct,* Meadowfield, Shildon, England.

Colbert, E. H., Cowles, R. B., and Bogert, C. M., 1946, Temperature tolerances in the American alligator and their bearing on the habits, evolution and extinction of the dinosaurs, *Bull. Am. Mus. Nat. Hist.* **86:**327–274.

Cowles, R. B., 1945, Heat-induced sterility and its possible bearing on evolution, *Am. Nat.* **79:**160–175.

Cowles, R. B., 1949, Additional speculations on the role of heat in evolutionary processes, *J. Entomol. Zool.* **41:**3–22.

Crain, I. K., 1967, Possible direct causal relation between geomagnetic reversals and biological extinctions, *Bull. Geol. Soc. Am.* **82:**2603–2606.

Croft, L. R., 1982, *The Last Dinosaurs,* Elmwood Books, Chorley, Lancashire, England.

Cuppy, W., 1964, *How to Become Extinct,* Dover, New York.

Cuvier, G., 1825, *Discours sur les Revolutions de la Surface de la Globe, et sur les Changements qu'elles ont Produites dans la Regne Animal,* Dufour et d'Ocagne, Paris.

Cys, J. M., 1967, On the inability of the dinosaurs to hibernate as a possible key factor in their extinction, *J. Paleontol.* **41:**226.

Darwin, C. R., 1859, *On the Origin of Species by Means of Natural Selection,* Murray, London.

De Laubenfels, M. W., 1956, Dinosaur extinction: One more hypothesis, *J. Paleontol.* **30:**207–218.

Desmond, A. J.,1979, Designing the dinosaur: Richard Owen's response to Robert Edmond Grant, *Isis* **70:**224–234.

Dott, R. H., Jr., 1983, Itching eyes and dinosaur demise, *Geology* **11:**126.

Erben, H. K., 1972, Ultrastrukturen und Dicke der Wand pathologischer Eischalen, *Akad. Wiss. Lit. Mainz, Abh. Math.-Naturwiss, Kl.* **6:**193–216.

Erben, H. K., Hoefs, J., and Wedepohl, K. H., 1979, Paleobiological and isotope studies of eggshells from a declining dinosaur species, *Paleobiology* **5:**380–414.

Fastovsky, D. E., and Dott, R. H., Jr., 1986, Sedimentology, stratigraphy, and extinctions during the Cretaceous–Paleogene Tertiary transition at Bug Creek, Montana, *Geology* **14:**279–282.

Ferguson, M. W. J., and Joanen, T., 1982, Temperature of egg incubation determines sex in *Alligator mississippiensis, Nature* **296:**850–853.

Flanders, S. E., 1962, Did the caterpillar exterminate the giant reptile?, *J. Res. Lepidopt.* **1:**85–88.

Fleming, J., 1826, The geological Deluge, as interpreted by Baron Cuvier and Professor Buckland, inconsistent with the testimony of Moses and the phenomena of nature, *Edinb. Philos. J.* **14:** 205–239.

Fremlin, J., 1979, Dinosaur death: The unconscious factor, *New Sci.* **81:**250–251.

Ganapathy, R., 1980, A major meteorite impact on the earth 65 million years ago: Evidence from the Cretaceous–Tertiary boundary clay, *Science* **209:**921–923.

Gartner, S., and Keany, J., 1978, The terminal Cretaceous event: A geologic problem with an oceanographic solution, *Geology* **6:**708–712.

Gartner, S., and McGuirk, J. P., 1979, Terminal Cretaceous extinction: Scenario for a catastrophe, *Science* **206:**1272–1276.

Ginsburg, L., 1964, Les regressions marines et le problème de renouvellement des faunes au cours des temps géologiques, *Bull. Soc. Géol. Fr.* **6:**13–22.

Gould, S. J., 1985, The paradox of the first tier: An agenda for paleobiology, *Paleobiology* **11:**2–12.

Hallam, A., 1984, Pre-Quaternary sea-level changes, *Annu. Rev. Earth Planet. Sci.* **12:**205–243.

Hallam, A., 1987, End-Cretaceous extinction event: Argument for terrestrial causation, *Science* **238:** 1237–1242.

Halstead, L. B., 1986, The physicists and the palaeontologists in the battle of the giants, *Guardian* **1986**(5 September)**:**11.

Hatfield, C. B., and Camp, M. J., 1970, Mass extinctions correlated with periodic galactic events, *Bull. Geol. Soc. Am.* **81:**911–914.

Hays, J. D., 1971, Faunal extinctions and reversals of the earth's magnetic field, *Bull. Geol. Soc. Am.* **82:**2433–2447.

Hennig, E., 1922, Paläontologische Beiträge zur Entwicklungslehre, *Tübinger Naturwiss. Abh.* **4:** 27–35.

Hickey, L. J., 1981, Land plant evidence compatible with gradual, not catastrophic, change at the end of the Cretaceous, *Nature* **282:**529–531.

Hoffman, A., 1985, [untitled letter], *Science* **230:**8.

Hoffman, A., and Nitecki, M. H., 1985, Reception of the asteroid hypothesis of terminal Cretaceous extinctions, *Geology* **13:**884–887.

Hsü, K. T., 1980, Terrestrial catastrophe caused by a cometary impact at the end of the Cretaceous, *Nature* **285:**201–203.

Hsü, K. T., 1984, A scenario for the terminal Cretaceous event, *Init. Rep. DSDP* **73:**755–764.

Hsü, K. T., He, Q., McKenzie, J. A., Weissert, H., Perch-Nielsen, K., Oberhänsli, H., Kelts, K., LaBrecque, J., Tauxe, L., Krähenbühl, U., Percival, S. F., Jr., Wright, R., Karpoff, A. M., Petersen, N., Tucker, P., Poore, R. Z., Gombos, A. M., Pisciotto, K., Carman, M. F., Jr., and Schreiber, E., 1982, Mass mortality and its environmental and evolutionary consequences, *Science* **216:**249–256.

Hut, P., Alvarez, W., Elder, W. P., Hansen, T., Kauffman, E. G., Keller, G., Shoemaker, E. M., and Weissman, P. R., 1987, Comet showers as a cause of mass extinctions, *Nature* **329:**118–126.

Hutchinson, H. N., 1910, *Extinct Monsters and Creatures of Other Days,* Chapman and Hall, London.

Huxley, T. H., 1870, On the classification of the Dinosauria, with observations on the dinosaurs of the Trias, *Q. J. Geol. Soc. Lond.* **26:**32–51.

Jaekel, O., 1911, *Die Wirbeltiere; Eine übersicht über die fossilen und lebenden Formen,* Borntraeger, Berlin.

Jakovlev, N. N., 1922, Vymiranye i Evo Prichiny kak Osnovnoy Vopros Biologii, *Mysl* **2:**1–36.

Jastrow, R., 1983, The dinosaur massacre: A double-barrelled mystery, *Sci. Digest* **1983**(September)**:**151–153.

Jepsen, G. L., 1964, Riddles of the terrible lizards, *Am. Sci.* **52:**227–246.

Kauffman, E. G., 1984, The fabric of Cretaceous marine extinctions, in: *Catastrophes and Earth History,* (W. A. Berggren and J. A. van Couvering, eds.), pp. 151–246, Princeton University Press, Princeton, New Jersey.

Kauffman, E. G., 1986, High-resolution event stratigraphy: Regional and global bio-events, in: *Global Bio-events* (O. Wallisser, ed.), pp. 279–335, Springer-Verlag, Heidelberg.

Keith, M. L., 1983, Violent volcanism, stagnant oceans and some inferences regarding petroleum, stata-bound ores and mass extinctions, *Geochem. Cosmochim. Acta* **47:**2621–2637.

Koch, N. C., 1967, Disappearance of the dinosaurs, *J. Paleontol.* **41:**970–972.

Krassilov, V. A., 1981, Changes of Mesozoic vegetation and the extinction of dinosaurs, *Palaeogeogr. Palaeoclimatol. Palaeoecol.* **34:**207–224.

Kuhn, O., 1937, *Die fossilen Reptilien,* Borntraeger, Berlin.

Lewin, R., 1983, Extinctions and the history of life, *Science* **221:**935–937.

Lewin, R., 1985*a,* Catastrophism not yet dead, *Science* **229:**640.

Lewin, R., 1985*b,* Catastrophism not yet dead, *Science* **230:**8.

Loomis, F. B., 1905, Momentum in variation, *Am. Nat.* **39:**839–843.

Lyell, C., 1832, *Principles of Geology,* Murray, London.

Maddox, J., 1985, Periodic extinctions undermined, *Nature* **315:**627.

Mantell, G. A., 1831, The geological age of reptiles, *Edinb. New Philos. J.* **11:**181–185.

Marsh, O. C., 1882, Classification of the Dinosauria, *Am. J. Sci. (3)* **23:**81–86.

Marsh, O. C., 1895, On the affinities and classification of the dinosaurian reptiles, *Am. J. Sci. (3)* **1:** 483–498.

Marshall, H. T., 1928, Ultra-violet and extinction, *Am. Nat.* **62:**165–187.

Matthew, W. D., 1921, Fossil vertebrates and the Cretaceous–Tertiary problem, *Am. J. Sci. (5)* **2:** 209–227.

Mayr, E., 1982, *The Growth of Biological Thought: Diversity, Evolution and Inheritance,* Harvard University Press, Cambridge, Massachusetts.

McLaren, D. J., 1970, Time, life and boundaries, *J. Paleontol.* **44:**801–815.

McLean, D. M., 1978, A terminal Mesozoic "greenhouse": Lessons from the past, *Science* **201:** 401–406.

McLean, D. M., 1982, Deccan volcanism and the Cretaceous–Tertiary extinction scenario: A unifying causal mechanism, *Syllogeus* **39:**143–144.

Moodie, R. L., 1923, *Paleopathology,* University of Illinois, Urbana, Illinois.

Müller, A. H., 1968, *Lehrbuch der Paläozoologie,* Vol. III, Part 2, *Reptilien und Vögel,* Gustav Fischer, Jena.

Müller, L., 1928, Sind die Dinosaurier durch Vulkanausbrüche ausgeratet worden? *Unsere Welt* **20:** 144–146.

Neruchev, S. G., 1984, *Uranium and Life in the History of the Earth.*

Newell, N. D., 1952, Periodicity in invertebrate evolution, *J. Paleontol.* **26:**371–385.

Newell, N. D., 1962, Paleontological gaps and geochronology, *J. Paleontol.* **36:**592–610.

Newell, N. D., 1967, Revolutions in the history of life, *Spec. Pap. Geol. Soc. Am.* **89:**63–91.

Nopcsa, F., 1911, Notes on British dinosaurs. Part IV: *Stegosaurus priscus,* sp. nov., *Geol. Mag. (5)* **8:**143–153.

Nopcsa, F., 1917, Über Dinosaurier, *Centralbl. Mineral. Geol. Paläontol.* **1917:**332–351.
Nopcsa, F., 1922, On the geological importance of the primitive reptilian fauna in the uppermost Cretaceous of Hungary; with a description of a new tortoise (*Kallikobotion*), *Q. J. Geol. Soc. Lond.* **79:**100–116.
Oelofson, B. W., 1978, Atmospheric carbon dioxide/oxygen imbalance in the late Cretaceous, hatching of eggs and the extinction of biota, *Palaeontol. Afr.* **21:**45–51.
Officer, C. B., and Drake, C. L., 1985, Terminal Cretaceous environmental events, *Science* **227:** 1161–1187.
Officer, C. B. Hallam, A., Drake, C. L., and Devine, J. D., 1987, Late Cretaceous and paroxysmal Cretaceous–Tertiary extinctions, *Nature* **326:**143–149.
Orlov, J. A., 1964, *Osnovi Paleontologii* (*Amphibians, Reptiles, and Birds*), Izdatelistvo, Moscow.
Ostrom, J. H., 1970, Terrestrial vertebrates as indicators of Mesozoic climates, *Proc. N. Am. Paleontol. Conv. D* **1970:**347–376.
Owen, R., 1842, Report on the British fossil reptiles, *Rep. Br. Assoc. Advan. Sci.* **1841:**60–204.
Pennant, T., 1771, *Synopsis of Quadrupeds,* Chester.
Perch-Nielsen, K., McKenzie, J., and He, Q., 1982, Biostratigraphy and isotope stratigraphy and the "catastrophic" extinction of calcareous nannoplankton at the Cretaceous/Tertiary boundary, *Spec. Pap. Geol. Soc. Am.* **190:**353–371.
Prinn, R. G., and Fegley, B., Jr., 1987, Bolide impacts, acid rain, and biospheric traumas at the Cretaceous–Tertiary boundary, *Earth Planet. Sci. Lett.* **83:**1–15.
Raup, D. M., 1972, Taxonomic diversity during the Phanerozoic, *Science* **177:**1065–1071.
Raup, D. M., 1986, *The Nemesis Affair,* Norton, New York.
Raup, D. M., and Sepkoski, J. J., Jr., 1984, Periodicities of extinctions in the geologic past, *Proc. Natl. Acad. Sci. USA* **81:**801–805.
Raymond, P. E., 1939, *Prehistoric Life,* Harvard University Press, Cambridge, Massachusetts.
Reid, G. C., Isaksen, I. S. A., Holzer, T. E., and Crutzen, P. J., 1976, Influence of ancient solar-proton events on the evolution of life, *Nature* **250:**177–179.
Renard, M., and Rocchia, R., 1984, Extinction des espèces au Secondaire: La terre dans un nuage interstellaire?, *Recherche* **15:**393–395.
Rocchia, R., Renard, M., Boclet, D., and Bonte, P., 1984, Essai d'evaluation de la durée de la transition Crétacé/Tertiaire par l'evolution de l'anomalie en iridium; implications dans la recherche de la cause de la crise biologique, *Bull. Soc. Géol. Fr. (7)* **26:**1193–1202.
Ruderman, D. A., 1974, Possible consequences of nearby supernova explosions for atmospheric ozone and terrestrial life, *Science* **184:**1079–1081.
Russell, D. A., 1971, The disappearance of the dinosaurs, *Can. Geogr. J.* **83:**204–215.
Russell, D. A., and Tucker, W., 1971, Supernovae and the extinction of the dinosaurs, *Nature* **229:** 553–554.
Russell, L. S., 1965, Body temperature of dinosaurs and its relationships to their extinction, *J. Paleontol.* **39:**497–501.
Schatz, A., 1957, Some biochemical and physiological considerations regarding the extinction of the dinosaurs, *Proc. Penn. Acad. Sci.* **31:**26–36.
Schindewolf, O. H., 1958, Über die möglichen Ursachen der grossen erdgeschichtlichen Faunenschnitte, *Neues Jb. Geol. Paläontol. Monatsh.* **1958:**457–465.
Schopf, T. J. M., 1979, Evolving paleontological views on deterministic and stochastic approaches, *Paleobiology* **5:**337–352.
Schuchert, C., 1924, *Historical Geology,* Wiley, New York.
Signor, P. W., III, and Lipps, J. H., 1982, Sampling bias, gradual extinction patterns, and catastrophes in the fossil record, *Spec. Pap. Geol. Soc. Am.* **190:**291–296.
Simpson, G. G., 1952, Periodicity in vertebrate evolution, *J. Paleontol.* **26:**359–370.
Sloan, R. E., 1964, Paleoecology of the Cretaceous–Tertiary transition in Montana, *Science* **146:** 430.

Sloan, R. E., 1970, Cretaceous and Paleocene terrestrial communities of western North America, *Proc. N. Am. Paleontol. Conv. D* **1970:**427–453.

Sloan, R. E., Rigby, J. K., Jr., Van Valen, L. M., and Gabriel, D., 1986, Gradual dinosaur extinction and simultaneous ungulate radiation in the Hell Creek Formation, *Science* **232:**629–632.

Smit, J., and Hertogen, J., 1980, An extraterrestrial event at the Cretaceous–Tertiary boundary, *Nature* **285:**198–200.

Smit, J., and Kyte, F. T., 1984, Siderophile-rich magnetic spheroids from the Cretaceous–Tertiary boundary in Umbria, Italy, *Nature* **310:**403–405.

Spotila, J. R., Lommen, P. W., Bakken, G. S., and Gates, D. M., 1973, A mathematical model for body temperatures of large reptiles: Implications for dinosaur ecology, *Am. Nat.* **107:**391–404.

Stechow, E., 1954, Zur Frage der Ursache des grossen Sterbens am Ende der Kreidezeit, *Neues Jb. Geol. Paläontol. Monatsh.* **1954:**183–186.

Stromer von Reichenbach, E., 1912, *Lehrbuch der Paläozoologie, Part 2,* B. G. Teubner, Leipzig.

Surlyk, F., and Johansen, M. B., 1984, End-Cretaceous brachiopod extinctions in the Chalk of Denmark, *Science* **223:**1174–1177.

Swain, T., 1976, Angiosperm–reptile coevolution, *Linn. Soc. Symp. Ser.* **3:**107–122.

Swinton, W. E., 1939, Observations on the extinction of vertebrates, *Proc. Geol. Assoc.* **50:**135–146.

Tappan, H., 1968, Primary production, isotopes, extinctions and the atmosphere, *Palaeogeogr. Palaeoclimatol. Palaeoecol.* **4:**187–210.

Terry, K. D., and Tucker, W. H., 1968, Biological effects of supernova, *Science* **159:**421–423.

Thaler, L. S., 1965, Les oefs des dinosaures du Midi de la France livrent le secret de leur extinction, *Sci. Prog. Nat.* **1965**(2)**:**41–48.

Tsakas, S. C., and David, J. R., 1987, Population genetics and the Cretaceous extinction, *Génét. Sélect. Evol.* **19:**487–496.

Tschudy, R. H., Pillmore, C. L., Orth, C. J., Gilmore, J. S., and Knight, J. D., 1984, Disruption of the terrestrial plant ecosystem at the Cretaceous–Tertiary boundary, Western Interior, *Science* **225:**1030–1032.

Tucker, W. H., 1977, The effect of a nearby supernova explosion on the Cretaceous–Tertiary environment, *Syllogeus* **12:**111–124.

Turco, R. P., Toon, O. B., Ackerman, T. P., Pollack, J. B., and Sagan, C., 1983, Nuclear winter: Global consequences of multiple nuclear explosions, *Science* **222:**1283–1292.

Urey, H. C., 1973, Cometary collisions and geological periods, *Nature* **242:**32–33.

Valentine, J. W., 1969, Patterns of taxonomic and ecological structure of the shelf benthos during Phanerozoic time, *Palaeontology* **12:**684–709.

Valentine, J. W., 1974, Temporal bias in extinctions among taxonomic categories, *J. Paleontol.* **48:** 549–552.

Van Valen, L. M., 1984, Catastrophes, expectations, and the evidence, *Paleobiology* **10:**121–137.

Van Valen, L. M., and Sloan, R. E., 1972, Ecology and the extinction of the dinosaurs, *Proc. 24th Int. Geol. Congr.* **7:**214.

Van Valen, L. M., and Sloan, R. E., 1977, Ecology and the extinction of the dinosaurs, *Evol. Theory* **2:**37–64.

Vogt, P. R., 1972, Evidence for global synchronism in mantle plume convection, and possible significance for geology, *Nature* **240:**338–342.

Von Huene, F., 1956, *Paläontologie und Phylogenie der niederen Tetrapoden,* G. Fischer, Jena.

Von Zittel, K. A., 1902, *Text-book of Paleontology,* Vol. 2, Macmillan, London.

Von Zittel, K. A., 1932, *Text-book of Paleontology,* Vol. 2, 2nd ed., Macmillan, London.

Wieland, G. R., 1925, Dinosaur extinction, *Am. Nat.* **59:**557–565.

Wieland, G. R., 1942, Too hot for the dinosaur!, *Science* **96:**359.

Wilfarth, M., 1949, *Die Lebensweise der Dinosaurier,* E. Schweizerbart, Stuttgart.

Williston, S. W., 1925, *The Osteology of the Reptiles,* Harvard University Press, Cambridge, Massachusetts.

Wolbach, W. S., Lewis, R. R., and Anders, E., 1985, Cretaceous extinctions: Evidence for wildfires and search for meteoritic material, *Science* **230:**167–170.

Wolfe, J. A., and Upchurch, G. R., 1986, Vegetation, climatic and floral changes at the Cretaceous–Tertiary boundary, *Nature* **324:**148–152.

Woodward, A. S., 1898, *Outlines of Vertebrate Palaeontology for Students of Zoology,* Cambridge University Press, Cambridge.

Woodward, A. S., 1910, Presidential Address to Section C, *Rep. Br. Assoc. Advan. Sci.* **1909:**462–471.

Yayanos, A. A., 1983, Thermal neutrons could be a cause of biological extinctions 65 Myr ago, *Nature* **303:**797–800.

Index